Agricultural Science at Your Fingertips

Agricultural Science at Your Fingertips

Volume-I

Jagan Singh Gora

Chet Ram
S.M. Haldar
R.P. Sharma

V. Bhati
Deepak Singh
Dushyant Parihar

Published by:

SCIENTIFIC PUBLISHERS (INDIA)
5A, New Pali Road, P.O. Box 91
Jodhpur 342 001, India
E-mail: info@scientificpub.com
Website: www.scientificpub.com

ISBN: 978-93-89832-92-1 (PB)
eISBN: 978-93-89832-93-8 (EBook)

Printed in India

PREFACE

In India, the agriculture is the backbone and participating near about 14% in national GDP. As the spectrum covered by the agricultural streams is too wide, the students had to study different book, assimilate their contents and prepare suitable materials. Often this task is not easy and in case of less time and shortage of book it is really impossible. Keeping these facts in mind we had made an effort to make available a book suitable for preparation and revision for competitive examination. All the chapter of this book had been properly covered in systemic way as detailed in bullet form. The important things are highlighted with bold letters which will give more visibility and attractive to readers. 150 or more multiple choice questions given in the end of every chapter are very relevant to the present competitive examinations and most of the previous year question asked in the different examinations had been incorporated. This book fulfills the need of a good reference-cum text book for the student community. The volume I is covering the ten disciplines of agriculture namely recent general agriculture, soil science, agri-meteorology, agronomy, entomology and nematology, plant pathology, environment and ecology, agro-forestry, statistics, economics and extension educations.

I express my sincere thanks to scientists of ICAR Institutes/ SAUs for their input in term of material collection compilation and correction. Every effort has been made to ensure that the information collected from various sources like individuals, institutes and publications are accurate. I hope that this book will be useful as reference material for the teachers, and student searching their career in ICAR, SAUs, State agricultural department for competitions like ARS, NET, SRF, JRF, UPSC, PSCs, and banking as well as entrance examinations for B.Sc., M.Sc. and Ph.D. degree in Agricultural subjects.

We hope that this effort will help the readers in their endeavors. All the suggestions in making the book further improvement are always welcome.

Author

(Jagan Singh Gora)

Scientist (Fruit Science)

ICAR- CIAH, Bikaner, Rajasthan 334006

E-mail: jagangora@gmail.com

Dr WS Dhillon

Former ADG Hort. Sci., ICAR, New Delhi

Former Director PHPTC, Govt. of Punjab

Former Director PAU-Fruit Research Station

FORWARD

The students of agricultural sciences are playing the important role in the development of agriculture and nation building because agriculture is the foremost basis of livelihood for about 58% of India's population. India ranks second worldwide in farm outputs. India is leading in the production of food grains 295.67 MT, horticultural crops 314.67 MT and milk 176.35 MT at global level, and contributed 17.2% to country's GDP. The Indian agriculture is very rich in crops biodiversity, and awards Master's and Doctoral degree in more than 20 disciplines through SAUs and Deemed Universities.

To understand the basics of different agricultural disciplines, the author presented this book in a scientific and systematic manner. The author has given proper emphasis to important things with bold letters for attracting the reader which is the beauty of this book. To fulfill agriculture students need, the book **"Agricultural Science At Your Finger Tips"** by **Jagan Singh Gora, ARS Scientist (Fruit Science)** is a welcome effort.

Jagan Singh Gora is keenly involved in fruit crop research particularly citrus. He himself has qualified many prestigious exams in the field of Agriculture like JRF, SRF, NET (Horticulture, 2011 and Fruit Science, 2012) and ARS, and also awarded Netaji Subhas ICAR International Fellowship.

I personally congratulate to Jagan Singh Gora for bringing out a systematic designed valuable book with all the latest information. I am confident that this book will be widely accepted among the students. I extent my best wished to Jagan Singh Gora for the success of his book.

Place: Ludhiana

Dated: November 10, 2020,

WS Dhillon

1. **Recent General agriculture:**

   Jagan Singh Gora, Scientist, ICAR- CIAH, Bikaner; Chet Ram, Scientist ICAR-CIAH, Bikaner and Dushyant Parihar, Babasaheb Bhimrao Amedkar University (Central University), Lucknow.

2. **Soil science:**

   R. P. Sharma, Sr. Scientist and P. C. Moharana, Scientist, ICAR-NBSSLUP, Nagpur; M.K. Jatav, Principle Scientist ICAR- CIAH, Bikaner; Shankar Lal Bijjarnia (Ph.D.) SKRAU, Bikaner; Ghous Ali, Scientist, ICAR- CSWRI, Bikaner and Dr. Roop Chand Balai, Sr. Scientist ICAR- CIAH, Bikaner

3. **Agricultural Meteorology:**

   Ved Prakash Jyani, Scientist ICAR-RCER Patna; N. K. Pareek, Assoc. Professor, SKRAU, Bikaner and Tanuja Poonia, Ph.D. SKRAU, Bikaner

4. **Agronomy:**

   Jagdish Prasad, Assoc. Professor, MPUAT, Udaipur; Tanuja Poonia (Ph.D.) SKRAU, Bikaner; C.R. Kantwa, Scientist, KVK, Ratlam and Ghous Ali, Scientist, ICAR- CSWRI, Bikaner; Madan Lal Reager, Assoc. Professor, SKRAU, Bikaner and Raghuveer Singh Bhamu, Scientist, IIFSR, Meerut}

5. **Entomology:**

   S. M. Haldar, Assoc. Professor, CAU Imphal and former Scientist, ICAR- CIAH, Bikaner and Sandeep Kudi (Ph.D.), SKNAU, Jobner

6. **Plant Pathology:**

   RP Ghasolia, Assoc. Professor, SKNAU, Jobner; Kartar Singh, Scientist, ICAR–NBPGR- RS, Jodhpur; A. K. Meena, Assistant Professor, SKRAU, Bikaner and Ramya Shree Devi, Scientist, ICAR- CIAH, Bikaner.

7. **Environment and ecology:**

   Sheetal K.R., Scientist, ICAR- CAZRI –RRS Bikaner; Birbal, Principle scientist, ICAR- CAZRI -RRS Bikaner and Vinod K. Yadav, Technical Officer, ICAR-National Research Centre on Camel

8. **Agro- Forestry:**

   Pawan Kumar Poonia, Assistant Professor, CCS HAU, Hisar and Vidya Bhati, Assistant Professor, Maharishi Markandeshwar (Deemed University), Mulana Ambala

9. **Agricultural Statistics:**

   Deepak Singh, Scientist and Raju Kumar, Scientist, ICAR- IASRI, New Delhi

10. **Economics:**

    Vikash Pawariya, Assistant Professor, AU, Jodhpur, Vikram Yogi, Assistant Professor SKRAU, Bikaner.

11. **Agricultural Extension:**

    Madan Singh & Sharuti, Scientist, ICAR- IVRI, Izatnagar; Annu Gora (Ph.D.) MPUAT, Udaipur; Mukesh Choudhary, Technical Assistant, AU Kota and Arvind Kumar Jhajharia, Assistant Professor, SKRAU, Bikaner

CONTENTS

Agro- Forestry:			
	Introduction and history of agro-forestry; Traditional Agro-forestry Systems in India; Systems for soil conservation or amelioration; Benefits from Agro-forestry; Classification of Agro-forestry System; Agro-forestry Diagnosis & Design (D & D) and MCQ	353–436	4.91
Entomology	Introduction and history of entomology; Factors for insect dominance; Insect taxonomy; Insect structure (external and internal); Beneficial insect; Polyphagous pests; Insect of crops; IPM; HPR; Chemical insecticides; Insect hormones; Insect pheromones; Biological control; Insect ecology; Mites; Rodent pests and MCQ	437–464	18.66
Plant Pathology	Terminology; Important famines; History of plant pathology; Plant pathogenic organisms (Eukaryotic - Algae, Fungi, Protozoa, Nematodes, Phanerogamic Parasites, Insects; Prokaryotic – Mycoplasma, Bacteria, BGA, Spiroplasma, Phytoplasma; Acelluar- Virus, Viroid, Virusoid, Satellite Viruses and Prions), Important diseases of crops, Forecasting and its integrated management, and MCQ	465–486	14.73
Agricultural Statistics	Data; Mean; Mode; Medium; Quartiles; Dispersion; Test of Significance; Regression and Correlation; Probability; Design of Experiments and MCQ	487–518	4.91
Economics	Economic history, Utility, Supply and Demand , Tax and Inflation; Farm management- Farming types, System of farming, Cost of cultivation, Relationship between Factor-Product and Enterprise; Agricultural Finance and Cooperation; Agricultural Marketing, and MCQ	519–546	4.91
Agricultural Extension	Extension Education, Objective and Principle of Agricultural Extension, Rural Sociology and Psychology, Communication; Diffusion and Adoption; Programme Planning and MCQ.	547–584	4.91

Chapter
1

RECENT INDIAN AGRICULTURE

INDIAN AGRICULTURAL GEOGRAPHY

- Total geographical area of India is **329 Mha and Rajasthan is the largest state (Areawise) in India (9.61%).**
- Net cultivated area of India is **143 Mha (42.57%).**
- Highest net sown area is in **MH** followed by **M.P.**
- Gross cultivated area in India is **195 mha.**
- Highest gross sown area in **U.P. > M.P. > MH.**
- Cropping intensity of India is **138%.**
- Total irrigated area in India is **64.7 Mha**.
- Waste land area in India is **178 Mha.**
- Barren land area in India is **41 Mha** and barren and uncultivated land is highest in **Rajasthan** state.
- Degraded land area in India is **63.85 Mha.**
- Poverty percentage of India is **21.9%.**
- India's position in Global Hunger Index is **102th Rank out of 117 (2019).**
- GDP growth rate of India in 2019-20 (4.2%) and 2020-21 (–10.2%) are projected by **RBI** (Fourth Monetary report).
- The Gross Value Added (GVA) share of agriculture and allied sectors during 2019-20 was **15.4%.**
- Gross value added (GVA) growth rate of India is **2.9 %** in 2018-19.
- Contribution of Agriculture & Allied sectors to GDP in India is **17.2%.**
- Power house of Nutrients/Nutri-Cereals/Climate resilient crops are **Millets.**
- Total Forest area in the country- **70.82 Mha (21.54%** of total geographical area).
- State with highest forest area in the country is **MP > AP > Chhattisgarh**.
- Share of Agriculture and allied sectors to India's national employment is **52%.**

- **Service sector (61.5%)** is contributes highest in national GDP followed by Industrial sector (23%).
- The **First World Agriculture Prize (2018)** awarded was given to **Dr. M. S. Swaminathan** for his contributions to Indian Agriculture.
- World food prize (2018) won by **Lawrence Haddad** and **Dr. David Nabarro**
- Best Horticulture state awarded to **Nagaland (2018).**
- At Present, India is having **first position** in cattle population, buffalo population, milk production, pulses, tea, jute & allied fibres.
- **14 products** received **Geographical Indication (GI)** tag from Government of India, 2019.
- Kisan Call Centre (KCC-Toll free No.1800 1801551) was started on **21st Jan., 2004.**
- Integrated Scheme of Oilseeds, Pulses, Oil palm & Maize (ISOPOM) was started on **1st Apr., 2004.**

Land Utilization of Country and Some States

Classificatio	India (mha)	Chhattisgarh (mha)	Madhya Pradesh (mha)	Rajasthan (mha)
Total Geographical area	**329**	13.7	30.8	**34.22**
Net cropped area	143	4.83	15.07	17.39
Net irrigated area	68	1.33	9.58	7.88
Gross irrigated area	97	1.52	10.30	10.15
Cropping intensity	**138 %**	134 %	130 %	125 %
Wasteland	63.85	–	5.84	10.56
Forest area	79.42	5.56	7.75	3.22
Per capita agriculture land	**0.12 (ha)**	–	–	0.38 (ha)
Size of operational holding	1.41 (ha)	–	2.5 (ha)	3.96 (ha)
Average rainfall (mm)	1190	1400	1089.3	575
Rainfed area	60%	75 %	70 %	70 %
Total fame holding	107 (millions)	–	6.64 (millions)	–
Agro-climatic/sub ACZ	**15 (ICAR)**	3 (SACZ)	11 (SACZ)	10 (SACZ)
Forest area (%)	24.16%	40.58 %	25.16 %	9.43 %

*According to ISFR (India State Forest Report), 2015

Net Sown Area, Gross Cropped Area, Net Irrigated Area, Gross Irrigated Area of India and Same States

Class	Area (mha)				
	India	**Rajasthan**	**UP**	**MP**	**Maharastra**
Net sown area	140.1	17.52	16.56	15.35	17.34
Net irrigated area	68.1	7.88	14.38	9.58	3.24
Total cropped area	198.3	24.23	26.15	23.18	23.47
Gross irrigated area	96.50	10.17	20.96	10.30	24.28

Source: Directorate of Economics and Statistics, Ministry of Agriculture and Farmers Welfare

ABOUT THE ICAR AND ITS INSTITUTES

- Indian Council of Agricultural Research (ICAR) is registered under **Society Act, 1860.**
- ICAR was established on the recommendation of **Royal Commission on Agriculture** (1928) on **16 July, 1929.**
- ICAR is a registered society came in existence in **1929.**
- Present ICAR, formerly known as Imperial Council of Agricultural Research and it was established on **16th July, 1929 (ICAR Day).**
- Department of Agricultural Research and Education (DARE) came into existence under the Ministry of Agriculture in **1973.**
- Establishment of Agricultural Research Service and Agricultural Scientists' Recruitment Board in **1975.**
- No. of ICAR institutions are **65.**
- No. of National Research Centres is **14.**
- No. of Directorates/Project Directorates is **13.**
- No. of non-educational institutes are **4**: EEI- Anand (Gujarat), EEI- Hyderabad, EEI- Jorhat, (Assam) and EEI- Nilokheri (Haryana).
- No. of universities with agriculture faculty are **4** namely AMU at Aligarh, BHU at Varanasi, NU at Lumani (Nagaland) and Shantiniketan at West Bengal.
- No. of Central Agricultural Universities is **3.**

  1. CAU- Imphal (1993).

  2. RLBCAU-Jhansi, Uttar Pradesh (2010).

  3. Dr. RPCAU Pusa Samastipur, Bihar (2016).
- No. of State Agricultural Universities is **64.**
- No. of All India Coordinated Research Projects is **59.**
- No. of Network Projects is **20.**

- First KVK in India was established at Pondicherry in **1974.**
- Total number of Agricultural Technology Information Centres (ATIC) under ICAR is **44.**
- All India Coordinated Research Project on Maize was started in **1957**
- The G.B. Pant University of Agriculture and Technology, Pantnagar (first Agricultural University) was started in **1960** in India.
- Number of Deputy Director Generals (DDGs) under the ICAR organization are **8.**
- Number of Assistant Director Generals (ADGs) under the ICAR organization are **24.**
- Total number of Agricultural Technology Application Research Institutes **(ATARIs)** of ICAR is **11 (Eleven).**
- The chairman of National Commission for Farmers **Dr. M.S. Swaminathan.**
- Present Director General (DG) of ICAR (as on February, 2020) is **Dr. Trilochan Mohapatra.**
- Present total number of Krishi Vigyan Kendra (KVK's) is 716 (February, 2020).
- The Department of Agriculture was established in **1880.**
- Central Institute of Horticulture (CIH) was sanctioned during 2006 at **Nagaland.**
- **FIRM-** Farm Innovation Resources Management.
- **KSHAMTA-** Knowledge Systems and Homestead Agriculture Management in Tribal Areas.
- **NARI-** Nutrisensitive Agricultural Resources and Innovations.
- **VATICA-** Value Addition and Technology Incubation Centres in Agriculture.
- **SAMPADA-** Scheme for Agro Marine Processing and Development of Agro Processing Clusters.
- **APEDA-** Agricultural and Processed Food Products Export Development Authority.
- **FICCI-** Federation of Indian Chambers of Commerce & Industry.
- **FSSAI-** Food Safety and Standards Authority of India.
- **NAFED-** National Agricultural Marketing Cooperative Federation.
- **PPV&FR-** Protection of Plant varieties and Farmers Right Act.

ICAR Institutions, Deemed Universities, National Research Centres, National Bureaux & Directorate/Project Directorates

DEEMED UNIVERSITIES - 4

S.N.	Name of Institutes	Situated at
1.	ICAR-Indian Agricultural Research Institute,	New Delhi
2.	ICAR-National Dairy Research Institute	Karnal
3.	ICAR-Indian Veterinary Research Institute,	Izatnagar
4.	ICAR-Central Institute on Fisheries Education	Mumbai

ICAR- INSTITUTIONS – 65

S.N.	Name of Institutes	Situated at
1.	ICAR-Central Island Agricultural Research Institute	Port Blair
2.	ICAR-Central Arid Zone Research Institute	Jodhpur
3.	ICAR-Central Avian Research Institute	Izatnagar
4.	ICAR-Central Inland Fisheries Research Institute	Barrackpore
5.	ICAR-Central Institute Brackish water Aquaculture	Chennai
6.	ICAR-Central Institute For Research on Buffaloes	Hissar
7.	ICAR-Central Institute For Research on Goats	Makhdoom
8.	ICAR-Central Institute of Agricultural Engineering	Bhopal
9.	ICAR-Central Institute For Arid Horticulture	Bikaner
10.	ICAR-Central Institute of Cotton Research	Nagpur
11.	ICAR-Central Institute of Fisheries Technology	Cochin
12.	ICAR-Central Institute of Freshwater Aquaculture	Bhubneshwar
13.	ICAR-Central Institute of Research On Cotton Technology	Mumbai
14.	ICAR-Central Institute of Sub Tropical Horticulture	Lucknow
15.	ICAR-Central Institute of Temperate Horticulture	Srinagar
16.	ICAR-Central Institute on Post Harvest Engineering and Technology	Ludhiana
17.	ICAR-Central Marine Fisheries Research Institute	Kochi
18.	ICAR-Central Plantation Crops Research Institute	Kasargod
19.	ICAR-Central Potato Research Institute	Shimla
20.	ICAR-Central Research Institute For Jute and Allied Fibres	Barrackpore
21.	ICAR-Central Research Institute of Dryland Agriculture	Hyderabad
22.	ICAR-National Rice Research Institute	Cuttack
23.	ICAR-Central Sheep and Wool Research Institute	Avikanagar, Rajasthan
24.	ICAR- Indian Institute of Soil and Water Conservation	Dehradun
25.	ICAR-Central Soil Salinity Research Institute	Karnal

S.N.	Name of Institutes	Situated at
26.	ICAR-Central Tobacco Research Institute	Rajahmundry
27.	ICAR-Central Tuber Crops Research Institute	Trivandrum
28.	ICAR-ICAR Research Complex For Eastern Region	Patna
29.	ICAR-ICAR Research Complex For Neh Region	Barapani
30.	ICAR-Central Coastal Agricultural Research Institute	Ela, Goa
31.	ICAR-Indian Agricultural Statistics Research Institute	New Delhi
32.	ICAR-Indian Grassland and Fodder Research Institute	Jhansi
33.	ICAR-Indian Institute of Agricultural Biotechnology	Ranchi
34.	ICAR-Indian Institute of Horticultural Research	Bengaluru
35.	ICAR-Indian Institute of Natural Resins and Gums	Ranchi
36.	ICAR-Indian Institute of Pulses Research	Kanpur
37.	ICAR-Indian Institute of Soil Sciences	Bhopal
38.	ICAR-Indian Institute of Spices Research	Calicut
39.	ICAR-Indian Institute of Sugarcane Research	Lucknow
40.	ICAR-Indian Institute of Vegetable Research	Varanasi
41.	ICAR-National Academy of Agricultural Research & Management	Hyderabad
42.	ICAR-National Institute of Biotic Stresses Management	Raipur
43.	ICAR-National Institute of Abiotic Stress Management	Malegaon, Maharashtra
44.	ICAR-National Institute of Animal Nutrition And Physiology	Bengaluru
45.	ICAR-ICAR-National Institute of Natural Fibre Engineering and Technology	Kolkata
46.	ICAR-National Institute of Veterinary Epidemiology and Disease Informatics	Bengaluru
47.	ICAR-Sugarcane Breeding Institute	Coimbatore
48.	ICAR-Vivekananda Parvatiya Krishi Anusandhan Sansthan	Almora
49.	ICAR-Central Institute For Research on Cattle	Meerut, Uttar Pradesh
50.	ICAR-National Institute of High Security Animal Diseases	Bhopal
51.	ICAR-Indian Institute of Maize Research	Ludhiana
52.	ICAR- Central Agroforestry Research Institute	Jhansi
53.	ICAR-National Institute of Agricultural Economics and Policy Research	New Delhi
54.	ICAR- Indian Institute of Wheat and Barley Research	Karnal
55.	ICAR- Indian Institute of Farming Systems Research	Modipuram
56.	ICAR- Indian Institute of Millets Research	Hyderabad

S.N.	Name of Institutes	Situated at
57.	ICAR- Indian Institute of Oilseeds Research	Hyderabad
58.	ICAR- Indian Institute of Oil Palm Research	Pedavegi, West Godawari
59.	ICAR- Indian Institute of Water Management	Bhubaneshwar
60.	ICAR-Indian Institute of Rice Research	Hyderabad
61.	ICAR- Central Institute For Women in Agriculture	Bhubaneshwar
62.	ICAR-Central Citrus Research Institute	Nagpur
63.	ICAR-Indian Institute of Seed Research	Mau
64.	ICAR-Indian Agricultural Research Institute	Hazaribag, Jharkhand
65.	ICAR-National Institute For Plant Biotechnology	New Delhi

NATIONAL RESEARCH CENTRES - 14

S.N.	Name of Institutes	Situated at
1.	ICAR-National Research Centre for Banana	Trichi
2.	ICAR-National Research Centre for Grapes	Pune
3.	ICAR-National Research Centre for Litchi	Muzaffarpur
4.	ICAR-National Research Centre for Pomegranate	Solapur
5.	ICAR-National Research Centre on Camel	Bikaner
6.	ICAR-National Research Centre on Equines	Hisar
7.	ICAR-National Research Centre on Meat	Hyderabad
8.	ICAR-National Research Centre on Mithun	Medziphema, Nagaland
9.	ICAR-National Research Centre on Orchids	Pakyong, Sikkim
10.	ICAR-National Research Centre on Pig	Guwahati
11.	ICAR-National Research Centre on Seed Spices	Ajmer
12.	ICAR-National Research Centre on Yak	West Kemang
13.	ICAR-National Centre for Integrated Pest Management	New Delhi
14.	Mahatma Gandhi Integrated Farming Research Institute	Motihari

NATIONAL BUREAUX – 6

S.N.	Name of Institutes	Situated at
1.	ICAR-National Bureau of Plant Genetics Resource	New Delhi
2.	ICAR-National Bureau of Agriculturally Important Micro-organisms	Mau, Uttar Pradesh
3.	ICAR-National Bureau of Agricultural Insect Resources	Bengaluru
4.	ICAR-National Bureau of Soil Survey and Land Use Planning	Nagpur
5.	ICAR-National Bureau of Animal Genetic Resources	Karnal
6.	ICAR-National Bureau of Fish Genetic Resources	Lucknow

DIRECTORATES/PROJECT DIRECTORATES - 13

S.N.	Name of Institutes	Situated at
1.	ICAR-Directorate of Groundnut Research	Junagadh
2.	IISR-India Institute Soyabean Research	Indore
3.	ICAR-Directorate of Rapeseed & Mustard Research	Bharatpur
4.	ICAR-Directorate of Mushroom Research	Solan
5.	ICAR-Directorate on Onion and Garlic Research	Pune
6.	ICAR-Directorate of Cashew Research	Puttur
7.	ICAR-Directorate of Medicinal and Aromatic Plants Research	Anand
8.	ICAR-Directorate of Floricultural Research	Pune, Maharashtra
9.	ICAR-Directorate of Weed Research	Jabalpur
10.	ICAR-Project Directorate on Foot & Mouth Disease	Mukteshwar
11.	ICAR-Directorate of Poultry Research	Hyderabad
12.	ICAR-Directorate of Knowledge Management in Agriculture (DKMA)	New Delhi
13.	ICAR-Directorate of Cold Water Fisheries Research	Bhimtal, Nainital

Important and Recent Varieties of Agricultural Crops

S. N.	Varieties	Crops	Remarks
1.	PUSA Arhar 16	Pigeon pea	Determinate, extra early maturity, 120 days
2.	Hybrid GTH -1	Pigeon pea	World's first **cytoplasmic male sterility**
3.	**Karan 4 (C0-0238)**	Sugarcane	Wonder variety of sugarcane
4.	Pusa Double Zero Mustard 31	Mustard	**First canola type**, erucic acid less than 2% and glucosinolates < 30 pp
5.	Yugank	Cotton	**World earliest maturity**
6.	**HDSCW 18**	Wheat	**first variet** of the country bred specifically for **conservation agriculture**
7.	PUSA 109, PUSA 1592 and Punjab Basmati-3	Rice	BLB of disease resistant
8.	DRR Dhan 45	Rice	**High zinc rice variety**
9.	CR Dhan 310	Rice	High protein

S. N.	Varieties	Crops	Remarks
10.	CR Dhan 510 & CR Dhan 511 CR Dhan 801 (Subhas)	Rice	MR to blast, BB, SBR & LF, 4.6 t/ha Shallow lowland(for submergence and drought prone areas)
11.	**Pusa Vivek QPM9**	Maize	Provitamin-A, lysine & tryptophan rich hybrid)
12.	**PUSA Breakfast Cereal**	Maize	Quality Protein Maize (QPM)
13.	Pusa Super Sweet Corn 1	Maize	hybrid with enhanced sweetness with a good grain (9.3 t/ha) and fodder (16.2 t/ha) yield
14.	Rattan, Nirmal, Prateek	Grass Pea	Low neuro-toxin
15.	NARI-6	Safflower	**Spineless**
16.	Virat (IPM 205-7)	Moong	short duration < 60 days
17.	Pusa Majesty	Papaya	**Nematode resistant and papian rich**
18.	**Solapur Lal**	Pomegranate	iron, zinc & vitamin-C rich
19.	**Pusa Beta Kesari 1**	Cauliflower	β-carotene rich
20.	**Arka Rakshak**	Tomato	India's first **Triple disease resistant hybrid** (Leaf curl, Bacterial wilt & early blight)
21.	Pusa Asita	Carrot	**Black coloured**
22.	**Swarna Vaideh**	Makhana	Released by ICAR
23.	Arka Samrat	Tomato	Bacterial wilt resistant tomato hybrid
24.	Pusa Seedless Cucumber-6	cucumber	**Parthenocarpic and gynoecious**
25.	**Bhu Sona**	Potato	β-carotene rich
26.	Pusa Utkarsh	Long melon	-
27.	Pusa Raunak	Round melon	-
28.	Pusa Madhurima Pusa Sunahari	muskmelon	-

Recent in 2020 released varieties by ICAR			
29.	Pusa Manohari	Mango	Amrapali x Lal Sundari , regular and **tolerant to malformation**
30.	Pusa Deepshikha	Mango	Amrapali x Sensation, **regular bearer**
31.	Pusa Seedless Pummelo-1	Pummelo	**India's first seedles** cultivar developed by **mutation**, white fleshed, juice recovery (41.13%)
32.	Pusa Purple seedless	Grape	Raised by **embryo rescue, extra early ripening (75-80 DAFB)** under sub-tropical region
33.	Pusa Aditi	Grape	Hybrid was released commercial cultivation in NCR region.
34.	Pusa Shanti	Gladiolus	-
35.	Pusa Pickling Cucumber-8 (DG-8)	Cucumber	Off season (Nov. -March)
36.	Pusa Shreyash	Summer squash	Suitable for winter season
37.	Pusa Snowball Hybrid-2 (KTH-DH-1)	Cauliflower	First doubled haploid based F1 hybrid and more uniformity
38.	Pusa Purple Cauliflower-1 (KTPCF-1)	Cauliflower	Intense purple colour and the pigmentation is present deep inside the bracts below the curd, anthocyanin concentration (43.7mg/ 100 g)
39.	Pusa Rakshit (DTPH-60)	Tomato	First F1 hybrid of IARI for protected cultivation, Average fruit yield/plant is 15 kg/m^2, TSS (5.10 Brix) and lycopene (6.0 mg/100 g).
40.	Pusa Vilayati Palak	Spinach	Short duration and rich in nutrients
41.	Pusa Sunehari	Musk melon	First orange fleshed variety of Sarda melon and suitable for protected cultivation.
42.	Pusa Hybrid-3 (DCH 976)	Cauliflower	Help farmers to earn more (2nd fortnight of December in Delhi condition with yield. of 37-39 t/ha).

Programme Launched by ICAR and GOI, for Agricultural Development in India

S. N.	Programmes	Full forms	Year
1.	FASAL	Forecasting Agricultural output using Space, Agro meteorological and Land based observations	2010
2.	PMJDY	Pradhan Mantri Jan Dhan Yojana	28th August, 2014
3.	Make in India	-	25th Sep., 2014
4.	SBA	Swachh Bharat Abhiyan	2nd October, 2014
5.	PMFBY	Pradhan Mantri Fasal Bima Yojana	11th October, 2014
6.	SAGY	Sansad Adarsh Gram Yojana	
7.	PDUSJY	Pandit Deendayal Upadhyay Shramev Jayate Yojana	16th October, 2014
8.	RGM	Rashtriya Gokul Mission	
9.	DBTL Scheme	PAHAL-Direct Benefits Transfer for LPG	1st Jan., 2015
10.	NITI Aayog	The National Institution for Transforming India	
11.	Beti Bachao, Beti Padhao Yojana	—	22nd January, 2015
12.	PMSSY	Pradhan Mantri Sukanya Samriddhi Yojana	
13.	SHC Scheme	Soil Health Card Scheme	17th Feb., 2015
14.	KVP	Kisan Vikas Patra	3rd March, 2015
15.	PRASAD	Pilgrimage Rejuvenation and Spiritual Augmentation Drive	9th March, 2015
16.	SDY	Swadesh Darshan Yojana	
17.	PMMY	Pradhan Mantri Mudra Yojana	8th April, 2015
18.	PMGKY	Pradhan Mantri Garib Kalyan Yojana	
19.	MUDRA Yojana	Mico Units Development and Refinancing Agency	April, 2015
20.	PMSBY	Pradhan Mantri Suraksha Bima Yojana	9th May, 2015
21.	PMJJBY	Pradhan Mantri Jeevan Jyoti Bima Yojana	
22.	APY	Atal Pension Yojana	
23.	AMRUT	Atal Mission for Rejuvenation and Urban Transformation	24th June, 2015

S. N.	Programmes	Full forms	Year
24.	PMAY	Pradhan Mantri Awas Yojana	25th June, 2015
25.	SCM	Smart City Mission	
26.	PMKSY	Pradhan Mantri Krishi Sinchai Yojana	1st July, 2015
27.	PMGSY	Pradhan Mantri Gram Sinchai Yojana	
28.	DI	Digital India	
29.	Skill India	-	16th July, 2015
30.	DDUGKY	Deen Dayal Upadhyaya Grameen Kaushalya Yojana	25th July, 2015
31.	Sagarmala Project	-	31st July, 2015
32.	PMKKKY	Pradhan Mantri Khanij Kshetra Kalyan Yojana	17th September, 2015
33.	PKSY	Pardarshi Kisan Sewa Yojana	
34.	Startup India, Stand-up India	-	16th January, 2016
35.	PMJAY	Pradhan Mantri Jan Aushadhi Yojana	March, 2016
36.	PMUY	Pradhan Mantri Ujjwala Yojana	1st May, 2016
37.	CHAMAN	Coordinated Horticulture Assessment and Management using Geoinformatics	2016
38.	PMKSY	Pradhan Mantra Kisan SAMPADA Yojana	May, 2017
39.	OG	OPERATION GREEN	Feb, 2018
40.	PM-JIVAN	Prime Minister-Jaiv Indan Vatavaran Anukool Fasal Awasesh Nivaran	Feb 28, 2019
41.	PM-AASHA	Pradhan Mantri - Annadata Aay Sanrakshan Abhiyan	Sep, 2018
42.	PM-KISAN	Pradhan Mantri Kisan Samman Nidhi	24 Feb, 2019
43.	KUSUM	Kisan Urja Suruksha evam Utthaan Mahabhiyan	Feb, 2019
44.	GOBAR – DHAN	Galvanizing Organic Bio-Agro Resources	February, 2018

ICAR Programmes			
45.	NICRA	National Innovations on Climate Resilient Agriculture	Feb. 2011
46.	KIRAN	Knowledge Innovation Repository of Agriculture in North East	2012
47.	AGRI UDAAN	-	2017
48.	ARYA	Attracting and Retaining Youth in Agriculture,	
49.	READY	Rural and Entrepreneurship Awareness Development Yojana	July, 2015
50.	MGMG	Mera Gaon-Mera Gaurav,	
51.	Farmer FIRST	Farmer's Farm, Innovations, Resources, Science and Technology (FIRST)	
52.	NAHEP	National Agriculture Higher Education Project	Sep, 2017

ICAR AWARDS

S. N.	Awards name	Areas
1.	Sardar Vallabhai Patel	Outstanding ICAR Institute
2.	Chaudhary Devi Lal	Outstanding AICRP
3.	Rafi Ahmad Kidwai	Outstanding Research in Agril. & allied sciences
4.	Fakruddin Ali	Outstanding research in Tribal areas
5.	Hari Om Ashram Trust	Recognizes published original research, fundamental or applied, in crop science, horticulture, resource management and animal science
6.	Jawaharlal Nehru	Outstanding Doctoral thesis research in agricultural and allied sciences
7.	Vasantrao Naik	Outstanding research and application in Dry land farming
8.	Lal Bahadur Sastry	Young scientist
9.	Bharat Ratan C. Subramanyam	Outstanding teacher
10.	Punjab Rao Deshmukh	Outstanding woman scientist award
11.	Chaudhary Charan Singh Award	Excellence in Journalism in Agricultural Research and Development
12.	NG Ranga Farmer	Award Diversified agriculture
13.	Jagjivan Ram Kisan Puraskar	Innovative farmer

S. N.	Awards name	Areas
14.	Swami Sahajanand Saraswati	Extension work
15.	ICAR Award for Outstanding Multidisciplinary Team Research in Agriculture and Allied Sciences	Outstanding interdisciplinary team research
16.	Dr. Rajendra Puraskar	Technical bulletins in Hindi
17.	NASI-ICAR	Farm implements
18.	Haldhar award	Organic Farming
19.	Pandit Deen Dayal Upadhyay Krishi Vigyan Puraskar	Best KVK
20.	King Baudoin International Prize	Green Revolution

Some Important International Years declared by UNO

S. N.	Year	Purpose
1.	2004	International year of rice
2.	2005	International year of micro credit
3.	2006	International year of desert and desertification
4.	2007	International year of water (Theme-More Crop Per Drop)
5.	2008	International year of **Potato and Sanitation**
6.	2009	International year of **Fiber, Reconciliation, Astronomy and Gorilla**
7.	2010	International year of **Biodiversity and Rapprochement of Cultures**
8.	2011	International Year for **Chemistry, Forests and Youth**
9.	2012	International Year of Cooperatives and International Year of Sustainable Energy for All
10.	2013	International Year of Water Cooperation
11.	2014	International year of Family farming & Farmer Producer organizations
12.	2015	International Year of **Soil and Light and Light-based Technologies**
13.	2016	International Year of pulses
14.	2017	International Year of sustainable tourism for development
15.	2018	Contrary to common practice, the United Nations has not declared
16.	2019	International Year of **Indigenous Languages**
17.	2020	International Year of **Plant Health**
18.	2023	International Year of **Millets** (Proposed by Govt. of India to FAO)

Some Important Day's Celebration Related to India Agriculture

S. N.	Name of Days	Day of celebration
1.	**World Pulses Day**	10 February
2.	World Forest Day	21 March
3.	World Water Day	22 March
4.	Meteorology day	23 March
5.	National Technology Day	11 May
6.	World Honey Day	20 May
7.	World Anti- Tobacco Day	31 May
8.	National Tomato Day	1 June
9.	World Environment Day	5 June
10.	Mango Day	7 July
11.	**ICAR Day**	16 July
12.	Ozone Day	16 September
13.	World Animal Day	4 October
14.	World Egg day/UN International Day for Natural Disaster Reduction	13 October
15.	World Food Day	16 October
16.	National Agriculture Education Day (Babu Rajendra Prasad birthday)	3 December
17.	**World Soil Day**	5 December
18.	International Tea Day	15 December (started from 2005)
19.	**Kisan Diwas** and Farmers awareness week (Dec 23-29)	23 Dec on Ch. Charan Singh Choudhary birthday

AGRICULTURAL PRODUCTION IN INDIA

Food Grain

- **Total Food grain production** of India was **283.37 Mt** during 2018-19 (3rd Advanced Estimate).
- Government of India sets **food grain production target** with **290.25 Mt** for 2018-19 (Kharif- 144.55 Mt; Rabi- 145.70 Mt).
- Total **Pulse production** is **23.22 Mt** during 2018-19.
- **Chick pea** has highest production and largest area under cultivation among Pulses the crop while **Soyabean** in oil seeds crop.
- Total Oil seeds production (9 Oilseeds crops) is 29.88 Mt.
- The **Chickpea** production is highest among all pulses in India.

- **Madhya Pradesh** and Gujarat state are having highest pulses **production** and **productivity,** respectively.
- **Cereals** are deficient in **"Lysine"** amino acid; **Pulses** are deficient in **"Methionine".**

Area, Production, Productivity of Major Cultivated Crops in India

Group/ commodity	(Area: million ha, Production: million tonnes and Productivity / Yield: kg/ha)					
	Area		Production		Productivity / Yield	
	2017-18	2018-19	2017-18	2018-19	2017-18	2018-19
Food grain	128.0	122.7	285.01	**283.37**	2153	2292
Rice	43.2	43.4	112.76	115.63	2550	2665
Wheat	30.6	29.1	99.87	101.20	3216	4308
Jowar	5.1	4.0	4.80	3.80	889	934
Maize	9.9	9.1	28.75	27.84	2664	3046
Bajra	7.5	6.7	9.21	7.5	1311	1114
Pulses	29.5	28.3	25.42	23.22	779	849
Gram	9.6	9.7	11.38	10.30	973	1067
Tur	5.4	4.3	4.29	3.70	885	861
Oilseeds	26.2	25.5	31.45	31.50	1225	1234
Groundnut	5.3	4.9	9.25	7.0	1424	1436
Mustard	6.0	6.2	8.4	8.8	1324	1346
Cotton	10.8	12.3	32.05	30.1	519	414
Sugarcane	4.4	5.1	380	400	70(t/ha)	75(t/ha)

{Source: Economic Survey (GOI), 2018-19}; Notes: #Lakh bales of 170 kg. Each * 3rd Advance Estimates, - 2018-19

Per Capita Availability of Edible Items in the Country

Year / Item	Milk (gm/day)	Eggs (No./ annum)	Edible oils (kg/ annum)	Coffee (g/ annum)	Tea (g/ annum)	Sugar Kg/an.
1950-51	—	5.0	2.5	67	362	—
2002-03	—	38.0	7.2	67	623	—
2010-11	281	53	13.6	90	715	17.0
2017-18	**375**	**74**	**19.5**	—	-	19.5

Leading State in Production of Agricultural Crops

Crops	Production	Productivity
Rice	WB> UP>Punjab	Punjab
Wheat	UP>Punjab>Haryana	Haryana
Pulses	MP> Raj	Haryana
Oil seed	MP>AP	TN
Ground nut	Gujarat	TN
Mustard	Rajasthan	UP
Cotton	MH	GUJ
Jute	WB	GUJ
Coffee	Karnataka	Karnataka
Tea	Assam	—
Rubber	Kerala	—
Potato	UP	—
Onion	MH	GUJ
Sugarcane	UP	TN
Soybean	MP	AP

Average, Maximum, and Minimum Yield of Major Crops (Kg/ha)

Crops	All-India average (Kg/ha)	Maximum (Kg/ha)	Minimum (Kg/ha)
Wheat	3145	Punjab (5017)	Andhra Pradesh (500)
Rice	2416	Punjab (3952)	Madhya Pradesh (1474)
Maize	2676	Tamil Nadu (5372)	Assam (898)
Jowar	957	Andhra Pradesh (1661)	West Bengal (280)
Groundnut	1764	Gujarat (2668)	Himachal Pradesh (600)
Rapeseed & Mustard	1185	Gujarat (1723)	Tamil Nadu (241)
Soybean	1012	Andhra Pradesh (1612)	Uttar Pradesh (577)
Gram	960	Andhra Pradesh (1439)	Tamil Nadu (653)
Tur (Arhar)	813	Bihar (1667)	Andhra Pradesh (542)
Sugarcane	70522	West Bengal (114273)	Jammu & Kashmir (1000)
Cotton* Bale weight @170 kg each	510	Punjab (750)	Maharashtra (358)

Horticultural Crops (2019)

- Total area under horticultural crops was **25.87 Mha.**
- Total production of horticultural crops was **314.67 MT.**
- The productivity of horticultural crops was **12.10 tonnes/ha.**
- In India, total fruit production is **96.75 MT** and vegetables crops (**187.47MT**).
- India's shares in total world **fruits and vegetables** production are **13% and 10.7%,** respectively.
- **Recommended Dietary Allowance** (RDA) of **fruit and vegetables** are **120gm/day** and **300 gm/day** according ICMR. Whereas, **Per Capita Availability** (PCA) is **125.8gm/day** and **286.2gm/day**, respectively in India during 2019 (Ministry of Agriculture, ICICI Securities).
- The contribution of horticulture to Gross Value Addition (GVA)/GDP of Agriculture is 38%.
- Largest producer of **Kiwi fruit** in India is **Manipur.**
- Largest producer of **Lychee and Mesta** in India is **Bihar**.
- India's rank is **second** production of fruits and vegetables after China.
- In India, the leading producing state of **fruits, vegetable and plantation** crops are **Andhra Pradesh, West Bengal and Kerala**, respectively.
- Overall horticultural crop production in UP {Area (2.51 mha) and production (39.7 MT)} among all Indian states.
- In India, the fruit crop with highest **area and production** is **mango and banana,** respectively.
- The vegetable crop with highest area and production is **potato.**
- **Assam** and **Karnataka** are producing highest states in India for **tea and coffee,** respectively.
- Type of dichogamy in fruit- **PDSD** (avocado), **Duodichogamy** (chestnut), **Heterodichogamy** (pistachonut and pecanut) and **Protoandry** (coconut, sapota, walnut and *A. murocata*).
- Self incompatibility – **Sporophytic** (mango, banana and cocoa) and **Gametophytic** (ber, loquat, pineapple, apple, pear, peach).
- Type of parthenocarpic – **Vegetative** (banana, fig, pineapple), **Stimulative** (litchi, bread fruit grape cv. **Black Corianth**) **Stenospermocarpy** (grape and mango cv. **Sindhu**).
- Inflorescence in fig and pomegranate is **hypanthodium**.
- **Polinizers required** – in date palm (3:100) 10% male plants, Aonla (1:197-307), apple (33;67) and papaya (1:20) but 50:50 in dioecious cultivar.
- **Vivipary** in fruit- grape fruit, cocoa and jack fruit.
- **Cauliflorus** bearing found in jackfruit and cocoa.

- **TPD, Brown blast and Ridley** terms are related to Rubber.
- **Monoluron** content of coconut oil has **anti-HIV** property.
- **Coconut water** contains 94.5%water, **Vit C**+Vit B+ sugar+ **pH (4.8-5.3).**
- **Peach** is very susceptible to **Fe deficienc** and **grape** is **universally Mg deficienc** .
- **One kg** fully ripe fresh dates provide approximately **3150 calories.**
- **Avocado** is the **Fruit of New World** and **pilu** is a **Khara jal/salt bush.**
- **Fig** is one of the **first fruit p eserved** by **drying.**
- **Citrus** is **micro nutrient loving** plant whereas **banana** is **water loving plant.**
- **Nagpur mandarin** is the **world finest cultiva** in mandarin group.

Animal Husbandry (Census 2019)

- India's position in livestock population is first (largest livestock population).
- India is share of world livestock population about 15%.
- The total Livestock population is **535.78 million** in the country showing an increase of 4.6% over Livestock Census 2012.
- Total Bovine population (Cattle, Buffalo, Mithun and Yak) is 302.79 Million in 2019 which shows an increase of 1.0% over the previous census.
- The total poultry has increased by 16.81% and the total poultry is 851.81 million during 2019
- About **36% of the total livestock** is contributed **by cattle**.
- India's three largest states with highest livestock population are Uttar Pradesh followed by Rajasthan and Andhra Pradesh plus Telangana.
- The livestock contributing overall GDP of country is 4.5% and 25.8% in agriculture sector.
- Total milk production of India was **176.35 Mt** (2017-18).
- Per capita availability of milk in India is **376 gm/person/day.**
- **India** is having **first position** in milk production in the world with share in total world milk production is **17%.**
- Egg production in India was **88.03 billion** (2018-19) with 3rd position after China and USA.
- Per capita availability of eggs in India is **69 eggs person/year**.
- Cattle population in India (The highest in world) is **192.49 millions.**
- Buffalo population in India (The highest in world) is **109.85 millions.**
- Uttar Pradesh is the top meat producing state in India while, Andhra Pradesh in egg production.
- In India, the **first livestock ensus** was conducted in 1919.

Gestation Period and Body Temperatures of Animals/Birds

Animal/Bird	Group is known as	Meat	Gestation period / incubation, days	Body temp, ^{0}F
1. Cow	Herd	Calf: Veal Cattle; Beef	283 – 285	101
2. Buffalo		Bufen	300 – 310	
3. Goat (Doe)	Flock	Chavon	148	101
4. Hen	Folk	Chicken	21	107
5. Sheep (Ewe)	Flock	Mutton	150	103
6. Swine (Sow)	Herd	Pork	114	101

Aquaculture (2018-19)

- India's fish production to the global fish production is 6.3% (2nd position in world production).
- Total fish production in India is 12.60 MT {includes Inland (7.4 MT) and Marine (5.2 MT)}.
- The availability of fish in India is 9.0 kg/capita.
- The contribution of fisheries to national GDP is 0.91% whereas in agricultural GDP is 5.23%.
- World Fish Centre headquarter is located at Penang (Malaysia).
- Govt. of India was launched **"Blue Revolution Scheme"** during 2015 with objective to increase production and productivity of fisheries upto 20 MT by 2022-23.
- National Fisheries Development Board (NFDB) headquarter is suited at Hyderabad.
- The contribution of fisheries in agricultural exports is 20%.

Irrigation In India

- **USA** is having the highest area under drip irrigation in the world.
- **Maharashtra** is having the highest area under drip irrigation in India.
- The highest area under sprinkler irrigation in **Haryana.**
- Water requirement of irrigated wetland rice is **1500 mm.**
- Watershed development program established in 2003 at Hariyali.
- Highest area under irrigation is in U.P. state.
- In India, the highly irrigated crop is **Sugarcane (93.7%).**
- 69.5 mha is the potential area for micro irrigation in India.
- Actual (as on 2015) area under **micro irrigation** system is **7.73 mha** in India {drip irrigation (3.37 mha) and sprinkler irrigation (4.36 mha)}.

- Highest area under **micro irrigation** in India **Rajasthan** state (1.68 mha).
- In India, the sources of irrigation tube well and Canals are 57% & 32%, respectively.
- **USA** is having highest micro irrigation system (sprinkler).
- 90% of irrigation in **Israel** is under micro-irrigation.
- Sprinkler method of irrigation can provide protection against **Frost damage.**
- National Mission on Micro Irrigation **(NMMI)** was started in **2010.**

Irrigated Area under different crops (mha)

Crops	2008-09	Remark
Wheat	25.5(91.3)	Rice crop have highest irrigated area (26.5 mha)
Rice	26.5(58.7)	
Jowar	0.7(8.9)	
Maize	2.0(25.2)	
Bajra	0.8(9.4)	
Barley	0.5(75.1)	
Total cereals	56.2(55.9)	
Total pulses	3.8(16.0)	Sugarcane has highest percent irrigation area (93.7%)
Total food grains	60.0(48.3)	
Cotton	3.3(35.3)	
Oilseed	8.2(27.1)	
Sugarcane	4.5(93.7)	

Note: Figures in parentheses represent percentage of irrigated area to total area under the crop.

- Currently 68 million ha or 45 % of net cultivated area is irrigated in India (2014-15)

Fertilizers and Organic Farming in India

- **India** is having 3rd **rank** in **fertilizer production** and 2nd **rank** in **consumption** after China at global level.
- First fertilizer consumption country in the world is **China.**
- IARI has developed **'Pusa Soil Testing and Fertilizer Recommendation Meter'** for facilitating the reach of soil testing at the farmer's doorstep.
- Total consumption of fertilizers in India in is- **26.79 million metric tonne** (2017-18).
- The ratio of NPK use in India is **6.2:2.5:1** (2017-18).
- The optimum use of NPK ratio as per recommendation is **4:2:1.**
- In India, mostly imported fertilizer is **Potash** from Jordan and Tunisia.
- The highest total fertilizer consumption state is **U.P.**

- India average fertilizer consumption rate is **144.14 kg/ha.**
- **Punjab** is having **highest fertilizer consumption** (266 kg/ha) in India while **Nagaland** is using **lowest fertilizer** and **Pondicherry in UT states.**
- **Fertilizer Control Order (1985)** Neem Coated Urea (45 Kg bag) is incorporated as regular fertilizer in 2010.
- Indigenous manufacturing of Neem coated Urea was started in **2014** in India and it save **20-30%** of urea. It is sold at 5% above the MRP to recover the cost of coating.
- 100 % Neem Coated production of urea made mandatory from 2015.
- **Nutrient Based Subsidy Scheme** was started in the 2010-11.
- India's rank is 9th in organically cultivated area (1.5 Mha).
- Australia (27.15 Mha) is having the largest area under organic farming.
- Sikkim is the **first organic state in India's** (Declared on Jan 18, 2016) and second organic state is Uttarakhand.
- The largest area under organic certification is **Madhya Pradesh** > Himachal Pradesh > Rajasthan.
- The total organic production of India is 1.35 Million tonnes.
- The India is the highest number of organic producers in the world (More than 30 per cent of world's organic producers are in India).
- National Organic Farming Research Institute (**NOFRI**) is located at Gangtok **(Sikkim)** and it was established in **10th February 2016.**
- National Centre of Organic Farming (**NCOF**) is situated at Ghaziabad (Uttar Pradesh).
- Organics International (IFOAM) is situated at Bonn, Germany.
- 19th **Organic World Congress was organized at** by ISOFAR, NCOF and TIPI at New **Delhi, India** during 2017 with theme of **"Innovative Research for Organic Agriculture 3.0"**.
- **Mycorrhiza** is increase the availability of **Phosphorus** in the soil.
- **"Law of Tolerance"** was proposed by **Sheilford.**
- **"Nutrient mobility"** concept was given by **Bray.**

Consumption of Major Fertilizers (In Lakh Tonnes) at National Levels

Fertilizer	2014-15	2016-17	2017-18
N	169.5	167.15	169.58
P	60.9	67.05	68.54
K	25.3	25.1	27.79
N+P+K	255.8	259.5	265.91

Per ha consumption (kg/ha) 144.14 128.0 * includes production and imports

Pesticides and Herbicides

- India uses pesticides about **300** grams/ha.
- **Taiwan** is the largest per hectare pesticide consuming country in the world.
- The major pesticides applications are found in **1st Paddy and 2nd Cotton.**
- The major areas of crops in which maximum herbicides applied **Rice & wheat.**
- The top pesticide consumption states are **Punjab & Maharashtra.**
- Total number of registered & banned pesticides is **287 and 28,** respectively in India.
- The first herbicide in world was **2,4-D** developed in 1945 by **Zimmerman and Hitchkock (USA).**
- **Insecticides Act** was started in **1968** by **Thakur committee** recommendation.

Weeds and Herbicides

Weed/ herbicide	Example
Absolute weed	*Cyprus rotundus* in lawn, Loranthus in mango, Dodder in Medicago
Contact herbicide	Paraquat, Diquat
Mimicry weed	Phalaris in wheat field and wild rice in the field
Non–selective herbicide	Diquat, Paraquat
Noxious weeds	Parthenium
Relative weed	Rice in wheat field
Selective herbicide	Simazine, Atrazine, 2, 4-D, MCPA, Butachlor, Pendimethalin, Isoproturon, Fluchloralin
Systemic herbicide	Propanil, 2, 4-D (For Broad leaves), Atrazine, Simazine

Agricultural Export and Import

- Directorate of Marketing Research and Inspection (**DMRI**), situated at **Nagpur Maharashtra.**
- **APEDA** is the nodal Agency to coordinate the efforts on the Agricultural export zones.
- Total no. of Agri Export Zones in India is 60 and comprising of 40 agricultural commodities.
- Highest exported crops from India are **Cereals.**
- Highest imported crops in India are **edible oilseeds and pulses.**
- At present MSP is announced for **23 crop commodities** only.
- The concept of 'Climate smart agriculture' was given by FAO.
- Target year fixed by Govt. of India for Doubling of Farmers Income is 2022 (75th Independence of India).
- **India** is the largest producer and consumer of black tea in world.

- **India** is the largest producer, consumer and exporter of spices in the world.
- First Mega Food Park is started in Madhya Pradesh.
- **National Food security Act/bill** was passed in the year of **2013** for **Rice, wheat and coarse grains (Rs. 3:2:1 kg per month).**
- **Rajasthan** is the top wool producing state in India.
- **Minimum Support Prices:** Statement Showing Minimum Support Prices Recommended by **CACP** and Fixed by Government for different commodities (as on 03-07-2019) (Rs. per quintal).
- Agricultural Produce (**Grading and Marketing**) Act 1937 provided by the Ministry of Agriculture and Farmer Welfare.

Three farm reform bills (2020):

1. The Farmer's Produce Trade and Commerce (Promotion and Facilitation) Bill.
2. The Farmer's (Empowerment and Protection) Agreement on price assurance and farm Services Bills.
3. The Essential Commodities (Amendment) Bill.

List of MSP of Agronomic Crops

Commodity	Variety	MSP for 2019-20 (Rs/q)	MSP for 2020-21 (Rs/q)	Increase over previous year (Rs/q)
KHARIF CROPS				
Paddy	Common	1750	1815	65
	Grade 'A'	1770	1835	65
Jowar	Hybrid	2430	2550	150
	Maldandi	2450	2570	150
Bajra		1950	2000	50
Maize		1700	1760	60
Ragi		2897	3150	253
Arhar (Tur)		5675	5800	125
Moong		6975	7050	75
Urad		5600	5700	100
Cotton	Medium Staple (24.5-25.5 mm)	5150	5255	105
	Long Staple (29.5-30.5mm)	5450	5550	100
Groundnut in shell		4890	5090	200
Sunflower seed		5388	5650	262
Soybean	Yellow	3399	3710	311
Sesamum		6249	6485	236
Niger seed		5877	5940	63

RABI CROPS				
Wheat		1840	1925	85
Barley		1440	1525	85
Gram		4620	4875	255
Masur (Lentil)		4475	4800	325
Rapeseed/Mustard		4200	4425	225
Safflower		4945	5215	270
Jute		3700	3950	250

Area and Production of Cereals/Sugarcane at World Levels (Major Producing Countries during 2017)

Commodity	Area (m ha)		Production (Mt)		World	
	1st rank	2nd rank	1st rank	2nd rank	Area (M ha)	Production (MT)
Maize	USA (33.5)	China (24.5)	USA (371)	China (259)	197.2	1134.7
Rice	India (43.8)	China (30.7)	China (212)	India (168)	167.2	770
Wheat	India (30.6)	China (24.5)	China (134)	India (98.5)	218.5	772
Sugarcane	Brazil (10.1)	India (4.38)	Brazil (759)	India (306)	25.9	1841.5

Crop/Animals improvement through Biotechnology

- Bacillus thuringiensis was discovered by **S. Ishiwata.**
- **Dolly** (sheep clone) was created as **Nuclear Transfer Technology** by **Wilmont** (1997).
- **Garima** is the buffalo clone.
- First plant genome sequenced was done in **Arabidopsis** (2000) and **Rice** (2001) was the first commercial crop genome sequenced.
- **Golden rice** is a rich source of **vitamin 'A'**, which is having beta-carotene about **37 PPM.**
- **Pea** was recently genome sequenced crop (2020).
- The first transgenic plant (**Tobacco)** was created by **Michael (1983).**
- **Bikaneri Nerma (BNBt)** was first public sector transgenic Bt-cotton variety in India.
- The Protection of Plant Varieties and Farmers Rights (PPV&FR) Authority at New Delhi (Chairman – KV Parbhu), established in **Nov, 2005.**
- **Genetic Engineering Approval Committee (GEAC)** is the final approval committee for transgenic crop release in India which is works under **Ministry of Environment and Forest.**

- Leading transgenic crops are **Soybean > Cotton > Maize > Canola.**
- In world, total area under GM crops is **189.8 mha** in 26 countries while in India **(11.07 mha)** during 2018.
- First genetically modified (GM) varieties of cotton *i.e.* **PAU Bt-1 and F-1861** were developed by Punjab Agricultural University, Ludhiana.
- In the past years, the government has approved field trials of GM crops for field trails include transgenic were **Rice, Cotton, Maize, Mustard, Brinjal & Chickpea.**
- Leading countries in transgenic cultivation are **USA (75 mha)** > Brazil > Argentina > Canada > **India (5th).**
- The first genetically engineered food product was **FlavrSavr (Tomato)** for enhancing the shelf life fruit through incorporation of **Polygalacturonase (PG gene).**
- First cultivated GM crop in the world was **Soyabean.**
- Transgenic papaya cultivars are **SunUp and Rainbow** using against Papaya Ring Spot Virus **(PRSV).**
- **TILLING** stands for Targeting Induced Local Lesions in Genome.
- **PBR 322** is the most common **plasmid vector** used in genetic engineering.
- **Nif gene** for nitrogen fixation is cereal crops like wheat, jowar *etc.*, is introduced by cloning in **Rhizobium meliloti.**
- **Eco RI** is a Restriction enzyme.
- **Yeast** is a good source for **Protein.**
- **Hirudin is obtained from transgenic plant of *Brassica napus.***
- **DNA fingerprinting** is based on occurrence of **VNTR's.**
- **Virus free plants** can be obtained through shoot tip culture/meristemic culture/ shoot tip grafting.
- Three dimensional shape of tRNA is **Clover leaf-like.**
- **Anticodon** occurs in **tRNA** and most abundant RNA of the cell of **rRNA**
- **Bacterial plasmid** contains DNA.
- **Okazaki segments** are formed during **Replication.**
- Plasmids are vectors for gene cloning because they self replicate in bacterial cells.

The important gene and their role in plants

S. N.	Gene	Role	S. N.	Gene	Role
1.	Attacin-E gene/ Lactoferin	Bacterial resistance	8.	Anticoat protein	**Virus resistance**
2.	**SAM ase**	Fruit ripening (ethylene inducer)	9.	Bar gene	Nematode resistance
3.	Rol C	**Plant architecture**	10.	Collagenase	**Rooting inducer**

S. N.	Gene	Role	S. N.	Gene	Role
4.	**Bt- gene**	Lepidopetrans insect resistance	11.	**DREB**	Drought tolerant
5.	Chitinase	Fungus resistance	12.	P5Cs	Salinity tolerant
6.	HAL	Salinity tolerance in citrus	13.	AKT1, AKT2 and SKOR	K^+ uptake
7.	HSP, LEA protein	**Heat tolerant**	14.	LHC genes	light regulator

GENERAL AGRICULTURAL IMPORTANT FACTS

King/Queen/of Agricultural Crops

S. N.	Kings	Crops	Queens	Crops
1.	King of cereals	Wheat	Queen of cereals, Drosophila of crop plants -	Maize
2.	King of pulse	Chickpea	Queen of pulse	Pea
3.	King of fodder crops	Berseem	Queen of fodder crops	Lucerne
4.	King of fruits	Mango	Queen of fruits	Mangosteen
5.	King of oilseeds	Groundnut	Queen of oilseeds, Poor man's substitute for ghee	Sesame
6.	King of spices	Black pepper	Queen of spices	Small Cardamom
7.	King of flowers	Rose	Queen of flowers	Gladiolus
8.	King of nut fruit	Walnut	Queen of nut	Pecanut
9.	King of coarse/camel crop	Sorghum	Queen of beverage crops	Tea
10.	King of vegetables, Poor man's friend	Potato	Queen of vegetables	Okra
11.	King of day	*Cestrum diurnum*	Queen of night	Cestrum nocturnum
12.	King of temperate fruits	Apple	King of arid/semi arid fruit	Ber
13.	King of forest	Teak	King of weeds	Congress grass (*Parthenium hysterophorus*)

- Bio-energy plant is **Jetropa (35%).**
- Dead pupae of silkworm are a **Brown gold.**

- A famine reserve is **Millets.**
- Glory of East, Queen of East, Autumn queen **is Chrysanthemum.**
- Poor man's food is **Pearl millet.**
- Poor man's meat, Wonder crop is **Soybean.**
- **Cowpea** is Vegetable meat.
- Example of Pseudo cereal is **Buckwheat.**
- PUFA content is highest in **Safflowe (74%) and Sunflowe (68%).**
- Pulse crop that doesn't fix N is **Rajma.**
- Oleoresin is extracted from **Chillies.**
- Pashmina wool is obtained from **Goat.**
- Ramie (*Boehmeria nivea*) belongs to family Utricaceae yields strongest vegetable fiber from **Stem.**
- In sorghum, commonly **HCN synthesizes** in **roots** but it is accumulates in **Leaves** under stress condition.
- **Oxytocin** hormone is responsible for milk let down.
- Statement "Every thing else may wait but not agriculture" was given by J. L. **Nehru.**
- Statement "Our salvation can only come through the farmer. Neither the lawyer, nor the doctor, nor the rich landlord are going to secure it" was given by **M.K. Gandhi.**
- Contribution of South West monsoon rainfall in India is **80-90%.**
- Recently, a pain killer for cattle has been banned by Drug Controller General of India **Diclofenac.**
- System of Rice Intensification (SRI) requires only **5 kg seed/ha** (this method was developed at Madagakar in 1983).
- **Methane** is the important green house gas releases from paddy field & Wasteland.
- The swine flu (Human disease) is caused by **H1N1 virus.**
- **GD-2329** variety was contributed much in green revolution.
- **Iron** is the deficient element in milk.
- **Quincunx** (diagonal system) is most suitable method of fruit plantation in **highly fertile** soil. Whereas, **most common method** of fruit orchard is a **square system** in the India.
- First Agri. Chemist of ICAR was **J.W. Leather.**
- Punjab was selected by the Indian government to be the first state to try the new crops for **Green Revolution.**
- **IARI, New Delhi** was the landmark for Indian green revolution.
- **Terrace** and **contour** system of planting is suitable for hills and undulated lands.
- The theme of Earth Day (2016) was **"Trees for the earth".**

- National Biodiversity Authority (NBA) was established by the Central Government in 2003 to implement India's **Biological Diversity Act (2002)** and **Biological Diversity Rules, (2004).**

National Committee/Mission on Agricultural Crops

S. N.	Name of Committee	Set up (years)
1.	National Cotton committee	1921
2.	National Lac cess committee	1931
3.	National Jute Committee	1936
4.	National Coconut Committee	1945
5.	National Tobacco Committee	1945
6.	National Oil Seeds committee	1947
7.	National Arecanut Committee	1949
8.	National Spices and Cashewnut Committee	1958
9.	National Horticulture Mission	2005
10.	National Bamboo Mission	2006-07
11.	National Food Security Mission (**Rabi crops**)	2007

National Board/Mission/Authorities on Agricultural Crops in India (2020)

S. N.	Name of Board/ Authorities	Established at and year	Chairmen/MD	Works under
1.	National Coffee Board	Bangalore, 1942	M.S Boje Gowda	Ministry of Commerce & Industry
2.	National Rubber Board	Kottayam, Kerala, 1947	Dr. Sawar Dhanania	Ministry of Commerce and Industry
3.	National Central Silk Board	Bangalore 1949	H. Hanumanthappa	Ministry of Textiles
4.	National Tea Board	Kolkata 1953	Prabhat Kamal Bezboruah	Ministry of Commerce
5.	Agricultural Scientists Recruitment Board (ASRB)	New Delhi, 1973	A.K. Misra	Ministry of Agriculture & Farmers Welfare
6.	National Tobacco Board	Guntur, A.P. 1976	Raghunadha Babu Yadlapati	Ministry of Commerce & Industries
7.	National Coconut Development Board	Kochi, 1981	G. Jayalakshmi	Ministry of Agriculture & Farmers Welfare
8.	National Horticulture Board (NHB)	Gurugram, 1984	B. Srinivas	Ministry of Agriculture and Farmers Welfare
9.	National Spices Board	Kochi, 1986	Subhash Vasu	Ministry of Commerce and Industry

S. N.	Name of Board/ Authorities	Established at and year	Chairmen/MD	Works under
10.	National Fisheries Development Board (NFDB)	Hyderabad, 2006	Rani Kumudini	Ministry of Fisheries, Animal Husbandry and Dairying
11.	National Biodiversity Authority	Chennai, 2003	Dr. V. B. Mathur	Ministry of Environment and Forests
12.	National Commission on Farmers	2004	M.S. Swiminathan	Government of India
13.	National Rain Fed Area Authority (NRAA)	New Delhi, 2006	Ashok Dalwai	Ministry of Agriculture & Farmers Welfare

Seed Rates of Hybrids/Others

S. N.	Crop	Seed rates/ha	S. N.	Crop	Seed rates/ ha
1.	Hybrid Cotton Variety	2.0	5.	Bt-Cotton Variety	1.5
2.	Hybrid Rice Variety	10 – 15	6.	Hybrid Maize Variety	20 – 25
3.	Hybrid Wheat Variety	65	7.	TPS (Potato)	120 g/ ha
4.	Tomato Hybrids	100-150 g/ha	8.	Papaya - Dioecious Gynodioecious	400-500 g/ha 250-300 g/ha

Organisations Chairman/MD/President and Its Headquarters

S.N.	Organisation	Suited at and established years	Chairman/President
1.	ICAR	New Delhi, 16th July 1929	President- Narendra Singh Tomar DG- Dr. Trilochan Mahopatra
2.	ASRB	New Delhi, 1 November, 1973	Dr. A.K. Mishra
3.	PPV&FR	New Delhi	Dr. K.V. Prabhu
4.	NAFED	New Delhi Oct 2, 1958	Bijender Singh
5.	APEDA	New Delhi 1986	Paban Kumar Borthakur
6.	FSSAI	New Delhi 2011	Ms Rita Teaotia

S.N.	Organisation	Suited at and established years	Chairman/President
7.	FCI	New Delhi 1965	D. V. Prasad
8.	IFFCO	New Delhi , 1967	Dr. U S Awasthi
9.	CACP	New Delhi 1965	Vijay paul sharma
10.	ICRISAT	Hyderabad, 1972	Dr Peter Carberry
11.	FAO	Rome, Italy Oct 24, 1945	DG- Qu Dongyu
12.	United Nations	Newyork, USA, 1945	Antonio Guterres
13.	WTO	Geneva, Switzerland 1 January 1995	President - Andry Rajoelina Director-General-Roberto Azevêdo

Some important Acts/Orders in India

- Prevention of Food Adulteration Act, 1954
- Vegetable Oil Products (Control) Order, 1947
- Fruit Products Order, 1955
- Oiled Meal and Edible Flour (Control) Order, 1967
- Meat Food Products Order, 1973
- Edible Oils Packaging (Regulation) Order 1988
- Milk and Milk Products Order, 1992
- Food Safety and Standard Act, 2006

International Institute in Agricultural Research

- Agricultural Research Centre for International Development **(CIRAD)** at **Paris (France).**
- Institute National de la Recherche Agronomique (**INRA**) National Institute of Agricultural Research **at Paris, France (1946).**
- Centre for International Forestry Research (**CIFOR**) is at **Bogor, Indonesia.**
- Centro International de Agricultural Tropical – International Centre for Tropical Agriculture (**CIAT**) is at **Cali, Columbia.**
- Centro International de law Papa, International Centre for Potato **(CIP)** is at **Lima, Peru.**
- Centro International de Majoramiento de Maizy Trigo **CIMMYT** International Centre for the Improvement of Maize and Wheat is at **Mexico.**
- International Rice Research Institute **(IRRI), Manilla, Philippines.**

Indians Who Secured World Food Prizes/ Padam Bushan

S. N.	Persons	Fields	Years
1.	M.S. Swiminathan	for Green revolution	1987 (1st Indian)
2.	Vergese Kurien	for Milk revolution	1989 (2nd Indian)
3.	Mr. B R Barwale	Founder of MAHYCO	1998
4.	S.K. Khush	improved yield potential of rice	2000 (5th Indian)
5.	Modaduga V Gupta	for Aquaculture	2005
6.	Dr. Surinder K Vassal	for Quality Protein Maize	2008
7.	Gebisa Ejeta	for 1st sorghum hybrid for drought & srtiga weed	2009 (Ethiopia)
8.	K L Chadda	Golden revolution	2012

Concept and Related Persons

S. N.	Concepts	Related Persons
1.	PURA concept	A.P.J Abdul Kalam
2.	Bt. Cotton in India	C. D. Mayee
3.	Hybrid cotton	C. T Patel
4.	Crop Ideotype	Donald
5.	Super rice	G.S. Khush
6.	Golden rice	Ingo Potrykus
7.	Golden Revolution	K. L.Chadha
8.	Evergreen revolution in India	M. S. Swaminathan
9.	Green revolution in India	M. S. Swaminathan
10.	Green Revolution	N.E Borlaug
11.	Rainbow Revolution	Nitish Kumar
12.	Super wheat	S. Nagarajan
13.	Nobelization of sugarcane	T. S. Vankataraman
14.	White Revolution	Varghese Kurien
15.	Green Revolution Term	William Guadd
16.	Hybrid rice	Yuan Long Ping
17.	Green house effect	J. B. Fourier
18.	Drip irrigation	Symcha Blass
19.	Proteto (Proteto rich potato)	Ashish Dutta

Founder/ First in Agriculture

S. N.	First/founder	Name
1.	ICAR's first DG	Dr. B.P Pal
2.	President of ICAR	M. Habibullah
3.	KVK in India	Pondicherry
4.	Agricultural University	G.B university of Agriculture and technology, Pantnagar (1960)
5.	Aromatic Rice Hybrid	Pusa RH 10 (IARI)
6.	Biotechnological crop introduced in India	Bt, Cotton
7.	Commercial bio insecticide	**Sporeine**
8.	Commercial Fertilizers	SSP
9.	Country to introduce zero tillage	USA
10.	First Bt cotton variety	**Boll guard (cry1 Ac gene used)**
11.	First hybrid var. in India	Pusa Meghdoot (bottle gourd)
12.	Fungicide	Bordeaux mixture (inorganic in nature)
13.	Hormone	Secretin
14.	Hormone artificially produced by culturing bacteria	Insulin
15.	Hybrid cotton	H-4 (Surat Farm, Gujarat)
16.	Hybrid of Mustard	Pusa Jai Kisan (Bio- 902)
17.	Hybrid Pigeon Pea	ICPH-8 (ICRISAT Hyderabad
18.	Hybrid rice released in India	APHR-1 and APHR-2
19.	Hybrid variety of Bajra	HB-1
20.	Mango hybrid for commercial cultivation	Mallika (Neelam × Dashehari)
21.	Man-made cereal	Triticale
22.	Organochlorine compound	DDT
23.	Organophosphate compound	Parathion
24.	Plant Parasitic Bacteria	Fire Blight of apple
25.	Plant Parasitic Nematode	Anguina tritici
26.	Rice hybrid variety of India	Jaya
27.	Rice variety introduced in India	IR 8
28.	Sorghum hybrid	CSH 1 and CSH 2 (IARI)
29.	Super Rice variety for saline/alkaline conditions	Lunishree
30.	Systemic fungicide	Carboxin
31.	Transgenic plant	Tobacco and tomato

Terms related to Revolution in India

S. N.	Revolution	Field
1.	Black revolution	Petroleum products
2.	Blue revolution	Fisheries
3.	Brown revolution	Food processing / Fertilizers
4.	Evergreen revolution	Reduction in wastage of food grains, fruits and vegetables
5.	Food chain revolution	Food grains Production
6.	Golden revolution	Horticulture
7.	Gray revolution	Manures and Fertilizers
8.	Green revolution	Food grains
9.	Parbhani revolution	Okra
10.	Pink revolution	Prawn
11.	Rainbow revolution	Overall development of agriculture sector
12.	Red revolution	Meat/Tomato
13.	Round revolution	Potato
14.	Silver revolution	Eggs/Poultry
15.	White revolution	Milk
16.	Yellow revolution	Oil Seeds

Disciplines of Science and Its Fathers

S. N.	Science	Father
1.	Agricultural Chemistry	J.V. Liebig
2.	Agroclimatology	Koppen
3.	Agrometeorology	D.N. Walia
4.	Agrometeorology in India	LA Ramdas
5.	Antibiotics	Fleming
6.	Bacteriology	Leeuwenhoek
7.	Biochemistry	J.V. Liebig
8.	Biodynamic farming	Rudolf Steiner
9.	Biology	Aristotle
10.	Blood circulation	William Harvey
11.	Blood groups	Karl Landsteiner
12.	Botany	Theophrastus
13.	Co-operative movement in India	F. Nicholson
14.	Crop rotation	Norfork
15.	Cytology	Robert Hooke
16.	Cytoplasmic inheritance	Carl Correns

S. N.	Science	Father
17.	DNA finger printing technique	Alec Jeffrey
18.	Ecology	Reiter
19.	Economic ecology	M.S. Swaminathan
20.	Endocrinology	Thomas Addison
21.	Eugenics	Francis Galton
22.	Evolutionary ideas	Empedocles
23.	Experimental genetics	Thomas Hunt Morgan
24.	Extension education	A.Seaman and Knapp (in USA) & Leagans (in India)
25.	Fermentation	**Louis Pasteur**
26.	Field plot experiment	J.B.Boussingault
27.	Forest pathology	Robert Haring
28.	Fruit and vegetable preservation	M.Nicholas Apart
29.	Genetic engineering	**Paul Berg**
30.	Genetics	**G.J. Mendel**
31.	Golden revolution in India	Dr. K.C. Chadha
32.	Golden rice	Dr. I. Potrykus
33.	Green revolution	Dr. N.E. Borlaug
34.	Green revolution in India	M.S. Swaminathan
35.	Hybrid cotton	C.T. Patel
36.	Hybrid rice	Y.L.Ping
37.	Hybrid Rice in India	E.A. Siddiqe
38.	Immunology	Edward Jenner
39.	Indian Ecology	R. Mishra
40.	Indian mycology	E.J. Butler
41.	Indian Ornithology	Salim Ali
42.	Indian Palaeontology	Birbal Sahani
43.	Indian Plant Breeding	Dr. B.P. Pal
44.	Indian plant pathology	E.J. Butler
45.	Indian Remote Sensing	Dr. P.R. Pishroty
46.	Indian rust	Dr. K.C. Mehta
47.	Medicinal Bacteriology	Robert Koch
48.	Medicine	Hippocrates
49.	Microbiology	Loius Pasteur
50.	Microscopy	Marcello Malpighi
51.	Modern Agronomy	Pietro de Crenscenzi

S. N.	Science	Father
52.	Modern botany	C. Linnaeus
53.	Modern Cytology	Swanson
54.	Modern Genetics	T.H.Morgan
55.	Mutation	Hugo De Vries
56.	Mycology	A.M. Micheli
57.	Natural farming	**Masanobu Fukuoka**
58.	Nematology	N.A.Cobb
59.	Nitrogen Fixation	Winogradsky
60.	Organic farming	Albert Howard
61.	Paleontology	Leonard da Vinci
62.	Parasitology	F. Platter
63.	Pedology	Dokuchaev
64.	Plant Anatomy	Grew
65.	Plant Pathology	Anton de Bary
66.	Plant Pathology in India	E. J. Butler
67.	Plant Physiology	Stephan Hales
68.	Plant Tissue Culture	G.Haberlandt
69.	Polygenic Inheritance	Kolreuter
70.	Pure Culture Technique	Oscar Brefeld
71.	Radiation Genetics	H. J. Muller
72.	Safeners	O. L. Hoffman
73.	Sociology	AugusteCompte
74.	Soil Conservation	H.H.Bennett
75.	Soil Microbiology	S.N. Winogradsky
76.	Soil Science	Dokuchaev
77.	Soil Testing	M.L. Troug
78.	Statistics	R.A.Fisher
79.	Super Rice	Dr.Khush
80.	Taxonomy	Carolus Linnaeus
81.	Tillage and weeds	Jethro Tull
82.	White Revolution	Dr.V.Kurien
83.	Zoology	Aristotle

Field of Agriculture and Their Technical Name

S. N.	Name of Studies	Subject
1.	Apiculture	Bee keeping
2.	Cereology	Crops cycles
3.	Citriculture	Citrus
4.	Cryology	Snow, ice & frozen
5.	Cryptology	Codes and ciphers
6.	Eremology	Deserts
7.	Floriculture	Flowers
8.	Nephology	Clouds
9.	Olericulture	Vegetables
10.	Pomology	Fruits
11.	Siphonapterology	Fleas
12.	Tsiology	Tea
13.	Viticulture	Grape
14.	Xylology	Wood

Growth of Plants (Tropisms Means Response to Light)

S. N.	Tropisms	Direction of growth
1.	Gravitropism	Growth toward gravity
2.	Heliotropism	Response to Sunlight
3.	Phototropism	Growth toward light
4.	Thigmotropism	Growth toward solid objects
5.	Thigmotropism	Plant movement toward solid object

Classification of Plants Based on Its Adaptability

S. N.	Condition	Plants
1.	Calcium sensitive plants	Calcifuges
2.	Plant require large quantity of Ca	Calciphytes
3.	Gypsum loving plants	Gypsophytes
4.	Light loving plants	Heliophytes
5.	Non-vascular thallus plants	Bryophytes
6.	Plant grow under desert condition	Xerophytes
7.	Plant grow well under acidic condition	Acidophiles
8.	Plant grows on another plant for physical support	Epiphytes
9.	Plant grows on ice or snow	Cryophytes
10.	Plant prefer sandy soils	Psammophytes

S. N.	Condition	Plants
11.	Plant shows the presence of sub-surface water	Phreatophytes
12.	Plants able to grow on rocks	Petrophytes
13.	Plants grow on warm and moist climate.	Phanerophytes
14.	Plants growing in hot and dry conditions (remains in dormancy under adverse condition)	Therophytes
15.	Plants growing under extremely cold climate	Chamaephytes
16.	Plants grows on rock surface	Lithophytes
17.	Plants grows on two contrasting condition *e.g.*, Land & water	Amphipytes
18.	Plants prefer alkali soils	Basophiles
19.	Plants prefer saline conditions	Halophytes
20.	Plants prefer shady condition	Sciophytes
21.	Plants roots capable of penetrating into rock fissures	Chasmophytes
22.	Plants suitable for grassland condition	Hemicryptophytes
23.	Plants tolerant to high acidic soil condition.	Oxylophytes
24.	Seeded vascular plants	Spermatophytes
25.	Seedless vascular plants	Pteridophytes
26.	Water loving plants	Hydrophytes

Synonyms of Fruit Crops

S. N.	Fruit Name	Synonyms Name
1.	Banana	Adam's fig, Tree of wisdom, Tree of paradise, Kalptaru, Apple of paradise
2.	Guava	Apple of tropics
3.	Avocado	Butter fruit
4.	Kiwi fruit	China miracle, Horticulture fruit, Wonder of Newzeland, National symbol of Newzeland, Chinese Goosberry
5.	Cashew	Dollar earning crop, plough crop, gold mine of waste land
6.	Mandarin	Fancy fruit
7.	Cocoa	Food of good
8.	Makahana	Gorgon nut
9.	Aonla	India Gooseberry, Malacca tree, Myrabalan
10.	Jamun	Indian Black berry , Black plum, java plum
11.	Loquat	Japan plum, Japanese maldar
12.	Coconut	Kalpavriksha, Tree of heaven
13.	Walnut	King of nuts
14.	Apple	King of temperate fruits

S. N.	Fruit Name	Synonyms Name
15.	Papaya	Melon tree
16.	Jack fruit	Monkey jack
17.	Mango	National fruit of India, Bathroom fruit, King of fruits
18.	Water chestnut	Panifal, Singhara
19.	Ber	Poor man's fruit, King of arid fruits, Chinese date
20.	Tea	Queen of beverage crop
21.	Mangosteen	Queen of fruits
22.	Peanut	Queen of nuts
23.	Sapota	Sapodilla plum, Bully
24.	Oil palm	Small holders irrigated crop
25.	Phalsa	Star apple
26.	Carambola	Star fruit, Kamrakh, five corner fruit

Common Name of Different Vegetables

S. N.	Vegetables Name	Synonyms Name
1.	Tomato	Wolf apple, Vilayati Baigan, Poor man's orange (India) and love of apple (England)
2.	Capsicum	Bell pepper, Sweet pepper
3.	Bitter gourd	Bitter cucumber, Balsam Pear
4.	Summer squash	Bush Squash, Chappankaddu, Field pumpkin. Vilayatikaddu
5.	Pumpkin	Butternut squash, Vegetable of immense value
6.	Brinjal	Egg Plant , Aubergine
7.	Winged Bean	Goa Bean, Four angled Bean, Vegetable of 20th century
8.	Broad Bean	Horse Bean, Faba Bean
9.	Drumstick	Horse raddish tree, Sahijan, Muringa
10.	Dolichos Bean	Indian Bean, Hycainth Bean
11.	French Bean	Kidney Bean, Haricot Bean, Snap Bean, Navy Bean
12.	Okra	Lady finger
13.	Chekurmanis	Multivitamin Greens, Madura kheera, Vegetable of 21st century
14.	Yam Bean	Potato Bean, Misrikand
15.	Spinach beet (Palak)	Swiss chord, garden Beet, Beetleaf
16.	Cassava	Tapioca, Yuca
17.	Cowpea	Vegetable meat, Black eyed pea
18.	Cylon spinach	Water leaf
19.	Ash gourd	White gourd, Wax gourd
20.	Musk Melon	Wholesome food
21.	Lima Bean	Baby Potato, Double Bean, Butter Bean

National Flowers of The Different Countries

S. N.	Flowers Name	National of the Country
1.	Lotus	India, Egypt
2.	Rose	England, Iran, Maldives, Laos, Bulgaria, Persia, Turkey, U.S.
3.	Cattilya Orchid	Argentina, Brazil
4.	Water lily	Bangladesh
5.	Poppy	Belgium, Italy
6.	Blue eyed grass	Bermuda
7.	Narcissus	China
8.	Clover	Denmark
9.	Corn flower	Germany
10.	Lily	Italy, Canada
11.	Chrysanthemum	Japan
12.	Daisy	Litiva, Russia
13.	Tulip	Netherlands, Holland
14.	Crescent	Pakistan
15.	Jasmine	Pakistan, Paraguay, Philippines, Indonesia
16.	Carnation	Spain
17.	Golden rose	USA
18.	Daffodil	Wales

Flower for Different Purposes

S. N.	Flower	Purpose
1.	African Marigold	Vulgar mind
2.	Amaryllis	Pride
3.	Carnation (White)	Women`s love
4.	Daffodil	Regards
5.	French Marigold	Jealously/ Sorrow
6.	Iris	Message
7.	Lily	Purity
8.	Narcisuss	Self esteem
9.	Pansy	Thoughts
10.	Rose	Love
11.	Stock	Luxury
12.	Sweet pea	Departure

Terms and Associated Crops

S.N.	Practice	Crops
1.	Pinching (for loose flower production)	Carnation, Chrysanthemum
2.	Disbudding (for cut flower production)	Carnation, Chrysanthemum
3.	Scooping	Cauliflower
4.	Blanching	Cauliflower, Asparagus
5.	Bahar treatment	Citrus, Guava, Pomegranate
6.	Scuffing (soil stirring)	Coffee
7.	Nipping	Cotton
8.	Ginning	Cotton
9.	Notching	Fig
10.	Curing	Garlic, Onion Potato
11.	Scoring, Hilling	Gladiolus
12.	Topping	Gram, Coffee
13.	Girdling	Grape
14.	Pegging	Groundnut
15.	Retting	Jute
16.	Stripping	Jute, Oil palm
17.	Cincturing	Litchi
18.	De-tasseling	Maize
19.	Smudging	Mango
20.	Pinching	Marigold (To avoid the rotting under water logging condition)
21.	Ablation	Oil palm
22.	Tapping	Papaya, Rubber, Coconut
23.	Earthing –up	Potato, Pineapple
24.	Dapog seedling	Rice
25.	Wintering	Rose
26.	Mulching	Strawberry, Lettuce
27.	Wrapping, Propping	Sugarcane
28.	Arrowing, Rationing	Sugarcane
29.	Trashing	Sugarcane
30.	Skiffing, Tipping	Tea
31.	Sod culture	Temperate fruits
32.	De-suckering	Tobacco, Banana
33.	Curing	Tobacco, Tea
34.	Staking	Tomato, French Bean (pole type)
35.	Thinning	Turnip, Beet Root,
36.	Tapping	Rubber

Toxins Produced by Horticultural Crops

S. N.	Toxins	Related Plants
1.	Amygladin	Almond
2.	Oxalic acid	Amaranthus
3.	Arecotine	Arecanut
4.	Marmelosin	Beal
5.	Momordicin	Bitter guard
6.	Solosodine	Brinjal
7.	HCN glycosides	Cassava
8.	Apiin	Celery
9.	Monoluron (anti- HIV)	Coconut
10.	Sinirgin	Cole crops
11.	Ca oxalate	Colocasia
12.	Cucurbitacins	Cucurbits
13.	Ficin	Fig
14.	Haemaglutine	French bean
15.	Naringin	Grape fruit
16.	Resins	Mango
17.	Trypsin	Pea and beans
18.	Prunacin	Peach
19.	Bromelin	Pineapple
20.	Choline esterase	Pumpkin
21.	Soponine	Tomato, Asparagus
22.	Diosgenine	Yam

Toxins Produced by Agronomy Crops

S. N.	Toxins	Related plants
1.	Saponin and plant estrogens	Alfalfa
2.	Gossypol	Cotton
3.	Antivitamin E Factor	Field pea
4.	Aflatoxin	Groundnut
5.	Neurotoxin	Lathyrus
6.	Lathryrogenes	Lathyrus sativum
7.	Erucic acid	Rapeseed and mustard
8.	Eicosenoic acid	Rapeseed and mustard
9.	Polyphenolics	Safflower
10.	HCN glucocides	Sorghum
11.	Goitrogen	Soybean
12.	Polyphenolics	Sunflower
13.	Glucocides	White clover

Post Harvest Management

- In horticultural crops, **Post-harvest loss** is 20-40% and less than 2% **processing** in India.
- **Highest carbohydrate source** among fruit is **Banana** (36.4%) followed by **Durian** (34.2%) and **Beal** (30.6%) in **fresh form**. While **Apricot** (72.8%) and date palm (67.39%) as **dried/processed form.**
- Cassava and potato is the richest source of carbohydrate in vegetable crops.
- **Walnut (64.5%)** and almond (58.9%) are the richest source of **fat.** While **cashew nut** (21.2%) and almond (20.3%) in **protein** among fruit.
- **Vitamin A** (Mango 4800 IU, carrot), **Vitamin B_1** (Cashew nut 630 mg/100g, palak 0.26mg/100g), **Vitamin B_2** (Beal 1191 mg/100g, palak 0.56 mg/100g), **Vitamin C** (Barbados cherry 1400mg/100g, chilli).
- **Iron** in fruit - **Karonda 39.1%** and date palm 10.6%; in vegetables – Amaranthus and palak.
- **Canning** is done for prevention the spoilage in processed fruit and vegetables at > **100°C.** Whereas, **blanching** for removing the **field heat** and **inactivation of enzymatic activity** mostly in **vegetables** at **82-93°C.**
- **BHA** stands for Butylactic Hydroxy **Anisol.**
- **BHT** stands for Butylactic Hydroxy **Toluene.**
- **Class -1 preservatives** *e.g.,* – Salt, Sugar, spices, vinegar and ascorbic acid
- **Class –II** preservatives *e.g.,* Sodium Benzoate (used in **coloured product**) and KMS **(colourless product).**
- Use of low temperature in storage as **cellular storage** (>15°C), **refrigerator storage** (0-5°C) and **freezing storage** (-18 to -40°C).
- Presentation of bacterial contamination by temperature as **Pasteurization** (below 100°C), **Boiling/Cooking** (at 100°C), and **Canning** (for **fruits at 100°C** and for **vegetables at 116-120°C**).
- **Nicin** is an antibiotic produced by **Streptococcus lactic.**
- **Water soluble vitamins** – B and C; **Fat soluble vitamins** are A, D, E and K.
- **Caramelisation** - conversion of sugar to carbon and water.
- **Mailard reactions**- reaction of amino acid and reducing sugar bringing about suitable flavour that result browning of food and impart flavour.
- **Ethylene absorbent** – $KMnO_4$, methyl bromide.
- **Water absorbent** - $CaCl_2$, sorbitol and xylitol
- **O_2 absorbent** – Ferrus oxide; and **CO_2 absorbent** – hydrated lime and charcoal.
- Use of **AOA** and **AVG** for prevention of **ethylene biosynthesis.**
- **CA Storage** first time reported by **Kids and West** (1927).
- Pusa Zero Energy Cool Chamber **(PUSA- ZECC)** was invented by **Susanta K. Roy and D.S. Khuridiya** and it worked on the **direct evaporative cooling principle.** It reduce temperature by 10-15°C and maintain high relative humidity (95%) that can increase shelf life and retain quality of horticultural produce.

- **HACCP** plan includes 12 logically sequenced steps.
- **Citrus degreening – treat with ethylene @ 20 PPM and stored at 26.7°C + 80-90% RH**
- **Staphylococcal** food poisoning is most commonly occurred in food born diseases.
- **Lye peeling** is done in potato with **1-2% NaOH** for 30 second to 2 minutes.
- **Drying and dehydration** temperature for **vegetables** is **60-66°C** and for **fruits (66-71°C).**
- **Jelly** –TSS (65%), pH (3.2), Pectin (1%).
- **Jam** - Acid (0.5-0.6%), TSS (68%), pH (3.0) End temperature (105°C).
- **Curing temperature** – Onion (40°C), Potato (20°C) and Sweet Potato (29°C).
- TSS is measured by **Refractometer** and Pectin by **Jelmeter.**
- **Highest perishable** fruit is **Apricot** than Banana while in vegetable is **Asparagus** and broccoli.
- **Highest ethylene production** – Apple (25-2500µl/l) and passion fruit (466-530 µl/l).

Fermented Beverages

S.N.	Beverages	Sources	Remarks
1.	Cider	Apple	Fermented organism – Yeast
2.	Feni	Cashew nut	It prepared from cashew apple
3.	Neera	Palmyra palm	
4.	Toddy	Palmyra palm	5% alcohol
5.	Wine	Grape	Alcohol - 7-20% (It is the oldest wine in fermented beverages)

Unfermented Beverages

S. N.	Product	Fruit juice%	Acidity%	TSS%
1.	Barley water	25	9.0	30
2.	Chutney	40	1.0	50
3.	Cordial	25	1.5	30
4.	Crush	25	1.0	55
5.	Glazed fruits (candid, crystalized)	Not less than 25	-	Not less than 70
6.	Jam	45	0.5 -0 .7	68
7.	Jelly	45	0.5 - 0.7	65
8.	Natural juice	100	-	-
9.	Nectar (All fruits)	20	0.3	15
10.	Orange and pineapple Nectar	40	-	15

S. N.	Product	Fruit juice%	Acidity%	TSS%
11.	Preserve	55	-	68-70
12.	RTS	10	0.3	10
13.	Sharbat	-	-	65
14.	Squash	25	1.0	40-50
15.	Sweetened juice	85	-	10
16.	Syrup	25	1.3-1.5	65
17.	Tomato K*etc.,*hup	-	-	28
18.	Tomato Sauce	-	1.2%	30

Some Important By Product of Fruit and Vegetable Commodities

S. N.	By product	Source
1.	Amachur	Immature mango (keri)
2.	Anardana	Pomegranate
3.	Chali	Arecanut
4.	Cidar	Apple
5.	Cocopith	coconut
6.	Fenny	Cashew apple (cashew nut)
7.	Gulgand	Rose
8.	Jaggery, Neera	Palmyrah palm
9.	Jam	Apple
10.	Jelly	Guava
11.	Kalipak	Arecanut
12.	Kangi	Carrot
13.	Kottpak	Arecanut
14.	Liquor/ Dibbis	Date palm
15.	Mouth watering	Banana
16.	Paddaneera	Palmyrah palm
17.	Palmolin	Oilpalm
18.	Papain	Papaya
19.	Perry	Pear
20.	Peta	Ash gourd
21.	Preserve	Aonla
22.	Rose water	Rose
23.	Sugar	S/cane, Beetroot
24.	Toddy	Palmyrah palm, coconut
25.	Tooty –Fruity	Papaya, Bottle gourd

Storage Criteria of Horticultural Crops

S. N.	Crops	RH %	Storage life (weeks)	Storage temperature (⁰C)
1.	Apple	85-95	16- 32	-1 - 0
2.	Banana	80-85	1-2	13
3.	Brinjal	85-90	2-3	10-12
	Cucumber	—	2	—
	S. Potato	—	13-20	—
4.	Carrot, radish	95	20-24	0
5.	Cole crops	>90	4-6	0-1
6.	Garlic, onion	<60	20-35	0
7.	Grape, fig, litchi, pomegranate			0-2
8.	Green chilli	—	—	7-8
	Ripe chilli	—	—	5-6
9.	Mango			8-9
10.	Pea, Turnip	95		0
11.	Pear		12- 26	0-1
12.	Pineapple, Jackfruit	80-85	5-6	11-13
13.	Potato	80	35	3-4
14.	Rose (dry), Chrysanthemum, Carnation	—	-0.5 to 0	—
15.	Rose (preservative)	—	-0.5 to 2	—
16.	Gladiolus	—	2.5	—
17.	Dahlia, Delphidium	—	4	—
18.	Vanda	—	13	—
19.	Poinsettia	—	10-15	—

Isolation Distance Kept in Seed Production

S. N.	Crop	Foundation seeds	Certified seeds
1.	Beet	1600	1000
2.	Brinjal	200	100
3.	Carrot	1000	800
4.	Chilli	400	200
5.	Chrysanthemum, Marigold	400	-
6.	Cole crops	1600	1000
7.	Cucurbits	800	400

S. N.	Crop	Foundation seeds	Certified seeds
8.	Okra	400	200
9.	Onion	1600	1000
10.	Other beans	50	25
11.	Pea	20	40
12.	Potato	5	5
13.	Radish	1600	1000
14.	Sweet pepper	400	200
15.	Tomato	50	25
16.	Turnip	1600	1000

- **World food price 2020** –Eminent Indian-American soil scientist **Rattan Lal** at Ohio State University and lauding his research in **soil science**, saying he is helping millions of small farmers around the world with his work on **increasing food production and recycling of nutrients**; and recipient of **$250,000** World Food Prize. Dr. Lal said the "urgent task of feeding humanity is not fulfilled until each and every person has access to an adequate amount of nutritious food grown on a healthy soil and in a clean environment".
- The **World Food Prize** was **created** by Nobel Peace Prize laureate **Norman Borlaug in 1986** to recognise scientists and others who have improved the **quality and availability of food.** The foundation that awards the $250,000 prize is based in Des Moines. This award is **considered to be the Nobel Prize in Agriculture.**
- The **COVID-19 pandemic,** also known as the coronavirus pandemic, is an ongoing pandemic of coronavirus disease 2019 (COVID-19), caused by severe acute respiratory syndrome coronavirus 2 **(SARS-CoV-2).** The outbreak was **first identified in Wuhan, China, in December 2019.** The **World Health Organization declared** the outbreak a **Public Health Emergency** of International Concern on **30 January 2020,** and a **pandemic on 11 March 2020.**
- The usual **incubation period** (the time between infection and symptom onset) ranges from **01 to 14 days,** and is most commonly five days.
- Newly discovered plant gene could boost phosphorus intake - **CLE53 gene** which regulates cooperation between fungi and plants (mycorrhiza).

- **Kerala is the first state** of India which is **set to introduce** a **Minimum Support Price (MSP)** for **16 fruit and vegetable crops**, starting from **November 1, 2020**.

List of fruits (02) and vegetable (14) with MSP

S.N.	Crops	MSP (Rs/kg)	S.N.	Crops	MSP (Rs/kg)
1.	Banana Wayanadan banana	30.0 24.0	9.	Beans Bean Nadan (Vallipayar)	28.0 34.0
2.	Pineapple	15.0	10.	Tomato	8.0
3.	Tapioca	12.0	11.	Okra	20.0
4.	Garlic	139.0	12.	Carrot	21.0
5.	Cucumber	8.0	13.	Potato	20.0
6.	Bitter gourd	30.0	14.	Beet root	21.0
7.	Snake gourd	16.0	15.	Cabbage	11.0
8.	Ash gourd	9.0	16	Cluster bean	34.0

- **Giloy** (*Tinospora cordifolia*) is declared national medicine of 2020 by Ayush ministry.
- **Maru Sena:** is a biopesticide, released by CAZRI, Jodhpur. It is isolated from *trichoderma harzianum*.
- **Pusa waste Decomposer:** It degrade the stubble within 25 days to 3 months. It is contain the Itrd strain of fungi, which assist in producing enzymes essential to quicken the decomposition of bio-mass. It contain four capsules @ Rs. 5.0 for one acre of farm waste and will solve stubble burning problem, reduce pollution and make soil fertile.
- The prime minister issued a **commemorative coin of Rs. 75** and dedicated **17 biofortified varieties of 8 crops** to the nation on the occasion of 75th Anniversary of the **Agriculture and food organization (FAO)** dated 16 October, 2020.
- Recently, Three institute were merged in ICAR namely National Institute of Plant Health Mamagement (NIPHM), Hyderabad, National Institute of Agricultural Extension Management (MANAGE), Hydrabad, CCS National Institute of Agricultural Marketing (NIAM), Jaipur.

Chapter

2

SOIL SCIENCE

- Soil science is the study of soil as a natural resource on the surface of the Earth including soil formation, classification and mapping; physical, chemical, biological, and fertility properties of soils; and these properties in relation to the use and management of soils.
- According to **Buckman and Brady**, "Soil is a dynamic natural body on the surface of the earth in which plants grow, composed of mineral and organic materials and living form"
- Soil as a natural body composed of mineral and organic constituents, having a definite genesis and a distinct nature of its own **(V.V. Dokuchaev - father of soil science)**.
- Soil is Latin word **Solum means Floor/ground.**
- **Organic matter** is the **store house of nutrient** in the soil,**source of energy** for micro organism in the soil and **act as chelate.**

ROCKS

- Earth crust consists of 106 elements and out of which 08 elements makes composition 99 per cent earth crust. Oxygen, silicon, aluminum, iron *etc.,* in the decreasing order of their amounts.
- **Lithopile:** is the major element group.
- **Petrology:** (Greek words petros means Rock and logos means "subject matter") is the branch of geology that studies the **origin, composition, distribution and structure of rocks.**
- **Petrography:** deals with the description of rocks and petrogenesis.
- **Petrogenesis:** is a study of the origin of rocks.
- Rocks are basically divided into 3 types.
- **Igneous Rocks:** Granite, Basalt (Deccan Trap), Gabbro, Pumice, Rhyolite and Tracheolite. Acidic igneous rocks contain 60-75% quartz *e.g.* granite. Basic igneous rocks contain less than 50% of quartz *e.g.* Basalt.
- **Sedimentary rocks:** Limestone, sandstone, Shale, **Dolomite** and **Conglomerate.**
- **Metamorphic rocks:** are formed from Igneous and sedimentary rocks *e.g.* **marble, gneiss**, Schist, graphite, slate.

- **Important conversions:** Gneiss is formed from Granite. Marble from Limestone, Graphite from Coal, Slate from Shale and quartz from sandstone.
- 74 per cent sedimentary rock is found in the earth crust while igneous rocks percent is 18 per cent only.
- Granite is 83 percent part of igneous rocks.
- Granite and Sandstone are **acidic nature** of rocks.

Soil Origin, Weathering, Morphology, Factors and Processes of soil formation

- Rocks are the chief sources of soil parent materials over which soils are formed.
- Soil is formed from weathering of rocks.

Weathering = Disintegration + Decomposition

- Disintegration breaks consolidated rocks into unconsolidated parent materials, which on further breaking and chemical decomposition forms soil.
- **Weathering** is the process of disintegration and decomposition of rocks, soil and minerals through contact with the Earth's atmosphere, waters and biological organisms.
- **Physical weathering:** involves agents such as temperature, Water, wind, Plant and animals and processes such as exfoliation (surface peeling off of rocks), alternate wetting and drying, freezing and thawing.
- Water, on freezing, expands 9% by volume.
- **Chemical weathering reactions:** Solution, hydration, Hydrolysis, carbonation, Oxidation and reduction.
- **Biological weathering:** due to man, animals, micro organism and roots of higher plants *etc.*,
- Chemical weathering is not possible without presence of water (Hydrolysis).
- Most resistant mineral to weathering is **quartz** while least resistant is **gypsum**> calcite> dolomite.
- **Stokes's Law** the relation between the radius of a particle and its rate of fall in a liquid is known as Stockes's law.
- Geo-chemical weathering is the weathering taking place at the layers down below whereas pedo-chemical or pedogenic weathering is the one taking place at the surface and subsurface layers.
- Soil found at the site of formation- sedentary soils, whereas soils found far away from the site of formation are called Cumulous or transported soils
- Agent of transportation & name of the soil formed: River water-alluvium, lake water- lacustrine, seawater-marine, wind-aeolian if coarse and loess if fine particles, gravity-colluvium, ice-moraine.
- Parent material is an end product of weathering.

Stages of soil formation- Infancy, Youth, maturity and old age

- Soils wherein there is continuous deposition of materials always remains young *e.g.* Desert soils and alluvial deposits.
- The study of origin, formation and geographic distribution of soils in nature is known as **Pedology**, whereas the study of soil in relation to crop growth is known as **Edaphology.**
- **Soil profil** : is the vertical section of soil through all its horizons and it extends up to its parent materials.
- A pedon is the smallest volume that can be recognized as a soil individual and it is 3-dimensional.
- The horizontal layers in a soil profile are called horizons.
- Typically there are 4 horizons - O, A, B and C. **O horizon is the organic horizon** found in the forest soils. A horizon is below O horizon out of which A_2 layer is called illuvial horizon. B-horizon is found below A horizon and is mostly called illuvial horizon. Elluvial layer is the one wherein soil materials are removed whereas illuvial layer is one wherein soil materials removed from other layers are deposited.
- Mineralization and immobilization of nutrient is occurred in A horizon and B horizon respectively.
- C-horizon consists of **unconsolidated parent material** sand **least weathering horizon.**
- Water stagnation in soil is due to hard pan of calcium carbonate ($CaCO_3$) or gypsum *etc.*,
- Below the C-horizon is found the **R layer**, which is known as the **Bed Rock**.
- **Solum** = A+B horizons
- **Regolith** = A+B+C horizons
- The study of the soil in the field condition with the help of various morphological characteristics is called **Soil Morphology**.
- **Dokuchaev (father of soil science)** first described about the factors of soil formation, which were later, modified by **Jenny**.
- The five factors of soil formation are **cl, o, r, p, t** - Climate and organism (**active factors**), Relief or topography, parent material and time (**passive factors**).
- There are various processes of soil formation out of which two are important like Laterisation and Podzolisation.
- **Laterisation** occurs in warm humid tropical conditions whereas **Podzolisation** occurs in the cold humid temperate conditions.
- In Laterisation, Silica (SiO_2) is removed from the A horizon and sesquioxides are left out in the A-horizon whereas in Podzolisation, sesquioxides are leached from the A horizon and silica is left out in upper layers.
- Laterite is used for preparing bricks; Laterites are very well weathered soils whereas laterite soils are still undergoing some weathering to become laterites.

Nature and composition if soil

- Soil is a 3-phase, particulate, disperse, porous, open and heterogeneous system
- Ideal soil contains 50% solid matter (45% mineral matter and 5% organic matter) and 50% pore space (25% air and 25% water).
- Out of the 3 phases, **Solid phase is the constant phase** in terms of composition.
- There are basically 3 soil separates viz., sand, silt and clay. There are two systems of classifying the sizes of these separates.

Classification of soil particles based on siz

S.No.	Particles names	Size (diameter in mm)	
		USDA (mm)	ISSS (mm)
1.	Gravel	72.0	72.0
2.	Sand – very course	1 – 2	–
	– course sand	0.5 – 1.0	0.2 – 2.0
	– medium sand	0.25 – 0.50	–
	– fine sand	0.10 – 0.25	0.02 – 0.20
	– very fine sand	0.05 – 0.10	–
3.	Silt	0.002 – 0.05	0.002 – 0.020
4.	Clay	<0.002	< 0.002

- India follows International system of particle size classification. It is given by Atterberg.

Physical properties of soil

- **Soil texture** and **soil structure** are both unique properties of the soil that will have a profound effect on the behavior of soils, such as water holding capacity, nutrient retention and supply, drainage, and nutrient leaching.
- **Soil texture** is the relative proportions of sand, silt, or clay in a soil.
- The **soil textural groups are 12** based upon these relative proportions *e.g.* sand, loamy sand, sandy loam, loam, silt loam, silt, sandy clay loam, clay loam, silty clay loam, sandy clay, silty clay and clay.
- Soils with the **finest texture are called clay** soils, while soils with the coarsest texture are called sands. However light textured or coarse textured soils are easy to plough whereas heavy textured or fine textured soils are difficult to plough.

To determine the soil texture knowing percent sand silt and clay using the table belo

Soil classification	Clay Soil	Loam soil	Sandy soil
percent clay	40-100%	7-27%	1-10%
percent silt	0-40%	28-50%	1-15%
percent sand	0-45%	23-52%	85-100%

- Gravel is neither a soil separate nor one of the soil textures.
- Particles greater than 2mm are not considered as soil.
- Texture is determined by **International Pipette and Bouycous Hydrometer** method.
- Stoke's law is applied in the determination of soil texture.
- Silt has intermediate characteristics between sand and clay.
- Loamy texture and granular or crumb structure (Sandy loam) is the best soil for agriculture cultivation whereas especially black soil for cotton and light soil for groundnut, tubers and roots and bulb crops
- Clay is called the active fraction of the soil. It is made up of alumino-silicates.
- Soil texture is a static/natural property of soil which **can't be changed**.
- **Density of soil (D)**: is the weight per unit volume of a substance is called density. It's expressed in **gm/cc or lb/cft or mgm^{-3}.**

  D = Mass (M)/ Volume (V)
- Soil density is measured by **Pycnometer**
- **Bulk density:** The mass (weight) per unit volume of a dry soil (volume of solid and pore space) is also called apparent density.
- Bulk density (A) is approximately half that of particle density. So bulk density is always lower than particle density value of **Bulk Density is 1.4 to 1.8 Mg/m^3**. Bulk density increases on compaction whereas it decreases on addition of organic matter.
- Bulk density of organic peat (0.5 gm/cc), sand soil (1.7 gm/cc) and compact sub soil (2.0 gm/cc).
- Poor soil indicates high BD and low BD is favorable for crop growth.
- **Particle density (True Density)** is the weight per unit volume of the solid portion soil or without pore space.
- The particle density (T) of soils is around **2.65 Mg/m^3** due to dominance of quartz whose density is also of the same value.
- Macro pore is found more in sandy soil while micro pore is more in clay soil.
- Maximum pore space texture is clay and structure is **granular** and crumb.
- **Porosity** percentage pore space is calculated by the formula **100x (T-A/T).**
- Values: Sandy soil +30% clay soils =50-60% and loams = 40-50%
- **Macropores** are more in coarse textured soil and are primarily **filled with air** while **micropores** are more in fine textured soils and are mostly **filled with water**.
- **Soil structure** is the arrangement of primary soil particles and their aggregates into certain definite pattern is called soil structure.
- Primary soil particles are sand, silt and clay.
- **Peds:** natural aggregate of soil; **Clods** are artificial aggregate by tillage and broken peds are called **fragment**

Type of soil structure

1. **Basic types** - Granular, Blocky, Prismatic and Platy
2. **Subtype** - Columnar
3. **Structureless types** - Massive and Single Grained
4. **Pseudo-types** - Oxic and Vertic.

Examples of Soil Structure Types

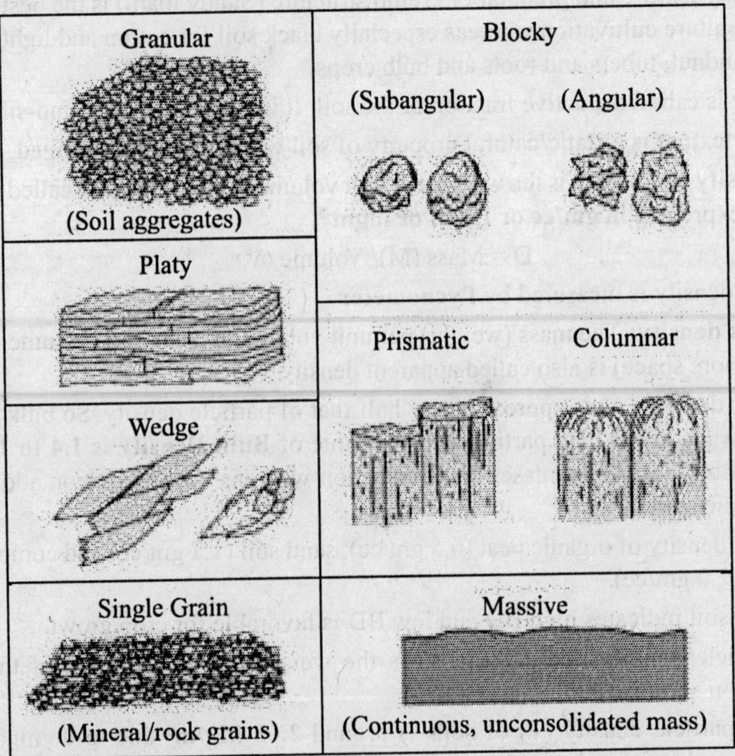

Source: http://weble.ugent.be/WLES/doc/UGent.WLES.Data.SoilData.SoilStructure.html

Type of soil structure and its remarks

S. No.	Type of structure	Remarks
1.	Plate like	• Horizontal dimension is much developed than vertical axis resulting a flattened, compressed and len like peds.
	Platy	• Thick plate like
	Laminar	• Thin plate like
2.	Prism like	• Vertical axis is more developed than other, flatted, and give pillar like shape
	Columnar	• Top of such ped is round
	Prismatic	• Clean cut

S. No.	Type of structure	Remarks
3.	Block like	• All three dimension are about same size, peds are cube like flat or rounded face
	Angular	• Cube and sharp blocky
	Sub angular	• Rounded (face and edge)
4.	Spheroid	• All axes are developed equally with the same length, curved and irregular face
	Granular	• Simple/fine aggregate and less porous
	Crumb	• When granular are especially porous

- Grade denotes the durability or stability of the aggregates – structureless, weak, moderate, strong and very strong.
- Class of soil structure denotes size of the aggregates: very fine, fine, medium, coarse and very coarse.
- Structure of the soil can be easily changed by management practices.
- Physical property of soil is most important for plant growth due to amount and nature of porosity.

SOIL COLOUR

- The colour of the soil is a result of the light reflected from the soil. It is an easily observable soil property and gives an immediate indication of the soil condition.
- Soil colour is inherited from its parent material and that is referred to as **litho-chromic**, *e.g.* red soils developed from red sandstone.

Determination of soil colour

- The soil colours are determined by using **Munsell colour chart**. In this chart, different colour chips are systematically arranged by three variables namely Hue, Value and Chroma.
  - **Hue** - it indicates the **dominant spectral** colour (red, yellow, blue and green).
  - **Value** - it indicates **lightness or darkness** of a colour (the amount of reflected light).
  - **Chroma** - it represents the **purity** of the colour (**strength of the colour**).
- Soil colour is usually due to 3 main pigments: **Black**—from organic matter, **Red**—from iron and aluminium oxides, **White**—from silicates and salt.
- **High denitrificatio risk and methane emission hazard** is problem of grey and green colour of soil.

Soil Plasticity and cohesion

- Plasticity is the capacity of the soil to change its shape under moist conditions. It exhibits more in clay soil.
- Cohesion is the capacity to stick together.
- Plastic soils are cohesive.
- **Redox potential:** is regulated by the concentration of reduced and oxidized substance. formula – Eh= E0
- **Redox potential (Eh):** The redox potential is a measure (in **volts**) of the affinity of a substance for electrons. Its electronegativity compared with **hydrogen** (which is set at **0**).
- **Formula** – Eh= E_0+RT/nF In (Oxidation.)/(Reduction)
- Eh of well drained soil (400- 700 mv)and water logging (-250 to 350 mv)

SOIL COLLOIDS

- Colloids are **inorganic (clay) and organic (humus),** which have very small particles size (**0.001mm or smaller 1 micron**) containing large surface area per unit mass.
- The two phases are the dispersed phase (clay and humus) and dispersion medium/ water).
- Soil colloids also carry electrostatic charges (- and +). Most of the organic and inorganic soil colloids carry a **negative charge.**

Types of Soil Colloids

1. Crystalline silicate clays - Phylosilicates (tetrahedral and octahedral crystal sheets)
2. Non-crystalline silicate clays –Dominantly amorphous clays (allophane and imogolite)
3. Iron and aluminum oxides –Dominantly gibbsite (Al-oxide) and goethite (Fe-oxide)
4. Organic (humus) colloids –Non-crystalline colloids dominated by long C-chain molecules

- Colloidal property increases with decrease in diameter.
- When colloidal particles are coagulated by adding oppositely charged ions called **flocculatio** . Maximum flocculation cation mineral are Al^{+3}>Ca^{2+}. Whereas minimum flocculation of $Na^+ > K^+$.
- Colloids exhibit the property of sacrificial activity such as the capacity to hold solid, gases, salts and ions.
- Soil colloids have high exchange capacity, which increases with silica sesquioxides ratio.

Silicate Clay Mineral Groups:

Group	Layer type	Cation to anion exchange capacity (EEC/AEC)	Remarks
Kaolinite	1:1	0.5	Mineral clay
Illite	2:1	2.3	Non expanding
Vermiculite	2:1	—	Limited expanding
Smectite/Montmorillonite	2:1	6.7	Most prominent expanding typeclay minerals with **swelling & shrinking characteristics**
Chlorite	2:1:1	—	—

SOIL WATER

- Water present in soil pores is called soil water.
- It is an important component of the soil which influences soil organisms and plant growth.
- It serves as a solvent and carrier of nutrients for plant growth.
- It regulates soil temperature and helps in chemical and biological activities of soil.
- It is essential for soil forming processes and weathering.
- **Water has maximum density at 4^0 C.** One molecule of water is attached to four molecules in the neighborhood.
- The diameter of water molecule is **3^0A ($3x10^{-10}$m).**
- Water has high dielectric constant of 80. Its surface tension is **$7.2x10^{-2}$N/m.**
- Structure of water molecule is **hexagonal lattice** and the **angle is 105^0.**
- **Adhesive**: An attraction between two unlike/dissimilar molecules. *e.g.* (water + soil).
- **Cohesive**: An attraction /bonding between two similar molecules *e.g.*(water + water)
- **Heat of solution** : when **ions are hydrated a large amount** of energy to released is known as heat of solution
- **Heat of wetting:** when **clay particles are hydrated in a certain amount** of energy to release is called heat of wetting.
- **Adsorbed water:** always less free energy than water in the pool **(zero potential)**. Free energy of soil water potential is measured by **pF Scale.**
- **pF Scale:** The logarithm of centimeter height of a water column to give the necessary suction is called pF scale. p means logarithm and F means free energy.
- Saturated soil having pF value lowest (0) and height (1.00 cm)

Physical classification of soil water:

1. **Gravitational water (free water):**
   - This form of water is loosely held in soil (in macro pores) and move downwards freely under the influence of gravity.
   - Water in excess of the field capacity is termed gravitational water.
   - The drainage or deep **percolation loss of water results from downward movement** of this gravity water.
   - It has a suction of **less than 1/3 atmosphere**.
   - The plants can't absorb it as it drains out of root zone in short period of time.
2. **Capillary water:**
   - Capillary water is held in the capillary pores (micro pores) with a suction ranging from **1/3 and 31** atmospheres. Capillary water is retained on the soil particles by surface forces (PF value –2.54 to 4.50).
   - It is held so strongly that gravity cannot remove it from the soil particles.
   - The **availability of capillary water to plant** roots depends on pore diameter which controls the pressure of water. The narrower the capillary pore, lesser is the availability.
3. **Hygroscopic water:**
   - This form of soil water is held with a high suction ranging from **31 to 10000** atmospheres. It is held tightly on the surface of soil colloidal particles.
   - Generally, it includes first two molecular layers of water on soil particles. **Only microorganism is uses this form of water.**

Biological classification of water:

1. **Available water-** it is held between **field capacity (-1/3) to permanent wilting coefficient (-15)**
2. **Unavailable water-** it is held at **> -15 bar** and it include whole hygroscopic water plus a part of capillary water below the wilting point.
3. **Superfluous water-** it is retained in the soil beyond the field capacity of soil moisture tension.

Soil moisture constants: These are of practical importance for irrigation and drainage management. These are also used to compare water retention capacity of different soils.

1. **Field capacity:**
   - If a soil is saturated, gravity water starts moving downwards. When all the gravitational water is drained away, and then the wet soil are almost uniformly moist.
   - The amount of water held by the soil at this stage is known as the field capacity of that soil. It is the capacity of the soil to retain water against the downward pull of the force of gravity.

- At this stage only micro/capillary pores are filled with water and plants absorb water for their use.
- At field capacity water is held with a suction of 1/3 atmosphere and PF = 2.54.

2. Wilting coefficient

- As the soil water content decreases, a point is reached when the water is so firmly held by the soil particles that plant roots are unable to extract water at a rate sufficient to meet the transpiration needs.
- The plant begins to wilt. At this stage even if the plant is kept in a saturated atmosphere it does not regain its turgidity and wilt unless water is applied to the soil.
- The stage at which this occurs is termed the **wilting point** and the percentage amount of water held by the soil at this stage is known as the **wilting coefficien** .
- Water at wilting coefficient is held with a force of 15 atmospheres (pF=4.2).

3. Hygroscopic coefficient: Þ

- The hygroscopic coefficient is the maximum amount of hygroscopic water absorbed by 100 g of dry soil under standard conditions of humidity (50% relative humidity) and temperature (15°C).
- This tension is equal to a force of **31 atmospheres (pF=4.5).**
- Water at this tension is not available to plant but may be available to certain bacteria.

Energy concept of Soil water

- The energy related phenomenon is retention and movement of water in soils, its uptake and translocation in plants and its loss to the atmosphere *etc.*,
- The more strongly water is held in the soil, the greater is the heat (energy) required. In other words, if water is to be removed from a moist Soil, work has to be done against adsorptive forces.
- Conversely, when water is absorbed by the soil, a negative amount of work is done. The movement is from a zone where the free energy of water is high (standing water table) to one where the free energy is low (a dry Soil). This is called energy concept of soil water.
- The difference between the energy status of soil water and pure free water is known as **soil water potential.**

- Soil moisture is measured by gravimetric method. And its drying temperature is 105-110^{0}c.

$$\text{Formula moisture (\%)} = \frac{\text{weight of wet soil} - \text{weight of dry soil} \times 100}{\text{Weight of dry soil}}$$

- **Tensiometer** is used to measure in situ soil moisture. And its range is **0.0 to 0.85** bar.

- More water means less tension and less water is held at more tension. So water moves from low tension to high tension.
- Darcy's law in soil deals to hydraulic gradient.

S. No.	Characters	pF value	Height (cm)	Atm. Pressure
1.	Field capacity	2.53	346	1/3
2.	Wilting point	4.18	15850	15
3.	Hygroscopic water	4.50	31623	31
4.	Oven dry soil	7.00	10^7	10000

- **Soil Water Movement** mainly through saturated flow, unsaturated flow and water vapour.
  1. **Saturated flow:** This occurs when the soil pores are completely filled with water. This water moves at water potentials larger than – 33 kPa. Hydraulic conductivity can be expressed as V = kf.
     Where, V = Total volume of water moved per unit time f = Water moving force k = Hydraulic conductivity of soil.
  2. **Unsaturated Flow:** It is flow of water held with water potentials lower than- 1/3 bar. Water will move toward the region of lower potential (towards the greater "pulling" force).
  3. **Water Vapour Movement:** The movement of water vapour from soils takes place in two ways:
     (a) Internal movement: the change from the liquid to the vapour state takes place within the soil, that is, in the soil pores and
     (b) External movement: the phenomenon occurs at the land surface and the resulting vapour is lost to the atmosphere by diffusion and convection.

Entry of Water into Soil

1. **Infiltratio** : Infiltration refers to the downward entry or movement of water into the soil profile. Vegetative covered soil surface has more infiltration rate than bare soil. It occurs in upper layers of soil.
2. **Percolation**: The downward movement of water through a column of soil is called percolation. Sandy soils have greater percolation than clayey soil, and it in lower loyers of soil.
3. **Permeability**: It indicates the relative ease of movement of water with in the soil. The characteristics that determine how fast air and water move through the soil are known as permeability.
4. **Drainage:** Water moving below the root zone of the plant is unavailable to plant and move the chemical deeper into the profile and to recharge the underlying aquifer.
5. **Hysteresis:** It is difference in energy status of soil when soil are alternately wetted and dried is called hysteresis. **Desorption** is upper solid line developed curve due to drying of saturated soil whereas lower dotted line developed curve due to wetting an initially dry soils called **Sorption.**

- **Consumptive use water:** quantity of water lost by evapotranspiration (ET) plus that contained in plant tissue is called consumptive water use.
- **Water use efficiency (WUE):** amount of water (evapotranspiration, plant tissue and drainage loss) required to produce a unit of dry weight material is known as WUE. It is **mathematically define as WUE= DW/ET**
- Maximum water movement occurred in single grain structure and lowest in massive soil structure.

SOIL AIR

- **Soil air** can be defined as the air that fills the soil pore spaces not occupied by water whereas **soil aeration** is the exchange of CO_2 and O_2 gases between the soil pore space and aerial atmosphere.
- Two type of pores are involve in aeration
  1. **Crumb pore** – pores within the crumb
  2. **Intercrumb pore** – pore between the crumbs.

Gaseous Composition of Soil Air

1. **Oxygen**
   - Once the supply of oxygen is virtually exhausted, the soil environment becomes anaerobic
   - Typically wet soils have very low oxygen contents.
   - Drastic reductions in the oxygen content of soil air may occur following a heavy rain.
2. **Carbon dioxide**
   - Carbon dioxide typically increases in soil air.
   - When it is as high as 10% in soil air it may become toxic to plants.
3. **Other Gases**
   - Soil air is usually higher in water vapour than in the atmosphere.
   - Under waterlogged conditions, the concentrations of **methane (CH_4), and hydrogen Sulphide (H_2S)** is particularly higher in soil air.
   - **Ethylene (C_2H_4) gas,** a product of anaerobic microbial metabolism is toxic to plant roots, even in very small concentrations.

Composition of gases in air and soil

S. No.	Name of gas	Atmosphere (%)	Soil (%)
1.	Oxygen (O_2)	20.97	20.60
2.	Nitrogen (N_2)	78.03	78.60
3.	Carbon dioxide (Co_2)	0.03	0.25 - 0.3
4.	Argon	0.94	0.90
5.	Hydrogen	0.01	—

- Mechanisms of gas exchange or soil aeration through **mass flow and diffusion.**
- **Diffusion:** is the principle process that drives gas exchange soil air and the atmosphere. Diffusion is defined as random thermal movement of molecule from higher concentration to lower concentration zone. **Numerically it is explained by Fick's law of diffusion.**
- **Mass flow:** The mass flow occurs due to total pressure gradient of gas. The pressure gradient causes movement of entire mass of air from a zone of high pressure to that of low pressure. Such a flow of air occurs within the soil or from atmospheric air to soil air or vice-versa. When the soil temperature is higher than the atmospheric temperature during mid-day then the soil gases will expand and move out of the pore space into the atmosphere. When the soil is cooler than the atmosphere during night, then the atmospheric gases enter the soil. When the atmospheric pressure is high, the atmospheric gases will enter in to the soil.
- **Oxygen Diffusion Rate** (ODR): It is the best and most reliable measurement of aeration capacity. It determines the rate at which O_2 in soil air is replenished. ODR decreases with soil depth. ODR should be above **40 x 10^{-8} g/cm²/minute for good growth** of most of the crops. However, the root growth is drastically reduced when the ODR decreases to about **20 x 10^{-8} g/cm²/minute.**
- **Fick's law:** diffusion is a function of the concentration gradient, the diffusion co-efficient of the medium and the cross- sectional area. **Formula- DQ= DA (dc/dx)^{dt}**
- Standard electrode potential is 0.55-0.75 for ODR
- Ideally 2/3rd of soil pores are filled with water and 1/3rd with air.

SOIL TEMPERATURE

- **Solar Radiation:** The primary source of energy to heat soils is the solar radiation from the sun.
- Factors influencing the amount of solar radiation:
  - **Albedo** - This is the fraction of incident radiation that is reflected by the land surface.
  - **Aspect** - The angle at which the sun's rays strike the soil
  - **Rain/Irrigation water** – affect the soil temperature.
  - **Soil Cover** - depends on whether the soil is bare, or is covered with vegetation or mulch.
- **Soil Thermal Properties:**
  - **Specific heat** (heat capacity): This is the amount of energy required to raise the temperature of a substance by 1°C. Unit - (cal/g) or joules per gram (J/g). Specific heat of water (1.0 cal/gm), humus/Organic matter (0.44 cal/gm), dry soil (0.2 cal/gm) and mineral matter (0.11 cal/gm)
  - **Thermal Conductivity**: The amount of heat transferred through a unit cross-sectional area of unit thickness in unit time and unit temperature gradient.

- **Thermal Diffusivity**: Temperature change that takes place in a portion of a given soil as heat flows into it from adjacent layer.
- **Mode of Heat Transfer**
  - **Conduction** – Flow of heat through matter unaccompanied by any motion
  - **Convection** – The transfer of heat by moving matter.
  - **Radiation** – The transfer of heat through vacuum / space
- **Density**
  - **Liquid** - 0.998 g cm^{-3} or 1000 kg m^{-3}
  - **Solid**-0.910 g cm^{-3}
  - **Vapor** - 1.73×10^{-5} g cm^{-3}
- Loss of heat in the form of **conduction,** radiation, and convection.
- **Fourier's law** deals with heat conduction in soils.
- Optimum temperature for plant growth and development is 25-35°c
- Sandy soils absorb more heat than clayey soils because the specific heat of water (heat required to heat a substance) is 4-5 times than of clay soil particles.
- Soil temperature is used at family level categorization in soil taxonomy.

Chemical properties

- pH is the negative logarithm of H ion concentration. **Sorenson gave pH scale.**
- There are two types of acidity in soil active acidity and potential acidity.
- pH measure only active acidity. Potential acidity forms the bulk of the soil acidity. It is greater than active acidity.
- Soil pH is also soil reaction.
- Soil with pH less than 6.5 are acidic, 6.5 to 7.5 are neutral and above 7.5 are alkaline.
- One unit change in pH changes H ion concentration by 10 times, 2 units by 100 times and so on.
- **Electrical conductivity**:
- Measure of soluble salts in mmhos/cm or dS/m in solubridge or conductometer.
- Ion exchange:
- Most important process occurring in soil Ion exchange is a reversible process. Soil colloids are the seat of ion exchange.
- CEC is measured at pH 7 & expressed as meq/100 g of soil. CEC varies greatly with nature and amount of clay and organic matter.
- Kaolinite has 3-10, Illite – 10-30, Montmorillonite – 80-150 and Organic matter – greater than 200 CEC.

Base saturation:

$$BS = \{(Na^{+} + K^{+} + Ca^{++} + Mg^{++})/CEC\} \times 100$$

- Fertile soils are saturated with Ca^{++} and Mg^{++} ions. If soil is saturated with more than 15% exchangeable sodium, then that soil is called Alkali soil. If soil is saturated with H^+ ions. Then the soil is said to be base unsaturated or acidic.

Organic matter (OM):

- OM on decomposition by humification process gives humus. Humus is amorphous in nature. In India OM is very less because of tropical and sub-tropical climate.
- In hilly and high altitudes, OM is above 1% in mangroves it is 10-30%.
- CN ratio of OM is 10:1 whereas that of Indian soils is 5:1 to 25:1 with an average of 14:1
- Organic matter decomposition stages: First sugars & starches followed by proteins followed hemicelluloses, cellulose and fatty acids and finally lignin and waxes.

Biological properties:

Bacteria > Actinomycetes > Fungi > Algae

- The smell of soils after fresh showers is due to **Actinomycetes**.
- Bacteria occur in neutral to alkaline pH. Fungi in acid pH and Algae in shade areas.
- **Symbiotic N fixe -Rhizobium** in Legumes, Non-symbiotic or **Free living is Azotobacter**.
- N fixation in paddy field by **algae or Azolla**. It fixed N around 50-150 Kg.
- In plant N is fixed as $R\text{-}NH_2$, which is converted to ammonia. The ammonia is converted to nitrite first by **Nitrosomonas or Nitrosococcus**, followed by Nitrite to nitrate by **Nitrobacter**. This process is called **Nitrificatio** .
- N-Serve and actinomycetes are nitrification Inhibitors.
- Optimum condition for nitrification process is field capacity and pH should be above 5.

Mineralogical Properties:

- There are primary, secondary, accessory and amorphous minerals.

1. **Primary mineral**: Order of occurrence

Feldspars> Quartz> Mica> Limestone> Hornblende and augite> Olivine and serpentine

- Serpentine is hydrated silicate of Mg.

2. **Secondary minerals**:

- 1:1- One silica and one alumina layer. Kaolinite, Halloysite and Dickite

- 2:1 - Two silica and one alumina layer: Montmorillionite **(expanding)**, Vermiculite **(Slightly expanding)** and Illite **(Non-expanding and 15% of silica is replaced by Al^{3+} and K^{+} ions)** Illite is also called as hydrous mica.
- 2:1:1 or 2:2 – Chlorite. The crystal unit is composed of one 2:1 unit (like mica and Montmorillionite) and one octahedral unit, Brucite {(Mg_3 (OH)6}.

3. **Accessory minerals**: B- Tourmaline, F-Topaz, P-Apatite, Ti-Rutile and Anatase.
4. **Amorphous clay mineral**: Allophane. Found in Soil order Andosols.

- Negative charge is due to exposed surface of clay and isomorphous substitution.

SOIL FORMATION, CLASSIFICATION, SURVEY, LAND CAPABILITY CLASSES AND SOIL OF INDIA

- A soil forming process may be defined as a complex or sequence of events including both complicated reactions and simple rearrangement of matter which intimately affect the soil. These processes are also known as **soil building processes or pedogenic processes**.
- Soil formation occurred due to **destructive** – weathering means change the consolidated rocks and minerals into unconsolidated material (parent material) and **constructive** – it is the second phase of soil formation in which development of soil profile.
- Soil formation factor according to **Dokuchaev (1889) –S= f(cl,o,r,p,t)**. cl- climate, r- relief/topography, p- parent material, o - organism, t- time, f- function and s- soil formation. Again **Jenny classifie** these factor based on function as
  1. **Active factor** - S= f (cl,b) cl - climate and b- biosphere.
  2. **Passive factor** = r,p,t.
- Fundamental soil forming processes:
  1. **Humification:** It is the process of transformation **or decomposition of raw organic matter into humus**. In this process the soluble organic substances regroup themselves in to **large molecules by polymerization and become poorly soluble**. The characteristics are influenced by the nature of vegetation residue and the way it becomes decomposed and **synthesized into new organic compounds**.
  2. **Eluviation:** Eluviations means washing out. Mobilization and translocation of clay, salt, humus from one area of soil body to other areas.
  3. **Illuviation:** The process of deposition of soil or accumulation of the eluviated constituent at a depth below soil surface
- India, from 1969, followed USDA's Comprehensive system of soil classification called soil Taxonomy. It is type of multi-category classification having 6 categories *i.e.,* soil order, suborder, great group, sub-group, family and series.
- There **12 orders, 63 sub order** and the lastly added 11th and 12th orders are Andisols and Gelisols respectively, Andisols are found in volcanic cruption areas and contains high content allophane. Gelisols are found in arctic regions.

- The largest orders in India are **Inceptisols** followed by Entisols.
- **Series:** (200+ in India and 12000 in USA) it is the **lowest category** in the system. The series is collection of soil **individuals essentially uniform** in differentiating characteristics like colour, texture, structure, consistency, pH and EC and in arrangement of horizons.

Soil orders with soil characteristics

S. No.	Order	Soil Characteristics
1.	Entisol	Soils with **little horizon development** or beginning of soil formation means **recent formed order in Thar desert of Rajasthan.**
2.	Inceptisol	Soils with altered horizons, Fe, Al oxides are absent
3.	Gelisols	Newly introduced soil order and find on higher altitude of Jammu and Kashmir.
4.	Histosols	Soils containing > 30 % organic matter to a depth of 40 cm
5.	Spodosols	Soils with spodic horizon within 2 m.
6.	Andisols	Soils that have andic soil properties in 60% or more of the thickness between the soil surface and 60 cm or lithic or paralithic contact. This order having **highest number of sub order (07)** among all 12 ordered of soil.
7.	Oxisols	Highly weathered soils of tropical and subtropical regions with oxic horizon
8.	Vertisols	Soils with more than 30% clay in all horizons and crack when dry, slickensides or pressure faces are present in Bss horizons
9.	Aridisols	Dry soils, little profile, light in colour and low organic content, **lowest no. of sub order(2)**
10.	Ultisols	Soils in warm humid regions with argillic horizon and low base saturation(<35% at 2 m depth below the surface
11.	Mollisols	Extensively important for agriculture due to dark colored, base rich soils of grass land areas, nitric, or calcic horizon
12.	Alfisols	High base status (>35%) soils of the humid and sub humid regions with an ochric epipedon and argillic (or nitric) horizon.

SOIL SURVEY

- Soil survey is a study of soil morphology in the field, corroboration of diagnostic soil properties in laboratory, classification of soil area in well defined unit, plots, and boundaries on a map. Predication of adaptability of soil to various uses is called soil survey.
- There are two types of soil survey- Reconnaissance soil survey and detailed soil survey.
- In RSS_1 1^9 = 1 mile. Toposheets of 1:50,000 to 1:1,00,000 or aerial photo of 1:25,000 or 1:50,000 is used. Observations are once at 3-6 Kms.

- DSS: 1⁹ = 8 mile or 1⁹ = 16 mile. Cadastral maps of 1:5000 to 1:8000 or Aerial photo of 1:10,000 is used Observation are once at ¼ - ½ Km.
- Forest area is surveyed by RSS.
- **Survey report and map are end product** of any soil survey

Land capability classification (LCC)

- Land capability classification (LCC) may be defined as **a system of grouping land into various classes based on inherent limitations imposed on sustained use by soil attributes, topography, drainage and climate.** The guiding principle underlying LCC is "use land according to its capability and treat it as per its need".
- The capability classes falls in two groups, one **suited for cultivation** and other **not suited for cultivation.**
- Land **suitable for agriculture** and other uses which include **class I to class IV** lands.
- Land not suitable for agriculture but very well **suited for forestry, grass land and wild life which include class V to class VIII lands.**
- On map, the capability classes are indicated in different colours as **Green (I),** Yellow (II), Pink (III), Blue (IV), Dark green (V), Orange (VI), Red (VII), **Purple (VIII).**

Remote sensing:

- Remote sensing is an art and science for obtaining information about an object or feature without physically coming in contact with that object or feature. Two type devices are used in remote sensing
- **Active sensing:** when the sensing device direct Electro Magnetic Radiation (EMR) at an object and detected the amount of that energy which is reflected back *e.g.* **Radar, Radio, Laser Imaging Radar (UV) and Camera etc.**
- **Passive sensing:** when the sensing device detected EMR originating from another source *e.g.* **Sun.**
- **Objective of remote sensing:** Soil survey, detect of water resources, crop management, nutrient management and land use planning.
- **Feature of devise:** Aerial photography lens (88 mm), black & white ratio scale (1: 85000), aerial photography for large scale ratio (1:1200), focal length lens (6 inch)

The spectral region used in satellite based remote sensing

S. No.	Regions	Wavelength	Property
1.	Visible	0.4-0.7 μm	Reflectance
2.	Reflective infrared	0.7-3.0 μm	Reflectance
3.	Thermal infrared	3.0-15 μm	Radioactive temperature
4.	Microwave	0.1-30 cm	Brightness temperature (active) Backscattering temperature (passive)

- **Visible Spectrum:** Blue, green, and red are the **primary colors or wavelengths** of the visible spectrum.
- Visible spectrum of colours:
  1. Violet: 0.400 - 0.446 μm
  2. Blue: 0.446 - 0.500 μm
  3. Green: 0.500 - 0.578 μm
  4. Yellow: 0.578 - 0.592 μm
  5. Orange: 0.592 - 0.620 μm
  6. Red: 0.620 - 0.700 μm

SOIL OF INDIA

- Soil can be defined simply as a mixture of small rock debris/particles and organic materials/humus which develop on the earth surface and support to plants growth. The classification soil in the ancient time was based on two things; whether the soil is fertile or sterile. Thus the classification was: **Urvara [fertile] and Usara [sterile]**
- **Soil Classification – Agencies involved** When the **Soil survey of India** was established in **1956,** they studied soils of India and their characteristics. **The National Bureau of Soil Survey and the Land Use Planning,Nagpur** an institute under the control of Indian Council of Agriculture Research did a lot of studies on Indian soil.

Alluvial soil:

- Mostly available soil in India **(about 43%)** which covers an area of **143 sq.km.**
- Indus-Ganga-Brahmaputhra plain, Narmada-Tapi plain *etc.,* are examples.
- The old alluvium is termed as **Bhangar** and latest alluvium termed as **Khadar.**
- Highly fertile.
- Humus, lime and organic matters are present.
- **Colour:** Light Grey to Ash Grey.
- **Texture:** Sandy to silty loam or clay.
- Rich in potash but Poor in phosphorous.
- Wheat, rice, maize, sugarcane, pulses, oilseed *etc.,* are cultivated mainly.

Red soil:

- Seen mainly in low rainfall area and also known as **Omnibus group.**
- Porous, friable structure.
- Absence of lime, kankar (impure calcium carbonate).
- **Deficient i** : lime, phosphate, manganese, nitrogen, humus and potash.

- **Colour:** Red because of Ferric oxide. The lower layer is reddish yellow or yellow.
- **Texture:** Sandy to clay and loamy.
- Wheat, cotton, pulses, tobacco, oilseeds, potato *etc.*, are cultivated.

Black soil/Regur soil:

- **Regur** means **cotton** – best soil for cotton cultivation.
- Mostly black soil is occupied by Deccan.
- Matured soil.
- High water retaining capacity.
- Swells and will become sticky when wet and shrink when dried.
- **Self-ploughing** is a characteristic of the black soil as it develops wide cracks when dried.
- **Rich in:** Iron, lime, calcium, potassium, aluminum and magnesium.
- **Deficient in** Nitrogen, Phosphorous and organic matter.
- **Texture:** Clay.
- **Colour:** Deep black to light black.

Laterite soil:

- It derived from Latin word 'Later' which means Brick. Become so soft when wet and so hard when dried.
- Found in the areas of high temperature and high rainfall. So formed as a result of high leaching.
- Lime and silica will be leached away from the soil.
- Organic matters of the soil will be removed fast by the bacteria as it is high temperature and humus will be taken quickly by the trees and other plants. Thus, humus content is low.
- **Rich in**: Iron and Aluminum
- **Deficient i** : Nitrogen, Potash, Potassium, Lime, Humus
- **Colour**: Red colour due to iron oxide.
- Rice, Ragi, Sugarcane and Cashew nuts are cultivated mainly.

Desert / arid soil:

- Seen under Arid and Semi-Arid conditions. Deposited mainly by wind activities.
- **High salt content, lack of moisture and humus.**
- Kankar or Impure Calcium carbonate content is high which **restricts the infiltratio** of water.
- Nitrogen is insufficient and Phosphate is normal. **Texture:** Sandy and **Colour:** Red to Brown.

Peaty / marshy soil:

- Areas of heavy rainfall and high humidity. A large quantity of dead organic matter/humus which makes the soil alkaline.
- Growth of vegetation is very less. Heavy soil with black colour.

Forest soil:

- Humus content is less and thus the soil is acidic. Found in regions of high rainfall.

Mountain soil:

- In the mountain regions of the country. It is immature soil with low humus and acidic.
- Largest area is occupied by alluvial soils 75 Mha.
- Black soils 9Vertisols mostly), 72 Mha, are characterized by Gilgai (shining surface), Micro- relief (ups and down), self-churning, smectite (expanding clays) and they are the Russian equivalent of Chernozems.
- 2/3rd of TN is of red soil, the parent material for red soils is **mostly granite**.
- Laterite is older than lateritic soils. Laterites are the oldest or most weather soils.

SOIL REACTION AND BUFFERING

pH: pH is define as the **logarithm of the reciprocal of hydrogen ion (H^+) activity**. It is expressed in 'g' ions l^{-1} or ml^{-1} (mole/liter). pH generally range in number between 0 and 14. Any pH below 7.0 denotes acidic condition while any number above 7.0 denotes alkaline condition.

$$pH = \log 1/a^{H+} \text{ or } -\log AH^+$$

- The concentration of H^+ at 7 pH (1×10^{-7}), 0 pH (1×10^{-0}) and 14 pH (1×10^{-14}).

Effect of pH on nutrient availability and uptake

- pH does not directly affect plants. It does affect the availability of different nutrients and toxic elements to plants. This is mostly due to the fact that pH changes the form of the nutrients and that forms may be relatively insoluble.
- Low pH causes nutrient deficiency, mainly calcium, phosphorus and nitrogen. Extreme pH reduces the population of some useful organisms, such as bacteria and earthworms.
- Highest resistant crops at higher pH in **cereal (Barley) legume (Barseem), field crops (sugar beet) and fruit crops (date plam than ber).**
- Optimum range of pH for microbial activity – **Fungus (4.5- 6.5), Bacteria (6.5-8.0) and Actinomycetes (> 8.0).**
- The power to resist a change in pH is known as buffer action.

Soil classification based on pH, EC and ES

Type of soil	pH	EC(dS/m)	ESP (%)
Saline	< 8.5	> 4	< 15
Alkali	> 8.5	< 4	> 15
Saline-Alkali	< 8.5	> 4	> 15

PROBLEMATIC SOIL OF INDIA

The soils which possess characteristics that make them uneconomical for the cultivation of crops without adopting proper reclamation measures are known as problem soils. Different kinds of problem soils are physical problem soils, chemical problem soils, biological problem soils and nutritional problem soils.

1. Physical problem soils

A. **Slow permeable soils/**Impermeable (due to very high clay content, infiltration rate <6cm/day, so more runoff which eventually leads to soil erosion and nutrient removal)

B. **Highly permeable** soils (Excessive permeability of the sandy soils results in poor water retention capacity)

C. **Heavy clay soils** (plastic and sticky due to more clay content up to 40percent)

D. **Soil surface crusting** (Prevent germination of seeds and retards root growth Results in poor infiltration and accelerates surface runoff creates poor aeration in the rhizosphere).

E. **Sub soil hard pan** (commonly in red soil).

F. **Shallow soils** (15-20 cm depth).

2. Chemical problem soils

A. **Salt - affected soils** - The salt-affected soils occur in the arid and semiarid regions where evapo-transpiration greatly exceeds precipitation. The accumulated ions causing salinity or alkalinity include sodium, potassium, magnesium, calcium, chlorides, carbonates and bicarbonates. Problem soils occupy 10 Mha. Saline soils- 7.2 Mha, alkali soils- 2.8 Mha and Problematic soils are found highest in UP. The salt affected soils can be primarily classified as saline soil and sodic soil.

I. Saline soils:

- Saline soils defined as soils having a conductivity of the saturation extract greater than 4 dS m-1 and an exchangeable sodium percentage less than 15. Saline soils defined as soils having a conductivity of the saturation extract greater than 4 dS m-1 and an exchangeable sodium percentage less than 15. The pH is usually less than 8.5. This soil is called **solanchak, white alkali soils because of surface crust of white salts** and are said to cause boron toxicity
- Formation process by which the saline soil formed is called **Salinization**. Saline soils occur mostly **in arid or semi arid regions**

- **Major production constraints**-presence of salts leads to alteration of osmotic potential restricted water and nutrient uptake by plants, reduced microbial activity

Management

- Construct drainage channels of 60 cm deep and 45 cm wide and allow leaching of salts
- Proportional mixing of good quality (if available) water with saline water
- Addition of extra dose of nitrogen to the tune of 20-25% of recommended level, organic manures like, FYM, compost, green manuring (**Sunhemp, Daincha, Kolingi**) and or green leaf manuring also counteracts the effects of salinity.
- Crop choice/crop management like barley, beet root and date palm *etc.*,
- Soil/cultural management planting the seed in the centre of the raised bed / ridge may affect the germination as it is the spot of greatest salt accumulation. A better salinity control can be achieved by using sloping beds with seeds planted on the sloping side just above the water line.
- Alternate furrow irrigation is advantageous as the salts can be displaced beyond the single seed row. Application of straw mulch had been found to curtail the evaporation from soil surface resulting in the reduced salt concentration in the root zone profile within 30 days.

II. Alkali / Sodic soils:

- Alkali soil is having a conductivity of the saturation extract less than 4 dS m-1 and an exchangeable sodium percentage greater than 15. The pH is usually between 8.5 – 10.0.
- Most alkali soils, particularly in the arid and semi-arid regions, and it rich $CaCO_3$ in the profile. OH- ions in soil solution is released during hydrolysis of $CaCO_3$ that's why its keeps higher pH. Alkaline soil is also called **solanetz, sodic and black alkaline.**
- Development of alkaline due to salt blowing by wind especially in arid region, use of excess **sodium nitrate fertilizer** and excess mineral in soil composition*e.g.* **Ca^{2+}, Mg^{2+}, Na^{2+}** and formation of alkaline is known as **solodisation**

Management

Physical Amelioration

- Deep ploughing is adopted to break the hard pan developed at subsurface due to sodium and improving free-movement water and aeration.
- Improving the aeration and to remove further accumulation of salts at root zone through proper management of drainage.
- Sand filling which reduces heaviness of the soil and increases capillary movements of water.

- Profile inversion – Inverting the soil benefits in improvement of physical condition of soil as that of deep ploughing.

Chemical Amelioration- The chemical amendments can be broadly grouped as follows:

1. Direct Ca suppliers: **Gypsum, calcium carbonate, phospho-gypsum,** *etc.,*
2. Indirect Ca suppliers**: elemental Sulphur, sulphuric acid, pyrites, FeSO4** *etc.,*

- Among them gypsum is the most commonly used chemical amendment.
- **Neutralizing value of some liming material are lime (Cao) –**(**179** CCE %),**Dolomite** [CaMg (CO_3)] –(108.7 CCE %) Calcite $(CaCO_3)$- (100 CCE %)

Other chemical sources

- Distillery spent wash: Distillery spent wash is acidic (pH 3.8-4.2) with considerable quantity of magnesium
- **Distillery effluen** : Distillery effluent contains macro and micronutrients
- **Pulp and paper mill effluent** : Pulp and paper effluents contain lot of dissolved solids and stabilized organic matter and if properly treated can safely be used for irrigation with amendments viz. pressmud @ 5 tonnes ha-1, fortified pressmud @ 2.5 tonne ha^{-1} or daincha as in situ green manure.

Crop choice: Rice, date palm, Ber, tamarind, sapota and wood apple,

III. Saline-alkali/sodic soil:

- A soil is having a conductivity of the saturation extract greater than **4 dS m^{-1}** and an exchangeable sodium percentage **greater than 15**. The pH is variable and **usually above 8.5** depending on the relative amounts of exchangeable sodium and soluble salts.
- When soils dominated by exchangeable sodium, the pH will be more than 8.5 and when soils dominated by soluble salts, the pH will be less than 8.5.
- Formation: These soils form as a result of the **combined processes of Salinization and alkalization.** If the excess soluble salts of these soils are leached downward, the properties of these soils may change markedly and become similar to those of sodic soil.
- The reclamation / management practices recommended for the reclamation of sodic soil can be followed for the management of saline-sodic soil.

B. Acid soils:

- Soil acidity refers to presence of **higher concentration of H^+ in soil solution** and at exchange sites. They are characterized by low soil pH and with low base saturation.
- The ranges in soil pH and associated degree of acidity are as follows:
  1. Very strong (3-4)
  2. Strong (4-5)
  3. Moderate (5-6)
  4. Slight (6-7).

- Acid soils occupy approximately 60% of the earth land area and are arise under humid climate conditions from carbonaceous less soil forming rocks in all thermal belts of the earth.
- World wide – 800 M ha
- India - 100 M ha
- Assam 95% and 30% of geographical area of Jammu and Kashmir are acidic.

Sources of soil acidity

- Leaching due to heavy rainfall
- Acidic parent material (granite)and alumina silicate minerals
- Acid forming fertilizers (Ammonium sulphate and ammonium nitrate)
- Humus and other organic acids (carboxylic (-COOH)and phenolic (-OH)]
- Carbon dioxide and hydrous oxides
- Acid rain

Production constraints

- Increased solubility and **toxicity** of Al, Mn and Fe
- **Deficiency** of Ca and Mg
- **Reduced availability** of **Mo** and P
- Reduced microbial activity

Two type of acidity soils:

- **Active acidity**: acidity develops due to (H^+) and (Al^{3+}) ions **concentration of the soil solution.**
- **Exchange acidity**: acidity develops due to adsorbed (H^+) and (Al^{3+}) ions **concentration on the soil colloids.** It is also called **potential or reserve acidity.**
- Ratio of active acidity and potential acidity in soil is 1:2

Management

- Soil amelioration – Lime
- Tolerant crops – Tea, Potato, Rice, Strawberry and Apple *etc.,*
- Gypsum is not considered as a liming material because gypsum dissociated into calcium (Ca^{2+}) and sulphate (SO_4^{-2})

  $CaSO_4 \longrightarrow Ca^{2+} + SO_4^{-2}$

  Now sulphate ions reacts with moisture and produce strong acidic mineral (H_2SO_4) which also increase soil acidity instead of reducing soil acidity.

  $SO_4^{-2} + H_2O \longrightarrow H_2SO_4$

- **Acid Sulphate soils (ASS):**Soil with sufficient sulphides (FeS_2 and others) which become strongly acidic when drained are termed acid sulphate soils or as the Dutch refer to those soils **cat clays** It is two types are PASS and AASS.
  1. **Potential acid sulphate soils (PASS):** ASS which has not been oxidized by exposure to air is known as potential acid sulfate soils (PASS). They are **neutral in pH (6.5–7.5)**, contain unoxidised iron sulfides, it is generally soft, sticky and saturated with water and are usually gel-like muds but can include wet sands and gravels have the potential to produce acid if exposed to oxygen.
  2. **Actual acid sulphate soils (AASS):** When PASS is exposed to oxygen, the iron sulfides are oxidized to **produce sulfuric acid (H_2SO_4)** and the soil becomes strongly acidic (**usually below pH 4**). These soils are then called actual acid sulfate soils (AASS). They contain oxidized iron sulfides, vary in texture and often contain **jarosite** (a yellow mottle produced as a byproduct of the oxidation process).
- **Hydrogen sulphide (H_2S)** often formed in lowland rice soils causing **akiochi disease**
- $SAR = \sqrt{Na^+ / (Ca^{++} + Mg^{++}) / 2}$
- $\text{Gypsum requirement (GR)} = \dfrac{\text{ESP (Initial)} - \text{ESP (Final)} \times \text{CEC}}{100}$
- $\text{Leaching requirement (LR)} = \dfrac{D_{dw} \times 100}{D_{jw}}$

Denotes

dw = drainage water

iw = irrigation water

- Amendments are required in salt affected soil based on pH *e.g.*Gypsum should be applied when the soil pH > 9.0; whereas Sulphur and Iron pyrite at 8.0-9.0; limestone at < 8.0

Salt resistance and sensitive crops

Tolerant	Moderate tolerant	Sensitive
Date palm	Sunflower	Pecan nut
Coconut	Barley	Plum
Sugar beet	Maize	Apple
Alfalfa	Sorghum	Grape
Onion	Oats	Citrus
Turnip	Sweet potato	Beans
Lettuce	Lime bean	

Salinity hazard and total soluble salt concentration of irrigation water

Salinity class	Electrical conductivity (dS/m)	Total dissolved salts (ppm)	Remarks
Low	<0.25	<150	Safe for irrigation
Medium	0.25 - 0.75	150 – 500	Moderate to used with leaching
High	0.75 - 2.25	500 – 1500	Not safe
Very High	>2.25	>1500	Not safe

Nutrient toxicity and deficiency with pH rang

Categories	pH	Toxicity	Deficiency
Acidity	< 7.0	Al, Mn	Ca, Mg, K
Alkaline	>8.5 (8.6-10.0)	Na,Bo	Fe, Zn, Mn, P
Saline	<8.5	Na, Cl	Ca, K

- Half Kg soil sample is taken and analysed for pH, total soluble salts by EC, Organic Carbon by Walkley and Black Method, Available N by alkaline permanganate method, available P by Olsen's or Bray's Method, available K by Neutral Normal Ammonium Acetate method.
- **2×10^6** is the weight of one hactare soil upto 15 cm soil depth.
- The concentration of boron is safe in water when it is < 0.04 ppm.
- The effect of its concentration on the quality of water is determined by the **residual sodium carbonate (RSC)**. The water containing RSC less than 1.25 meq L -1 is safe for plant growth.

  $RSC = (CO_3^{--} + HCO_3^{-}) - (Ca^{++} + Mg^{++})$

Guideline for irrigation water quality (FAO)

Parameter	Permissible/ safe	Moderately safe	Moderately unsafe	Unsafe
RSC (meL^{-1})	< 1.25	—	1.25-2.50	>2.50
SAR	< 10	10-18	18-26	>26
pH	6.5-8.4	—	—	0-5, >9.5
B ($mg\ L^{-1}$)	< 2.0	2.0-2.5	2.5-3.0	>3.0
Cl ($mg\ L^{-1}$)	< 140	140-350	—	>350

SOIL FERTILITY, MANURES AND FERTILIZERS

- **Fertility Versus Nutrition** - The term "fertility" refers to the inherent capacity of a soil to supply nutrients to plants in adequate amounts and in suitable proportions. The word "nutrition" refers to the interrelated steps by which a living organism assimilates food and uses it for growth and substitution of tissue
- **Essential nutrient** - in its absence the plant is unable to complete a normal life cycle (Emanuel Epstein)

- There are **17 essential elements** for plant growth. There are divided into macro and micronutrients based on the amount of plant uptake. Recently added essential nutrient is **Nickle** in year**1987.**
- Ni **Plants'** Ni requirement is the lowest of all essential elements at < 0.5 mg per kg of dry weight.
- A criterion of essentiality of **elements** was given by **Arnon and Stout (1939)** while Criteria of essentiality of **nutrient** were given by **Arnon (1954).** He proposed three criteria of essentiality of nutrient as given below-
  1. A deficiency of the given element makes it for the plant impossible to complete its life cycle.
  2. The deficiency is specific for the given element and not replaceable by another element.
  3. The element is a constituent of an essential metabolite or it is required for the action of an enzyme system.
- **Nicholas** (1961) suggested the term **"functional or metabolism nutrient,"** which is defined as "any mineral element that functions in plant metabolism irrespective of whether or not its action is specific but it can disrupt the metabolic functions thus inhibit the plant growth.
- He proposed **20 functional nutrients** which are **Vanadium, Silicon, Cobalt, and Sodium plus essential nutrients.**
- **Cation** (+) form of elements – N, K, Fe, Mn, Cu, Zn, Ca and **Anion** (-) form of P, B, S, and Mo while **Nitrogen and Mn are found in both form** as NH_4^+ and No_3^-;Mn^{+2} and MnO_4^- respectively

Classification of essential plant nutrients (17

	Elements	Remarks
1. Frame work elements (available from air)	C, H, O	They account for **approximately 96%** of dry plant weight, mostly in the form of carbohydrates.
2. Mineral nutrients (plants taken from soil)	**A. Macro nutrients-** **a.** Primary nutrients (N,P,K) **b.** Secondary nutrients (Ca, Mg, S)	They account for approximately **3.5%** of dry plant weight
	B. Micro nutrients - Chlorine (Cl), Iron (Fe), Manganese (Mn), Zinc (Zn), Boron (B), Copper (Cu), Molybdenum (Mo) and Nickel (Ni)	It contribute about **0.04%** of dry plant weight

Difference between Macro and micro nutrients

Macronutrients	Micronutrients (Trace elements)
It required in relatively large amount. (Inexcess of 10 mmole Kg-' of dry matter).	Required in very small amount (less than 10 mmole Kg -' of dry matter).
The macronutrients include Carbon, Hydrogen, Oxygen, Nitrogen, Phosphorous, Sulphur, Potassium, Calcium and Magnesium.	These include Iron, Manganese, Copper, Molybdenum, Zinc, Boron, Chlorine and Nickel.
They are non toxic in slightly excessive quantity.	They show toxic effects even in slight excess
Some macronutrients contribute in the development of osmotic potential in the cells.	They do not play any role in the development of osmotic potential in the cells

Essentiality of nutrient in plant nutrition

- **Deficient** When the concentration of an essential element is lower to limit yield severely and distinct deficiency symptoms are visible. Extreme deficiencies can result in plant death. The moderate or slight deficiencies, symptoms may not be visible, but yields will still be reduced.
- **Critical range:** The nutrient concentration in the plant below which a yield response to added nutrient occurs. Critical levels or ranges vary among plants and nutrients, but occur somewhere in the transition between nutrient deficiency and sufficiency.
- **Sufficient** The nutrient concentration range in which added nutrient will not increase yield but can increase nutrient concentration. The term luxury consumption is often used to describe nutrient absorption by the plant that does not influence yield.
- **Excessive or toxic:** When the concentration of essential or other elements is higher to reduce the plant growth and yield. Excessive nutrient concentration may cause an imbalance with others essential nutrients, which also may lead to reduce the yield.

Deficiencies symptoms of nutrients on plant leave

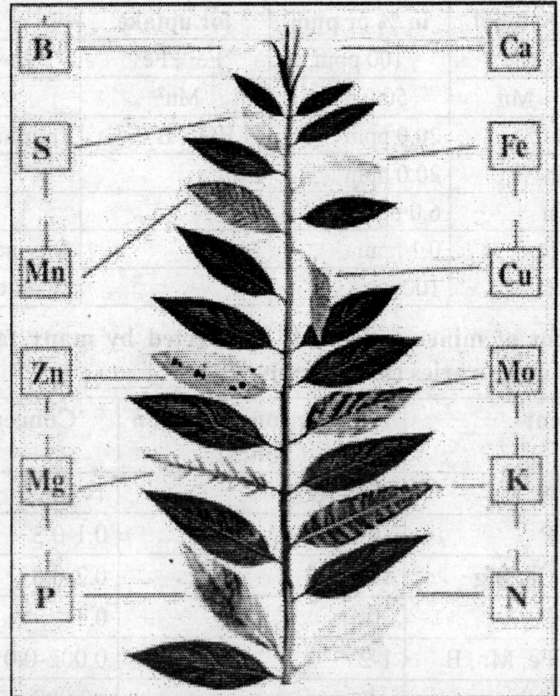

Source: https://in.pinterest.com/pin/971793044349458299/

Essential Nutrients for plant growth and their principal forms for uptake

Nutrient	Chemical Symbol	Concentration in % or ppm	Principal forms for uptake	Source of nutrients
Carbon	C	6.0	CO_2	Carbonate and Air
Hydrogen	H	45.0	H_2O	Air and water
Oxygen	O	45.0	H_2O, O_2	Air and water
Nitrogen	N	1.5	NH^{+4}, NO^{-3}	Organic Matter
Phosphorus	P	0.1-0.3	$H_2PO^-_4$, HPO^{2-}_4	Pica
Potassium	K	1.0	K^+	Micas, Feldspar
Calcium	Ca	0.5	Ca^{2+}	Dolomite, Apatite, Calcite
Magnesium	Mg	0.2	Mg^{2+}	Mica, Hornblende, lime stone
Sulfur	S1	0.1	SO^{2-}_4, SO_2	Organic sources – protein, amino acids Inorganic sources – Gypsum, pyrites

Nutrient	Chemical Symbol	Concentration in % or ppm	Principal forms for uptake	Source of nutrients
Iron	Fe	100 ppm	Fe^{2+}, Fe^{3+}	
Manganese	Mn	50.0 ppm	Mn^{2+}	
Boron	B	20.0 ppm	$H_3BO^-_3$, $B_40_7^{-2}$	Tourmaline
Zinc	Zn	20.0 ppm	Zn^{2+}	
Copper	Cu	6.0 ppm	Cu^{2+}	
Molybdenum	Mo	0.1 ppm	MoO_2^{-4}	Olivine
Chlorine	Cl	100 ppm	Cl^-	Apatite

The plant content of mineral elements is affected by many factors and their concentration in crops varies considerably

Nutrient		Relative concentration (Times)	Concentration (%)
Primary	N, K	400- 1000	1- 1.5
	P	30	0.1-0.3
Secondary	Ca, Mg	100-200	0.2-0.5
	S	30	0.1
Micro	Fe, Mn, B	1-2	0.002-0.01
	Others	<1	<0.002

Nutrients are classified based on mobility in soil and plants

Base of classification	Nature of elements	Example
	Mobile	NO_3^-, SO_4^{2-}, BO_3^-, Cl^- and Mn^{2+}
Mobility in soil	**Less mobile**	NH_4^+, K^+, Ca^+, Mg^{2+}, Cu^{2+}
	Immobile	$H_2PO_4^-$, HPO_4^{2-}, Zn^{2+}
Mobility in plants	**Mobile**	N, P and K
	Less mobile	Zn
	Immobile	Ca and B

Specific function of nutrients in plant

Function	Example
Energy storage and bonding nutrients	N, S and P
Act as regulatory and carriers	Ca, Mg and K
Catalyzed and activator	Fe, Mn, Zn, and Cu
Energy and exchange nutrients	H and O

Deficiency symptoms on plants leav

Type of leaves	Nutrients
Old leaves	N, P, K, Mg and Mo
New leaves	S, Fe, Mn and Cu
Old and new leaves	Zn
Terminal buds	Ca and Boron

Nutrient deficiency indicato plants

Nitrogen	Maize , Sorghum ,Leguminous plants
Phosphorus	Tomato ,Maize, Lucerne, Cereals, Duranta
Potassium	Maize, Lucerne , Cotton, Potatoes, Banana, Cucurbits
Sulpur	Lucerne , Clover , Cereals, Tea
Zinc	Maize, Tomatoes , Potatoes, Beans, Citrus
Copper	Citrus
Iron	Ornamental plants, Ixora, Acacia, Eucalyptus, Gooseberry, Securmanis
Boron	Lucerne, Coconut, Guava
Manganese	Citrus
Molybdenum	Cauliflower, Cabbage
Sodium	Sugar beet and Turnip
Source- http://agritech.tnau.ac.in/agriculture/agri_min_nutri_chart.html	

Plant nutrient interaction Mulder's chart

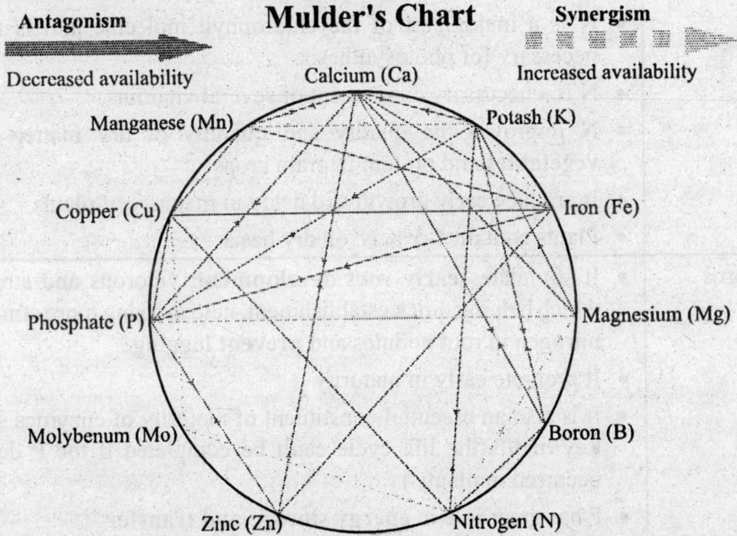

Source: -http://www.omafra.gov.on.ca/english/crops/hort/news/hortmatt/2016/07hrt16a3.htm

- **Ballast elements** are Al and Si.
- **Metal nutrients** *e.g.* K, Ca, Mg, Fe, Mn, Zn, Cu and **non metal nutrients** *e.g.* C, H, O, N, P, S, Cl
- **Law of minimum** was given by ***Father of Agriculture Chemistry*** **Leibig** and ***Law of Diminishing Returns***was given by **Mitscherlich.**
- Nutrient such as nitrate, chloride and sulphate are not absorbed by the soil colloids and remain mainly in the soil solution.
- Diffusion mechanism enables nutrient movement without the movement of water.
- It is the mechanisms predominant are supplying most of the P and K to plant roots.
- Nutrient uptake by the plants through mass flow, diffusion and root interception.
- Transportation of ions into the plants system through two methods like active transport (it need energy for water absorption) and passive transport.
- **Chelates** – it driven from a Greek word which means **"CLAW"**. Chelate is type of chemical compound in which a metallic ion is firmly combined with a molecule by means of multiple chemical bonds *e.g.* **DTPA, EDTA, EDDA and CIT.**
- **Nitrigin**- The fertilizer preparation with Rhizobium culture is called Nitrigin
- **Azotobacterin**- These fertilizer prepared with azotobacter culture

Functions of nutrients in plants

Nutrient	Functions and deficiency in plant
	Major - Primary nutrients
Nitrogen	• It is **basic constituent of life** • N is a major part of the chlorophyll molecule and is therefore necessary for photosynthesis. • N is a necessary component of several vitamins. • N improves the quality and quantity of dry matter in **leafy vegetables** and protein in grain crops. • It produce early growth and delay in maturity of plants • Plants contain 1-5 % N_2 on dry basis
Phosphorus	• It stimulates **early root development**, vigorous and strengthens which helping quick establishment, help in fixing more atmospheric nitrogen in root nodules and **prevent logging.** • It promote early in maturity • It is also an essential constituent of majority of enzymes so called **key of life**(the life cycle can't be completed if the P deficiency occurred in plants) • P has great role in **energy storage and transfer** • P is closely related to cell division and development • P compounds act as **energy currency** with in plants

Potassium	• K regulate **osmotic regulation; and opening and closing of stomata** • K deficiency greatly reduce quality and crop yield so this serious loss called **hidden hunger** • **K imparts increased vigor and** resistance to stress (cold and water), diseases, insect pests, *etc.,* as well as **quality of produce.** • K play vital role in conversion of soluble sugar into starch **(translocation of sugar)** resulting **grain filling**
Secondary nutrients	
Calcium	• Ca deficiency manifests itself in the **failure of terminal buds** of the plants • Ca is a main **constitution of cell wall** and increases stiffness of plants • Ca has an essential role in cell division and elongation • Ca **encourages seed production**
Magnesium	• Mg is a **major constituent of chlorophyll**; chlorophyll formation usually accounts for **15-20 percent** of the total Mg content of plants • Mg also Serves as a **structural component of Ribosomes** • Mg act as a **cofactor** for certain enzymes ***e.g.* RuDP carboxylase.**
Sulphur	• S is essentially required for the synthesis of **Cystine, cysteine and methionine as well as biotin, thiamin and glutathione** • S occurs in **volatile compounds** responsible for the characteristics **taste and smell** of plants *e.g.* mustard, garlic and onion *etc.,* • S promotes **nodule formation** on roots in **legume plants** • S also activates a number of **proteolytic enzymes** such as papainases
Micronutrients	
Iron	• Fe help in the **formation of chlorophyll** production • Its deficiency caused **Chlorosis between the veins of young leaves** • Fe is the **structural component of porphyrin** molecule like cytochrome, hemes, hematin, ferrichrome and **leghemoglobin.** These substances are involved in oxidation, reduction, ferrichrome, and photosynthesis.
Boron	• B primary role is associated with **calcium metabolism** • It act as a **regulator of K/Ca ratio** in the plants and help in nitrogen absorption through help in formation of nodules in legume plants • It is necessary for the **translocation of sugar, starch, phosphorus** and **pollen tube growth** as well as fruit or seed setting *etc.,* • B regulates carbohydrate metabolism • B is concerned with precipitating **excess cations, buffer action** and regulatory effect on their nutrient elements

Manganese	• Mn is **needed for citric acid cycle.** • Excess Mn concentration favour the **breakdown of Auxin (IAA)** • Its deficiency caused **intervienal Chlorosis**
Molybdenum	• Essential for nitrogen assimilation by plants and nitrogen fixation by bacteria. This means that it is needed for the production of nitrogen-based proteins.
Chlorine	• Stimulates photosynthesis.
Copper	• Cu helps in the utilization of iron during chlorophyll synthesis • Lack of Cu caused Fe to accumulate in the nodes of plants • It also act as electron carier in enzymes which bring about oxidation –reduction reactions in plants
Zinc	• Plays an important role in the synthesis of proteins, enzymes and growth hormones. • It help in the synthesis of tryptophan **(auxin Precursor)** • Zn stabilizes ribosomal fraction • It is associated with water uptake and water relation in the plants
Nickel	• Key component of selected enzymes involved in N metabolism and biological N fixation.
Beneficial element	
Silicon	• Strengthens cell walls, energy transfer & drought resistance • Reduces water loss & prevents fungal infection.
Cobalt	• Cobalt is essential for **microorganism fixing atmospheric nitrogen.** • Cobalt is the form of **vitamin B_{12}** • It also act as **electron career of oxygen** • Cobalt is also known as **animals protein factor**
Sodium	• Na replaces K in certain functions in halophytes plants and help to maintain turgor and growth • C4 plants require sodium as an essential nutrient
Vanadium	• It is essential to fixing atmospheric nitrogen for **Blue Green Algae**

MANURE AND FERTILIZERS

Manure - is originally as a organic, bulky and concentrate in nature; capable to plant nutrient supply and improving fertility of the soil by adding organic matter and nutrients without any hazardous effect. Manures are produced from animals, plants and other waste matter. Use of bulky organic manures has several advantages:

- They supply plant nutrients including micronutrients
- They improve soil physical properties like structure, water holding capacity *etc.*,

- They increase the availability of nutrients
- Carbon dioxide released during decomposition acts as a CO_2 fertilizer and
- Plant parasitic nematodes and fungi are controlled to some extent by altering the balance of microorganisms in the soil.

Kinds of manure-

1. **Bulky organic manure**–It is having less nutrient content but easy to decomposed and absorbed by plants *e.g.* plants and animals product like FYM, compost and crops (green manuring)

   A. **FYM** - It is a mixture of cattle dung, urine, litter or bedding material, portion of fodder not consumed by cattle and domestic wastes like ashes *etc.,* collected and dumped into a pit or a heap in the corner of the backyard. It is allowed to remain there and rot till it is taken out and applied to fields. Well rotten Farm Yard Manure contains O.5.% N., 0.2 % P2 O5 and 0.5 % K2 O

   Average amount of Moisture and Nutrient Content of Different Animals

Animals	Dung : Urine Moisture		'Nutrients' (kg/t)		
	Ration	(K)	Nitrogen (N)	Phosphorus (P_2O_5)	Pottasium (K_2O)
Dairy Cattle	80 : 20	85	4.53	1.22	3.40
Feeder Cattle	80 : 20	85	5.40	2.13	3.22
Poultry	100 : 0	62	13.54	6.48	3.17
Swine	60 : 40	85	5.84	3.22	4.94
Sheep -	67 : 33	66	10.42	3.17	9.83
Horse	80 : 20	66	66.75	2.03	5.98
Source - www.soilmanagementindia.com/organic-manures					

   B. **Compost** is the well rotten and decomposed remnants of plants and animal residue is known as compost.

   C. **Green manure** crops are grown in the field and incorporated into soil through ploughing in the same field along with growing others vegetation at the flowering or pod stage. The green manure crops are **especially legumes** *e.g.* **sannhemp, dhaincha and guar.**

2. **Concentrated** - Concentrated organic manures have **higher nutrient content** than bulky organic manure *e.g.* oilcakes, blood meal, fish manure *etc.,* it is also known as organic nitrogen fertilizer. These organic fertilizers are supplying **available nitrogen for a longer period due to relatively slow release of nutrients.**

- **Oil cakes**- it is the solid material of oilseeds after extracted which is dried as cake and can be used as manure. It is a two types:

- **Edible oil cakes**; *e.g.,* Groundnut cake, Coconut cake *etc.,*
- **Non edible oil cakes**; *e.g.,* Castor cake, Neem cake, Mahua cake *etc.,*,

Oil-cakes	Nutrient content (%)		
	N	P_2O_5	K_2O
Non edible oil-cakes (not fit for feeding livestock)			
Castor cake	4.3	1.8	1.3
Cotton seed cake (undecorticated)	3.9	1.8	1.6
Karanj cake	3.9	0.9	1.2
Mahua cake	2.5	0.8	1.2
Safflower cake (undecorticated)	4.9	1.4	1.2
Edible oil-cakes (safely fed to livestock)			
Coconut cake	3.0	1.9	1.8
Cotton seed cake (decorticated)	6.4	2.9	2.2
Groundnut cake	7.3	1.5	1.3
Linseed cake	4.9	1.4	1.3
Niger cake	4.7	1.8	1.3
Rape seed cake	5.2	1.8	1.2
Safflower cake (decorticated)	7.9	2.2	1.9
Sesamum cake	6.2	2.0	1.2
Source- agritech.tnau.ac.in			

Other Concentrated Organic Manures

Blood meal and meat of dead animals is dried and converted into meat meal which is a good source of nitrogen. Average nutrient content of animal based concentrated organic manures is given as follows.

Organic manures	Nutrient content (%)		
	N	P_2O_5	K_2O
Blood meal	10 – 12	1 – 2	1.0
Meat meal	10.5	2.5	0.5
Fish meal	4 – 10	3 – 9	0.3 – 1.5
Horn and Hoof meal	13	–	–
Raw bone meal	3 – 4	20 - 25	–
Steamed bone meal	1 – 2	25 - 30	–
Source- agritech.tnau.ac.in			

FERILIZER

The fertilizer is industrially manufactured or synthesis chemicals that, having definite chemical composition with higher analytic value, and capable to supply of specific nutrient in larger amount.

Type of fertilizers

1. **Simple fertilizers**: **contains only one main nutritional element**, such as nitrogen, phosphate, or potassium; it is also called Straight fertilizer *e.g.* Urea and Ammonium Sulphate
2. **Mixed fertilizers**: **contains at least two main** nutritional elements obtained via mixing or chemical processes are called mixed fertilizer. e. g. DAP
3. **Complex fertilizer** – Such type of fertilizer contain all primary nutritional elements. *e.g.* NPK (18:18:18)
4. **Special Purpose Fertilizers** – This type of fertilizer are formulated especially to target certain plants' or very specific requirements such as blueberry food (Blue Hydrangea Food) sometimes called rhododendron or azalea fertilizer)
5. **Slow release fertilizer**- A synthetic fertilizer, usually encased in capsules, that releases its nutrients over a specific period of time much the same as a controlled-release drug releases its medicine. The basic objective of slow release fertilizer (SRF) is Loss of nutrients due to volatilization and leaching; and reduction of plant toxicity and stress as well as Reduction in pollution of soil, water and atmosphere. *e.g.* Neem Coated Urea
6. **Liquid fertilizers**: – These types of fertilizer come in a variety of formulations and even include organic fertilizer, complete fertilizer as well as special purpose fertilizer. *e.g.* Nitrosol and African Violet Food.
7. **Fertilizer with insecticide**: These types of fertilizer that are prepared and combined with an insecticide. One such example is **Wonder 4:1:1 (21) + Karbaspray**

Difference between manure and fertilizer

Characteristics	Manures	Fertilizer
Origin	Plant or animal origin	Chemical synthesized or manufactured
Nature	Organic in nature	Inorganic in nature
Type	Natural product	artificial product
Conc. of nutrients	less concentrated	More concentrated
Material	Supply organic matter	Supply inorganic matter
Nutrient availability	slowly available	May or may not be readily available
Nutrients	Supply all the primary nutrients including Micronutrient	Supply specific type of nutrients one, two or three. micro nutrients may or may not be present
Effect on Soil health	Improves physical condition of soil	Do not improve the physical condition of soil

Effect on plant growth	No bad effect when applied in large quantities.	Adverse effect on plant whenever there is deficiency or excessive application

Fertilizer and its characters

Type of fertilizer	N	P	K	Remarks
NITROGEN FERTILIZERS				
Anhydrous Ammonia	82			highest N content fertilizer
Urea $[CO(NH_2)_2]$	46	-	-	Non-proteined organic form of N, amide form of N_1. 84 is urea acid equivalent
Ammonium sulphate [(NH4)2 SO4]	20.6	-	-	When close to seeds affects seed germination
Ammonium nitrate [(NH4)(NO3)]	33.5	-	-	(half as nitrate form and other half as Ammonical form), acidulating and explosive.
Calcium Ammonium Nitrate (CAN) $[NH_4NO_3 + CaCO_3\ MgCO_3]$	25.			Kisankad (1/2 ammonical and ½ nitrate)
Ammonium phosphate, 20/20 $[(NH4)_3PO_4]$	16	20		
Ammonium phosphate, 22/49	18	49	-	
Sodium nitrate 1	15.5	-	-	1st nitrogenous fertilizer
Calcium nitrate	15.5	-	-	
PHOSPHATE FERTILIZERS				
POTASH FERTILIZERS				
Potassium sulphate $[K_2SO_4]$	-	-	48	
Potassium chloride (muriate) [KCl]	-	-	60	
Potassium nitrate [KNO3]	13.8	-	38.7	
COMPLEX FERTILIZER				
Suphala: Nitro phosphate	20:20:2, 15:15:15, 18:18:9			
Diammonium Phosphate: (DAP) $[(NH_4)_2HPO_4]$	18:46:0			

Per hectare Fertilizer Consumption in Kg

S. No.	N	P	K	Total NPK(kg/h)

State	Pondicherry (365.81) Punjab (188)	Pondicherry (90.7) Jharkhand (60.91)	Pondicherry (52) Kerala (34.40)	509
India	84.54	33.44	10.36	128.34
Source: Department of Agriculture & Cooperation year 2012-2013				

Acidity and basicity of fertilizer

- **Equivalent acidity:** The amount of $CaCO_3$ required neutralizing the acid residues caused by acidic fertilizers in the soil *e.g.* 100 kg $(NH_4)_2SO_4$ produce acidity which need 110 kg of $CaCO_3$ to neutralize it. Therefore equivalent acidity of $(NH_4)_2SO_4$ is 110; anhydrous NH_3 has more equivalent acidity.
- **Equivalent Basicity:** Residual basicity caused by application of basic fertilizers, expressed in term of $CaCO_3$ of equivalent of basic residues left by a fertilizer material (in kg/100 kg of fertilizer salt).

Fertilizers	Equivalent acidity	Equivalent Basicity
CaCN	-	**63**
$NaNO_3$	-	29
KNO_3	-	29
$Ca(NO_3)$	-	**21**
Anhydrous Ammonia	**148**	-
NH_4Cl	128	-
$(NH_4)_2SO_4$	110	-
Urea	80-84	-
DAP	77	-
Ammonium Nitrate	60	-

- Rock phosphate is the 1 tone of cattle dung can give only 2.95 Kg of N, 1.59 Kg of Phosphoric acid and 2.95 Kg of potash.
- Night soil is also called Powdrette.
- Nitrate form of nitrogen is highly mobile in nature.
- Potassium fertilizers are totally import in India
- NPK grade in GROMOR is 29:29:0
- Phosphorus fertilizers are best for basal dose purpose
- Ammonium (NH_4) form of nitrogen fertilizers in rice is applied in reduced zone because it prevents the leaching and volatilization loss. While nitrate fertilizer applied in oxide zone (7-8 mm soil depth).
- Starter solutions- It refers to the application of solution of N, P2O5 and K2O in the ratio of 1:2:1 (legume) and 1:1:2 (to young plants at the time of transplanting, particularly for vegetables).

- **Biuret concentration of urea should not exceed to 1.5**
- Active Index of N_2 fertilizer (AI) = $\frac{CWIN\% - HWIN\% \times 100}{CWIN\%}$
- Metamorphic acid and Orthosphoric acid form of P is HPO_3 and H_3PO_4 respectively.

Conversion factors for plant nutrients:

P_2O_5	×	0.44	=	P
P	×	2.29	=	P_2O_5
K_2O	×	0.83	=	K
K	×	1.20	=	K_2O
CaO	×	0.71	=	Ca
Ca	×	1.40	=	CaO
MgO	×	0.60	=	Mg
Mg	×	1.66	=	MgO
SO_2	×	0.50	=	S
S	×	2.00	=	SO_2

Biofertilizers - 'Biofertilizer' is a substance which contains living microorganism which, when applied to seed, plant surfaces, or soil, colonizes the rhizosphere or the interior of the plant and promotes growth by increasing the supply or availability of primary nutrients to the host plant. A biofertilizers provide that is the nutrient availability in the soil, symbiotically, cost effective and eco friendly.

Type of biofertilizers

1. **Rhizobium –** It belongs to bacterial group and it is symbiotic nitrogen fixation @ 40-250 kg N /ha/year in legume crop. *e.g. R. leguminosarum* (Green pea, Lentil), *R. japonicum* (Soybean), *R. melliloti* (*Medicago, Trigonella*), it fixed highest nitrogen among all *Rhizobium sp.* @ 100-150 kg N /ha/year.
2. **Azotobacter** - It is well known free living **nitrogen fixing aerobic bacterium.** It is used as a Bio-Fertilizer for all non leguminous plants especially rice, cotton, vegetables *etc.,*
3. **Azospirillum**- It belongs to bacteria and to **fix the considerable quantity of nitrogen** in the range of 20- 40 kg N/ha in the rhizosphere in non- leguminous plants such as millets, Oilseeds, cotton, cereals, *etc.,*
4. **Cyanobacteria**- A group of one ornumerous **celled aquatic organisms** and also known as **blue-green algae (BGA)**
5. **Azolla** - Azolla is a free floating **water fern** and **fixes atmospheric nitrogen** in association with nitrogen fixing blue green algae Anabaena azollae. Azola dose required **0.1-0.4 kg/m².**
6. **Arbuscular Mycorrhiza (Fungi):** is a type of mycorrhiza which penetrates the cortical cells of the roots of a vascular plant.

Groups	Examples
Nitrogen (N2) fixing Biofertilizer	
Free-living	Azotobacter, Clostridium, Anabaena, Nostoc,
Symbiotic	Rhizobium, Frankia, Anabaena azollae
Associative Symbiotic	Azospirillum
P Solubilizing Biofertilizers	
Bacteria	*Bacillus megaterium var. phosphaticum*
	Bacillus circulans, Pseudomonas striata
Fungi	*Penicillium sp, Aspergillus awamori*
P Mobilizing Biofertilizers	
Arbuscular mycorrhiza	*Glomus sp., Gigaspora sp., Acaulospora sp.,*
Ectomycorrhiza	*Laccaria sp., Pisolithus sp., Boletus sp., Amanita sp.*
Orchid mycorrhiza	*Rhizoctonia solani*
Biofertilizers for Micro nutrients	
Silicate and Zinc solubilizers	*Bacillus sp.*
Source- http://agritech.tnau.ac.in/org_farm/orgfarm_biofertilizertechnology.html	

- Nitrogen occurs in soil mainly in two form of **nitrate (NO^-_3)** and **ammonium (NH^+_4) ions**
- Soil (0-15 cm depth) contain nitrogen in the range of 0.2-0.4 percent by weight
- Transformation of nitrogen in soil by two pathway like **mineralization (aminization, ammonification and nitrification) and immobilizati**
- **Mineralization** is process by which nitrogen in organic compounds becomes converted into the inorganic ammonium and nitrate ions carried out by different soil microorganism
- Organic N → Amine → Ammonium (NH^+_4) → Nitrite (NO^-_2) → Nitrate (NO^-_3)
- Ammonium form of nitrogen is convert into nitrite through Nitrosomonas (obligate autotrophic bacteria) while nitrite to nitrate through Nitrobacter
- Optimum temperature required for ammonification and nitrification are 40-60°C and 30°C respectively.
- **Immobilization** nitrogen immobilization means the change of mineral nitrogen to organic form by soil microorganism
- **Denitrificatio** is defined as the microbial reduction of nitrate and nitrite with the release and loss of molecular nitrogen, in some instances of NO_2 (means loss of nitrogen).
- **Some facultative** aerobic bacteria are responsible for denitrification like Pseudomonas, Bacillus and Paracoccus
- The total P content in the range of 0.02 to 0.15 per cent
- **Carbon :** Phosphorus ratio for mineralization is 200: 1 or less whereas 300: 1 or above for immobilization

- P are absorbed by plants mostly as primary (**$H_2PO^-_4$ at below 7.2**); and secondary (**HPO^{2-}_3 at above 7.2**) but at **pH 7.2** there are almost equal in amount of $H_2PO^-_4$ and HPO^{2-}_3
- The Q/I concept has been developed by **Beckett** for the use of predicting the status of K in soil.
- **Hidden hunger** - it is evident that K deficiency doesn't immediately show the visible symptoms and reduced the plant growth rate initially, is known as hidden hunger.
- **Glutathione is precursor of Phytochelatins**; its functioning as a detoxification the certain heavy metals in higher plants. Phytochelatins are composed of two or more repeating gamma-glutanyl cystine units with a terminalglycine residue.
- S requirement in decreasing order for plants families as **cruciferae > leguminosae > gramineae**
- **Submerged soil** is soil that saturated with water for a sufficiently long time in a year is called submerged soil. It reduced oxygen level, aerobic microbial activity and altered chemical status of soil also.
- Nutrient availability are depends on Specific zone such as **Fe^{3+}, NH^{4+} and NO^{3-}** available in oxide zone whereas **Fe^{2+}, Mn^{2+} and N_2** in reduced zone.
- **Puddling** is essential practice for submerged soil to get successful rice cultivation.
- **Radioactivity** concept was discovered by **Henri Becquerel (1896)**

SOIL EROSION AND CONSERVATION

- **Soil degradation** is defined as the loss in soil productivity due to physical, chemical and biological deterioration.
- Causes are excessive pressure on land to meet the growing demands of population. They are in the form of over exploitation of natural resources like overgrazing, excessive deforestation, faulty methods of agricultural practices, **shifting cultivation or Jhuming.**
- **Soil erosion** is the **detachment, transportation and deposition** of soil mass from one place to another through action of wind, water in motion or by the beating action of rain drops, is known as soil erosion.
- There are 2 types of soil erosion Normal erosion (geological erosion) and Accelerated erosion.
- In normal erosion, rate of soil loss rate at which soil is formed.
- Accelerated erosion is one where rate of soil of loss is far greater than the rate at which it is formed.
- Area affected by soil degradation is 187.9 Mha (57.1%) of the total geographical area.
- Deterioration in the form of water erosion (148.9 Mha), Wind erosion (13.5 Mha), chemical deterioration – 13.8 Mha, Physical deterioration like water logging (11.6 Mha) and Biological deterioration.

- **Water erosion**: Erosion by water – splash erosion, sheet, rill, gully, stream bank and landslide erosion.
- **Splash erosion** is the splashing effect of raindrops on soil particles.
- **Sheet erosion** is not clearly recognized but can be seen as muddy run-off.
- **Rill erosion** is leads to formation of finger like rills and gully erosion is the severe form of rill erosion wherein grooves form deep channels called pullies. Land becomes unfit for cultivation.

Size of gullies:

Symbol	Description	Specification
G_1	Very small	Up to 3 m deep and 18m bed width. Side slopes vary
G_2	Small	Up to 3m deep and greater than 18m bed width. Side slope very
G_3	Medium	Depth between 3 and 9m. Bed width greater than 18m. Side slopes uniformly sloping between 8 and 15%
G_4	Deep and Narrow	3m-9m deep. Bed width less than 18m. Side slope vary. Depth greater than 9m. Bed width varies. Side slope vary, mostly steep or even vertical, with intricate and active branch gullies.

The mechanism of water erosion is detaching effect of raindrops and surface flow of water carries the detached soil particles to far places.

Wind erosion:

- Wind erosion normally occurs in arid and semi-arid regions.
- 3 types of soil movement are seen in wind erosion: saltation, suspension and surface creep.
- **Saltation:** is the most important process in wind erosion and control of wind erosion is mainly based on elimination of movement in saltation. Particles of **size 0.1 to 0.5mm** are lifted.
- Major portion of soil carried by wind is moved in a series of bounces called saltation.
- **Suspension**: Very fine particles **less than 0.1mm** in diameter is carried into suspension over long distances. They are kicked up into air by action of particles in saltation.
- **Surface creep**: Particles **larger than 0.5mm but smaller than 1.0mm** are pushed and spread along the surface by impact of particles in saltation to form a surface creep.
- The mechanism of wind erosion is Initiation, transportation and deposition.

Factors affecting soil loss:

- Universal soil loss (USLE) was given by **Wischmeier & Smith's**.
- **A= RKLSCP** where A denotes soil loss, R-Rainfall erosivity, K-soil erodibility, L-slope length, S-slope%, C-crop management factor, P-Soil Management factor.
- Soil loss is calculated by comparing soil loss with an ideal plot of 22m long (72 ft) and 9% slope.

Rainfall erosion index = Kinetic energy of the storm x Max 30 minute mtensity

- Intensity of the rainfall is more important than duration and frequency of rainfall in causing soil loss.
- Soil texture, structure, infiltration, permeability, organic matter content *etc.*, affects soil erosion.
- Lateritic soils are less erodible than black soils.
- Speed and extent of run-off is decided by slope% and length.
- Velocity of water flow is proportional to square root of slope% or vertical drop.
- Hence if land slope% is increased 4 times then velocity is doubled.
- If velocity is doubled, erosive power is increased by 4 times *i.e.,* erosive power is proportional to square of the velocity.
- Loss of plant nutrients increases with increase in degree of slope.
- Soil loss is less when land is left undisturbed under a natural cover and soil loss increases steeply when vegetation is removed and land is cultivated.
- Legumes and grasses are stronger in preventing soil loss.
- Monoculture of cereals should be avoided.

SOIL AND WATER CONSERVATION MEASURES

- The key to soil and water conservation is to follow land capability classification
- Land capability classes are based on intensity of hazards and limitations, as class increases, the intensity of limitations increases.
- **Class V has no erosion hazard** but it is placed in class V only due to the fact that it limitations are practically difficult to remove.
- Class VIII includes bad land rocks, sandy beaches, marshes, deserts, river washes, mine spoils and other nearly barren lands.
- The colours for the various classes from I to VIII are Green, yellow, Brown, Pink, Grey, Orange, Red and Purple.
- Sub class indicates dominant limitation and is denoted by **E, W, S, C. where E for erosion, W for wetness, S for soil limitations and C for climate.**
- If two limitations are found, the dominant limitations are written first.
- The order of priority in case of equal extent of limitations is E, W, S, and C.
- Arid lands start with class IV, semi-arid with class III or IV, sub-humid where crop yield is affected frequently by drought with II or III, Humid climate with

occasional dry spells with II and humid climate with well distributed rainfall with CLASS I.

- Soil and water conservation measure are divided into Agronomic and Engineering measures.

Agronomic measure:

- The principles are to intercept raindrops and stop splash effect, to increase intake rate and to stop overland flow.
- **Contour farming**: It is farming across the slopes along the contour bunds within 6% slope, Important examples of soil loss: Maize + cow pea (Contour cultivation) < maize (up and down cultivation).
- Potato (contour cultivation) < Potato (up and down cultivation)
- **Mulching:** synthetic and natural.
- **Selection of crops**: crops selected should provide maximum cover of soil. Legumes are very effective particularly cowpea and mung. Crops can be divided into erosion resistant (groundnut) and erosion permitting crops (urd).
  1. Erosion resistant crops *e.g.* Urd > Maize > Paddy > Groundnut.
  2. Erosion permitting crops *e.g.* Jowar > Black gram > Groundnut.
- **Strip cropping:** Alternate strips of erosion resisting and erosion permitting crops.
- Wind strip cropping is growing alternate rows of tall and short crops across the direction of wind.
- **Mixed cropping:** Better protection and yield than strip cropping.
- **Engineering measures**: To increase the opportunity time and to divide the long slopes into short ones so that the velocity of flow is reduced.
- Gully control measures: Check dams, sodding, Gully plugging with brush woods, wire meshes, sand bags, boulders, live hedges (Vetiver), bricks masonry items, *etc.,* earthen plugging.
- In case of small gullies, diversion check bunds and in case of medium gullies, check dams at vertical interval of 1.2m and terracing the side slopes are done.

SOIL SCIENCE OBJECTIVE QUESTIONS (MCQ)

1. The equation of soil formation, S = f(cl. o. r. p. t.) was given by

   A. Dokuchaev B. Glinka C. Jenny D. Kellog

2. Mineral at the 8th of Jackson's weathering sequence is

   A. Muscovite B. Kaolinite C. Vermiculite D. Quartz

3. The stage of development of tropical soils when only the most resistant materials have survived

   A. Initial B. Juvenile C. Virile D. Senile

4. In anorthite the amount of isomorphous substitution is
   A. 25% B. 50% C. 30% D. 75%
5. Concept of active and passive soil forming factors was given by
   A. Jenny B. Dokuchaev C. Joff D. Marbut
6. The reaction $2Fe_2O_3 \rightarrow 4FeO + O_2$ is an example of:
   A. Oxidation B. Reduction C. Hydration D. Hydrolysis
7. What is soil development stage for goethite and hematite?
   A. 11 B. 12 C. 3 D. 13
8. The wind blown materials consisting primarily of silt with some fine sand and clay are called
   A. Loess B. Aeolian C. Alluvial D. Colluvial
9. A group of soils which have developed from the some parent material in same climate but under different topographical condition is called
   A. Soil catena B. Cat clays C. clay micelle D. Podzol soils
10. The diagnostic horizon in Alfisol is termed
    A. Argilic B. Spodic C. Mollic D. Oxic
11. Oxisols are the most weathered soils and the one just preceding these are
    A. Ultisols B. Alfisols C. Entisols D. Inceptisols
12. A soil formation process which humus and sesquioxides become mobile and remains behind
    A. Pedoturbation B. Calcification C. Laterization D. Podsolization
13. The identifiable morphological feature of Vertisols is
    A. A horizon B. Crumb structure
    C. Clay skins D. Slickensides
14. An argillic sub-surface horizon with ESP of more than 15 is called:
    A. Cambic horizon B. Natric horizon
    C. Agric horizon D. Oxic horizon
15. In Vertisol, horizon formation is inhibited by
    A. Cumulization B. Haproidization C. Desiccation D. Mineralization
16. In the taxonomic system of soil classification the numbers of soil orders are
    A. 15 B. 12 C. 10 D. 20
17. The sub-surface horizon having an illuvial accumulation of free subsquoxides and organic matter is
    A. Argillic B. Natric C. Spodic D. Mollic
18. A mineral soil horizon in which clay, silt, and humus derived from an overlaying cultivated and fertilized layer have accumulated is called
    A. Agric horizon B. Hydro horizon C. Paleo horizon D. Hygro horizon
19. Soils, which are developed from volcanic eruption and have a dark in colour surface layer with low bulk density are named as

A. Edaposols B. Andisols C. Aldisols D. Acidic soils

20. The most commonly present soil type in desert soils, which have highest infiltration rate and low in organic matter is called as

A. Edaposols B. Andisols C. Aldisols D. Ardisols

21. The soil order covering the maximum area in the world is

A. Alfiso B. Inceptisol C. Entisol D. Vertisol

22. A silicate clay enriched horizon formed by illuviation of clay is termed as

A. Natric B. Oxic C. Kandic D. Argillic

23. Podzolic soils are produced from

A. Acid igneous rocks B. Basic igneous rocks

C. Alluvium D. Sandstone

24. Collection of soil individuals essentially uniform in differentiating characteristics is termed as soil

A. Family B. Type C. Series D. Phases

25. An illuvial horizon of clay accumulation is denoted by

A. Ap B. Bw C. Bt D. Bh

26. Gypsic horizon contains at least

A. 5% or more $CaSO_4$ B. 10% or more $CaSO_4$

C. 2% or more $CaSO_4$ D. 1% or more $CaSO_4$

27. Base saturation of argillic horizon should be

A. <35% B. <20% C. >35% D. >50%

28. A horizon enriched with Fe and Al oxides and dominance of 1:1 type clay mineral is termed as

A. Kandic B. Alillic C. Oxic D. Spodic

29. Removal of silica from soil and accumulating of sesquioxides occurs in the pedogenic process

A. Gleization B. Laterization C. Podzolization D. Ferruginisation

30. The most important group of soil-forming minerals are

A. Micas B. Amphiboles C. Quartz D. Felspars

31. A colour or structural B-horizon is indicative of the endopedon:

A. Argillic B. Agric C. Cambic D. Odic

32. In 1:50,000 scale fo base map distance between field observations is

A. 50 m B. 500 m C. 100 m D. 1000m

33. As per USDA soil taxonomy, shallow soil depth for series level classification is

A. <10 cm B. 10-25 cm C. 25-50 cm D. 50-75 cm

34. The process associated with the formation of Hitosols:

A. Padulization B. Leucinization C. Melonization D. Rubification

35. Grassland soils are

A. alfisol B. Mollisol C. spodosoll D. inceptisol

36. The process of soil formation occurring under reducing conditions endangered by poor drainage and lack of O_2 and due to presence of organic matter under water logged conditions is known as

A. Gleization B. Podzolization C. Salinization D. Calcification

37. Well drained soils with mean annual soil temperature > 22⁰C <28⁰C is termed as

A. Mesic B. Thermic C. Hyperthermic D. Megathermic

38. Ustic moisture regime indicates soils are dry for

A. 90 days B. >90 days < 180 days

C. 45 days D. >45 days < 90 days

39. In general the limit of soil moisture control section (SMCS) in coarse sandy soil is:

A. 10 – 30 cm B. 20 – 60 cm C. 30 – 90 cm D. 40 – 120 cm

40. Percent pore space of a soil having bulk density Pb and partical density Ps is given by:

A. (1 – Ps/Pb) × 100 B. (1 – Pb/Ps) × 100

C. (Pb – Ps/Pb) × 100 D. (Ps– Ps/Pb) × 100

41. The cations viz. Ca, Mg and Al have the ability to bind the soil aggregates. This process is known as

A. Flocculation B. Aggregation C. Deflocculation D. None of these

42. At field capacity the soil moisture is retained at tension of

A. 1/2 bar B. 1/3 bar C. 1 bar D. 15 bar

43. 1 hectare cm of water is equivalent to how many litres of water

A. 1,000 B. 10,000 C. 100,000 D. 110,000

44. If K is intrinsic permeability and f is fluidity then hydraulic conductivity is given by:

A. kf B. f/k C. fk D. k/f

45. Tensiometers can be used to measure soil moisture potential up to (bar)

A. 0-0.85 B. 0.9-1.0 C. 1.1-1.2 D. 1.3-1.4

46. The water that occurs as a thin film around the soil particles and is held tenaciously with a tension of 31 bar or more is called as

A. Hygroscopic water B. Capillary water

C. Gravitational water D. None of the above

47. If w and θ be the mass wetness and volume wetness, respectively of a soil then θ is given by

A. $w\rho_b/\rho_w$ B. $w\rho_w/\rho_b$ C. $\rho_w/w\rho_b$ D. $\rho_b/w\rho_w$

48. The domain reflectometry of soil moisture estimation is based on estimation of

A. Dielectric constant B. Roughness of soil surfaces

C. Thermal conductivity D. Electrical capacity

49. Which law follows heat flow through soil?
A. Darcy's law B. Fourier's law C. Ohm's law D. Fick's law

50. The average annual soil loss tonnes/ha in India is of the order of
A. 12.4 B. 14.4 C. 16.4 D. 18.4

51. The soil water erosion is mainly caused by rain drops is
A. Splash B. Sheet C. Rill D. Gully

52. Which of the following mineral is not a polymorph:
A. Quartz B. Tridymite C. Crystobalite D. Leucite

53. Negative ions absorption increase with
A. Increasing CEC B. Decreasing CEC
C. Independent of CEC D. Decreasing pH dependent

54. Which one of the following has the highest cation exchange capacity on weight basis?
A. Illite B. Kaolinite C. Humus D. Montmorillonite

55. The amount of charge reaching the central cation Na^+ from an anion in cubic coordition is:
A. – 1/8 B. –1/6 C. +1/8 D. +1/6

56. The negative ions adsorption is maximum in
A. Montmorillonite B. Illite
C. Kaolinite D. Vermicullite

57. The inter layer spacing of K-saturated air dried montmorillonite is:
A. 10 A^0 B. 12 A^0 C. 14 A^0 D. 17 A^0

58. The tetrahydral rotation is minimum in:
A. Moscovite B. Biotite C. Beidilite D. Illite

59. Ligand adsorption of P by soil results in:
A. Increase in CEC B. Increase in AEC
C. Decrease in pH D. Increase in surface area

60. In soils, Gapon's equation is widely used equation to describe
A. C : N ratio B. CEC of soils
C. Bulk density of soil D. None of the above

61. Which of the following is not recommended as amendment to reclaim alkali soils:
A. Gypsum B. Lime C. Iron sulphate D. Pyrite

62. Potential source of acidity in acid soils is:
A. H^+ B. Ca^{++} C. Al^{+++} D. Fe^{+++}

63. Rock phosphate is used in
A. Alkaline soil B. Saline soil C. Acidic soil D. Normal soil

64. Highest limit of exchangeable sodium percentage in saturated paste of saline soil is:

A. < 5 B. < 10 C. < 15 D. < 20

65. A soil which has pH < 8.5, ESP < 15 and EC > 4 dSm^{-1} at 25°C in saturated condition is called:

A. Saline B. Saline alkali C. Alkali D. Normal

66. A solution which has pH value 6.0, the OH^- ions (gram ions L^{-1}) concentration in the solution is:

A. 10^{-6} B. 10^{-8} C. 10^{-10} D. 10^{-12}

67. The best salt tolerant crop is:

A. Wheat B. Barley C. Maize D. Sorghum

68. Sodium adsorption ratio (SAR) is calculated as

A. Na+/ Ca+Mg/2 B. Na+/ Ca+Mg

C. √Na+/ Ca+Mg/2 D. Na+/ Ca+Mg

69. Residual sodium carbonate is calculated as

A. $(CO_3^{2-} + HCO_3^-) - (Ca^{++} + Mg^{++})$ B. $(Ca^{++} + Mg^{++}) - (CO_3^{2-} + HCO_3^-)$

C. $(Ca^{++} + Mg^{++}) - (CO_3^{2-} - HCO_3^-)$ D. $(CO_3^{2-} - HCO_3^-) - (Ca^{++} + Mg^{++})$

70. Permissible limit of RSC (me L^{-1}) in irrigation water up to which gypsum application is not recommended:

A. 2.5 B. 3.5 C. 4.5 D. 5.5

71. The nutrient element that is not a constituent of any biosynthesized compound in plant is:

A. P B. Ca C. K D. Mg

72. The element reportedly having beneficial effect on rice plants is:

A. Se B. Co C. Na D. Si

73. Internal cork disease of apple is due the deficiency of:

A. Zn B. Cu C. B D. Mn

74. Khaira disease is due to..............deficiency

A. Zn B. Mo C. Cu D. Mn

75. The 'Pahala blight of sugarcane' is caused by the deficiency of:

A. Zn B. Mn C. Cu D. Fe

76. The biochemical process whereby ammonical nitrogen is released from nitrogen containing organic compounds is called as

A. Nitrate reduction B. Nitrification

C. Nitrogen fixation D. Ammonification

77. Detected intermediate involved in the process of nitrification is:

A. Hyponitrite B. Hydroxylamine C. Hydrated nitrite D. None of them

78. How many electrons are required for conversion of NO_2^- to NH_4^+

A. 4 B. 6 C. 8 D. 10

79. To prevent denitrification losses of N2 in paddy field, Nitrogen should be applied in the
    A. Oxidized zone B. Reduced zone
    C. Hydronized zone D. Calcified zone
80. Which formed of Nitrogen adsorbed by plant is toxic?
    A. NH_4^+ B. NO_3^- C. NO_2^- D. Amide
81. Total Nitrogen content in Indian soil
    A. 0.2-0.5% B. 0.03-0.05% C. 0.002-0.005% D. 0.2-0.005%
82. C: P ratio for net immobilization should be
    A. >300 : 1 B. <200 : 1 C. <300 : 1 D. >200 : 1
83. The microorganism which is responsible for oxidation of sulphur is
    A. Nitrobacter B. Clostridium C. Nitrosomonas D. Thiobacillus
84. At what temperature the soil samples are oven dried
    A. 100°C B. 105°C C. 110°C D. 115°C
85. In a pot containing 10 kg soils to which a dose of 10 ppm Zn is to be applied. The amount of Zn in milligrams to be applied in the pot:
    A. 10 B. 100 C. 1000 D. 10000
86. The pH of DTPA extractant (Lindsay and Norvill) for micronutrient is adjusted at:
    A. 7.3 B. 7.5 C. 7.8 D. 8.2
87. The pH of Olsen extractant for extraction of available phosphorus is adjusted at
    A. 8.0 B. 7.5 C. 8.5 D. 9
88. The concentration of $CaCl_2$ in DTPA extractant for micronutrient is:
    A. 0.1 M B. 0.01 M C. 0.001 M D. 0.005 M
89. The indicator used in determination of organic carbon by Walkley and Black method:
    A. Eriochrome Black-T B. Mixed indicator
    C. Bromocresol green D. Diphenylamine
90. Electric conductivity is a measure of:
    A. Na-salts B. Ca-salts C. Soluble salts D. Insoluble salts
91. Di-acid digestion of plants is carried out at:
    A. 80 – 100°C B. 180 – 200°C C. 280 – 300°C D. 380 – 400°C
92. One molecule of $K_2Cr_2O_7$ when react with H_2SO_4 gives three atoms of oxygen, the equivalent weight of $K_2Cr_2O_7$ is:
    A. 1/3rd of molecular weight B. 1/4th of molecular weight
    C. 1/6th of molecular weight D. 1/12th of molecular weight
93. Which of the following Concentration of $KMnO_4$ was suggested by Subbiah and Asija for determination of available nitrogen in soil:

A. 0.25 % B. 0.32 % C. 0.48 % D. 0.56 %

94. In order to prevent considerable loss of K during ashing of plant samples the temperature is not allowed to exceed:
A. 200ºC B. 400ºC C. 600ºC D. 800ºC

95. Hanway and Hiedel method is used to estimate.........
A. K B. P C. B D. Mo

96. In organic carbon estimation Diphenylamine (DPA) is replaced by.......................
A. DTPA B. EDTA C. Feroin D. Mureoxide

97. 1 ml of 0.025 N H_2SO_4 = ...mg N
A. 0.35 B. 0.70 C. 1.05 D. 1.40

98. CEC of soil is expressed as:
A. percentage B. parts per million
C. milliequivalent per 100 gm D. milligram per 100 gm

99. Composition of DTPA is
A. 0.005 *M* DTPA + 0.01 *M* TEA + 0.01 *M* $CaCl_2$
B. 0.0005 *M* DTPA + 0.1 *M* TEA + 0.01 *M* $CaCl_2$
C. 0.005 *M* DTPA + 0.1 *M* TEA + 0.01 *M* $CaCl_2$
D. 0.005 *M* DTPA + 0.1 *M* TEA + 0.001 *M* $CaCl_2$

100. At what temperature the plant samples are oven dried
A. 98°C B. 105°C C. 65°C D. 115°C

101. Molarity of concentrated H_2SO_4 (AR grade) is
A. 18 M B. 10 M C. 12 M D. 16 M

102. What is the role of TEA in DTPA extractant of micronutrient?
A. It act as buffer B. It clearly burn on flame
C. It act as reducing factor D. Both A and B

103. Which one of the following indicator is known as self indicator
A. Muraxide B. $KMnO_4$ C. Peranitrophenol D. All of the

104. Colorimetric estimation of Total P in Plant sample is done by
A. Nesslers reagent B. Barton's reagent
C. Fehling's solution D. Tollen 's reagent

105. Percentage recovery of organic carbon by Walkley-Black method is
A. 100% B. 90% C. 76% D. 60%

106. The DRIS approach was given by:
A. Bray B. Beaufils C. Jackson D. Nemeth

107. The fertility gradient approach is given by:
A. B. Murthy B. B. Ramamoorthy
C. B.V. Mehta D. A.B. Ghosh

108. In India, the oldest surviving long term manurial trial initiated in 1909 is located at

A. TNAU, Coimbatore B. 'UAS, Dharwad

C. APAU, Rajendra Nagar D. IARI, New Delhi

109. Which amino acid do not contains S

A. Cystine B. Cystein C. Methionine D. Arginine

110. Higher molecular weight substances released from roots as a result of active metabolic processes are:

A. Exudates B. Lysates C. Secretions D. Mucigel

111. Which of the following is P solubilizing bacteria

A. Pseudomonas striata B. *Nitrosomonas europaea*

C. *Thiobacillus ferooxidans* D. *Furarium oxysporum*

112. Host specificity in legume Rhizobium symbiosis attributed in

A. AAA B. Lectins C. nif genes D. nod

113. Organism most tolerant to soil moisture stress is:

A. Fungi B. Protozoa C. Actinomycetes D. Bacteria

114. Among the prominent green house gases the one whose emission is almost entirely dependent on soil related processes is

A. Methane B. Carbon dioxide

C. Chlofluorocarbon D. Nitrous oxide

115. The book "Soil conditions and plant Growth" is written by

A. Nyle C. Brady B. M. Alexander

C. E.J. Russell and E.W. Russell D. S.A. Waksman

116. A byproduct of coastal salt industry- schoentite is a rich source of plant nutrient

A. N B. P C. K D. S

117. A combination of many taxonomic unit likes soil series, phase or types is known as

A. Soil association B. Soil texture

C. Soil structure D. Soil genesis

118. Aspergillus niger test is used for

A. Cu and Mg B. S C. N D. P

119. Availability of manganese increase in

A. Neutral soil B. Alkali soil C. Acidic soils D. Saline soil

120. Carry over fertilizer effects are more pronounced in the case of

A. SSP B. DAP C. MOP D. MAP

121. Green ear disease of bajra is suppressed by increasing the concentration of

A. Mg B. Cu C. Ca D. S

122. Optical Density equals to

A. OD = LogT-2 B. OD = 2-Log T

C. OD = 1/2-Log T D. OD = LogT-1/2

123. The method of survey adopted for intensive cultivation is
A. Detailed survey B. Semi-detailed survey
C. Reconnaissance survey D. All

124. Nutrient absorbed more through mass flow is due to
A. Transpiration B. Respiration C. Both D. None

125. Nutrient which is having the lowest diffusion rate is
A. N B. K C. P D. S

126. Sackette and Stewart technique of nutrient deficiency test is used to study
A. K, P B. N, S C. Mg, Mo D. Fe, Mn

127. Physical mixture of straight fertilizers is called
A. Straight fertilizer B. Complex fertilizer
C. Mixed fertilizer D. All

128. Reclamation disease exists due to deficiency of
A. Cu B. Mg C. Ca D. Mn

129. Percentage of conditioners used in fertilizer mixture is
A. 1% B. 4.5% C. 10% D. 20%

130. S content in ammonium sulphate
A. 15% B. 20% C. 22% D. 24%

131. Deep placement of fertilizers increase the efficiency of
A. Ammonium sulphate B. Ammonium nitrate
C. Ammonium chloride D. All

132. Fertilizers have less losses in calcareous soil is
A. Ammonium sulphate B. Ammonium nitrate
C. Urea D. All

133. Maximum Ca/Mg ratio limit beyond which Mg deficiency exist, limit is
A. 3 : 1 B. 5 : 1 C. 6 : 1 D. 7 : 1

134. Formation of soil is the result of
A. Action of wind B. Chemical process
C. Bio-chemical process D. Physical process

135. Alternate drying and wetting improves
A. Soil structure B. soil texture C. Soil colour D. Soil temperature

136. More than 90% gaseous exchange in soil takes place by
A. Mass flow B. Diffusion
C. Partial diffusion D. All

137. Flesh of soil
A. Sand B. Silt C. Clay D. None

138. Buffering capacity is high for
A. Silt B. Clay C. Sand D. All

139. Approximate humus content in soil
A. 1.5% B. 2.5% C. 3.5% D. 4.5%

140. Which clay is sometimes used for medicinal purpose
A. Kaolinite B. Chloride C. Vermiculite D. Illite

141. Press mud is used for reclamation of
A. Saline soil B. Acidic soil C. Sodic soil D. Saline-alkali soil

142. Limonite is the source of
A. Iron B. Nitrogen C. Potash D. Phosphorus

143. Phosphate fixing capacity is high in
A. Sandy soil B. Red soil C. Alluvial soil D. Laterite soils

144. Khaira disease of rice is diagnosed by
A. Necrotic lesions B. Dark reddish brown pigmentation
C. Yellowish spots on leaves D. All

145. Nufrients absorbed by plants are carried upward through
A. Xylem B. Phloem C. Both D. None

146. The losses of ammonia usually become greater with........ in soil
A. Increase in soil PH B. Decrease in soil PH
C. Both D. None

147. Chlorosis in mid rib of plant leaves is caused by
A. Mn B. B C. Mg D. S

148. Improper development of wheat inflorescence is due to deficiency of
A. B B. Fe C. Zn D. P

149. Critical soil test level approach was given by
A. Bray B. Liebig
C. Cate and Nelson D. Arnon and Stout

150. Premature leaf fall is due to deficiency of
A. P B. Zn C. N D. Fe

ANSWERS

1. C	2. C	3. D	4. A	5. C	6. A	7. D	8. A
9. A	10. A	11. A	12. D	13. D	14. B	15. B	16. B
17. C	18. A	19. B	20. D	21. B	22. D	23. A	24. C
25. C	26. A	27. C	28. C	29. B	30. D	31. C	32. B
33. C	34. A	35. B	36. A	37. C	38. B	39. C	40. A
41. A	42. B	43. C	44. A	45. A	46. A	47. A	48. A
49. B	50. C	51. A	52. D	53. A	54 .C	55. A	56. D

57. B	58 .B	59. A	60. B	61. B	62. C	63. C	64. C
65. A	66. B	67. B	68. C	69. A	70. A	71. C	72. D
73. D	74. A	75. B	76. D	77. B	78. B	79. B	80. C
81. B	82. D	83. D	84 .B	85. B	86. A	87. C	88. B
89. D	90. C	91. B	92. C	93. B	94. C	95. A	96. C
97. A	98. C	99. C	100. C	101. A	102. D	103. B	104. B
105. C	106. B	107. B	108. A	109. D	110. A	111. A	112. B
113. C	114. D	115. C	116. C	117. A	118. A	119. C	120. A
121. C	122. B	123. A	124. A	125. C	126. A	127. C	128. A
129. B	130. D	131. A	132. B	133. D	134. C	135. A	136. B
137. C	138. A	139. C	140. B	141. C	142. A	143. B	144. B
145. A	146. A	147. A	148. A	149. C	150. C		

Chapter

3

AGRICULTURAL METEOROLOGY

- **Meteorology:** A branch of physics of the earth dealing with physical processes in the atmosphere that produce weather.
- **Agriculture Meteorology:** The study of those aspects of meteorology that have direct relevance to agriculture or a branch of applied meteorology which investigates the responses of crops to the physical conditions of the environment.
- The different aspect of agricultural meteorology is yield forecasting, disaster management and environmental pollutions.
- **Climatology:** The science dealing with the factors which determine and control the distribution of climate over the earth's surface.
- **Weather**: It is defined as "A state or condition of the atmosphere at a given place and at a given instant of time".

CLIMATE

It is defined as "The generally weather or summation of weather conditions over a given region during comparatively longer period"

Differences between Weather and Climate

S. No.	Weather	Climate
1	A typical physical condition of the Atmosphere at particular time	Generalized weather condition of the atmosphere of a place which represents characteristics of a region
2	Varies from one place to another place even in a small locality	Since it covers larger area so varies at larger distance
3	Changes according to time (every moment)	Change requires longer (years) time
4	Crop growth, development and yield are decided by weather in a given season	Selection of crops suitable for a place is decided based on climate of the region
5	Under aberrant weather conditions planners can adopt a short-term contingent planning	Helps in long-term agricultural planning

Koppen's Classification

- Koppen (1936) is based on annual and monthly means of temperature and precipitation and the types of native vegetation associated with them.
- Koppen divided the world climate into five major types labelled by capital letters A, B, C, D and E.
- Climates in the B category are termed 'dry' since the potential evapotranspiration during an average year exceeds the precipitation; the other four categories are considered 'moist', since the precipitation during an average year exceeds the potential evapotranspiration.

Five type of climate

A : Tropical rainy climate with no cool season; Mean temperature of each month more than 18°C throughout the year.

B : Dry climate with no temperature limitations.

C : Mesothermal (intermediate heat) middle latitude rainy climate; mean temperature of the coldest month less than 18°C but above -3°C and mean temperature of at least one month above 10°C.

D : Microthermal (small or little heat) middle latitude rainy climates with severe winter; the same as C except that the coldest month has mean temperature below -3°C.

E : Arctic or polar: mean temperature of the warmest month below 10°C; very short and cool growing season (tundra climate) or all months below freezing (ice cap or perpetual frost climate).

Thornthwaite's Rational Classification (1948)

It is the improved classification upon Koppen's method.

- Humidity index $I_h = 100 s/n$
- Aridity index $I_a = 100 d/n$
- Moisture index $I_m = Ih - 0.6\ Ia = (100\ s - 60\ d) / n$

On the basis of this moisture index, he determined the following climatic types by the criteria indicated beside each.

Climatic types	Moisture index
A - Perhumid	100 and above
B4 - Humid	80 to 100
B3 - Humid	60 to 80
B2 - Humid	40 to 60
B1 - Humid	20 to 40
C2 - Moist subhumid	0 to 20
Cl - Dry subhumid	-33.3 to 0
D - Semi-arid	- 66.7 to - 33.3
E - Arid	-100 to -66.7

Factors affecting the climate of a region are: Latitude, Altitude, Land and water, Winds and air masses, Low and high pressure belts, Ocean currents, Extent of forests *etc.,*

ATMOSPHERE

A relatively stable mixture of a number of gases as envelope surrounding the earth is called atmosphere. These gases are colorless, odorless and tasteless.

Sources of oxygen in the atmosphere are **water vapour** (through dissociation into hydrogen and oxygen by the action of sun's ultra-violet) and **green plants** (through photosynthesis)

Composition of the atmosphere

Principal gases comprising dry air in the lower atmosphere

S. No.	Constituents	Per cent by volume	Per cent by weight
1	Nitrogen	78.08	75.51
2	Oxygen	20.94	23.15
3	Argon	0.93	1.28
4	Carbon-dioxide	0.036	0.0456

Physical structure of the atmosphere

According to Petterssen, the atmosphere is divided into different spheres or layers on the basis of vertical temperature variation as detailed below:

1. Troposphere

- The term first advised by **Teisserence de Bort.**
- Troposphere literally means the **'region of mixing'**.
- It has been derived from the **Greek word 'Tropos'** meaning 'mixing or turbulence'.
- It is also called the **convective region,** for the convective activities cease at the upper limit of the troposphere.
- It is nearest to the earth or lower most part of atmosphere.
- As weather events like clouds, rains and thunderstorms *etc.,* occurs in this layer, this sphere is also known as **"Seat of weather phenomena"**.
- The average height of this layer is 18 kilometers (8 to 18 km.) above sea level.
- The average height/thickness of this lower most layer of the atmosphere is about 14 kilometers above the mean sea level; at the equator it is 18 km; and 7- 8 kilometers at the poles
- 85 percent weight of the total mass weight.
- Temperature decrease with increase in height/elevation @ **1°C/165m or 6.5°C/km or 3.6°F/1000 feet or 18.8 °F/mile (Normal lapse rate).**

- **Adiabatic lapse rate** = 10ºC /km.
- The temperature at the boundary is roughly 60ºC.
- At the top of the troposphere there is a shallow layer separating it from the next thermal layer of the atmosphere *i.e.* the stratosphere. This shallow layer is known as **tropopause.** This layer is called as **"Seat of weather phenomena"**.

2. **Stratosphere**
   - This sphere is found between 18-50 km.
   - This layer is called as **"Seat of photochemical reactions"**.
   - It is the dust free, cloudless and warm layer.
   - It contains 15 percent of the total mass/weight.
   - Temperature increase with increase in altitude.
3. **Ozonosphere**
   - There is a maximum concentration of ozone between 30 and 60 km above the surface of the earth and this layer is known as ozonosphere.
   - **Ozone gas** is made of three atoms of oxygen. The formation of ozone takes place in the upper stratosphere when oxygen molecules broken into two atoms by ultra violet radiation and the free unstable atoms combine with two other oxygen molecules
   - A property of ozone is that it absorbs ultra violet rays. It acts as a **filte for the ultra violet rays** of the sun.
   - The temperature of the ozonosphere is high (warm) due to selective absorption of ultra violet radiation by ozone.
4. **Mesosphere**
   - This is the middle part of the atmosphere.
   - It extends from 50-85 km above the surface of the earth.
   - Temperature ceases to rise and start decreasing and becomes about -87ºC.
   - Because of the preponderance of chemical process this sphere is called as **"Chemosphere"**.
   - It is **warmest layer** due to mainly **absorption of ultra violet radiation** by ozone.
   - Temperature increase with height at the rate of 5ºC / km.
   - The upper boundary of stratosphere is **mesopause.**
5. **Ionosphere/Thermosphere**
   - It lies beyond the ozonosphere/mesosphere at a height of about 80 km. above the earth's surface and extends up to 400 kilometers
   - Above the ozonosphere the temperature falls again.
   - According to some climatologists, the layer between 80 and 140 kilometers is known as "thermosphere"

- Ionized layers by means of **aurora, sound waves and satellites** *etc.,*
- It reflects the medium and high frequency radio ways.

6. **Exosphere**
   - The outermost layer of the earth's atmosphere.
   - It lies between **400 to 1000 kilometers.**
   - **Hydrogen and helium** gases predominate in this region.

RADIATION

It is a form of energy emitted by all objects having a temperature above absolute zero; its propagated in the form of electromagnetic waves.

- **Law of radiation-**
  1. Kirchoff's law $\{\varepsilon(\lambda) = \alpha(\lambda)\}$
  2. Planck's law – $\{E = h\nu\}$
  3. Stefan- Boltzman law (or Stefan's law) – $\{E = \sigma T4\}$

Ultraviolet radiation

- Ultraviolet radiation is one form of radiant energy coming from the sun. The various forms of energy, or radiation, are classified according to wavelength, measured in nanometers (one nm is a millionth of a millimeter).
- The **shorter the wavelength, the more energetic the radiation**. In order of decreasing energy, the principal forms of radiation are gamma rays, X rays, UV (ultraviolet radiation), visible light, infrared radiation, microwaves, and radio waves.
- **There are three categories of UV radiation as per wavelength**:
  - UV-A, between 320 and 400 nm
  - UV-B, between 280 and 320 nm
  - UV-C, between 200 and 280 nm - **most dangerous to human health** because of shorter in wave length and very energetic.
- **Atmospheric Variables** are Solar Radiation, atmospheric temperature, atmospheric pressure, wind, atmospheric humidity, precipitation and cloud

Solar radiation (SR)

- Solar radiation is the radiant energy emitted by the Sun in the form of electromagnetic waves.

Absorptions of Solar radiation

S. No.	Process	Per cent
1.	Earth surface	43
2.	Atmosphere	22
3.	Reflected back (Albedo)	35

About the Sun

- The sun is a **gaseous body** composed mostly of **hydrogen**.
- The gravity causes intense pressure and heat at the core initiating **nuclear fusing reactions**; that atoms of lighter elements are combined into atoms of heavier elements, which releases huge quantities of energy.
- The Sun is **source of energy in the universe**

Solar constant

The energy received in 1 minute on a surface of 1 cm^2 at the outer boundary layer of the atmosphere that would be incident on a plane perpendicular to the rays, at a distance of one **astronomical unit** from the Sun (roughly the mean distance from the Sun to the Earth).

- The units are **cal/cm^2/min**. "cal/cm^2 "is also known as **"Langley"**.

Albedo - Solar radiation that is reflected without any change in quality.

Solar spectrum - The energy from the sun is spread over a very broad band of wave lengths known as solar spectrum. It is also known as electromagnetic spectrum. **Example:** U.V. rays, light part, Near I.R., Far I.R. Radio weaves, micro waves, radar waves, *etc.,*

Distance between the Earth and the sun: differential distance due to Elliptical orbit

- Average distance : 14.9 x 10^7 Km (93 M miles)
- At perihelion (nearest position) : 14.7 x 10^7 Km (91.5 M miles)
- At aphelion (farthest position) : 15.2 x 10^7 Km (94.5 M miles)

Net radiation - The radiation balance between global and reflected solar radiation.

- Unit of solar radiation is **Watts (latest), j oule and Lux (oldest).**

Radiation Instruments

S. No.	Instrument	Radiation measured
1	Pyranometer / Solarimeter	Total incoming radiation (direct and diffused)
2	Shading pyranometer	Diffused solar radiation
3	Pyrheliometer	Direct component of solar radiation
4	Net Radiometer	Net solar radiation
5	Quantum sensor and Line Quantum sensor	Photosynthetically active radiation or PPED (photosynthetic photon flux Density- μmol/sec/m^2)
6	Spectro-Radiometer	Solar radiation in narrow bands
7	Infrared Thermometer	Infrared radiation (Terrestrial radiation)
8	Albedometer	Reflected radiation
9	Pyradiometer	Short and long radiation
10	Pyrgeometer	Long wave radiation or infrared radiation.
11	Albedometer or Inverted Pyranometer	Albedo or reflected solar radiation

Atmospheric Temperature

- It is relative term which indicates capacity to transfer heat by conduction.
- It indicates the relative degree of molecular activity or heat of a substance. It is an index of sensible heat.
- It is degree of hotness/coldness of a substance.

Temperature is expressed in three ways as given below

1. **Celsius (°C) scale -** Swedish astronomer **Anders Celsius** (International unit)
   - 0°C is the triple point temperature.
   - 100 °C is boiling point.
2. **Kelvin (°K) scale -**
   - Absolute zero :273.06 °K
   - Molecular activity cease
   - Indicates temperature in the upper atmosphere.
   - Studies involving energy exchange process.
3. **Fahrenheit (°F) scale -**
   - Triple point of water at 32 °F
   - Boiling point at 212°F
   - **Melting Temp**. of water is 273°K/32°C)
   - **Boiling Temp**. of water is 210°F/98°C

Conversion formula of temperature

- °C = (°F-32)5/9 or °C/5= (°F-32)/9
- °F = °C (9/5) + 32
- °K = °C+273

Specific heat: The specific heat (C_s) of a material is defined as the amount of heat absorbed or released in raising or lowering the temperature of a unit mass of the material by 1°C.

- Thermal conductivity of water – (1.42×10^{-3} cal cm^{-1} sec^{-1} $°C^{-1}$)
- Thermal conductivity of air - (0.062×10^{-3} cal cm^{-1} sec^{-1} $°C^{-1}$)
- Thermal conductivity of quartz - (20×10^{-3} cal cm^{-1} sec^{-1} $°C^{-1}$)
- Ratio of thermal conductivity of water, air and quartz is 333:23:1.
- The thermal conductivity are varies in order to sand>loam<clay>peat.

Potential temperature

The potential temperature of a parcel of fluid at pressure P is the temperature that the parcel would acquire if adiabatically brought to a standard reference pressure P0, usually 1000 millibars. The potential temperature is denoted θ and, for air, is often given by

$$\theta = T(P_o/P)^{R/cp}$$

Where: T is the current absolute temperature (in K) of the parcel, R is the gas constant of air, and cp is the specific heat capacity at a constant pressure. This equation is often known as **Poisson's equation.**

Some expressions of temperature

Mean daily temperature - Mean of 24 readings taken at hourly intervals as in the self-recording instruments or average of the readings (2/3/4) taken during the period of 24 hours.

$$= \frac{T_{max} + T_{min}}{2}$$

Mean monthly temperature - It is the average of the total daily mean temperature for the month divided by the number of the days in a month.

Mean annual temperature - It is average of 12 months temperature for the total of monthly means divided by 12.

Mean annual range - Temp. of warmest month – Temp. of coldest month.

Diurnal range - T_{max} of the day – T_{min} of the day.

Temperature anomaly - Deviation from the normal/average temperature

Cardinal temperatures

- It is the minimum and maximum temperatures that define limits of growth and development of an organism or plants, and an optimum temperature at which growth proceeds with greatest rapidity.

Cardinal temperature for crop growth and development

Season	Minimum (°C)	Optimum (°C)	Maximum (°C)
Cool (*Rabi*)	0-5	25-30	30-38
Warm (*Kharif*)	15-20	30-38	45-50

Lapse Rate

- The lapse rate is defined as the rate at which atmospheric temperature decreases with increase in altitude. In the troposphere, the average environmental lapse rate is a drop of about 6.5 °C/km.

Dry adiabatic lapse rate

- The rate at which temperature decrease with height in atmosphere; the dry adiabatic lapse rate is 10°C/km.

- The adiabatic lapse rate for dry air is 0.98°C per 100 m.
- If condensation occurs in the air parcel, latent heat is released, thereby modifying the rate of temperature change is known as wet adiabatic lapse rate
- The average adiabatic lapse rate is 6.5°C per kilometer height and it is assumed as 0.5°C per 100 m.

Temperature inversion/Negative lapse rate

- Condition in which the atmospheric temperature increases with increase in height/ altitude in contrast to the normal decrease with altitude. When this condition occurs, cold air underlies warmer air at higher altitudes.

Vant Hoff's Law or Temperature Quotient or Quotient 10/Q_{10}

- It is the ratio of the rate of reaction at a certain temperature to that at a temperature 10 °C lower. The rate of reaction may double or even be more for every 10 °C increase in temperature.

  Formula $Q_{10} = \frac{\text{Rate of reaction at } (t+10)°C}{\text{Rate of reaction at } t°C}$

- The optimum temperature for CO_2 uptake in wheat/barely is 24 t°C

Growing Degree Days (GDD)

- GDD or heat units is the relationship between plant growth development and maturity to air temperature.

  $$GDD = \sum \frac{(T_{max} + T_{min})}{2} = -T_b$$

  $T_{max.}$ =Maximum temperature of the day

  $T_{min.}$ =Minimum temperature of the day

  T_b = Base temperature of the crop

Base temperature (T_b)

- It is the minimum threshold temperature or lowest temperature at which no growth takes place, ranges between 4.5 to 12.5 °C, higher for tropical plants and lower for temperate plants.

Thermal Conductivity (TC)

- It is the property of a material to conduct heat. It is evaluated primarily in terms of **Fourier's Law** for heat conduction. It is often denoted ***k*, *λ*, or *κ*.**
- It is highest for sandy soil whereas, lowest for clay soil

- It is decrease with increase organic matter and porosity while increase with increase in soil moisture
- The temperature of soil at 80 cm depth almost stable irrespective of day & night time.

Mode of energy transfer

S. No.	Process		Activity
1.	Conduction	:	Molecular activity (Occurs in solid)
2.	Convection	:	Transfer of heat within liquid and gases
3.	Radiation	:	Electromagnetic wave (EMW) may occur in vacuum

Heat Wave

- The continuous spell of abnormally hot weather.
- Heat wave is declared when plains regions temperature is beyond at least 40° C and for hilly regions at least 30° C.

Hot day

- When, the maximum temperature remains **40°C or more** and minimum temperature is **5° C or more above normal**, it may be defined as hot day.

Cold wave

- A rapid fall in temperature (-5°C to -7°C) from its normal within 24 hours to a level requiring substantially increased protection to agriculture, industry, commerce, and social activities is known as cold waves.

ATMOSPHERIC PRESSURE

It is a defined as the weight of the air column per unit area, above the earth surface, extending to the limits of the atmosphere. It is measured by **Barometer/barograph**

- 1 hPa = 1 mbar = 0.1 kPa. Hence, 1013.25 mbar = 1013.25 hPa = 101.325 kPa
- 1 atmosphere = 1.013 bar = 76 cm height of Hg Column

Standard atmospheric pressure:

At a temperature of 15°C and at 45° latitude the standard normal pressure is 1013.2 millibars which is equivalent to 29.92 inches (or) 760 mm of mercury at the sea level, which is considered as standard atmospheric pressure.

- Fortin barometer has a fixed cistern (*i.e.,* it does not require the zero point setting).

Winds

- It is the **horizontal motion** which travels from a high pressure area to low pressure area.

- The direction of the wind is the direction from which the wind approaches the station (for example Northerly wind means wind approaching the station from North side). **Windward** refers to the direction a wind comes from and **leeward** is the direction towards which it flows.
- Atmospheric motion is characterized by **direction and speed**.
- Wind direction measured by **Wind Vane**
- Wind velocity measured by **Anemometer**
- It is an important parameter for weather forecasting
- The **"knot"** is the unit used for wind speed of Canadian marine forecasts and measured in nautical miles per hour. One nautical mile is equal in distance to 1.852 kilometers.
- 1 knot = 1.852 km/h, 1 km/h = 0.54 knot

Gales

- It is a very strong wind (34 to 47 knots)

Squall

- A sudden increase of wind speed by at least 3 stages on the **Beaufort Scale**, and speed rising to force 6 time or more for one minute is called a squall

Gust

- A rapid increase in the strength of the wind relative to the mean strength at the time.

Tornado

- The typical tornado first appears as a rotation in a huge thunder cloud, behind a shroud of heavy rain or hail. The sky usually turns **green, yellow or black**.
- The tornado descends as a violently rotating funnel cloud and sounds like the **rumble of a freight train or a jet.**
- The more violent tornadoes are the **most devastating storms on earth**.
- Its winds approaching 500 km/h and the path of destruction may reach over 100 km long and over 1 km wide.
- Tornado strength is measure by the **Fujita scale** (Dr. T. Fujita, a pioneer in tornado research at the University of Chicago) F0 is the least intense; F5-most intense.

Thunderstorm

- A Thunderstorm is a meteorological phenomenon in which one or more sudden electrical discharges manifested by a **sharp rumbling sound** (thunder) and **flash of ligh** (lightning) occurs from a cloud of vertical development.

Dust storm

- When the dust or sand particles are energetically lifted to great heights by a strong and turbulent wind. It is reduced the surface visibility below 1000 m.

Water spouts

- Tornado formed over water surface and due to high moisture content of air, the funnels are heavily laddened with water drops, so they look somewhat like a stream of water pouring from the base of the clouds.

Land breeze

- Wind occurs during night from land towards sea due to rapid cooling of land.

Sea breeze: Wind occurs during day from sea towards land due to rapid heating of land.

ATMOSPHERIC HUMIDITY

Expression of moisture content in atmosphere:

1. **Vapour pressure**
   - The pressure exerted against the walls of the container and the surface of the water by the water vapour molecules. This pressure is measured in **millibars.**
2. **Specific humidit**
   - It is the weight of water vapour per unit weight of air.
   - Unit g per kg of air
   - It is related to temperature and which is the **highest in summer** and **lowest in winter.**
3. **Absolute humidity**
   - It is the amount of water vapour present in a unit volume of air.
   - Unit: g per cm^3 of air
4. **Relative humidity**
   - It is expressed as the ratio of actual vapour pressure to the saturated pressure expressed in terms of percentage. Or ratio of actual moisture content at particular temperature to the moisture content in the saturated condition at the same temperature and **expressed in percentage**
   - It is a critical for plants and animals growth and development.
   - At dew point the RH is 100 %.
   - The minimum humidity at sun rise and maximum in the afternoon between 14 and 15 hours. It is just reverse to the temperature.

$$\text{RH (\%)} = \frac{\text{Water vapour present in the air}}{\text{Water vapour required for sturation}} \times 100$$

- Relative humidity is measured by **Hygrometer/psychrometer**.
- The moisture content of the atmosphere is about 0.035% of all fresh water.
- The evaporation of 1 kg of water at 20° C requires about 2.45 MJ, which is known as the **latent heat of vapourization.**
- When water condenses in the atmosphere the same amount (2.45 MJ) of latent heat is given off as is used for evaporation at the same temperature.
- When 1 kg of water freezes, about 0.34 MJ is released which is referred to as **latent heat of fusion.**
- Total consumption or release of water from solid to vapor phase or *vice-versa* is about 2.834 MJ/ kg.

5. **Water (mass) mixing ratio**
   - The mass of water vapor compared to the mass of the rest of the air parcel

Condensation - Change of state of water from vapour to liquid state is called condensation. It is the reverse process of evaporation/vapourization.

Condensation Nuclei - The particles (objects), present in the atmosphere, on which condensation of water vapour occurs are called **condensation nuclei**.

Forms of condensation on the basis of location

- Condensation near the ground - Dew, fog, mist, haze or smog and frost.
- Condensation in troposphere - Clouds and Precipitation

Dew

- Condensed moisture in the form of liquid droplets as a consequence of contact cooling is **dew** or it is droplets form of water that appears on thin and exposed objects due to contact cooling, mostly observed during early morning or late evening hours.
- The exposed surface cooled by radiating its heat, the atmospheric moisture condenses rate greater than it's evaporation, resulting in the formation of water droplets on grass, leaves, railings, car roofs, and bridges

Fog

- Microscopically small droplets of water condense and remain suspended in the air near the ground which reduces horizontal visibility are called **fog**.
- A phenomenon of small suspended droplets (10-50μm) in air and the horizontal visibility is one kilometer or less.

Mist

- It is a phenomenon of small suspended droplets (50-500 μm) in the air as a part of natural weather or volcanic activity.

Smog

- It is occurred when smoke and fog come together.
- Smoke and fog together reduce the visibility.

Haze

- Haze is usually an atmospheric phenomenon where dust, smoke and other dry particles obscure the clarity of the sky.

Snow

- It is the form of precipitation in which loose aggregate of branched ice crystal fall.

CLOUDS

- It is a visible mass of condensed watery vapour floating in the atmosphere, and always the signs of precipitation.
- Proportion of clouds in clear sky (0-5%), scattered cloud (5-55%), broken cloud (55-95%) and overcast cloud (95-100%).

Sky Conditions - Cloud cover is reported in terms of **Octa** where the sky is divided into 8 equal parts as given below.

S. No.	Sky conditions	Octa parts
1.	Clear sky	0 Octa
2.	Mainly clear	1-2 Octa of sky covered
3.	Partly cloudy	3-4 Octa of sky covered
4.	Generally cloudy	5-7 Octa of sky covered
5.	Cloudy	>7 Octa of sky covered

Basic type of clouds and its characters

S. No.	Clouds	Characters
1.	Cirrus **(Ci)**	Vein like fibrous or feathery form, bright orange/red in colour before sunrise and after sun set. The highest of this cloud ranging from 7 to 12 kilometers.
2.	Stratus **(St)**	Cloud looks like a sheet and uniform in grey colour It is found at the lowest height from the ground.
3.	Cumulus **(Cu)**	Cumulus means heap or globular mass (cauliflower shape). Cloud is wooly and bunchy with rounded top and flat base

Nimbus (Nb) - Used as prefix means associated with precipitation (**rain clouds**) This looks dark and ragged.

Alto (A) - means above normal height, used as prefix with basic forms

Cloud formation- When the air rises due to increase in temperature and less pressure, it expands and cools until temperature is equalized. If the cooling proceeds further till the saturation point, the water vapour condenses and cloud is formed.

PRECIPITATION

It is the water in liquid or solid forms, falling on to the earth surface. Various forms of precipitation are rainfall, drizzle, snow, hail, mist *etc.*,

Rainfall

- It is the amount of rain that falls on an area in a particular period of time and size of rain **drop is ≥0.5** *μm.*
- The amount of rainfall is measured by **Rain gauges**
- Ordinary the rain drop size varies from 0.5 – 4.0 mm in diameters.

Drizzle

- It is a light liquid precipitation consisting of liquid water drops smaller than those of rain generally **< 0.5** *μ***m** in diameter.
- These drops barely reaches ground.

Glaze

- When rain falls on any material or on ground having subfreezing temperature, it freezes into a sheet or wating of ice is known as graze or freezing rain.

Hail

- It is solid precipitation as **pieces of ice or form of balls** (hailstones) with diameters ranging from **up to 50 or more μm.**

Types of rainfall - There are mainly 3 types of rainfall

1. **Convectional rains**
   - The air near the ground gets heated due to heating of land mass which favours upward movement of air as the air becomes light in weight. This process is known as convection rain.
2. **Orographic rains**
   - Moisture laden wind coming from the sea or ocean is stopped by mountain barrier and brings rain. It is also known as **orographic rains or relief rains.**
3. **Cyclonic**
   - The rains occurs due to cyclones

Frontal precipitation

- Occurs when air masses of different properties collides

Intensity of Rainfall: It is the amount of rain water received on surface during unit time and it is classified in various class which is given below.

Descriptive Term used	Rainfall amount in mms
No rain	0.0
Very light rain	0.1- 2.4
Light rain	2.5 – 7.5
Moderate rain	7.6 – 35.5
Rather heavy	35.6 – 64.4
Heavy rain	64.5 – 124.4
Very heavy rain	124.5 – 244.4
Extremely heavy rain	> 244.5
Exceptionally heavy rain	When the amount is a value near about the highest recorded rainfall at or near the station for the month or season. However, this term will be used only when the actual rainfall amount exceeds 12 cm.

WEATHER PHENOMENA

Graupel

- It is a form when snow in the atmosphere encounters super cooled water. The size of graupel is typically **less than 5 mm** or some graupel can quarter size (coin).

Blizzard

- It is a severe storm condition characterized by strong winds, low temperatures and heavy snow. The strength of the wind differentiate between blizzaro and snowstorm.

Squally weather

- Squally weather is an occasional or frequent squalls with rain or persistent type of strong gusty winds (>20 knot) accompanied by rain. These conditions are coupled with onset and strengthening of monsoon or low pressure systems.

Jet stream

- The Jet stream is narrow, swift currents or tubes of air found at heights ranging from 7 to 8 mi (11.3–12.9 km) above the surface of the earth.
- They are caused by great temperature dissimilarity between adjacent air masses.
- The average wind speeds is 35 mi (56.3 km) per hour in summer and 75 mi (120.7 km) per hour in winter, even though speeds as high as 200 mi (321.9 km) per hours have been recorded.

El Nino effect and La Nino

- **El Nino**, Spanish for **"boy child"** is an **abnormal warming** of water responsible for drought, flooding as well as tropical cyclones and severe winter storms in Equatorial Pacific Ocean in every three to five years and can last up to 18 months.
- Its severity impacts have been blamed for the deaths of hundreds of people and billions of dollars in an estimated 15 countries especially in the Panama Canal region and in east coast Africa during 1997/98.
- **La Nino** means **"the little girl"**, it is just opposite to El Nino, and known for the **abnormal cooling** of the ocean temperatures in the same Pacific region.

Southern Oscillation (SO) and ENSO

- SO is a **"see-saw"** in the surface air pressure between eastern and western tropical Pacific.
- It was discovered by **Sir Gilbert Walker** in the early 1920's.
- It's periodicity about 2-5 years.
- This is the most common index of SO computed as the difference between standardized sea level pressure anomalies at Tahiti and Darwin.
- **ENSO (El Nino Southern Oscillation)** is an acronym designed to fact of stress that the coupled phenomena of El Nino and SO components at the same global Ocean Atmosphere.

Radar

- It is acronym for radio detection and ranging.
- It uses electro-magnetic waves in microwave region to detect location (range & direction), height (altitude), intensity (in case of weather systems) and movement of moving and non-moving targets.

Hydrological cycle (HC)

- It is continuous circulation of water between hydrosphere (ocean), atmosphere (clouds) to lithosphere (run off) is known as hydrological cycle.
- Hydrological cycle is the only 0.001 percent of the total earth water.
- 65 percent of water is loosed by evapotranspiration (ET).

Doldrums

- It is the low pressure areas near the equator is called doldrums

MONSOON

- "MONSOON" is the **Arabic** word **"MAUSIM"** means **Season**.

1. **North- East monsoon**
   - It is also known as **post monsoon season.**
   - The large indentation made by the **Bay of Bengal** into India's eastern significant precipitation occurred in **winter periods.**
   - Bay of Bengal monsoon first come in Assam.
   - The contribution of North- East monsoon - **15-20 per cent in India.**
2. **South - Western or summer monsoons**
   - It occurs during **j uly to September** and it is grand period of rain fall in India
   - The low pressure area over the northern and central India attracts the moisture-laden winds from the Indian Ocean, the high pressure area.
   - These winds, rich in moisture, are drawn towards the Himalayas and this Himalayas act like a high wall, blocking the winds from passing into central Asia and forcing them to rise as result the clouds rise their temperature drops and precipitation occurs.
   - This monsoon accounts for **80-85%** of the rainfall in India.
   - SW monsoon comes from **Arab Sagar** and **Bay of Bengal**
   - SW monsoon first come in **Kerala (1st j une)**
   - Both the monsoon clashed at Delhi state
   - The **Coolest month** of the year is January while May is the **hottest month.**

Artificial Rain

- Artificial rain making was initiated in the year of 1944, just after II World war by American scientists **V. j. Schaefer and E.L. Angmuir.**
- Generally **nimbus clouds,** dark brown in appearance, are elected for artificial precipitation.

Artificial rain making chemical & its characters

- **Warm clouds** - Positive, NaCl (Common salt)
- **Cold cloud** - Negative, Silver Iodide/dry ice (CO_2)
- The temperature of **super cooled (water) clouds** between -5°C and -15°C for artificial rain through silver iodide.

Acid Rain

- Rain water is slightly acidic, with pH (5-6). Water that evaporates from earth is neutral (pH 7) and it becomes weak acid **(sulphur dioxide and nitrogen oxides)** when mixed with carbon dioxide in the atmosphere.

Collision theories

- Alternative rain drop theories use collision coalescence and 'sweeping' as the drop growth generator. It was originally thought that atmospheric turbulence is caused by collision of significant proportions of cloud particles.

DROUGHT

- It is a condition in which crops is failed to grow and mature due to insufficient soil moisture is known as drought.

Type of drought

1. **Meteorological drought**: A situation when the seasonal rainfall received over the area is less than75% of its long term average value.
2. **Hydrological drought:** A period during which the stream flows are inadequate to supply established use of water under a given water management system.
3. **Agricultural drought:** It occurs when available soil moisture is inadequate for healthy crop growth and cause extreme stress and wilting.
4. **Socio-Economic drought**: Abnormal water shortage affects all aspects of established economy of a region. This in turn adversely affects the social fabric of the society creating unemployment, migration, discontent and various other problems in the society. Thus, meteorological, hydrological and agricultural drought often lead to what is termed as Socio-economic drought'.

CYCLONE

- Cyclone a spinning storm that rotates around a low pressure in center, with an anticlockwise in the northern hemisphere and clockwise in the southern hemisphere is called 'cyclone' in the Indian Ocean and southern Pacific.
- Cyclones are also known as **'Lows'** or **'Depressions'**. They originated over the **tropical ocean** and intensify is low.
- They are two types as tropical cyclone and extra-tropical cyclones.
- It is know as **Hurricance** in Western Atlanti and **Typhoon** in Western Pacific.

Classification of low pressure area

1. **Tropical depression (wind up to 34 knots)**
   - Low - <14 knots
   - Depression – 17-27 knots
   - Deep depression – 28-34 knots
2. **Tropical storms (wind 35-63 knots)**
   - Cyclonic storm – 35-47 knots
   - Severe cyclonic storm – 48-63 knots

3. **Hurricane or typhoon (wind of 64 knots or higher)**
   - One nautical mile = 1.15 statute miles = 1.85 km
   - One knots (kt) = 1 nautical mile per hours = 1.15 mih^{-1} = 0.515 ms^{-1}

Characters of cyclone

- Low pressure (20-70mb) & more temperature in centre
- The Initial temperature of cyclone is at least **above 27°C** and it is mostly observed in the **middle latitude.**
- The center of cyclone is known as **eye** and speed of wind in eye is **< 4 km/hrs.**
- Average diameter of cyclone's eye 23 km.
- Average speed of cyclone movement 18 km/hours.

Impact of cyclone

- Flooding, beach destruction, and property devastation to seaside communities due to strong cyclones.
- Terms are related to cyclone *e.g.*, TUTT cell, Tornado, dust devil, steam devil, waterspout *etc.*,

One standard atmosphere (atm) = 1,01,325 Pa = 1013.25 hPa = 1013.25 mb = 760 mm Hg

EVAPOTRANSPIRATION

It is a bio physical process and it occurs in crop's life cycle from germination to harvest. The evapotranspiration components are evaporation and transpiration which are most important segments of hydrologic cycle.

Latent heat: The heat released or absorbed per unit mass by a system during change of phase. In meteorology at 0°C the latent heat of vapourization, fusion and sublimation of water are about 600, 80 and 680 calories per gram respectively.

Evaporation (E): This is the phenomenon in which volume of liquid or solid water is lost to atmosphere from a water body.

- Pan evaporimeter instrument is used to measurement of Evaporation.

Potential Evapotranspiration (PET): PET according to **Thomthrwaite (1948)** is the evaporation from an extended surface of short green crop which fully shades the ground, exerts little or negligible resistance to flow of water and it is always well supplied with water.

FORECAST

The prediction of happening things in the upcoming future is known as forecasting.

Types of Weather forecast

- **Nowcasting:** A short ranging forecast of less than 24 hrs.
- **Short range forecasts:** Forecasts for the period of 1 to 3 days.
- **Medium range forecasts:** Forecasts for the period of 4 to 10 days.
- **Long range/Extended Range forecasts:** Forecasts for the period beyond 10 days.
- **Local Forecast:** It is valid for a radius of 50 km around the station and is updated 4 times in a day.

Methods used in weather prediction - Three methods are used for accurate weather prediction.

- Synoptic method
- Statistical methods
- Numerical methods

Kalpana-1 is located at longitude 74° E and Insat-3A located at 93.5° E both geostationary satellites are being used to monitor the weather of Indian region"

Long Range Forecasting of South-west Monsoon

- The first operational LRF was issued by IMD in June 1886. Landmark contributions in this field were made by Sir Gilbert Walker who was the Director General for 1904-1924.
- In 1988, a new LRF model termed as **Parametric and Power Regression Model** was developed by IMD scientists led by **Dr. V.R. Govarikar**.

Hysteresis: The relation between the matric suction and soil wetness is not generally a unique and single valued one. This relation can be obtained in two ways.

Desorption: It can be obtained by taking an initially saturated soil sample and applying increasing suction to gradually dry the soil while taking successive measurements of water content at any suction.

Sorption: It can be obtained by gradually wetting up initially dry soil sample while reducing the suction.

Determination of moisture percentage - by weighed samples is kept for drying at **105 °C** for about 48 to 72 hr in oven.

Weather in the Incidence of Pests and Diseases

Relation of weather and crop disease and pests

Crop	Pest/Disease	T min. (°C)	T max. (°C)	RH%	Mean rainfall (mm)	Remarks
Crop insect and pests						
Rice	Rice gall midge	19.8	35.2	89.94	4.5-62.5	High humidity, less SSH, high temperature are conducive
	Rice swarming caterpillar	22	32	65-70		Good monsoon activity coinciding with the vegetative phase of crops followed by 2-3 weeks dry spells.
Jowar	Shoot fly	24.4-25.5	34.5-38.8	40-75		
Cotton	Spotted bollworm	19-20	30-32	95-100	17-210	
Sugarcane	Shoot borer	24.4-31.1	37-8-41.4	16-73		
Potato	Colorado potato beetle	16	27		600-1500	
Diseases						
Rice	Blast disease	20	33	88		
Potato	Late blight	10			0.1 mm	At least four hrs. of dew in the night
Cucurbits	Alternaria blight	25	28	85		
Sugarcane	smut	25-30	36-40	100		
Apple	scab	0	<25			leaf wetness hours are nine (9) or more
Grapevines	Downy mildew	5	16	—	—	

Acronyms and abbreviations

- **AMS:** American Meteorological Society
- **AREP:** Atmospheric Research and Environment Programme
- **AWC:** Aviation Weather Center
- **AWIPS:** Advanced Weather Interactive Processing System
- **CAPE:** Convective Available Potential Energy

- **CAPS:** Center for Analysis and Prediction of Storms
- **CSIRO:** Commonwealth Scientific and Industrial Research Organisation
- **EMC:** Environmental Modeling Center
- **ERL:** Environmental Research Laboratories
- **EROS:** Earth Resources Observation Satellite
- **ESSA:** Environmental Science Services Administration
- **GARP:** Global Atmospheric Research Program
- **GCM:** General Circulation Model (OR global climate model)
- **GMT:** Greenwich Mean Time (also known as UTC)
- **GOES:** Geostationary Operational Environmental Satellite
- **GOES-NEXT:** Next Generation Geostationary Operational Environmental Satellite
- **GOOS:** Global Ocean Observing System
- **GPCP:** Global Precipitation Climate Project
- **GPS:** Global Positioning System
- **IPCC:** Intergovernmental Panel on Climatic Change
- **LANSAT:** Land Satellite
- **MetOp:** Meteorological Operational Satellites
- **NASA:** National Aeronautics and Space Administration
- **NCAR:** National Center for Atmospheric Research
- **NCEP:** National Centers for Environmental Prediction
- **NHC:** National Hurricane Center
- **NOAA:** National Oceanic and Atmospheric Administration
- **NSSFC:** National Severe Storms Forecast Center
- **NWC:** National Weather Center
- **NWS:** National Weather Service
- **TDWR:** Terminal Doppler Weather Radar
- **TORRO:** Tornado and Storm Research Organization
- **UARS:** Upper Atmosphere Research Satellite
- **UNCED:** United Nations Conference on Environment and Development
- **UNDP:** United Nations Development Programme
- **UNEP:** United Nations Environment Programme
- **UNESCAP:** United Nations Economic and Social Commission for Asia and the Pacific
- **UNFCCC:** United Nations Framework Convention on Climate Change
- **WIMS:** Weather Information Management System
- **WCRP:** World Climate Research Programme
- **WMO:** World Meteorological Organization

- **WPC:** Weather Prediction Center
- **WSR:** Weather Surveillance Radar
- **WWW:** World Weather Watch Programme

AGRICULTURAL METEOROLOGY OBJECTIVE QUESTIONS (MCQ)

1. Microclimate is also known as
   A. Plant climate B. Eco-climate C. Habitat climate D. All of the above
2. Blue colour of the clear sky and red colour of sunset is due to
   A. Reflection B. Refraction C. Dispersion D. None of the above
3. Lower part of the atmosphere contains water vapour (by volume)
   A. 0-1% B. 0-2 % C. 0-3 % D. 0-4%
4. Average value of solar constant
   A. 1.94 cal/cm^2/hr B. 1.94 cal/cm^2/sec
   C. 1.94 cal/cm^2/min D. 1.94cal/cm^2/day
5. Units of photosynthetically active radiation (PAR) is
   A. Photon B. Lux C. Watt D. Einstein
6. Equinox where day and nights are equal is observed on
   A. March 21 B. September 23 C. Both a and b D. None of these
7. Air temperature is measured continuously by
   A. Max. thermometer B. Min. thermometer
   C. Thermograph D. None of these
8. Lifting up of plants along with soil due to very low temperature is called
   A. Chilling injury B. Heaving C. Freezing injury D. Suffocation
9. Which among the following parameter is more important in weather forecasting?
   A. Rainfall B. Wind
   C. Temperature D. Atmospheric pressure
10. Range of RH conducive for crop growth
    A. 40-6% B. 80-100% C. 60-80% D. 50-60%
11. The contribution of PAR in crop production is
    A. 1-2% B. 5-10% C.10-20% D. 20-30%
12. Stage of crop where the utilization of solar energy is high
    A. Vegetative B. Germination C. Flowering D. Maturity
13. Conversion efficiency (Calories required for production of one gram of dry matter) in crop plants ranges between
    A. 3700-4100 cal/g B. 1700-3700 cal/g
    C. 2700-3700 cal/g D. 3100-4100 cal/g
14. Cloud which is commonly forming hallow around sun is
    A. Cumulonimbus B. Cumulus C. Cirrostratus D. Cirrus

15. Cloud which brings continuous precipitation is
    A. Cumulus B. Nimbostratus C. Cirrus D. Altocumulus
16. Indian agriculture faces cyclonic effect commonly during
    A. January- February B. June- September
    C. October- December D. March- May
17. According to IMD, range of rainfall variation is considered as normal is
    A. -19 to +19 B. -19 to +29 C. -19 to +10 D. -19 to +39
18. Low pressure belt at equator (Doldrums) are found in
    A. 10° N and S of equator B. 5° N and S of equator
    C. 15° N and S of equator D. None of the above
19. High pressure belts (Horse latitudes) are confined to
    A. 0-5° N and S B. 5-15° N and S C. 15-25° N and S D. 30-35° N and S
20. Movement of air from high pressure belt to equator is named as
    A. Trade wind B. Anti- trade wind
    C. Tropical easterlies D. Westerly's wind
21. Which is most destructive among weather abnormalities?
    A. Tornado B. Frost C. Drought D. Cold waves
22. Formation of cyclones starts when sea surface temperature exceeds
    A. 15°C B. 20°C C. 22°C D. 26°C
23. Plant growth is generally inhibited if the wind velocity is more than
    A. 5 km/hr B. 10 km/hr C.15 km/hr D.20 km/hr
24. Which among the following is having least albedo?
    A. Dry black soil B. Moist black soil
    C. Grey soil D. Moist ploughed field
25. Name the force which tends to turn the flow of air due to earth's rotation is
    A. Electromagnetic force B. Electrostatic force
    C. Coriolis force D. Gravitational force
26. 1° longitude equals the time span of
    A. 1 minute B. 2 minute C. 3 minute D. 4 minute
27. The most important weather elements are
    A. Temperature and Precipitation B. Temperature and RH
    C. RH and Wind D. Precipitation and wind
28. Agro-climatic zones are formed in India during
    A. V Five Year Plan B. VI Five Year Plan
    C. VII Five Year Plan D. VIII Five Year Plan
29. 'Isolated rain' denotes ----------- area receives rain used for forecasting of rainfall
    A. 1/4 B. 1/3 C. 1/2 D. 1/5

30. 'Scattered rain' denotes -------------- area receives rain used for forecasting of rainfall

A. 1/5 B. ¼ C. 1/3 D. 2/3

31. Satellite takes the picture in

A. Red region B. Infra red region C. Blue region D. Yellow region

32. The percentage of dependable rainfall in Hargreaves Moisture Availability Index is

A. 25% B. 50% C. 75% D. 80%

33. High and low pressure are developed mainly due to

A. Amount of solar radiation received

B. Absorption of heat by water and earth

C. Rotation of earth

D. All of these

34. Fall of pressure from periphery to the center of the storm ranged between

A. 20 and 70 millibar B. 20 and 30 millibar

C. 20 and 50 millibar D. 20 and 40 millibar

35. During summer season which cloud is responsible for causing rainfall

A. Cumulus B. Cumulonimbus C. Stratus D. Nimbostratus

36. Which of the following is the most important factor affects climate?

A. Latitude B. Topography

C. Ocean currents D. Distribution of land and sea

37. The aggregate of minute drops of water suspended in air at higher altitude is called

A. Snow B. Glaze C. Sleet D. Cloud

38. The Coriolis force is

A. Nil at equator B. Maximum at equator

C. Nil at poles D. Maximum at 66 N and S

39. Tundra climate is characterized by

A. Abundant precipitation B. Meager precipitation

C. Moderate precipitation D. Droughts

40. In taiga region average temperature of the coldest month is

A. Below 0°C B. Above 0°C C. 5°C D. 3°C

41. The average annual temperature in tundra climate is

A. -20°C B. -12°C C. -5°C D. 0°C

42. Sclerophyllus natural vegetation is found in

A. Mediterranean regions B. Equatorial regions

C. Monsoon regions D. Tundra regions

43. The equatorial rain forest is also known as
A. Savanna B. Canpos C. Llanos D. Selva
44. Zones separating different air masses are called
A. Anticyclones B. Cyclones C. Fronts D. Cols
45. Thunderstorms originate from
A. Altocumulus clouds B. Stratus clouds
C. Cumulonimbus clouds D. Nimbostratus clouds
46. The central low pressure core of a tropical cyclone is called
A. Col B. Trough C. Eye D. Depression
47. The typhoons derive their energy from
A. The falling rain B. Density contrast
C. The rising air D. Latent heat of condensation
48. Doldrum is a zone of
A. Intertropical divergence B. Intertropical convergence
C. Frontolysis D. Local wind
49. The pressure system with higher pressure at the center is called
A. Anticyclone B. Front C. Cyclone D. Depression
50. Continuous recording of air pressure is done by
A. Hygrograph B. Seismograph C. Thermograph D. Barograph
51. The height of midday sun at the equator is never less than
A. 66 B. 23 C. 90° D. 30°
52. Insolation intensity of depends on
A. Altitude B. Nature of earth's surface
C. Latitude D. Winds
53. The rays of the sun take about-------------to reach the earth's surface
A. 18 minutes B. 12 minutes C. 8 minutes D. 30 minutes
54. Excitation of phytochrome is done by
A. Red light B. Green light C. White light D. All
55. Formation of chlorophyll occurs in the light of spectral range
A. < 0.25 μm B. 0.3 μm – 0.338 μm
C. 0.4μm – 1.0 μm D. > 1.0 μm
56. Uniform grey cloud sheet is characteristic of
A. Cirrus B. Stratus C. Cumulus D. Cumulonimbus
57. The instrument used for measuring relative humidity is
A. Psychrometer B. Hygrometer C. Both D. None
58. Line joining the equal water table
A. Isotherms B. Isobars C. Isohyets D. Isohalines

59. The normal lapse rate is
    A. 6.5° C/km B. 65° C/km C. -6.5° C/km D. -65° C/km
60. Direct solar radiation is measured by
    A. Pyranometer B. Pyrheliometer C. Polymeter D. None
61. PAR is measured using
    A. Thermo sensor B. Albedo meter
    C. Quantum sensor D. None
62. The instrument used for measurement of total incoming solar radiation
    A. Solar meter B. Pyranometer C. Thermometer D. Quantum sensor
63. Ozone layer which restricts the entry of ultraviolet radiation is found in
    A. Troposphere B. Stratosphere C. Mesosphere D. Thermosphere
64. The study of correlation between agricultural crops and environment is called
    A. Autecology B. Agro-ecology
    C. Agrology D. Agrometeorology
65. Stevenson's screen is placed at a height of
    A. 3 feet B. 5 feet C. 2 feet D. 6 feet
66. Specific heat of water, soil and organic matter is
    A. 0.2, 1.0 and 0.5 B. 1.0, 0.2 and 0.5
    C. 1.0, 0.5 and 0.2 D. 0.5, 1.0 and 0.2
67. De Candole classified the climate based on
    A. Vegetation B. Rainfall
    C. Temperature D. Atmospheric pressure
68. Wind movement from Sea to Land during summer is
    A. Sea breeze B. Land breeze
    C. Ocean currents D. Warm air current
69. The most common way for air to be cooled in order that a cloud may form is by
    A. Emitting radiation B. Rising and expanding
    C. Sinking and contracting D. Reflecting radiation
70. Why do clouds represent a source of heat for the atmosphere
    A. They increase the absorption of solar radiation
    B. they conduct heat from the earth's surface
    C. cloud formation releases latent heat to the air
    D. melting ice crystals in the cloud absorb heat
71. The important difference between liquid droplets and ice crystals in a supercooled cloud is that ice crystals
    A. Are more pure
    B. are smaller
    C. increase the water vapor content of the cloud
    D. grow much more rapidly

72. The cloud droplets in a cloud are formed by water vapor molecules and

A. Molecules of air B. other water vapor molecules

C. hygroscopoic nuclei D. protons

73. The air mass type most likely to form thunderstorms is

A. mT B. Mp C. Ct D. Ca

74. After a cold front passes, which of these does not usually occur?

A. Marked temperature drop B. wind direction shift

C. clearing skies D. falling barometer

75. What causes the dissipating stage of a thunderstorm

A. Spreading downdraft cuts off air inflow

B. release of latent heat within the cloud

C. converging surface winds

D. converging winds aloft

76. A tornado is a small, very intense example of the

A. Cyclone B. anticyclone C. coriolis effect D. geostrophic

77. Deserts such as the Great Basin, Gobi, and Takla Makan are examples of

A. Chinook deserts B. subtropical deserts

C. rain shadow deserts D. monsoon deserts

78. The stability of an air layer refers to its

A. Albedo

B. average temperature

C. pressure as measured at its base

D. tendency to either sustain or suppress vertical motions

79. What determines when a tropical depression or storm is given hurricane status?

A. Wind speed B. central pressure

C. diameter D. water temperature

80. Why do surface wind speeds increase toward the center of a hurricane

A. Reduced friction B. stronger coriolis effect

C. warmer water D. conservation of angular momentum

81. Ozone concentration is reduced by

A. CO_2 B. O_2 C. CFC D. CH_4

82. The rate of decrease of air pressure is higher where the density is

A. Low B. High C. Constant D. None of these

83. Energy transfer in the atmosphere takes place through

A. Two modes B. Three modes C. Four modes D. All

84. The value of auto-convective lapse rate is

A. 10 ºC/km B. 5 ºC/km C. 34.1 ºC/km D. 15 ºC/km

85. Inter tropical convergence zone (ITCZ) is located between
   A. 10-20^0 N B. 20-30^0 N C. 30-60^0 N D. 5^0 N-5^0 S
86. Maritime tropical air mass is
   A. Hot and humid B. Cold C. Hot and dry D. Stable
87. Annual rate of increase of CO_2 is
   A. 5 ppm B. 1.5 ppm C. 10 ppm D. 4.5 ppm
88. Cotton crop needs frost free period of
   A. 100 days B. 200 days C. 300 days D. 50 days
89. Maximum diurnal variation of atmosphere is
   A. 10 ^{0}C B. 20^{0}C C. 16^{0}C D. 26^{0}C
90. Radio communication is made possible through
   A. Inosphere B. Troposphere C. Stratosphere D. Mesosphere
91. The immediate cause of wind is
   A. Temp. difference B. Rotation of earth
   C. Pressure difference D. Humidity difference
92. Coriolis force is maximum at
   A. Equator B. Tropic of cancer
   C. Poles D. None
93. The value of crop coefficient may be
   A. Equal to unity B. More than unity
   C. Both D. None
94. Generally cyclones moves
   A. Westward B. North- Westward
   C. South- Westward D. Both a and b
95. Oilseed zone is
   A. Gujarat plains and hills B. Western plateau and hills
   C. Island zone D. Southern plateau and hills
96. Water balance concept introduced by
   A. Koppen B. Thornthwaite C. De condole D. Hargreaves
97. An instrument used to measure the amount of cloudiness in the sky is called as
   A. Nephometer B. Nephelmeter C. Nephoscope D. None of these
98. Solar radiation produced by which process
   A. Adiabatic process B. Nuclear fusion
   C. Photoelectric effect D. Chaos process
99. What is solar constant on earth?
   A. 100 Langley B. 22 Langley C. 1.94 Langley D. 2.987 Langley
100. Which one of these wavelengths has maximum energy?
   A. 0-300 nm B. 400-500 nm C. 500-530 nm D. 1000 nm

101. Combination of Dry bulb and Wet bulb thermometer used for
A. Net radiation B. Gas composition
C. Cloud D. Relative humidity

102. Which phenomena is responsible for maintaining earth temperature for life?
A. Detoxification B. Procurement
C. Assortment D. Green house effect

103. In which part of atmosphere all types of weather phenomena occur?
A. Mesosphere B. Stratosphere C. Exosphere D. Troposphere

104. World meteorological organization (WMO) classified the cloud in
A. 5 parts B. 8 parts C. 10 parts D. 12 parts

105. Satellite which move around to North pole to South pole known as
A. Polar satellite B. Navigator satellite
C. Weather satellite D. Scientific satellite

106. What is the name of world first artificial satellite?
A. Sputnik 1 (Oct 4, 1957) B. TIROS
C. METSAT D. GOES

107. The 100 meter length of represent by 10 centimeter on the map called as
A. Trim B. Dapper C. Scaling down D. Mangitude

108. At which height the satellite move in space?
A. 36000 km B. 50000 km C. 60000 km D. 45000 km

109. The latent heat of condensation is released in air under the atmospheric condition of
A. Dry adiabatic lapse rate B. Wet adiabatic lapse rate
C. Isothermal lapse rate D. Inversion

110. In stratosphere temperature increases due to absorption of
A. Ultraviolet radiation B. Visible radiation
C. Infrared radiation D. Microwave radiation

111. Secondary circulation of atmosphere over earth surface is called
A. Esterlies B. Westerlies
C. Cyclones & anticyclones D. Land & sea breezes

112. Albedo values of agricultural field ranges between
A. 0.18-0.27 B. 0.29-0.35 C. 0.40-0.50 D. 0.51-0.65

113. Heat transfer in soil takes place mainly through the process of
A. Convection B. Advection C. Conduction D. Radiation

114. Heat flux which remains dominant over a dry barren field is
A. Sensible B. Latent C. Ground D. Photosynthetic

115. The average sea level pressure is equal to
A. 1103 mb B. 1013 mb C. 1030 mb D. 1003 mb

116. The circulation of the atmosphere occurs due to
   A. Kinetic energy B. Inertial frame of reference
   C. Thermal differences D. Gottfried Leibniz
117. 1 mb =.........Hectopascal
   A. 10 B. 01 C. 1013 D. 100
118. Kalpana Chawala-1 satellite was launched in
   A. 2001 B. 2002 C. 2003 D. 2004
119. The value of auto convection lapse rate is
   A. 34 ^{0}C/km B. 6.5 ^{0}C/km C. 10 ^{0}C/km D. 25 ^{0}C/km
120. How much time does a polar satellite require for an orbit of the earth?
   A. 90 minutes B. 60 minutes C. 80 minutes D. 100 minutes
121. Generally LAI (Leaf Area Index) is for field crops
   A. 1-2 B. 3-5 C. 5-7 D. 7-9
122. IRGA used to measuring
   A. Respiration B. Photosynthesis
   C. Respiration & Photosynthesis D. Transpiration
123. Polar orbiting satellite operates at an altitude of
   A. 550 km B. 1600 km C. 1800 km D. 3600 km
124. .... is the first Indian satellite placed in orbit
   A. Aryabhata B. Rohini C. GSAT-11 D. Chandrayaan
125. On surface weather map isobars are usually plotted an intervals of
   A. 2 mb B. 4 mb C. 6 mb D. 8 mb
126. Which type of EM waves used in radar
   A. UV B. IR C. Micro wave D. Visible
127. The negative value of Richardson no. (Ri) indicates
   A. Stable condition B. Unstable condition
   C. Neutral condition D. Semi neutral
128. Livestock weather safety index is baesd on
   A. Temperature index B. Humidity index
   C. Precipitation index D. Humidity and Temperature index
129. Wig- Wag instruments used to measure the
   A. Solar radiation B. Evaporation C. PFD D. Brightness
130. Oldest unit used to measuring the intensity of solar radiation
   A. Flux B. Lux C. Candela D. None of these
131. Distribution of radiation within plant canopy is estimated with the help of
   A. Beer s law B. Kirchoff s law
   C. Plank s law D. Stefan boltzman law

132. Change of phase from solid to vapour or vice versa is known as
A. Evaporation B. Condensation C. Precipitation D. Sublimation

133. Tornadoes most often move towards what direction
A. South East B. South West C. North East D. North West

134. The unit of pressure most commonly found on a surface weather map is
A. Milibars (mb) B. Hectopascals C. Pascals D. A & B

135. The combined albedo of the earth and the atmosphere is approximately……. percent
A. 10 B. 20 C. 30 D. 40

136. Which cloud type would most likely from in an unstable atmosphere?
A. Cumulonimbus B. Nimbus C. Altostratus D. Stratus

137. The first meteorological satellite was launched in the year
A. 1950 B. 1960 C. 1955 D. 1965

138. The cup anemometer rotates
A. North to South B. South to East C. West to East D. East to West

139. Which Sensor is used in Quantum sensor
A. Thermopile B. Silicon photo diode
C. A & B D. Silicon Triode

140. ……… is called as Umbrella of the earth
A. Troposphere B. Ozonosphere C. Ionosphere D. Exosphere

141. The trade wind belts lie between…..to…… latitude in both hemispheres
A. 10^0 to 20^0 B. 20^0 to 30^0 C. 5^0 to 30^0 D. 30^0 to 60^0

142. The belt of westerlies between……to……latitude in each hemisphere
A. 10^0 to 20^0 B. 20^0 to 30^0 C. 5^0 to 30^0 D. 30^0 to 60^0

143. Both wind speed and wind direction is measured using
A. Wind meter B. Aero vane C. Aero deck D. None

144. Temperature inversion is found in which season
A. Summer B. Winter C. Both D. None

145. The germination of seed is induced by the light of …….colour
A. Blue B. Red C. Green D. Yellow

146. The surface wind in Hadley cells, ferrel cells and polar cells are called……. winds
A. Trade, Easterlies and Westerlies B. Westerlies, Trade and Easterlies
C. Trade, Westerlies and Easterlies D. Westerlies, Trade and Easterlies

147. Clouds at lower or ground surface is called as
A. Smog B. Fog C. Smoke D. Smaze

148. A low pressure with wind speed < 34 knots, the system is called as
A. Depression B. Cyclone C. Trough D. Ridge

149. Which light is the most effective inhibitor of flowing in the case of long day plants?
   A. Blue B. Red C. Green D. Yellow
150. Aerosol consisting of liquid droplets is called as
   A. Fog B. Smaze C. Mist D. None

ANSWERS

1. D	2. C	3. D	4. C	5. D	6. C	7. C	8. B
9. D	10. C	11. A	12. C	13. A	14. C	15. B	16. C
17. A	18. B	19. D	20. A	21. A	22. D	23. B	24. B
25. C	26. D	27. A	28. D	29. B	30. D	31. B	32. C
33. D	34. A	35. B	36. A	37. D	38. A	39. B	40. A
41. B	42. A	43. D	44. C	45. C	46. C	47. D	48. B
49. A	50. D	51. A	52. C	53. C	54. B	55. B	56. B
57. C	58. C	59. A	60. B	61. C	62. B	63. B	64. D
65. B	66. B	67. A	68. A	69. B	70. C	71. D	72. C
73. A	74. D	75. A	76. A	77. C	78. D	79. A	80. D
81. C	82. B	83. B	84. C	85. D	86. A	87. B	88. B
89. C	90. A	91. C	92. C	93. A	94. D	95. A	96. B
97. C	98. B	99. C	100. A	101. D	102. D	103. D	104. C
105. A	106. A	107. C	108. A	109. B	110. A	111. C	112. A
113. C	114. A	115. B	116. C	117. B	118. C	119. A	120. D
121. B	122. B	123. D	124. A	125. B	126. C	127. B	128. D
129. A	130. B	131. A	132. D	133. C	134. D	135. C	136. A
137. B	138. C	139. B	140. B	141. C	142. D	143. B	144. B
145. B	146. C	147. B	148. A	149. B	150. C		

Chapter

4

AGRONOMY

TILLAGE

- Tillage is the **physical manipulation** of soil with the help of tools and implements for obtaining conditions ideal for seed germination, seedling establishment and subsequent growth of crops.
- The tillage word is derived from **'Anglo-Saxon'** words **Tilian** and **Teolian,** meaning 'to plough and prepare soil for seed to sown, to cultivate and to raise crops'.
- **Tilth** is a physical condition of soil resulting from tillage.
- Father of tillage /weed/ modern agronomy - **Jethro Tull** (1674-1741).
- Jethro Tull proposed a theory – **Plants absorb minute particles of soil.**
- Jethro Tull Wrote a book – **Horse Hoeing Husbandry.**
- Father of modern Agronomy- **Pietro De 'Crescenzi'** (1230-1307).
- Father of Indian Agriculture/agronomy/green revolution – **MS Swaninathan.**

Objectives of tillage

- To prepare a good seed bed which helps the germination of seeds.
- To create conditions in the soil suited for better growth of crops.
- To control the weeds effectively.
- To make the soil capable for absorbing more rain water.
- To mix up the manure and fertilizers uniformly in the soil.
- To aerate the soil.
- To provide adequate soil- seed contact, and permit water flow to seed and seedling roots.
- To remove the hard pan and to increase the soil depth.
- Soil and water conservation.
- Ease of planting.
- Control of insect- pest and diseases.

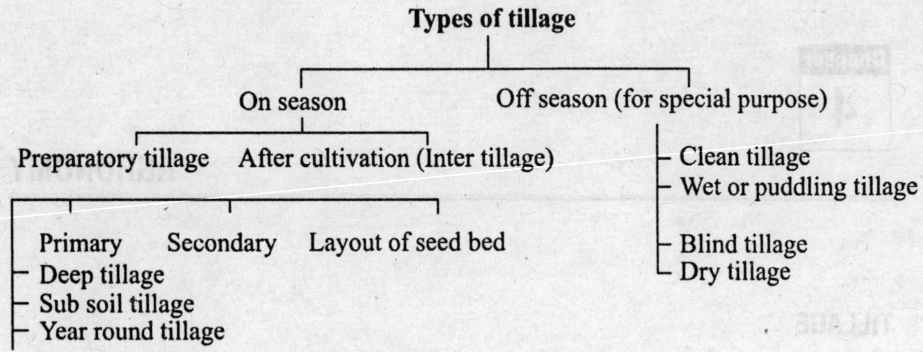

Tillage operations may be grouped into-

1. On season tillage
2. Off-season tillage

1. On-season tillage: it is done with the onset of the crop season for raising crops are known as on-season tillage. It may be further divided into two groups-

A. Preparatory cultivation
B. After cultivation.

A. Preparatory tillage:

- The tillage operations are carried out from the time of crop harvesting to the next crop sowing is known as preparatory cultivation.
- To bring out desirable tilth through deep opening and loosening of the soil and also incorporate or uproot weeds and crop stubble.
- It is having three distinct operations *viz.* primary tillage, secondary tillage and layout of seed bed.

a. Primary tillage or ploughing:

- This tillage operation (ploughing) is done after the harvest of crop to bring the land under cultivation.
- Implements - mould board plough, country plough, bose plough, tiller/cultivator, indigenous plough (V- shaped furrow open)

b. Secondary tillage:

- It is performed with lighter or finer operations of soil after primary tillage for cleaning of the soil, break down the clods and incorporate the manure and fertilizers.
- Implements- harrows, cultivators, guntakas and spade

c. Layout of seed bed and sowing:

- The field is properly laid out with the seedbed preparation for better irrigation and sowing or planting seedlings.
- Implements- leveling board, buck scrapers

B. After cultivation (Inter-tillage):

- These tillage operations are performed in the standing crop after the sowing or planting to prior the harvesting. It is also known inter cultivation or post seeding/ planting cultivation; It includes hoeing, earthing up, weeding, drilling or side dressing of fertilizers *etc.*,
- Equipments- Spade, hoe, weeders *etc.*, are used for inter cultivation.

2. Off-season tillage:

- These tillage operations carried out for the forthcoming main season crop. It may be summer tillage, post harvest tillage, winter tillage and fallow tillage.

1. Modern concepts in tillage: The conventional tillage involves primary tillage and secondary tillage from field prepration to crop harvest which results poor infiltration due continuous use of heavy ploughs create to hard pan in the sub soil and also more susceptible to run-off and erosion. It is high capital intensive and soil degradation tillage. Therefore, to avoid these ill effects; the modern concepts on tillage is in rule.

Types of modern tillage:

1. Minimum tillage
2. Zero tillage
3. Stubble mulch tillage or Stubble mulch farming
4. Conservation tillage

1. Minimum tillage:

- The aims minimum tillage is reducing tillage operations by the minimum necessity operation performed for ensuring a good seed bed, rapid germination, satisfacrary stand and favourble growing condition.
- The concept of minimum tillage was first time given in USA when the high price of oil due to higher tillage cost in 1974.

The advantages of minimum tillage over conventional tillage are as follows

- The cost and time for field preparation can be reduced by reducing the number of field operations (çombin application of seeds and fertilizer in the soil).
- Soil compaction is less and reduced water loss through runoff and erosion
- Water storage is increased in the plough layer.

2. Zero tillage (No tillage):

- It is an extreme form of minimum tillage, where the primary tillage is totally avoided and secondary tillage is restricted to the seedbed preparation in the row zone only. Under this concept, the new crop is planted in the residues of the previous crop without any soil tillage or seed bed preparation
- It is applicable for **Oxisols, Alfisols and Ultisol**
- Zero tillage is using in rice, wheat, blackgram crops.
- The Andra Pradesh is the first state to follow the zero tillage in India.
- The highest area under Zero-tillage under wheat cultivation mainly Haryana and Punjab_

Advantages:

- Zero tilled soils are having more number of earthworms and homogenous in soil structure
- It increased organic matter content due to less mineralization.
- It also reduced surface run-off due to the presence of mulch.
- It is reduced the cost of cultivation
- Save the time as go for direct seeding
- Less problem of *P. minor* as seed remain in lower surface

Disadvantages:

- It required the higher amount of nitrogen fertilizers for mineralization of organic matter.
- Perennial weeds may be a problem and much higher chance of pest infestation

3. Stubble mulch tillage or stubble mulch farming:

- The soil surface is covered by growing a crop or leaving the crop residues at all times with height of 12 to 15 cm.
- The crop residues on the soil surface interfere with seed bed preparation and sowing. Moreover, the traditional tillage and sowing implements are not suitable for stubble mulch farming.

4. Conservation tillage:

- In this system the organic residues are inverted into the soil and remain on surface also as protective cover with an objective *to conserve soil and soil moisture.* This system is also face the same problem of stubble mulch farming systems.

Tilth:

1. Tilth is a **physical condition of the soil** resulting from tillage. It is done to loose, roughly powdery, granular and crumby condition of soil which is become favourable for the crop production. The better tilth of the soil is improving the aeration and water retention, promote root spread and penetration, and also enhance the uptake and supply of nutrients in the roots.

According to **Russell** (1953) indicates two properties of soil:

1. **The size distribution of aggregates**: it is the relative proportion of different sized soil aggregates.
2. **Mellowness/friability**: it is property of soil in which the clods become more crumbly, smaller in size and less sticky when it wet.

Tith is done based on size of soil aggregates:

- Larger aggregate (>5mm) necessary for - Irrigated Agriculture
- Smaller (1-2 m) desirable for - Dry land Agriculture
- 1-5mm aggregate size is the best for agriculture.
- Crumbly & granular are the best soil structure for Agriculture.

Name of Implements used for specific conditions

S. No.	Implement	Remark/ functional area
1.	Sweep cultivator	Clogging, Ground nut harvesting
2.	Chisel plough	**Hard pan soil** with 60-70 cm depth
3.	Disc plough	Stone soil, effective pulverise the clods
4.	Rotary plough/ hoes	Light soil
5.	Pedda guntaka (Big blade harrow)	Control of grasses
6.	MB plough	Incorporation of manure/ fertilizer and plant residues, Deep ploughing inversion
7.	Country plough	Multi purpose

Depth of tillage acording to CRIDA:

S. No.	Tillage	Depth in cm
1.	Deep	25-30
2.	Moderate	15-20
3.	Shallow	5-6

No. of ploughing are required in

- Light soil : 1-3
- Heavy soil: 3-5

Puddling: It is a practice of ploughing which is done in the standing water for reducing deep percolation losses of water and to provide soft seedbed for planting of rice.

It destroy the soil structure and soil particle settled in order sand > silt > clay and to make an impervious layer which help to maintain standing water in rice field.

SEED AND SOWING

Seed: Seed is a matured ovule which consists of an embryo (germination), endosperm (food storage tissue) and seed coat (protective layer of seed) is called seed.

Type of seed

Monocot seed	Dicot seeds
Only one cotyledon present in embryo	Two lateral cotyledon are present in the embryo axis
Cotyledon is thin and small and lack of food materials	Cotyledons are fleshy and store the foods
Endosperm is mostly present and stores food	Endosperm mostly absent and lacks
Coleorhizae and coleoptiles are protect of radical plumule respectively	Coleorhizae and coleoptiles are absent
e.g. Gramineae family (Rice, wheat, barley, maize)	*e.g.* Pulses

Part of seeds

Ovule:

- It is the Small egg like structure in the seed consisting of the nucellus which contains the female gametophyte, one or two integuments and funiculus.
- After the fertilisation of the ovule egg cell develops into a seed.

The ovule consists of a number of components:

- **Funicle** (funiculus, funiculi) – it is also known as seed stalk which attached with the ovule to the placenta and hence ovary or fruit wall, at the **pericarp**.
- **Nucellus-** is the remnant of the megasporangium and main region of the ovule where the megagametophyte develops.
- **Micropyle**-is a small open pore in the apex of the integument where the pollen tube usually enters during the process of fertilization.
- **Chalaza**- is the base of ovule opposite to the micropyle, where integument and nucellus are joined together.

Embryo:

- It is the non-self-supporting immature organism formed from the zygote by cell division, differentiation, and a rudimentary plant within the seed is called embryo.

The main components of embryo-

1. **Cotyledons**:
   - It is the seed leaves and attached to the embryonic axis. It is classified in two types (monocotyledons and dicotyledons).
   - The cotyledons are the source of nutrients for the non-endospermic dicotyledons, in which case they replace the endosperm, and are thick and leathery. But cotyledon is the thin and papery in endospermic seeds.
2. **Epicotyl:** the embryonic shoot axis attachment above the cotyledon, it shows hypogeal germination in dicot plants.
3. **Plumule:** it is the tip of the epicotyls and feathery appearance due to the presence of young leaf primordia at the apex, and it emerged in tho form of shoot upon germination.
4. **Hypocotyl:** is the attachment of embryonic axis below the cotyledon, and it is promoted radical and it is being the stem-root transition zone.
5. **Radicle:** is the basal tip of the hypocotyl, which grows into the primary root.

Seed-coat:

- It is protective outer layer of the seed derived from the integuments.
- When two layers of the seed-coat are distinguishable, the terms ***testa*** for the outer coat and ***tegmen*** for the inner coat.

Types of seeds:

(i) **Breeder seed/Nucleus seed:**

- Breeder seed is produced by breeder as result of hybridization, selection and mutation.
- These seeds are high in genetic value.
- It is very little in quantity and often costly.

(ii) **Foundation seed:**

- It is the produced by National Seed Corporation, State Seed Corporation *etc.,*
- The foundation seed is relatively less pure as compared to breeders or nucleus seed.
- Physical purity of 98 % is permissible.

(iii) **Certified seed:**

- It is the proper labelled and sold to the farmers for commercial cultivation.

(iv) **Registered seed:**

- It is not taken in India.
- Working sample of seed must contain 2500-3000 seeds.
- Weight of working sample is 25 gram.
- Minimum seeds are required for germination test 400 seeds (According to ISTA).
- Seed germination requirement: 80% soil moisture and soil temperature of 10^0c and 20^0c for temperates crops and tropical crop respectively.

Classes of Seed and their tag colour.

S. No.	Types of seeds	Parents	Tags colour	Genetic purity (%)
1.	Nucleus /Breeder	Breeding lines	Golden Yellow	100
2.	Foundation seed	Breeder seed	White	99.5
3.	Registered seed	Foundation seed	Purple	99.5
4.	Certified seed	Foundation seed	Blue	99.0
5.	Truth Full Level (TFL)	—	Light green/ Opal green	

Characteristics of Good Quality Seeds

Seed health- free from pest and disease, good germination percentage with uniform seed size

Physical Purity of Seeds- free from inert matter like dust, stones, broken seeds, weed seeds and chaffy seeds

Genetic Purity- The characteristics of the progeny should exactly resemble its mother plant.

Moisture Content of the Seeds- high moisture increase chances pest infestation and attack by diseases and reduce seed vigor and viability. Seeds should be stored at a safe moisture level of 9-13%.

Seed Multiplication Ratio (SMR) - It is the number of seeds to be produced from a single seed when it is sown and harvested.

SMR = Seed Yield /Seed Rate

Seed Replacement Rate (SRR) - Seed replacement rate is the percentage of area sown out of total area of crop planted in the season by using certified / quality seeds other than the farm saved seeds.

SRR = X / Y × 100 Where,

X = Quantity of farmer saved seed

Y = Quantity of quality seeds of a particular variety reported to cover a given area .

Mandatory Practices under seed production programme

Isolation distance: It is the distance maintained between the seed crops and the different varieties of the same crop located in the adjacent area in order to maintain the genetic purity of the seeds.

Recommended isolation distance for field c op

Crop	Foundation Seed (m)	Certified Seed (m)	Crop	Foundation Seed (m)	Certified Seed (m)
Rice	3	3	Sorghum (OPV)	200	100
Wheat	3	3	Sorghum hybrid	300	200
Soybean	3	3	Sesame	100	50
Groundnut	3	3	Cotton	50	30
Green gram	10	5	Maize (OPV)	400	200
Black gram	10	5	Maize hybrid	400	200
Bengal gram	10	5	Pearl millet (OPV)	400	200
Cowpea	10	5	Pearl millet hybrid	1000	200
Pea	10	5	Castor	300	150

Bagging Technique - In order to protect the pollination through insects or wind covering the floral portion of a plant is practiced.

Rouging - It is the process of removing the plant varieties of other than the variety grown for seed production; it's a removing of off type plants.

Seed index - it the weight of 100 seeds and it is used for large size seeds crops like groundnut, gram, maize *etc.,*

Test weight - Weight of 1000 seeds (in case of small seeds like rice, wheat, mustard *etc.,*).

Name of crop	Test wt(g)	Name of crop	Test wt (g)
Tobacco	**0.05-1.2**	Barley	35-40
Canary grass	2	Sunflower	40-50
Rapeseed and mustard	3-5	Safflower	50-60
Pearl millet	5-9	Cotton	70-100
Basmati rice	**21**	Soybean	100-105
Lentil	17-38	Castor	100-150
Rice	25	Peas	120-220
Greengram	20-70	Chickpea(desi)	150-160
Sorghum	25-30	Maize	200-250
Wheat	40	Groundnut	200-250
Black gram	36-49	Gram kabuli	**260-450**

Difference between seed and grain

S. N.	Seeds	Grains
1.	A well programme is designed for seed production.	No designed programme for grain production.
2.	Varietal and genetic purity of the seed is clearly identified from its breeder seed	Varietal purity of the grain is not known.
3.	Quality standards like removal of off-types, weed plants or other crop plants and diseased plants are carried out.	Quality standards are not followed.
4.	Seeds have physical purity and good germination.	Physical purity and germination is not ensured.
5.	Scientific seed production follows the processing, treatment, packaging and labeling procedures.	Processing, labeling and tagging are not followed in grain production.
6.	Viability and vigor of seeds are maintained during seeds storage.	Grains are protected from pests and diseases only not done vigour and viability during storage.
7.	Seeds are certified and labeled properly.	It not required any certification labels and tags.

Classification of seed (based on seed gremination moisture):

1. **Orthodox seed**:
   - Orthodox seeds are mostly hard coated seeds which can be dried at low moisture content around 5%.

- These seeds are able to successfully store at low or sub-freezing temperatures for long periods.
- These seeds are no metabolism in dry phase but they may deteriorate by general ageing and ultimately lose their viability. *e.g.* hard seed coated (cereals, legumes) and soft seed coated (Picea and eucalyptus).

2. **Recalcitrant seed**:
   - These seeds are mostly larger in size and higher moisture content which can not survive drying below relatively high moisture content (20-35%) and do not tolerate low temperature during storage (tropical species)
   - These seeds rapidly lose their viability and cannot be stored for long periods therefore; they should be sown as soon as collection for better germination. *e.g.* mango, citrus, jamun *etc.*,

Seed Viability:

- It is capacity of seeds to germinate when provide the appropriate water and environment conditions to reactivation of their biochemical processes.
- Viability tests are done by cutting test or Tz staining.
- Tetrazolium test is commonly known as the **T_z test for seed viability**.

Machanism of the reaction:

- This test invented by **Lakon** (German scientist) and it is very reliable and widely utilized.
- Chemical - **2,3,5- triphenyl tetrazolium chloride or bromide.**
- Tz is used @ 0.1-1.0% solution at 30-35^{0}C temperature for 24-48 hours at pH 6-8.
- Red colour of seed is indicating the positive seed viability.

Photoperiodism:

- Photoperiodism can defined as the developmental responses of plants to the relative lengths of light and dark periods.
- "Photoperiod" was derived from:
  - Photo means light
  - Period means a specific length of time
- It is the relative **length of daylight and night.**
- It is essential for **flowering p ocess.**

Classification of plants in accordance to day length/ photoperiod:

1. **Short day plants (SDP):**
   - These plants are requiring short days and long nights (flower when the photoperiod is less than the critical length 10 hours daylight and 14 hours

darkness). *e.g.* chrysanthemums, poinsettias and Kharif crops (Except maize)

2. Long-day plants (LDP):

- These plants are requiring long days and short nights (flower when the photoperiod is greater than the critical length14 hour's daylight and 10hours darkness). *e.g.* sunflowers, radishes, lettuces, irises, Rabi crops *etc.*,

3. Day-neutral plants (DNP):

- These plants are relatively unaffected by the day length of light and dark, and produce flowering at any time of the year. *e.g.* dandelions, tomatoes, cotton, tobbaco, maize, tomato, chilli, sunflower, banana *etc.*,

Depth of sowing seed:

S. No.	Particulars	Size (cm)	Example
1.	Bigger	6.0	Groundnut, Sunflower & castor
2.	Medium	2.3	pearl millet
3.	Small	1.0	Tobbaco

- 3-5 cm is sowing depth of most of the field crops
- Seed sowing depth should be 3-4 times more of its diameter

Methods of crop establishment:

1. Broadcasting:

- Small sized seeds are broadcast and it is most primitive method of sowing. *e.g.* leafy vegetables, lucerne *etc.*,

2. Drilling:

- Seed sowing into the line by seed drills *e.g.* wheat, barley, sorghum, bajra *etc.*,

3. Planting:

- When individual seed is placed in the soil by mannul or machine is known as planting. *e.g.* cotton, maize, potato, sugarcane, potato, EFY *etc.*,

4. Transplanting:

- Seeds are sown first in nursery under appropriate condition in early stage and than transfer into main field where the crops finally cultivated. *e.g.* rice, tomato, brinjal, chilli, cole crops, tobbaco, cotton, cucurbits (now, recently for polyhouse cultivation for rootstocks benefits) *etc.*,

Why transplanting is needed?

- To extend the growing season by starting the plants in indoors before outdoor conditions are favourable
- Protecting young plants from diseases and pests

- Avoiding germination problems by setting out seedlings instead of direct seeding.
- To reduce the crop cost by growing the costly seeds in nursery for better germination and protection.

Type of sowing:

1. **Dry sowing:**
   - Dry sowing is adopted in black soil where sowing operation are difficult to carry out once rain comes.
   - Sowing is done 7-10 days before rain is received.
2. **Wet sowing:**
   - It is done when minimum amount of rainfall is 20 mm
   - It is the most common methods of sowing in India
   - Plastic mulch increase temperature - 1-5°c
   - GA_3 is most common hormon used to break dormancy @ 500 ppm for 12 hours
   - Gap filling should generally be done within 7-10 days after sowing and in hybrid cotton, it should be completed within 25 days after sowing

Seed treatment/Seed dressing:

- Seed treatment is the application of biological organisms and chemical ingredients to seed suface for suppress, control, or repel the plant pathogens, insects, or other pests that attack on seeds, seedlings or plants is known as seed treatment.

Approaches for seed treatment through biological or chemical methods-

1. **Protectant** - Protection of health of the host plant.
2. **Eradicant** - Curative therapy of an infested host plant.
3. **Disinfectant** - Destruction phase in the life cycle of the pest.
   - Pattern of seed treatment must be follow as **Fungicide > Insecticide > Rhizobium.**

Seed Priming:

- It is a treatment given to freshly harvested seed before sowing to improve speed and uniformity of seedling establishment especially under adverse climatic conditions such as sub and supra temperature and water stress; and to overcome the dormancy through stimulating physiological and biochemical activities.

Various priming methods are used to enhance seed quality:

1. **Hydropriming** - submerged in water
2. **Halopriming** - treating with inorganic salts. *e.g.* NaCl, KNO_3, $CaCl_2$
3. **Osmopriming** - soaking in osmotic solutions with a low water potential. *e.g.* Polyethylene glycol **(PEG)**
4. **Hormonal priming** - seed treat with plant growth regulators and hormones. *e.g.* Kinetin, ABA, auxins, gibberellins *etc.,*
5. **Biopriming** - inoculation with beneficial organisms to protect seed. Pseudomonas aureofaciens, Plant Growth-Promoting Rhizo-bacteria (PGPR)

Benefits of priming seeds:

- Emergence can occur before soil crusting becomes fully detrimental
- Crops can compete more effectively with weeds
- Priming allows growers to better control their water usage and scheduling.
- Priming can eliminate or greatly reduce the amount of seed-borne fungi and bacteria.

Types of seed germination:

- **Epigeal germination** - In dicotyledon plants like groundnut, beans where cotyledon emerge out of soil. *e.g.* neem, pinus, some cucurbits and beans.
- **Hypogeal germination** - cotyledon remains inside the soil; plumule grows or pushed upward by the elongation of epicotyls. *e.g.* tomato, brinjal, chilli, radish, carrot and *Pongamia pinnata.*

Dormancy - it is the condition of seed when it fails to germinate even under suitable external conditions due to unavailability of internal factors.

Double dormancy - it is the condition in which seed coat and embryo both are dormant at same time.

Quiescent - it is the state of alive seed when it is not able to germinate due to unavailability of external factors such as scarcity of water, unfavourable temperature *etc.,*

Rest period - it is the physiological condition of the bud of woody perennial plant beginning shortly after formation during which the bud fail to expand into flower or leafy shoot even if the growing condition is suitable.

Type of seeds Dormancy

1. **Innate (Primary)** - due to genetic factors and chemical present in seed
2. **Enforced** – due to deeper placement of seed during sowing.
3. **Induced (secondry)** - due to both environmental factors like temperature, water logging.

- Wild oat (*A. Fatua*) is having all three type of dormancy according Thurstone (1959).

Seed Dormancy Breaking Methods

1. **Scarification** – to break the physical strength of seed coat and improve the seed germination through mechanical, chemical and thermal treatments.
   - **Hot water treatment** – treat the seed with 75-100°C for 30 minutes
   - **Acid trestment** – treat the seed with Sulphuric acid (H_2SO_4) or HCl in the ratio of one part of seed to two parts of acid.
   - **Hormonal treatmnent** – GA_3 (200-500ppm), thiourea (1%) and KNO3 (1-3%) and cytokinine (100ppm).

PLANT POPULATION AND ITS IMPORTANCE

Plant population - Any group of individuals of one species that occupy a given area at the same time (***Curtis & Barnes***, Biologist).

- Plant height increases with increase in plant population due to competition for light
- Yield per plant decrease gradually as plant population per unit area increases.
- A population is "closed" if there is no **immigration and emigration.**
- **Genet** - is a single genetic individual.
- **Clone** - is a group of distinct individuals that are part of single genet *e.g.* aspen trees.
- **Ramet** - is a single member of a clone (a branch of genet).
- **"Plant ideotype"** concept was given by **Donald**

Asymptotic response - the yield increases with increase in the plant population, and then reaches a maximum and a further increase in plant population. It is followed where the entire dry matter is the economic product. *e.g.* fodder crops & dry matter in Tobbaco

Parabalic response - the yield increases with increase in plant population, and then reaches a maximum and it decreses with further increase in population. It is followed in cases where the economic yield is a fraction of the total dry matter or reproductive parts.

Plant demography - The study of changes in population structure through time.

Fitness:

- **The life time reproductive success of an organism.**
- This concept is much easier to apply to animals than to plants.
- The ability of plants to reallocate reproductive effort to vegetative modules during times of stress makes analysis of fitness difficult without following the reproduction among all its vegetative shoots

Plant density:

- **Number of individuals per unit area** *e.g.* trees per ha, or plants per m^2.
  - Seed production : Plants to seeds
  - Germination : Seeds to seedlings
  - Establishment : Seedlings to plants

CROPPING AND FARMING SYSTEM

Farming System - the Integration of two or more appropriate combination of farm enterprises like crop, livestocks, fishery, bee keeping, forestry *etc.,*

Cropping System - is refers to a set of crop systems, making up the cropping activities of a farm system.

Cropping pattern - is the yearly sequence and spatial arrangement of crops or of crops and fallow on a given area.

Cropping intensity- is the number of crops cultivated in a piece of land per annum is called cropping intensity. Highest cropping system in Punjab and Tamil Nadu is more than 100% (*i.e.* around 140-150%) while lowest in Rajasthan.

Crop rotation - is the practice of growing a series of dissimilar or different types of crops in the same area in sequenced seasons to maintain and propoer utilization of soil nutrient and avoiding the soil born pest and diseases.

Types of Cropping Systems

1. **Mono-cropping** - is growing of only one crop on a piece of land year after year. it is also known as monoculture.
2. **Double-cropping** - is a case where the land is occupied by two crops, which are grown in a year in sequence.
3. **Multiple-cropping** - Growing two or more crops on the same piece of land in one calendar year is known as multiple-cropping. It includes inter-cropping, mixed-cropping and sequence cropping.

A. Inter-cropping - the growing two or more crops simultaneously on the same field in a year in a definite row pattern.

i. Mixed inter-cropping - the growing of two or more crops simultaneously intermingled without a row pattern.

a. Parallel inter-cropping – the cropping system is having zero competition among them.

b. Companion cropping – production of both intercrops is equal to its sole cropping.

c. Multitier cropping – the growing crops are varying in the height with the proper utilization of space. These crops are having annidation effect on each others.

ii. **Row inter-cropping** - Growing two or more crops simultaneously where one or more crops are planted in rows. *e.g.* maize + greengram (1:1), maize + blackgram (1:1), groundnut + rredgram (6:1), wheat + mustard (9:1) and gram +mustard (4:1/6;2).

iii. **Strip Inter-cropping** - Growing two or more crops simultaneously in strips wide enough to permit independent cultivation but narrow enough for the crops to interact agronomically. *e.g.* Groundnut + redgram (6:4) strip.

iv **Relay Inter-cropping** - Growing two or more crops simultaneously during the part of the life cycle of each. A second crop is planted after the first crop has reached its reproductive stage of growth, but, before it is ready for harvest.

B. Sequence cropping - as growing of two or more crops in a sequence on same piece of land in a farming year. The succeeding crop is planted after the preceding crop has been harvested.

i. **Double triple** and quadruple cropping - Growing two, three and four crops, respectively, on the same land in a year in sequence.

- Double cropping - Rice: cotton
- Triple cropping - Rice: rice: pulses
- Quadruple cropping - Tomato: ridge gourd: Amaranthus greens: baby corn

ii. **Ratoon cropping** - The cultivation of crop re-growth after harvest, although not necessarily for grain. ***e.g.*** Bajra, Napier, Sorghum for fodder purpose.

C. Relay cropping – growing of succeeding (next) crops before harvesting the procedding (previous) crops. It known by another names such as **Paira** in WB and Bihar; and **Utera** in MP.

Others term of farming

1. **Alley Cropping** – it is row planting of trees at wide spacings with a companion crop grown in the alley ways between the rows. It is the type of agroforestry to meet out multiple requirements of food, fodder and fuel.
2. **Lay Farming** – in this system involve the rotation of legume forage with cereal for low risk in dry land farming.
3. **Precision farming** - the integrated crop management techniques as right place at right time and in right way to optimize use farm resources (soil, water, land labour and chemical inputs) for optimum profitability, sustainability and protection of resources as well as minimize the crop input cost.
4. **Plantation farming** – the extensive commercial cultivation of a single cash crop owned by a groupe or company on a large scale is known as plantation farming. *e.g.* coffee, tea, cucunut, rubber, cashewnut *etc.*,
5. **Jhoom farming** - the slashing and burning of forest areas and growing the crops by trible groups the northeastern Indian states of Arunachal Pradesh, Meghalaya, Mizoram and Nagaland. It is also called shift cultivation or land rotation.

6. **Tongya cultivation** - it is special arrangement between forest department and local farmers to allow for temporarily cultivation of need based agricultural crops during early stage forest establishement.
7. **Organic farming** - growing of agricultural crop without use of chemical pesticide, fertilizers, hormone and anti-biotics.
8. **Contingent farming** – growing of short duration crops for getting some benefit under aberrant weather conditions where the main crops is failed to cultivate.
9. **Diara cultivation**- the cultivation of specially cucubits vegetable in the bank or basin of the river when the flood level receded.
10. **Contract farming** – the land is given on the rent to the company for growing the crops.
11. **Integrated Farming System (IFS)** - Integration of two or more appropriate combination of enterprises like crop, dairy, fishery, poultry, bee keeping, piggery, *etc.*,, for each farm according to the availability of resources to sustain and satisfy the necessities of the farmer.

Crops grown for special purpose

1. **Avenue crops** - growing of crops on boundry of the farms or fencing side like pigeonpea, sisal; or in case of horticultural crops grown for shade purpose like neem, peepal, mango *etc.*,
2. **Alley crops** - growing of annual crops in between of tree rows *e.g.* ginger in coconut/areacanut orchards.
3. **Augment crops** - such crops are growing as to assist the main crops yield e.g Japanese mustard in wheat.
4. **Cash crops** - those crops are grown for direct sale in the market rather than for own use /livestocks feed *e.g.* tobacco, potato, onion, jute, sugarcan *etc.*,
5. **Catch/contingent crops** – some crop crops are cultivated to catch the forthcoming season when the main crop is failed.
6. **Guard crops** – these crops are protect the main crops from animals/ man/or restrict the wind speed.
7. **Smoother crops** - some crops are having the ability to suppress the weeds growth by providing the dense foliage *e.g.* moong, cowpea, urd *etc.*,
8. **Trap crops** – these crops are protect the main crops from the some especial pest and diseases *e.g.* marigold in anar for nematodes, bhindi in cotton, **marigold in totmato/ chilli (1:16)** *etc.*,
9. **Truck crops** – these crops are grown on large scale for fresh shipment to market distance *e.g.* Onion, potato, tomato and melons.

MINERAL NUTRITION

Plant nutrients

- **Criteria of essentiality**: Proposed by **Arnon and Stout** (1939) and it again refined by **Arnon** (1952)

An element is considered as essential only when

1. The plants are unable to complete the vegetative and reproductive stages of life cycle under the deficiency of that particular element.
2. Deficiency of an element can be corrected or prevented only by supply that particular element to the plant and it cannot be replaced by any other nutrients.
3. The element is directly involved in some physiological and biological process of plant.

Essential plant elements: 16+1 = 17 **(Ni)** - added by **Brown & Welch (1987).**

A. Absorbed by leaves (3) - C, H, O (Provide basic structure of plant).

B. Absorbed by roots (13) - and they also had known as *Mineral nutrients.*

Functional plant nutrient:

- It is proposed by **Nicholas (1961),**
- Total functional plant nutrients are 21 (Sodium, Cobalt, Vanadium and silicon in addition to above 17).
- Sodium, Cobalt, Vanadium and silicon are not required for all plant but they absorbed for special purpose by few plant *e.g.* sodium absorbed by (sugar beet, turnips and celery), silica by rice.

Classification of essential elements

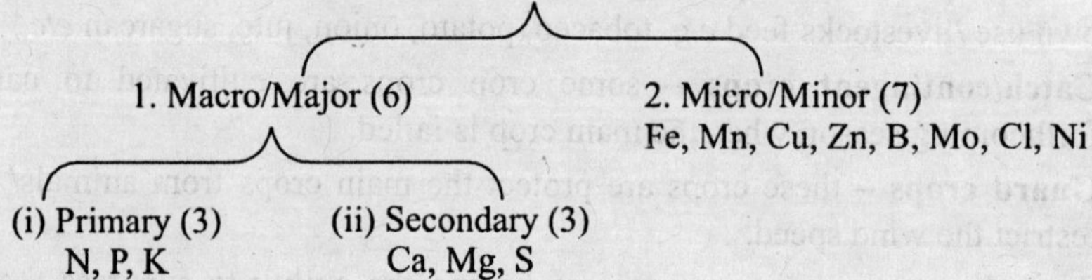

- Macro element do not cause toxicity to plant if applied in higher amount but micro element cause phytotoxcity after certain level

About the elements

Basic structural elements	C, H, O (forms 94% of the plant body)
Energy storage,transfer,bonding	N, P, S
Charge Balance	K, Mg, Ca
Catalyst and activators	Fe, Mn, Zn, Cu, B, Mo and Cl

Mobility in plant:

Highly mobile	N, P, K
Moderote mobile	Zn
Less mobile	S, Fe, Mn, Mo, Cl
Immobile	Ca, B

Identification of deficiency symptoms of nutrients in plan

Old leaves	N, P, K, Mg, Mo
New leaves	S, Fe, Mn, Cu
Old & new leaves	Zn
Terminal buds	Ca, B

Nutrient deficiency symptoms appea ed on plants:

Green viens on old leaves	Mg
Yellow viens on old leaves	N
Green viens on new leaves	Fe, Mn
Yellow veins on old leaves	S, Cu
'V' shaped yellowing in cereals	N
Hooked or cupped shaped leaves	Ca
Tea – yellow	S
Yellow Lucern, browning or hallow stem of cauliflower, Top sickness of tobacco	B
White bud disease of maize, little leaf in cotton, khaira disease of paddy, mottle leaf / little leaf or frenching of citrus	Zn
Die- back in citrus	Cu
Grey speck of oats, marsh spot of peas, pahala blight of sugarcane / streak disease	Mn
Whiptail in cauliflower, yellow spot of citrus	Mo

Indicator plants: Those weedy plants that grow in specific environment to allowing an assessment of biotic and abitotic stresses under adverse conditions by simple observation of vegetation.

Indicator plants for specific nutrient deficienci

S. No.	Nutrient element	Indicator plants
1.	Fe	Sorghum (CHS-1)
2.	Zn	Hybrid maize
3.	Mn	Rice (IR 8), Marigold, Wheat
4.	Cu	Wheat

5.	Mo	Tomato, Cauliflower
6.	B	Sunflower, Cauliflower, Maize
7.	K	Potato
8.	P	Toria

Bulky Organic manure: Plant nutrients available in small quantities that's why large quantity required fulfilling crop nutrient requirement,

The effect of bulky organic manures on soil:

- These manures provide nutrients to the plants.
- Increase organic matter content and improve the physical properties of soil.
- Provide food for soil micro- organisms.

Types of organic manures:

1. **Farm yard manure**: It refers to the decomposed mixture of dung and urine of farm animals along with the litter (bedding material) and left over fodder fed to the cattle. FYM contain 0.5, 0.2 and 0.5 % N, P_2O_5 and K_2O, respectively. Sheep & Goat Manure:
   - It contains 3: 1: 2 NPK
   - Poultry manure: It contains 3: 2.6: 1.4 NPK due to high vaccination poultry manure not allowed in organic farming
2. **Compost:** It made by farm waste like sugarcane trash, paddy straw, weeds and other plant and organic waste. The compost contain NPK (0.5, 0.15, 0.5 %)
3. **Night soil**: It is human excrement solid and liquid. It contains NPK (5.5, 4.0 2.0 %).
4. **Vermicompost**: It contains NPK (3.0, 1.0, 1.5%). Out of 350 earthworm species *Eisenia foetida, Eudrillius engenial* and *Perionyx excavates* are mainly used for preparation of vermicompost
   A. ***Eudrillius engenial* (African earthworm):** Produces higher production of vermicompost in short period of time
   B. ***Eisenia foetida* (Red worm):** Most commonly used in India and because it's extremely tough and adaptable. This worm is commonly known as the compost worm, manure worm, red worm and red wiggler.
   C. ***Perionyx excavates* – it is the Asian worm.**
5. **Green manuring**: It is practice of ploughing or turning undecomposed plant tissues into the soil for the purpose of improving soil structure as well as fertility of the soil.
   A. **Green manuring *in situ*:** Green manure crops are grown and buried in the same field along with either as a pure crop or as intercrop with the main crop. *e.g.* Sannhemp (*Crotolaria juncea*), Dhaincha (*Sesbania aculeata* & *Sesbania rostrata* – Nodules are present on stem), Guar (*Cyamopsis tetragonoloba*).

B. Green leaf manuring: In this practice, the green leaf and tender green twings collected from several shrubs and trees grown on bunds, waste lands and nearby forest areas. *e.g.* Glyricidia (*Glyricidia maculata*), Karanj (*Pongamia pinnata*), Wild dhaincha (*Sesbania speciosa*) and Wild cassia (*Cassia tora*) and apply in the field after decomposed.

Sewage: It is a liquid portion of human excreta.

Sludge: It is a solid portion of the sewage

Concentrated Organic Manures:

These manures are organic in nature and contain higher percentages of major plant nutrients like NPK as compared to bulky organic manure.

Oil cakes:

(i) **Edible cakes:**

- Suitable for cattle feeding *e.g.* Groundnut, safflower, rapeseed, sesame cakes

(ii) **Non edible cakes:**

- Not suitable for cattle feeding. *e.g.* Castor, karanj, mahua, neem cakes

Concentrate cakes:

Blood meal: It contains 10-12% N and 1-2% P_2O_5

Meat meal: It contains 10.5% N and 2.5% P_2O_5

Fish meal: It contains 4-10% N 3-9% P_2O_5 and 0.3-1.5% K_2O

Commercial fertilizers:

1. Nitrogenous Fertilizer:

Ammonium: Nitrogen in ammoniacal form

- Acidic residual effect *e.g.* Ammonium sulphate [(NH_4) SO_4], Ammonium Phosphate ($NH_4H_2PO_4$), Ammonium chloride (NH_4Cl), Anhydrous ammonia, Ammonium solution.

Nitrate: Nitrogen in nitrate form

- **Basic in nature**. *e.g.* Sodium nitrate or Chilean nitrate ($NaNO_3$), Calcium nitrate [$Ca(NO_3)_2$], Nitrophosphate.

Ammonium and nitrate: Nitrogen both in ammoniacal and nitrate form

- Acidic residual effect. *e.g.* Ammonium nitrate (NH_4NO_3), Calcium ammonium nitrate (CAN), Ammonium sulphate nitrate (ASN)

Amide: Nitrogen in amide form

- Organic compound *e.g.* Urea [$CO(NH_2)_2$], Urea phosphate, Urea sulphate, Calcium cyanamide ($CaCN_2$)

Slowly available:

- These fertilizers are slowly dissolve in water and are mainly decomposed by microbial activity after that they are available to plants for longer duration. *e.g.* Urea: Formaldehyde compounds, Oxamide, Urea-Z, IBDU

2. **Phosphatic Fertilizer:**

(i) Phosphatic fertilizer containing water soluble phosphoric acid or mono-calcium phosphate-Ca $(H_2PO_4)_2$
   - Used on neutral to alkaline soil but not on acidic soils.
   - Under acidic conditions, water soluble phosphoric acid gets converted into unavailable iron & Al phosphates. *e.g.* Single Super Phosphate (SSP), Double Super Phosphate (DSP), Triple Super Phosphate (TSP), Ammonium Phosphate.

(ii) Phosphatic fertilizer containing citric acid soluble phosphoric acid or dicalcium phosphate - $Ca_2H(PO_4)_2$
   - Suitable for acidic soil, because with low pH (basic reaction). *e.g.* Basic slug, Dicalcium Phosphate, Rhenania Phosphate, Raw and Steamed Bone Meal

(iii) Phosphoric acid- not soluble in water/citric acid
   - Well suited for strongly acidic soil or organic soil. *e.g.* Rock Phosphate, Raw Bone Meal, Steamed Bone Meal

3. Potassic Fertilizer:

(i) Fertilizer having K in the chloride form
   a. Muriate of potash (KCl)

(ii) Fertilizer having K in non chloride form
   a. Potassium sulphate (K_2SO_4)
   b. Potassium nitrate (KNO_3)

Multiple nutrient or complex fertilizer:

- A commercial fertilizer contains at least two or more primary essential nutrients (NPK).
- **Examples:** Mono Ammonium Phosphate (N 11%; P_2O_5 48%), Di Ammonium Phosphate (N 18%; P_2O_5 46%), Crude Ammonium Phosphate (N 16 & 20%; P_2O_5 20%), Urea Ammonium Phosphate, Nitrophosphate/Nitric-Phosphates

Straight fertilizer:

- Such fertilizer supply only one plant nutrient.

Mixed fertilizer:

- A mixture of two or more straight fertilizer materials.

Complete fertilizer:

- Those fertilizers that contain three major plant nutrients namely NPK.

Incomplete complex fertilizer:

- When the fertilizer contains only two primary nutrients.

Coated fertilizers:

- Fertilizers granules are covered with a thin layer of different materials in order to improve the behaviour and to modify the characteristics of the fertilizer. *e.g.* Lac Coated Urea, Neem Coated Urea, and Sulphur Coated Urea.

Nitrification inhibitors:

- Certain substances are toxic to the nitrifying bacteria when added to the soil/ urea. Thereby, it checks the rate of nitrification and ultimately reduces the leaching losses. *e.g.* **N- Serve, A.M., D.D., Thio- urea, K.N., MAST**.

Fertilizer application methods:

- **Broadcasting**: Application of fertilizer through broadcast in standing crops
- **Band placement**: Application of feritlizer in the side of crop rows
- **Point placement**: Application of feritlizer closed to plant around root zone either by making hole or depression followed closing or covering with soil
- **Basal Application**: application of fertilizer before or at the time of sowing *e.g* DAP fertilizer
- **Fertigation**: application of fertilizer fertilizer with irrigation water to the plant, through drip irrigation especially in horticultural crops.

WATER MANAGEMENT

- **Transpiration:** Transpiration is process of vaporization of liquid water from plant tissues or any living tissues are known as transpiration.
- The ability of trees to suck water from roots to leaves, against the gravitational pressure, above a hundred meters in height, without any mechanical pump was explained by many theories.
- Most acceptable **"Cohesion-Tension"** theory of the ascent was proposed by **Dixon and Joly in 1894**(Irish scientist).

The theory relies mainly on the

- Cohesive and adhesive properties of water
- The structural properties of trees
- Role of evaporation ("transpiration") from leaves

Types of transpiration:

1. **Stomatal transpiration – vaporization of** water through stomata is known as stomatal transpiration. It is contributed near about 85-90% of the total water loosed.
2. **Cuticular Transpiration** –loss of water in plants via the cuticle and its make up only about 5-10% of the total transpiration. The thickness of the cuticle and presence or absence of wax coating on the leaves surface determined this transpiration loss.
3. **Lenticular Transpiration** – The loss of water from plants as vapor through the lenticels and it is the only 2-3%.

Evaporation: - Evaporation is the process by which liquid water is converted into water vapor (vaporization) and removed from the evaporating surface such as lakes, rivers, pavements, soils and wet vegetation.

Evapotranspiration (ET): It is the sum of evaporation and plant transpiration loss of water from the earth parts.

Irrigation: It is defined as the artificial application of water other than rainfall to land to assist in the production of crops.

Importance of Irrigation:

- Protection from famine
- production of superior crops with higher yield
- Elimination of mixed cropping
- Economic development
- **National Mission on Micro Irrigation** was Central government sponsored scheme on Micro Irrigation has been introduced in **January, 2006** to increase the area under improved methods of irrigation for better water use efficiency to provide stimulus to agricultural growth.
- **CWR** stand for Crop Water Requirement
- **CCA** stand for Culturable Command Area

Irrigation schemes/projects classified based on area

1. Minor Irrigation (M.I.) Scheme: A scheme having CCA up to 2,000 hectares individually.
2. Medium Irrigation Scheme: A scheme having CCA more than 2,000 hectares and up to 10,000 hectares individually.
3. Major Irrigation Scheme: A scheme having CCA more than 10,000 hectares.

Terms related to irrigation/ Terminology:

Irrigation efficiency (IE): It is the ratio of the amount of water consumed by the crop to the amount of water supplied through irrigation (surface, sprinkler or drip irrigation).

Water management: It is planned use of water for better utilization in agriculture. It is includes irrigation and drainage.

Infiltration The downward entry of water from the surface into soil profile.

Infiltration rate (cm/hour): The rate at which water enters into the soil when the flow is non- divergent.

Accumulated/cumulative infiltration The total quantity of water that enters the soil in a given time.

Runoff: The flow of excess water from the field after saturation of soil during rain.

Life saving irrigation: It is balanced application of water to the crop protects and produces the yield, it is also known as contingency irrigation.

Viscosity: Internal force that brings about resistance to flow measured in poise.

- 1 poise = 1 dynes sec / cm^2
- 1/100th poise is called centipoises.
- Viscosity of water at 20° C is 1 centipoise and it increase more than 1% with each degree drop in temperature.

Percolation: The downward or lateral movement of water through saturated soil in response to gravity at a tension below 1/3 atm.

Water-use efficiency (WUE):

- It refers to the ratio of water used in plant metabolism to water lost by the plant through transpiration. It is having two types as given below-
  1. **Photosynthetic water-use efficiency** is the current net CO_2 assimilation rate, divided by the current transpiration rate. It is also known intrinsic or instantaneous water-use efficiency.
  2. **Productivity water-use efficiency** (integrated water-use efficiency)- it is the ratio of biomass produced to the rate of transpiration.

Consumptive Use of Water: It is the amount of water required by a crop for its vegetative growth to evapotranspiration and building of plant tissues plus evaporation from soils and intercepted precipitation. It is expressed in terms of depth of water.

Mathematically, Consumptive Use = Evapotranspiration + Water used in metabolic activity in plant

- It is expressed in terms of depth of water.
- Plant use <1% metabolic activity almost 99% loss through in Evapotranspiration

Productivity of water (PW): It is the ratio of output (physical, economical or social) to the amount of water depleted in producing the output.

Water requirement of a crop: It is the total quantity of water required from its sowing time to harvest

Net water requirement (NWR): NWR is the amount of water required to replenish evapotranspiration and deep percolation/seepage or root zone soil moisture deficit.

Gross water requirement (GWR): GWR is the actual amount of water supplied to meet crop evapotranspiration and or percolation/seepage observed under field conditions.

Command area (CA): CA is the area that can be irrigated by a canal system

Duty of water: The total volume of irrigation water required maturing a particular type of crop, including consumptive use, evaporation and seepage from ditches and canals, and the water eventually returned to streams by percolation and surface runoff.

Delta: It is total depth of water required for a crop during the entire period in the field and is denoted by Δ. The unit of delta is days.

Base period: It is the total duration from first irrigation water issued for land preparation for planting the crop to its last watering before harvesting.

Relation between duty and delta:

$$\text{Delta } (\Delta) = 8.64B/D$$

Where,

Δ - Delta in meter

B - Base period in days

D - Duty in hectares/cumec

Water Requirement of Different Crops

Crop	Water Requirement (mm)	Crop	Water Requirement (mm)
Rice	900-2500	Chillies	500
Wheat	450-650	Sunflower	350-500
Sorghum	450-650	Castor	500
Maize	500-800	Bean	300-500
Sugarcane	1500-2500	Cabbage	380-500
Groundnut	500-700	Pea	350-500
Cotton	700-1300	Banana	1200-2200
Soybean	450-700	Citrus	900-1200
Tobacco	400-600	Pineapple	700-1000
Tomato	600-800	Gingelly	350-400
Potato	500-700	Ragi	400-450
Onion	350-550	Grape	500-1200

Water (H_2O):

- Two hydrogen atoms bonded to an oxygen atom form a water molecule.
- It is an essential component of protoplasm.
- Water is dipolar in nature (104.5 °C).

Water Resources on the globe (%) and Constitution of water (%) by different plant parts

Water resources	Present (%) on globe	Plants and its parts	Constitution of water (%)
Oceans	97.39	Plant	35-95
Polar ice glacier	2.01	Leaves	40-95
Ground water	0.58	Roots	60-90
Lakes & river	0.02	Fresh fruit	70-90
Atmosphere water	0.0001		

- **2.6 % of water** (polar ice glacier, ground water, lake, river and vapour) is available only for plant
- Loss/gain of water measured by **Lysimeter**

Irrigation sources in India

- Cannal – 40%
- Wells – 30%
- Tubewell – 22%
- Tank – 8%

Important formula related to crop yield

1. **Mobilization index** = Crop yield / Straw yield
2. **Crop index** = Economic yield / Biological yield
3. **Harvest index** = Economic yield / biological yield
4. **Relative grain yield** = (Yield of treatment / Yield of control) X 100
5. **Evapotranspiration efficienc** = Water use efficiency/ Harvest index
6. **Water use efficienc** = Grain yield (ton) / Total water used (gallon)
7. **Irrigation requirement (IR)** = WR - (ER+S) here, ER means Effective rainfall and S means Surface runoff
8. **WUE** = Y/ET where Y - marketable yield
9. **Soil moisture (weight basis)** = Wet soil - dry soil/ Dry soil x100

Methods of Irrigation - Advantages and Limitations

Irrigation methods:

I. Surface

II. Sub-surface

III. Pressurized irrigation

I. Surface irrigation:

- Surface is grouped as Border, check basin and furrow irrigations.
- Border is again classified in to two as *straight* and *contour*.
- Check basins may be of rectangular, contour or ring,
- Furrow irrigation is classified as *deep furrow* and *corrugated furrows*.
- These may be again straight or contour according to direction and leveled and graded as per their elevation

(i) Border irrigation

- The land is divided into number of long parallel strips is called borders.
- These borders are separated by low ridges.
- The border strip should be having a uniform gentle slope in the direction of irrigation.
- Width of border strip is varies from 3-15 m

Length of beds based on slop

Slop	Soil	Length
0.25 - 0.60%	Sandy and sandy loam	60-120 m
0.20 - 0.40%	Medium loam soil	100-180 m
0.05 – 0.20%	Clay loam and clay soil	150-300 m

Advantages

- Border ridges can be constructed with simple farm implements like ridger or bund former.
- Labour requirement in irrigation is reduced as compared to conventional check basin method.
- Uniform distribution of water and high water application efficiencies are possible.
- Large irrigation streams can be efficiently used.

Limitations

- Water use efficiency is very low
- Water wasting is more
- Not suitable for those crops are sensitive to water lodging *e.g.* papaya

(ii) Check basin irrigation

- It is the most common method and filed is divided into smaller unit areas
- The size of the basin varies from $10m^2$ to 25 m^2 depending upon soil type, topography, stream size and crop.

Advantages

- Check basins are useful in the salts affected soil profile.
- Rainfall can be conserved and also soil erosion is reduced by retaining large part of rain

- High water application and distribution efficiency.

Limitations

- The ridges interfere with the movement of implements.
- More area occupied by ridges and field channels.
- The method impedes surface drainage
- Labour requirement is higher.
- Not suitable for crops which are sensitive to wet soil conditions around the stem.

(iii) Furrow irrigation

- The furrows are formed between crop rows.
- Water is applied by small running streams in furrows between the crop rows.
- Water infiltrates into the soil and spreads laterally to wet the area between the furrows.
- In heavy soils furrows can be used to dispose the excess water.
- **Surge irrigation**- it is the type of furrow irrigation where application of water into the furrows intermittently in a series of relatively short ON and OFF times of irrigation cycle. The irrigation efficiency of surge irrigation is 85 and 90%.

Advantages

- Water in furrows contacts only one half to one fifth of the land surface.
- Labour requirement for land preparation and irrigation is reduced.
- Compared to check basins there is less wastage of land in field ditches.

II. Sub-surface irrigation

- In subsurface irrigation, the water is applied beneath the ground by creating and
- Maintaining an artificial water table at the depth of 30-75 cm below the ground surface.
- Moisture moves upwards towards the land surface through capillary action
- The irrigation water should be of good quality to prevent soil salinity.

Advantages

- Minimum water requirement for raising crops
- Minimum evaporation and deep percolation losses
- No wastage of land
- No interference to movement of farm machinery
- Cultivation operations can be carried out without concern for the irrigation period.

Disadvantages

- Requires a special combination of natural conditions.
- There is danger of water logging
- Possibility of choking of the pipes lay underground.
- Higher cost.

Drip Irrigation System:

- Drip or **trickle irrigation** is one of the latest methods of irrigation.
- It is suitable where the water scarcity and salt affected soils.
- Water is **applied in the root zone** of the crop
- The laterals may be designed to operate under pressures as low as **0.15 to 0.2 kg/cm²** and as large as **1 to 1.75 kg/cm²**.

Advantages

- More water use efficient (90–95%).
- Reduced cost for fertilizers through fertigation
- Reduced labour demand:
- Low energy requirement
- Reduced salinity risk:
- Chemigation is means application of pesticides and other chemicals too periodically with drip irrigation system, such as chlorine or sulfuric acid.
- Water saving (> 60%)
- Better Weed management

Disadvantages

- High initial cost and need frequent maintenance
- Drippers are susceptible to **blockage or clogging**
- Interferes with farm operations and movement of implements and machineries
- Restricted root zone

Sprinkler Irrigation System

- The application of water to crop in the form of spray, somewhat as rain.
- It is highly useful particularly for sandy soils where absorb water too fast and undulating land tophography.

Types of sprinkler system based on the portability

1. **Portable system:** It has portable mainlines and laterals and a portable pumping unit

2. **Semi portable system:** A semi portable system is similar to a fully portable system except the water source and pumping plant are fixed.
3. **Semi permanent system:** A semi permanent system has portable lateral lines, permanent main lines and sub mains and a stationery water source and pumping plant. The main lines and sub-mains are usually buried, with risers for nozzles located at suitable intervals.
4. **Solid set system:** A solid set system has enough laterals to eliminate their movement.
5. **Permanent system:** It consists of permanently laid mains, sub-mains and laterals and a stationary water source and pumping plant and the sprinklers are permanently located on each riser.

Advantages

- Water saving can extent of 35-40% as compared to surface irrigation methods.
- Fertilizers saving.
- Suitable for undulating topography (sloppy lands) and sandy soil
- Reduces erosion
- Frost management
- Eliminating drainage problems

Disadvantages

- High initial cost
- Efficiency is affected by wind
- Higher evaporation losses in spraying water
- Not suitable for tall crops like sugarcane
- Not suitable for heavy clay soils
- Poor quality water can not be used

Cabligation: It is an automatted method of surface irrigation which can save labour & water as compared to other surface methods of irrigation is called cabligation.

SOIL CONSERVATION

Soil conservation: It is use of land within its capability and according to need so as to keep it permanently productive.

- Father of Soil conservation is **H. H. Benatt**
- Most erosive river are **Narmanda and Hirakund**
- Highly water erosive state is Uttar Pradesh
- Slope % in bench terracing is 16-33
- Best vegetative barrier for soil conservation is **Khus grass (Vetveria)**

Agronomic practices in soil and water conservation:

1. **Strip cropping**: It is a system under which erosion resistant crops are planted with eroding crops (ordinary farm crops) alternately in relatively narrow strips. Types of strip cropping

   1. Contour strip cropping 2. Buffer/spreader strips 3. Turn and border strips 4. Field strips 5. Correction strips 6. Filter strips.
2. **Terraces:** An embankment or ridge of earth constructed across a slope to control runoff and discharging soil erosion. A complete series of terraces occupying a slope and discharging runoff into one or more outlets channels called terrace system.

Types of Terraces:

1. Bench terrace, 2. Graded channel / runoff interception terraces, 3. Puerto Rico / stone wall terraces, 4. Zing terraces, 5. Broad base terraces.

Soil erosion:

- It is the detachment transportation and deposit of soil particles from one place to another through the action of wind and water is known as soil erosion.

Factors affecting erosion by water:

1. Climate
2. Topography
3. Vegetation
4. Soil

$$E = f(C, T, V, S)$$

1. Water erosion:

- Erosion by water (removal of soil from the land's surface by water in motion).

Types of water erosion:

(i) **Rain drop/Splash**:

- Results from soil splash caused by the impact of falling rain drops.
- The falling rain drops breakdown soil aggregates and detaches soil particles from the soil mass.
- Rain drop splash gradually removes the fine materials from the soil and leaves the less fertile sand and gravels.

(ii) **Sheet erosions**:

- Removal of a fairly uniform layer of soil from the land surface by the action of rainfall and runoff.

- It results commonly on the lands having a gentle slope.
- It results in the uniform **"Skimming the cream"** of the top soil with every hard rain.

(iii) **Rill erosion**:

- It is the removal of soil by running water with the formation of shallow cannels
- It is a transition stage between sheet erosion and gully erosion.

(iv) **Gully erosion**:

- It is the removal of soil by running water with the formation of wider channels .
- Gully erosion is **more spectacular** than other type of erosion.
- 'U' shaped gullies formed where both surface and sub surface soils are easily eroded. 'V' shaped gullies are formed where sub soils are resistant to erosion.

Sediment in channels is transported by

1. **Suspension**: Suspended sediment remains in flowing water for considerable period of time without contact with the stream bed.
2. **Saltation**: Occurs where the particles jump or bounces along the stream bed.
3. **Bed load**: Sediments rolled or pushed along the bottom by the water force.
   - About 10 to 54 % of the total sediment is transported by bed load movement.

2. Wind erosion:

- Wind picks up lighter particles, lifts them bodily from the surface and transports them long distances.
- Wind erosion is a serious problem in areas where land is bare and devoid of vegetation.
- It is a natural phenomenon in arid and semi arid zones.
- The **Thar Desert** is deformed mainly by blown in sand.

Processes:

- Soil loosening, Disintegration and Denudations
- **Initiation of movement** – Most erodible fractions 0.1 to 0.5 mm diameter.
- **Transportation-** Surface creep, Saltation and Suspension,
- **Eluviation-** is the transport of soil material from upper layers of soil to lower levels by downward precipitation of water across soil horizons, and accumulation of this material (illuvial deposit) in lower levels is called illuviation.
- **Abrasion-** Hitting clods or stones and breaking them.
- Avalanchin

- Deposition
- Soil consolidation and stabilization
- **Eluriation**. The conditioning of sludge from sanitary waste
- **Elutriation** is a process for separating particles based on their size, shape and density, using a stream of gas or liquid flowing in a direction usually opposite to the direction of sedimentation. This method is mainly used for **particles smaller than 1 μm.**

Types of soil erosion

A. **Suspension**:

- The fine particles of soil are kicked into the air, stay in suspension at higher height and carried long distances by the wind.
- The particles are **smaller than 0.1 mm in diameter**.

B. **Saltation**:

- The soil particles moved by saltation are between **0.1 to 0.5 mm in diameter** (fine sand).
- Sand sized particles rolling, jump up vertically into air and rise to a height of 30 to 60 cm.
- This erosion is contributed 50-70 % of total wind erosion.

C. **Surface creep**:

- Rolling of coarse grains, larger than **0.5 to 3.00 mm in diameter** and too heavy to be lifted, by wind along the surface of the ground.
- Generally 5 to 25 % erosion is carried by surface creep.

Sedimentation: It is an integral part of the erosion process means soil eroded from its original location and deposited at some other place is called sedimentation.

Wave erosion:

- Erosion caused by the combined action of wind and water.
- Wave erosion mainly occurs in canals and river banks.

Type of soil erosion based on particle size:

S. No.	Erosion type	Particle size (mm)	Loss (%)
1.	Suspension	< 0.1	3-5
2.	Saltation	0.1-0.5	50-75
3.	Surface creep	0.5-3.0	5-25

Some facts:

- Most erodible particle size is < 0.1 mm
- Date palm is most suitable tree to check the soil erosion in desert

- Casurina, Cashew, Coconut-most suitable trees to check the soil erosion in coastal area
- Trees useful for timber, fodder and fuels are babul, teak and eucalyptus
- Most suitable drought resistance grass is *Cenchrus ciliaris*.

DRY LAND AGRICULTURE

Dry land Farming/Dry farming:

- It is the practice of crop production in the regions where annual potential evepotranspiration is more than annual precipitation.

PEP > P

Dry land Agriculture: The growing of crop entirely under rainfed condition.

- Rainfed agriculture is contributing 95 mha (67%) out of 142 mha of net sowing area in the country.
- Dry land agriculture is covering 91% area of cereals and pulses. 80% oil seeds. 65% cotton, 50% rice and 19% wheat.
- Dry land agriculture contributes 42% of total food grain production in the country.

Dry land farming or agriculture can be grouped into three categories:

1. **Dry farming**
   - Cultivation of crops such in areas where rainfall is less than 750 mm/annum.
   - Prolong dry spell and more crops failure.
   - Growing days is less than 75 days.
2. **Dry land farming**
   - Cultivation of crops where rainfall is receiving above 750 -1150 mm/annum.
   - Single crop is taken and growing season is less than 200 days.
   - Wind & water erosion are main constraints.
3. **Rainfed farming**
   - Cultivation of crops in region where rainfall is receiving more than 1150 mm.
   - Inter cropping and double cropping
   - Constraint is water erosion
   - Growing season >200 days
   - Maximum area comes under rainfed farming (65%) in India
   - Dry spell in light soil for 10 days wheras 15 days in heavy soil.

Improved agronomical practices for dry land agriculture: There are two ways to increase and stabilize on dry land.

1. Soil moisture conservation practices for dry land
2. Improved crop production technology for dry land.

1. **Practices for soil moisture conservation:**
   i. **Sub surface moisture barriers-** Asphalt, bentonite, clay, wax, sodium carbonate, sodium chloride and pod sediments.
   ii. **Mulch**: Any material applied on the soil surface to check evaporation and improve soil water.
   iii. **Water harvesting-** Collecting and storing of water for subsequent use. *e.g.* Tanka, Nadi, Khadin
2. **Improved crop production technologies:**
   - Proper management of these following crop production technologies for better utilization under moisture constraint areas are selection of crops and their varieties, more seed rate, seed treatment, planting pattern, mixed cropping and inter cropping, crop rotation, soil fertility management, plant protection, alternative crop planning and mid season corrections, tillage practice (zero and minimum tillage), crop substitution and diversification, weed control, stabilizing yields through life saving and supplemental irrigation, mulching, Increasing water storage in the root zone and runoff control.
3. **Antitranspirant:**
   - Antitranspirant is any material applied to transpiring plant surface for reducing water loss from the plant.
   - Nearly 99% of the water absorbed by the plants is lost in transpiration.

Types of antitranspirant:

(i) **Stomatal closing type**:
- It causes the closing of stomata when applied on the leaves *e.g.* Phenyl mercuric acetate (PMA), Agroson GN (Fungicide), Atrazine (Herbicide), and CO_2.

(ii) **Film forming type**:
- They form a thin film on the leaf surface and retard the escape of water due to formation of physical barrier. *e.g.* Hexadeconol, Silicon, Alkyl Succinic Acid (ASN), Mobileaf.

(iii) **Reflectant typ** :
- These are white materials which form a coating on the leaves and increase the leaf reflectance (albedo) by reflecting the radiation, they reduce leaf temperature and vapour pressure gradient from leaf to atmosphere and thus reduce transpiration. *e.g.* Kaolin (5%) and Diatomaceous earth.

(iv) **Growth retardants**:
- These chemicals reduce shoot growth and increase root growth and thus enable the plants to resist drought. They may also induce stomatal closure. *e.g.* Cycocel (CCC).

Note: Antitranspirants generally reduce photosynthesis. Therefore, their use is limited to save the crop from death under severe moisture stress.

Definitions:

Wind break: Any structure obstructs the wind flow and reduces wind speed is called wind break; the best direction for wind break is **North- East side** which are checked the cold and hot waves, respectively. The wind break/shelterbelt provides the protection against hot and cold waevs to **4-5 times** to its height.

Shelter belt: Row of trees planted for protection of crops against wind

Windward side: The side of mouantain which faces the moisture laden winds and receive heavy rainfall with higher vegetation. *e.g.* West-Coast Mumbai.

Leeward side: The other side of the mountain which does not face the moisture laden so may not get rainfall and less vegetation (dry side) *e.g.* Pune.

Contingency Cropping (CC):

- It is a growing of suitable crops in place of normally sown highly profitable crops of the region due to aberrant weather conditions. These are short duration crops. *e.g.* Black gram, green gram, cowpea *etc.,*

Suitable varieties for dry land farming:

Crop	Varieties
Sesamum	Madhuri, Gowri, Purva-1
Castor	Kranthi, Haritha
Cotton	Mahanandi, Jayadher
Ground nut	TMV-2
Sorghum	CSH- 5, 9, 1, 112, CSV-11, 12
Sun flower	MFSH-1, 8, 17, Adarsha
Coriander	CS-4, 6
Pearl millet	HHB -45, 50, ICTP- 8203

Paraheliotropy:

It is a type of tropism by which a plant's leaves move to reduce injuriously intense illumination, usually by keeping their blades parallel to the light rays.

- The watering of crops in regions where normal rainfall ordinarily supplies most of the moisture. It is provided when crop is facing moisture stress due to dry spells.
- Effective weeding time is within 30 & 45 days after of sowing.
- Ground nut is the most common mono-cropping system in alfisols of Andhra Pradesh.
- Now days, the most common irrigation method in India is drip irrigation.

Watershed management:

- Any surface area from which rainfall water is collected and drains through a common point.
- It may be in few hectares or several thousands of hectares which is used during shortage of water to fulfill the water requirement is called watershed management

Objectives of watershed management:

- To effectively conserve soil, rainfall water and harness the surplus water to create water source in addition to ground water recharge.
- To promote sustainable farming and stabilize crop yield by adopting suitable cropping & its management.
- To cover non arable area through aforestation, horticulture and pasture land developed based on capability of land.
- To restore ecological balance.

Component of water harvesting

- **Catchment area-** a part of land which contribute the rain water.
- **Storage body** – where the runoff water is stored *e.g.* water pond.
- **Command area** – where the stored water is used

SUSTAINBLE AGRICULTURE

Sustainable Agriculture:

- A system of cultivation which includes the use of manure, crops rotation and minimum tillage or minimum dependence on synthetic fertilizer, pesticides and antibiotics to the land degradation through soil erosion, atmospheric and water pollution as well as prevent accumulation of pesticide residues in the farm produce with objective to conserve the ecological assets for future generation.
- The sustainable agriculture is the successful management of resources for agriculture to satisfy changing human needs while maintaining or enhancing the quality of the environment and conserving natural resources (International Agricultural Research).
- It is practiced in form of ecofarming or organic farming or natural farming or permaculture
  - **Ecofarming**: In this ecological balance is more important.
  - **Organic farming**: As organic matter is the source for nutrient management.
  - **Natural farming** : Without chemical input farming
  - Ozone layer damaging by **CFC, Methane, Nitrogen oxide**
  - CFC contain **CHLORINE**

- Major cause of land degradation is due to **Shifting cultivation** in N-E India
- Denudation of forest by Steep slope farming and over grazing of pasture

Management practices:

- Watershed management for dry land
- Conservation of genetic resource
- Minimum tillage
- Nutrient management *e.g.* - crop rotation
- Weed management
- Integrated Pest Management, Integrated Disease Management

Advantage of Sustainable Agriculture:

- Ecological balance
- Low cost of cultivation
- Clean environment
- Nutrition food without pesticide residues

Disadvantage of Sustainable Agriculture:

- Low yield
- Lack of time
- Lack of effective control of weeds, insects & diseases

WEED MANAGEMENT

Weed:

- Weed is a plant growing out of place and out of time
- It is a plant grown at place where it is not desired or an unwanted plant in the field is called weed.
- The first person to use the word weed in the present day meaning was **JethroTull**.

Classification of weeds

A. Based on their life cycle:

1. **Annual weeds:** Those Weed species that completes its life cycle (seed to seed) within one growing season or one calendar year *e.g. Amaranthus viridis, Argemone mexicana, Avena fatua, Chenopodium album, Cleome viscosa, Celosia argentia*
2. **Biennial weeds:** Two season required to complete life cycle in first season vegetative growth in second season go for seeding *e.g. Alternanthera philoxeroides,*

3. **Perennial weeds:** They survive more than two season regeneration occur from roots or vegetative parts *Cynodondoctylon dactylon, Cyperus rotundus*

B. Based on their morphology:

1. **Grasses weed:** *Echinochloa colonum, Digitaria sanguinalis, Panicum antidotale, Chloris barbata, Cynodon doctylon*
2. **Seed weeds:** *Cyperus rotundus*
3. **Broad leaves:** *Celosia argentia, Portulaca oleracea, Eclipta alba*

Important weeds in India

S. No.	Common name	Botanical name
1.	Nut grass (sedge)	Cyprus rotundus
2.	Bermunda grass (Grass)	Cynodon dactylon
3.	Barnyard grass	*Echinochloa colonum*
4.	Jungli rice	*Echinochloa crusgalli*
5.	Jhonson grass	*Sorghum halepense*
6.	Water hyacinth (BLW)	*Eichornia crassipes*
7.	Fat hen or lambs Quarter	*Chenopodium album*
8.	Field bind weed	Convolvulus arvensis
9.	Pig weed	Amaranthus sp.
10.	Night shade weed	Solanum sp.
11.	Prickle –pear	Opuntia sp.

Special mechanical/ physical method for weed control in particular crops

S. No.	Practices	Remark
1.	Mowing	Lawn
2.	Chiselling	Plantation crops particularly in tea
3.	Digging	Perennial weeds
4.	Mulching	Horticultural crops
5.	Burning	Uncultivated field
6.	Hoeing	Row growing crops (field crops)

Integrated approaches of weed management

1. **Eradication:**
   - Weed eradication is complete removal of all live plant parts and seeds of weed from an area.
2. **Weed control:**
   - It is the process of limiting weed infestations so that crops could be grown profitably and other activities of man conducted efficiently.

3. **Weed management**:
   - It is a system approach whereby whole land use planning is done in advance to minimize the invasion of weeds in aggressive forms and give crop plants a strong competitive advantage over the weeds.
4. **Integrated weed management:** concept was developed by Japanese scientist Noda, 1977.
   - A weed management system that uses all suitable control methods in a compatible manner, to reduce weed population and maintain them at levels below those causing economic injury.

Weed control

1. **Cultural**
   - Field preparation
   - Fertilizer application
   - Planting method
   - Irrigation and drainage
   - Varieties
   - Cropping system
   - Planting density
2. **Physical**
   - Hand weeding
   - Dredging and chaining
   - Hand hoeing
   - Burning and flaming
   - Digging
   - Mulches
   - Mowing
   - Inter-cultivation
   - Cutting
3. **Chemical**
   - **Herbicide** is a **Latin** word; **herba** means plant and **caedere** means to kill.
   - A chemical used for killing or inhibiting the growth of plants is known as herbicide.

a. **Based on chemical structure**:

(i) **Inorganic herbicide:**
- Do not contain carbon atoms in their molecules.
- They were the **first chemicals** used for weed control before the introduction of the organic compounds.
- *e.g.* Arsenic acid, Arsenous acid, Arsenic trioxide, Sulphuric acid, Inorganic salts –Sodium Arsenate, Sodium chlorate, Borax, $CuSO_4$, Ammonium sulphate, Copper nitrate.

(ii) **Organic herbicides:**
- Organic herbicides contain carbon atoms in their molecules.
- They may be oils or non oils.
- Majority of the present day herbicides are organic compounds which are non oils.
- They can be further divided into 18 groups.

b. Classification based on selectivity

(i) **Selective**:

- Kill only targeted plants or weeds while crops are not affected *e.g.* Simazine, Atrazine, 2,4–D, MCPA, Butachlore, Alachlor, Fluchloralin and Pendimethalin *etc.*,

(ii) **Non- selective**:

- Kill all kind of vegetation in applied field when it come in contact with irrespective of whether it is a crop or weed *e.g.* Paraquat, diquat, glyphosate *etc.*,
- The selectivity depends on dosage of herbicide.

c. Classification based on translocation

(i) **Systemic**:

- These enter in the plant system and kill the target weeds. *e.g.* Atrazine, Simazine, Propanil *etc.*,

(ii) **Contact**:

- Kill plants when they come in contact with the plants.
- They kill the part of the plant that is in contact with herbicide. *e.g.* Paraquat, Diquat (Plough herbicide).

d. Biological: Any biological process or living organism used to control weed

- *Bio herbicides / Myco herbicides* are pathogens cultured artificially and made available in sprayable solution or formulations just like a chemical herbicide.
- Example as
  - **De- vine**: *Phytopthora palmivora* (soil born) -Strangler vine / milk weed
  - **Collego**: *Colletotrichum gloesporiodes*-Jointvetc.,h
  - **Bipolaris**: *Bipolaris sorghicola*- Jhonson grass
  - **Product F**: *Cercospora rodmanii*- water hyacinth
  - *Fusarium oxysporium*- Orobanche in sunflower.
- *Parthanium hysterophorus*: *Zygogramma bicolorata* (Leaf eating insect)
- *Cyprus rotundus*: *Bactra vertuana* (Moth).
- *Lantana*: *Crocidosema lantana* (Moth)
- Water hyacinth: *Neochatina bruchi* (Beetle)

Biological control of weeds:

S. No.	Weeds	Biological control	Remarks
1.	*Lantana camera*	*Crosidosema lantana* / Octotoma / Uroplata	It was introduced from Australia
2.	Opuntia (prickle-pear)	*Cactoblastis cactorum*/ *Dactylopius tomentosus*	*Tetranychus sp.* (Spider mite)
3.	Water hyacinth	*Rhizoctinia* blight	Fungus

4.	Aligator weed (*Alternanthera sp.*)	*Agasicles hygrophyla*	Flea beetle as predator
5.	Cascuta (Dodder)	*Melanagromyza cuscutae*	Parasite weed of *Medicagosativa*
6.	Nut grass (Motha)	*Bactrocera racemosa (insect)*	
7.	Parthenium grass (Congress grass)	*Diacrisia obliqua*	Worldwide problematić weed
8.	Anoda	*Alternaria/ paccinia (fungus)*	weed of malvaceae family
9.	Hyacinth	*Cercospora roclmanii*	aquatic weed

Herbicide formulation:

1. **Soluble powder**:
   - They form a homogenous solution when dissolved in water which can be applied by spraying. *e.g.* Sodium salts of 2,4-D, TCA, are soluble powder.
2. **Soluble concentrates**:
   - Herbicides formulations which are in the form of soluble liquids are called water soluble concentrates (WSC). *e.g.* 2,4-D Amine, Dicamba, Diquat.
3. **Wettable powder**:
   - When the herbicide materials are of low solubility in water, they can be grouped into fine powder for suspension in water. *e.g.* Atrazine 80%, Simazine 50%, Diuron 80% and Isoproturon 70% WP.
4. **Liquid suspension**:
   - If the active ingredient is not soluble in water, it is solubilised in organic solvents when the product of a.i. and solvent is added to water for spraying, it forms a liquid suspension. *e.g.* Atrazine, Nitrolin.
5. **Emulsifiable concentrat** :
   - The a.i. is dissolved in solvent and an emulsifier is added to it. Since the emulsifier helps in uniform distribution of the chemicals in water, no stirring is necessary while spraying. *e.g.* 2,4-D ester, alachlor, nitrofen, diallate.
6. **Granules**:
   - Small pellets or granules are made with inert clays. The solution of the toxicant sprayed on these granules in desired quantity and dried. The granules are then packed for subsequent use. *e.g.* Butachlor, 2, 4-DEE.

Mode of action of herbicides:

- Herbicides may act on one or more following process and dislodge the normal functioning of plants resulting in death of plants.

1. Inhibition of photosynthesis:

- Herbicides cause damage to the photosynthetic apparatus or to the process itself resulting in death of the plants due to starvation.
- The important reactions that are inhibited are:
  - (a) **Hill reaction**: The breakdown of water in the presence of chlorophyll by sunlight with the help of enzymes is called photolysis or hill reaction. Inhibit photolysis by deactivation of the water splitting enzymes. *e.g.* Trizines, Urease, Carbonates.
  - (b) **Inhibition of electron transfer**: A photo system II, electrons are released due to the breakdown of water during photolysis. These electrons are carried through a series of intermediate compounds and they are finally accepted by ferredoxine, NADP and NAD. *e.g.* Triazines, Carbomates, Ureas, Uracils, Anilides. Interfere with electron carriers of the electron transport chain.
  - (c) **Inhibition of CO_2**: Oxidations inhibits CO_2 fixation and export of photosynthesis.

2. Inhibition of respiration:

- Food materials are oxidized during respiration and released energy (stored in ATP molecules). The combined action of oxidation and conversion of ADP to ATP is called **oxidative phosphorylation**. These ATP molecules supply energy for carrying out several metallic reactions.
- Example: Dinitrophenol, Dinitroanilines and Anilides. The energy released during oxidation is not used to convert ADP to ATP.

3. Inhibition of biosynthesis:

- Several materials like proteins, fats, waxes, vitamins *etc.,* are synthesized in the plants. Herbicides inhibit some of these biochemical activities and interfere with the normal functioning of the plants.

4. Inhibition of protein synthesis:

- Herbicides like butachlor, alachlor belonging to amide group inhibit protein synthesis resulting in improper root and shoot development.
  - Trifluralin: Inhibits RNA synthesis.
  - Paraquat: Inhibit thymidine incorporation in broad bean. This indicates inhibition of DNA synthesis.

5. Inhibition of lipid synthesis:

- Some of the thiocarbamate herbicides like triallate, diollate have been known to inhibit wax production in plants.

6. Inhibition of vitamin synthesis:

- **Dalapon** is known to inhibit the synthesis of pantothenic acid.

7. Inhibition of cell division: Herbicides suppress cell division by several ways-

- Herbicides interfere with the synthesis/ active transport of precursors into the nucleus that are required for DNA synthesis during interphase.

- They modify the chemical and physical properties of DNA and its complexes.
- They also interfere with spindle formation and inhibit cell wall formation. *e.g.* Carbamate herbicides (Asulam and Barban), Dinitroanilines.

8. Inhibition of germination:

- One of the major metabolic processor that place during germination is the production of hydrolytic enzymes such as amylase, protease, lipases, phosphatises, esterase *etc.,*
    - **Carbonate** - Simple sugar
    - **Protein** - Amino acids
    - **Lipids** - Long chain fatty acids
    - These enzymes are regulated by **gibberillic acid**.
    - Alachlor, propachlor belonging to amide group act on germinating seeds by inhibiting gibberillic acid production.
    - Inhibition of germination mainly by inhibition of amylase activity.
    - Natural herbicide: Coffeaarabica seeds, Ethanolic extract (1,3,7 trimethyl xanthine).

9. Abnormal tissue development:

- They are suspected to interfere with metabolism of nucleic acid at cellular level.
- Rapidly proliferating roots and stems consume most of the available photosynthates and nutrients and the developing leaves and secondary roots are starved. This results in premature ageing of leaves leading to death.
- *e.g.* Phenoxy group (2,4-D, 2,4- DB, MCPA), Benzoic acid group: Chloramben, Diacamba, 2,3,6 –TBA, Picloram (auxin type herbicide)

Selectivity of herbicides:

- The differential response of plants to herbicides is called selectivity of herbicides. In other words, herbicides harm or kill weeds whereas crop plants are not affected due to selectivity.

Control of aquatic weeds:

- **Chaining** - The floating aquatic weed break down in pieces removed by chain
- **Dredging** - Removing of weeds along with their roots & rhizome with the help of mechanical force is called dredging

Fundamental principle of herbicide selectivity:

- It is that more toxicant reaches the site of action in active form inside target plants than in non- target plants.

Methods of herbicides application:

A. Soil application:

1. **Soil surface:**
   - Herbicides are sprayed on the soil surface to form a uniform herbicide layer.
   - The herbicides thus sprayed, due to their low solubility may penetrate only few cm into the soil. Weeds germinating in the top layer are killed due to incidental absorption of herbicides. *e.g.* Trizines, Ureas and Amides.
   - The herbicides for surface soil application have to be soil active, less soluble and less volatile. After application, the surface soil should not be disturbed.
2. **Soil incorporation:**
   - Volatile herbicides are incorporated into the soil to reduce losses.
   - Example: Aniline, carbamate groups.
3. **Sub-surface:**
   - Herbicides are applied to the lower layers of the soil.
   - *Cyperus rotundus* and *Cynodon dactylon* are to be controlled.
4. **Band:**
   - Herbicides can be applied as narrow bands.

B. Foliage/ foliar application:

Time of application of herbicides:

1. **Pre- plant application:**
   - Any time before crop is planted or sowing.
   - Herbicides are incorporated into the soil because herbicides are volatile in nature.

     e.g. Basalin.
2. **Pre- emergence:**
   - The application of herbicide after sowing of the crop but before emergence of crop and weed is called pre emergence herbicide
   - Time of application is 1-4 day after sowing.
   - Only selective herbicide is used for this purpose.
   - *e.g.* Atrazine, Simazine, Alachlor *etc.*,
3. **Post- emergence:**
   - The application of herbicide after sowing of the crop to kill the unwanted plants is called post emergence application
   - Time of application is between 30-35days after sowing.
   - *e.g.* Isoproturon, 2,4-D, Sulfosulfurn, Chlodinafop *etc.*,

Terminology:

Acid equivalent: The theoretical yield of parent acid from an active ingredient.

Active ingredient: The chemical in a product responsible for the herbicidal effects.

The amount of commercial product to be used = $\frac{\text{Recommended does} \times 100}{\text{a.i.\%}}$

Catch crop: A crop employed to stimulate germination and growth of some specific, parasitic weed which is then destroyed before planting the main crop.

Facultative weed: Weed of a wild community origin, escaping sometimes to the crop field.

Noxious weed: a plant arbitrarily defined by law as being especially undesirable, troublesome and difficult to control. *e.g. parthenium hysterophorus*

Objectionable weed: Troublesome weed whose seeds once mixed with crop seeds are extremely difficult to separate. *e.g.* wild oat in wheat, dodder in lucerne

Obligate weed: Weed of cropland incapable of surviving in a wild community.

Parasitic weed/Crop bound weed: A weed that partially or fully depends for its growth or some nutrients drawn from its host plant.

Satellite weed: Weed that has become an integral part of a crop ecosystem.

Stale seed bed: A seed bed left unsown to let a heavy flush of weeds appear and be destroyed with herbicide before seedling the crop, without any further tillage.

Trap crop: A crop employed to induce germination of specific parasitic weed without succumbing to it. *e.g.* Cotton (Marigold), Tomato (Marigold) and Ground nut (Sun flower) *etc.*,

Some facts:

- Most common weeds in the world (Country)
  - *Cyperus rotundus* (92)
  - *Cynodon dactylon* (80)
  - *Echnichloa colonum* (61)
- Most of the weeds are C_4 plants in nature.
- Most of the weeds belong to Gramineae family.
- Gramineae weed is known as Grasses.
- The weeds belong to family Cyperaceae is known as Sedges.
- Dicot weeds are also known as Broad Leaves Weed (BLW).
- Up-land type of rice cultivation have critical period of weed competition entire the period of crop.
- Orobanche cuased 40 percent loss in tomato.
- Parthnium is imported from USA (1960).
- Far Red light colour induced Dormancy.

- Red light colour is Promoting germination.
- 2,4-D, MCPA are effective herbicide for dicot weed while Dalapon, Fluchloralin for monocot
- Nodulation is decreased in legume (*Vicia faba*) due to application of Alachlor herbicide.
- 2-4-D and Dalapon are highly soluble herbicides.
- Concentration of herbicide with fertilizer should be having 0.5-2.0%.

HARVESTING AND STORAGE

Harvesting: Removal of entire plants or economic parts of crop after maturation from the field is called harvesting.

Stubble: The portion of the stem that is left in the field after harvesting is known as stubble.

Desiccation:

- It is an important harvesting management practice to hasten harvest especially for hybrids which stay green even at harvest maturity. *e.g.* Sunflower.
- Desiccation is done by use Paraquat and sodium salt weedicides

Harvesting maturity:

- Generally occurs seven day after physiogical maturity.
- Harvesting cost percent involed in the total production cost is 20-50 %.

Criteria of harvesting:

S. No.	Crops	Maturity indices
1.	Mango	Tapka
2.	Banana	Angularity of finger
3.	Sapota	Milky sap turn watery
4.	Water melon	Dull sound on tapering Yellow spot on ground touching portion
5.	Rice	32 days after flowering Milk % less than 1% Moisture content < 20 % At least five hills
6.	Sorghum	40 days after flowering Moisture content < 28%
7.	Pearl millet	28-35 days after flowering
8.	Maize	25-30 Days after tasselling

9.	Wheat	moisture should be 15%
10.	Sugarcane	brix ratio b/w top and medium is 1 Brix 18-20 % Sucrose 15 % 35-40 days after flowering
11.	Redgram	80-85 % pod turn brown
12.	Cotton	Bolls fully opened

Dhurin is present in sorghum is high up to 30 days after sowing.

Picking is the harvesting of green gram, black gram, cotton is known as Picking.

Threshing is the practice for separating fruits or seeds from the plant or ears.

Winnowing is the practices for separating grain or seed from chaff.

Moisture content for safe storage (percent) must be in

Cereals - 12-14%

Spices - 10%

Oil seed crops - 06%

Some facts:

- CAP stand for Cover and Plinth.
- **Pyramidal** is the best storage house (Stock) shape.
- Drying curing process is involved in hay making.
- Most suitable crops for hay making are **Maize than Sorghum**.
- Legume crops are not suitable for hay making.
- The process of silage making is called ensiling.
- Silage is more nutritious conserve than hay.
- Silage is the fermentation product.

CROP PRODUCTION

- The word **"cereals"** derived from a **roman** word **Ceres** means **Goddess of grain.**

Protein content in the agronomic crops

S. N.	Crops	Protein content (%)	Crops	Protein content (%)
		in pulses		
1.	Soyabean	40-42	Pea	22.5
2.	Moong and lentil	25	Arhar	21-25
3.	Urd	24	Gram	21.1

4.	Cowpea	23.4		
		In cereals		
5.	Bajra	11-12	Sorghum	10-12
6.	Wheat (**Gluten**)	11-12	Maize (**Zein**)	10
7.	Barley	11.5	Rice (**Oryzenin**)	6-7
		In Oilseeds		
8.	Ground nut	26		

Oil and poly-unsaturated fatty acid (PUFA) in oilseeds

S. N.	Crops	Oil content	Oleic acid (%)	Linoleic acid (%)	Linolenic acid (%)
1.	Oil palm	**60**	39	10	-
2.	Sesamum	46-52	**47.3**	39.3	-
3.	Mustard	46-48	12	24.3	12.7
4.	Ground nut	45	71.4	11.1	-
5.	Linseed	38	19.5	24.1	47.4
6.	Rape seed	37	9.9	13.5	9.8
7.	Safflower	33	12.4	**78.5**	<3.0
8.	Sunflower	29	19.1	67	<0.05
9.	Soybean	20	24	49	8

CEREAL CROPS

RICE / PADDY/ DHAN

Scientific nam : *Oryza sativa*
Family : Gramineae
Sub family : Oryzineae
Origin : Asia (South – Eastern) Indo – Myanmar region

Importance of rice as

- Rice is the **staple food** of India.
- Milling process of paddy reduces vitamin B complex content.
- The lemma & palea together known as **Hull.**
- Outer glumes of spikelet known as **Floret.**
- Promote of plumule (shoot) by Coleoptile.
- Promote of radical (root) by Coleorhiza.

- **Short Day Plant**
- **Semi aquatic** plant.
- Favourable soil P^{H} for paddy cultivation is 5.5 to 6.5.
- Cardinal temperature for paddy is 30-32°c.
- Dwarfing gene of rice is **Dee Gee Woo gene.**
- It is a selection from **TN-1 variety by Dr. T. I. Chong at IRRI Manila, Taiwan.**
- Highest productivity of rice in world - **Japan (58 q/ha),** India (17.5 q/ha).
- Field experiment on rice growth was done by **Matsushima.**
- Hybrid variety of paddy developed by which mechanism- **CMS (Cytoplasmic Male sterile Line).**
- Approximately 500 litre of water required to produce 1.0 kg rice grain.
- **Major/mimicry weed** in paddy is *Echinochloa colonum* (wild rice) and *Eichinocloa crusgulli* (Barnyard Grass).
- Most widely used post emergence herbicide in paddy is Propanil.
- **Khaira disease** occurs due to Zn deficiency.
  - Its occurrence in **nursery seedlins**
  - It was first time observed by **Y. L. Nene (1966)** at Pantnagar
- Oxygen transport in sub-merged soil through **Aerenchymataus cells.**
- Most suitable crop rotation in paddy is Paddy - Potato - Wheat
- Indian Institute of Rice Research located at Hyderabad (Rajendera Nagar)
- Number of hybrid varieties in paddy is 12.
- Rice have only two cultivated species out of 24
  1. Oryza sativa
  2. Oryza glaberrima

Cultivated rice species and their characters

S. No.	Species	Climate	Growing country
1.	*Oryza sativa sub. indica*	Tropical	India
2.	*Oryza sativa japonica*	*Temperate*	Japan, (No-lodging)
3.	*Oryza sativa javanica*	Tropical	Indonesia
4.	*Oryza glaberrima*	Tropical	S. Africa

Sowing time of paddy in India

Season	Local name	Sowing time	Harvesting time
Kharif	Aus/Autum (West Bengal, Bihar)	May –Jun	Sep - Oct
Rabi	Aman (Aghani), Dalua (Orissa)	June-July	Nov - Dec
Spring	Boro (West Bengal)	Nov-Dec	March–April

Methods of sowing

- In upland area - Direct seed (100-105 days)
- In low-land area - Transplanting (130-135 days)

Seed rate:

A. Under dry/ rainfed upland conditions:

Method		Seed rate (kg/ ha)
Broadcasting	:	100
Drilling	:	60

B. Under wet /lowland conditions:

Method	Grain size	Seed rate (kg/ ha)
Transplanting	Bold	40-45
	Fine	30-35

Required area under nursery sowing

Time of sowing/ method		Area (m^2/ ha)
Normal sowing	:	500
Late sowing	:	750-1000
Dapog method	:	30-40

Days required for transplanting

Time of sowing/ method		DAS
Wet bed method (days)	:	20-25
Depog method	:	11-14
Short duration varieties	:	21-28
Long duration varieties	:	28-35

- No. of leaf /plant required for transplanting - 3-4.

About Dapog method

- Dapog method introduced from **Philippines.**
- Dapog method follows in Andhra Pradesh state.
- 3-4 seedling seedlings/hill required for transplanting.
- Puddling is most important practice in paddy.
- Seed rate - **1.5 kg/m^2 (45-60 kg/ha).**
- Depth of ploughing by MB plough is 25-30 cm.
- Optimum population (hills/m^2) is 50.

- Best fertilizer for rice quality production is Ammonium fertilizer.
- Application zone of fertilizer in rice is **reduced zone (3-4 cm).**
- Weed growth suppressed because of flood condition.
- Flooding increase the availability of **P, K, Ca, Fe and silica** nutrients.

Varieties of rice and their characters

S.No.	Varieties	Remarks
1.	Jaya	Miracle variety in India, Parents - TN -1 x T- 141 It was Developed by **Dr. Shastry** (1966).
2.	IR-8	First high yield dwarfest variety of the world, Parents –Dee Gee Woo gene x Pete), made by **Henry Beachell** (1966) at IRRI, Salinity tolerant
3.	Pusa Basmati-1	World first high yielding dwarf variety under quality rice (IARI)
4.	Padma	It is a back crossed variety of Jaya
5.	Jaganath	**Mutant variety** its parent T-141
6.	Phalguna	Iron toxicity variety under acid soil
7.	PHB -71	Only single hybrid variety of private sector
8.	Bala	Suitable variety for direct seeding
9.	TKM-6	BLB tolerant
10.	Lunishree	Salinity tolerant
11.	Jalmagna	Suitable variety under waterlogging condition (3-4 cm)
12.	PRH-10	First hybrid variety of basmati rice in the world
13.	TN-1	Introduced in 1964 -65 from IRRI by G.V. Chalam with help of 2.0 kg seed
14.	APHR-1	**First hybrid variety** of rice in 1994 at Hydrabad

- Javanica rice is also known as **blue rice.**
- **Golden rice** – it is transgenic rice, rich in beta carotene (vit. A).
- Aroma in basmati rice is due to **di-acetyl-1-propaline.**
- Methane gas (CH_4) is emited in low land rice field
- Browning in rice due to iron toxicity.
- Akoichi disease in rice is due to sulphide toxicity (H_2S).
- Harvest index of rice is **40%**

Processing of rice

- **Milling** – removing of aleurone layer and embryo during polishing, it cuased the vit. B loss particulary thiamine.

- **Polishing –** removing of very fine bran clinging to milled rice and improve the smoothing of surface
- **Parboiling –** it is pre milling hydrothermal treatment is given to unhulled rice, it improve the quality of rice as nutritive value and cooking.

WHEAT

Scientific name	:	*Triticum spp.*
Family	:	Graminae / Poaceae
Sub family	:	Triticineae
Origin	:	South – Western Asia (Asia Minor – Turkey)

Importance

- King of cereals
- Grain straw ratio is 1:1.5
- Central ziz-zag of axis is called **Rachis.**
- **Lemma -** Outer bracts of floret which closed the flower.
- **Palea -** Inner bracts of floret is called palea.
- Seed consist (%)

  Starchy endosperm - 83-87

  Germ - 2.5

  Bran - 14
- **Germ** is consist of embryo (radical and plumule)
- **Bran** is consist of three organelles are as Pericarp, Testa and Aleuron layer.
- Bran is used in dairy & poultry feed -
- Type of photoperiodism – Long Day Plant
- Bread Wheat was developed by **Dr. N. E. Borlaug** (Father of green revolution in world).
- Dwarfing gene of Wheat is **Norin-10** and it was developed by **S.C. Salman (USA, 1948).**
- Norin gene was discovered by Dr. N. E. Borlaug
- Sonalika, Kalyan Sona are late sowing & bold grain varieties.
- Kalyan Sona and Sonalika are susceptible to Rust and Karnal bunt respectively
- Source of dwarf gene in

  Japan - Norin

  Tibet - Tom Thumb

  S. Rhodesia - Olsen dwarf
- **Triticale** (First man made cereal) developed by **Rimpu.**

- **DT-46** is the variety of triticale and its parent is **Wheat X Rye**.
- Most of Indian varieties have which gene - Rht -1 and Rht-2.
- Moisture content for harvesting 25-30% and for best storage (10-12%).
- When seed is dropped by hand in furrow is called **Kera method**.
- When seed is dropped by nai/orani in furrow is called **Pora method.**
- **Perling index** – it is used for determine the hardiness of grain (higher value of PI)
- **Harvest index** of wheat is 40-45%
- Polyploids series of wheat

  Diploid - Einkorn

  Tetraploid - Einkorn

  Hexaploid - Dinkale

Wheat species and their characters:

S.No.	Species	Characters
1.	Triticum aestivum /vulgare	It is also known -Common/Bread wheat 2n = 42 (Hexaploid) 87% growing areas, **Chapatti making and bakery product**
2.	T. durum	Other name -Macaroni / durum 2n=28(Tetraploid), 12 % growing area **Suji making** (Sevya) Drought resistant
3.	T. diccoum	Emmer wheat/samba 1% growing area Best for **Uppuma making**

Optimum temperature

- for Germination - 20-25 °c
- for Ripening - 14-15 °c
- Best crop rotation with Wheat is Rice.
- Best mixed crop with wheat is Mustard
- Row spacing - 20-22.5 cm

Depth of sowing seed: sowing depth decided by coleoptile length

Particulars	Depth in cm
Normal seed	4-6
Tall variety	8-9
Triple dwarf	3-4

Time of sowing:

- Tall wheat – 2nd Fortnight October
- Dwarf - Third week of Novmber month

Requirement of seed:

Particulars	Seed rate (kg/ha)
Normal sowing	100
Late sowing/ Bold seeded	125
Dibbling	25 -30 (1-2 seed/hole)
Zero tillage	140-150 (20-25% more)

- Best zone for P & K placement is below the seed

Requirement of Zinc/hectare:

- Mixed with ploughing - 25-30 kg/ha
- Foliar spray - 5 kg/ha

Important wheat varieties and their characters

S. No.	Varieties	Remarks
1.	Lok-1	Suitable var. for late sowing
2.	Sujata	Suitable for dry land
3.	Raj-4037	Suitable for warm climate
4.	PBW-502	Suitable under irrigation
5.	DBW-17	Best for bread making
6.	Raj -3077	Late sowing /saline resistant
7.	Raj -3765	Late sowing, resistant to water lodging
8.	Raj -4120	Highest yielder(48-60q/ha), resistant to UG-99 rust race
9.	HD 2329	Resistant to PM
10.	Kalyan Sona	Susceptible to Rust
11.	Sonalika	Susceptible to Karnal bunt
12.	C-306	Tall variety, suitble for rainfed /restricted area
13.	Single gene var	Lerma Rojo – 64- A, Sonalika, Sujata
14.	Double gene	Kalyan Sona, Sonara-63 and 64, Janak, Arjun (HD-2009), HD-2329, Choti Lerma
15.	Triple gene	Hira, Moti, Lal Badshah, Jyoti, Sangam, Jawahar
16.	Mutant varieties	Sharbati, Sonara, Pusa Lerma

- Irrigation stages of wheat were discovered by **R.B.L. Bhardwaj.**

- Two most important irrigation stages are – **CRI (21 DAS) and late jointing (60-65).**
- ***Phalaris minior*** is the Mimicry weed of wheat.

Some facts about *Phalaris minor*:

- It produce number of seed/plant - 10,000-30,000
- Effective time of weed control - 30 -35 (DAS)
- *Phalaris minor* has become resistant to *Isoproturon* herbicide
- *Phalaris minor* is resistant to isoproturon, reported by - ***Dr. R. K. Malik*** in Haryana
- It is the first time in world that reported *Phalaris minor* resistant to **Isoproturon**
- Alternate herbicides (clodinafop, fenoxaoprop and sulfosulfuron) were recommended to tackle the resistant *P. minor* in 1997-98.

Irrigation stages of wheat

1. **CRI** – 21 DAS, it is most critical stage; this stage is given by R.B.L. Bhardwaj.
2. **Tillering** – 10-45 DAS
3. **Jointing stage** – 60-65 DAS
4. **Flowering (booting) stage** – 80-85 DAS (second most important stage of irrigation)
5. **Milking stage** – 100-105 DAS
6. **Dough stage** – 115-120 DAS

Diseases in wheat crop:

S. No.	Disease	Causal organism	Remark
1.	Brown/leaf rust	*Puccinia recondita tritici*	-
2.	Yellow/strip rust	*P.recondita striiform*	-
3.	Black/stem rust	*P. graminis tritici*	Epidemic form of rust
4.	Loose smut	*Ustilago nuda tritici*	Internally seed born
5.	Karnal bunt	*Neovossia indica*	Seed + soil born
6.	Powdery mildew	*Erysiphe grannulris tritici*	White powdery growth on leaves
7.	Flag smut	*Urocystis tritici*	
8.	Ear cockle	*Anguina tritici*	Caused by first plant parasitic nematode
9.	Tundu	*Anguina tritici + corynebacterium tritici*	-
10.	Alternate leaf blight	*Alternaria triticina*	-

Black Rust control:

- Grow resistant varieties - HD 2733, 2428, 2189, PBW – 343
- Avoid late sowing and its occurring time is last week of January
- Use less N_2 & more K
- Spray of Mancozed/Zenab

MAIZE / MAKKA

Scientific nam	:	*Zea mays* L.
Family	:	Gramineae
Origin	:	Central America and Mexico

Importance

- It is populary known as **Queen of cereals**.
- Adaptability and productivity is much more so called **Miracle crop.**
- **Tryptophan and lysin** essential amino acids are deficient in maize protein **zein.**
- Yellow colour of maize is due to **Zeaxanthin**.
- Pop corn maize has the **popping** quality.
- Maize is non tillering plant.
- **Heat and moisture contents** are most important factors for popping.
- Maize contain calcium mineral in high amount whereas, phosphorus in low amount.
- Classification of maize based on kernel was given by **Kipps (1959).**
- Male flower of maize is known as **Tassel.**
- Removal of male flower is known as **Detasseling.**
- Male flower are situated at Top end of stem as a terminal panicle (cluster).
- Style filament of female flower cluster is called **Silk.**
- Female inflorescence developed into **Ear/cob.**
- **Idea of hybrid maize** was first given by **G. H. Shull & E. M. East** (1910) by **single cross technique.**
- **Monoecious** in nature but it is a **cross pollinated** due to **Protandry.**
- It is the classical example of **C4 plants.**
- Day neutral plants.

Classification of maiz :

S. No.	Maize species	Scientific name	Remarks
1.	Pop corn	*Zea mays* var everta	Grains are hard corneous interior structure
2.	Flint corn	*Zea mays* var indurata	Most common growing maize in India
3.	Sweet corn	*Zea mays* var. saccharata	Sweetest
4.	Dent corn	*Zea mays* var indentata	Mostly grown in USA
5.	Soft corn	*Zea mays* var amylaceae	
6.	Pod corn	*Zea mays* var tunicata	Primitive type of maize
7.	Waxy corn	*Zea mays* var certina	Starch similar to tapioca starch for making adhesive for articles

Growing season of maize in India

S. No.	Location	Season	Sowing time
1.	North India	Kharif	Onset of monsoon(June)
2.	South India	Kharif Winter	April October

Seed rate

- **For composite var.** - 18-20 kg/ha (Kharif) and 25-30 kg/ha (Rabi).
- **Hybrids var.** – 20- 25 kg/ha.
- **For fodder purpose** – 40-50 kg/ha.

Planting distance and irrigation

- Spacing (60x20 cm).
- No. of plant/ha - 90,000 (Rabi) and 65000-70000 (Kharib).
- No. of irrigation required for production - 4-6.

  Temperature requirement for
  - Germination - 21^{0}c
  - Growth - 32^{0}c
- Depth of sowing seed- 3-5 cm.
- Maize is sensitive to Water logging.
- **White bud** of maize is due to deficiency of Zinc.
- Which is the common herbicide used in maize - Simazine, Lasso.
- **Pythium stalk** rot is caused due to over plant population (>75,000/ ha).
- Most serious pest of maize is Stem borer/Dead heart.

- Shoot fly infected mostly at seedling stage.
- White grub insect is most prevalence in sandy soil.
- Single cross technique is grown mostly in USA.
- Double cross technique in India.

Varieties

- Deccan, Ranjeet, Ganga1,2,3,4,5, Hi-starch, Vikram, Kisan, Amber Sona, Himalaya 1,2,3 and Vijay
- Highest yielding var - Amber (5-6 t/h) due to open pollination.
- Highest lysine rich var - Protina , Shakti, Rattan
- Lysine rich var is called - **Opaque -2 composite.**
- **QPM** stand for - Quality Protein Maize, by using **Opaque -2 gene**.
- **Varieties of QPM** - Shaktiman -1, 2 and HQPM-1.
- Rabi maize is only grown in Bihar state.

SORGHUM/ JAWAR

Scientific name : Sorghum vulgare

Family : Graminaeae

Origin : Africa

Importance

- Sorghum third major food grain crop of India.
- Sorghum plants can **tolerant high temperature** throughout their life cycle better than any other cereal crop.
- Its can also tolerant **water logging condition** better than any other field except rice.
- Highest drought & water logging resistant crops among field crop.
- Most efficient & exhaust nutrient & water uptake by sorghum roots.
- **No tillering** roots crop.
- Supporting roots of sorghum are called **Prop roots.**
- Upper most internode which bears the inflorescence is commonly known as **Neck.**
- **Leucine (7-17%)** is containing the highest amount among proteins in sorghum and **Lysine (1-3%)** with lowest amount.
- It is major crop of Kharif season in N- India. Whereas, in Rabi season in Western & Southern India.
- Optimum temperature for germination is 26-30 °c.
- 7-10^{0}c at least minimum temperature for growth is required.
- Photoperiod is **Short Day Plant.**

- CSH-6 hybrid var. of sorghum is suitable for mixed cropping.
- Sowing in row is common in Black cotton soil.
- Seed rate for seed purpose (12-15 kg/ha) and for fodder (40-50 kg/ha)
- Number of plants per hectare - 1,50,000
- Depth of seed sowing - 3-4 cm
- Sorghum is most sensitive to nutrient deficiency of **Iron.**
- Thinning is the most common practice required for cropping.

  Thinning is required twice (DAS)
  - First thinning at - 10-15
  - Second thinning at - 20-25
- **Composite varieties** - Swarna (CSV-1) CSV - 2, 3, 4, 5, 6, 10
- **Hybrid varieties** - CSH- 1,2,5,9,10,11,12,13,14
- Rabi season varieties - CSH -12R, CSH-13R
- Rajasthan chari-1, 2, 3 are fodder purpose varieties in Rajasthan.
- First hybrid of sorghum is **CSH -1**(1964)
- Male sterile variety is **Combine Kafi 60**
- Atrazine (500-1000 g/l) is most common herbicide in sorghum.
- Rust (*Puccinia purpurea*) is common problem in South India.
- Causal Organism of Leaf blight is *Heliminthosporium turcicum.*

BAJRA (Pearl millet)

Scientific name : Pennisetum glaucum/ americanum
Family : Gramineae
Origin : Africa

Importance

- Bajra is popularly known as **"Poor's man food"** because it rich nutrient and cheaper in price

S. No.	Nutrients	Per cent
1.	Protein	11.6
2.	Fat	5.0
3.	CHO	67
4.	Mineral	2.7

- Ancestor of cultivated bajra - ***Pennisetum purpureum***
- Tropical climate is required.
- Most **droughts tolerant** among cereal and millet but sensitive to water logging & acidic soil.

- Classification given in 5 sections by **Staph (1934).**
- Tillers arise from Basal nodes and that tillers roots are known as **Brace/prop roots.**
- No. of spilket/spike - 800-3000.
- Length of spike - 15-60 cm.
- Perfect flower is appearing first.
- Required rainfall - 40-75cm.
- Temperature required for Rabi crops in South India - 20-28^{0}c.
- Best inter cropping - black gram/ green gram.
- Optimum population (plant/ha) - 175000- 200000.
- Spacing - 45x10-15
- Sowing depth - 2-3 cm

Transplanting method of sowing

- Seed rate - 2.0 kg
- Area req. for nursery (m^2) - 500-600 m^2.
- Best sowing time - 20 July -10 Aug.
- First hybrid of bajra- **HB-1**(1965).
- Parent of HB- 1 - **Tift 23A(ms) x Bil 3B**
- Varieties suitable for Rajasthan - RHB-30, RCB-2, RAJ-171 PUSA- 605 and Pusa Moti.
- NPK (kg/ha) - 100-120:40-60 :30-40
- Foliar spray done under Barani condition.
- Pre-emergence herbicide *e.g.* – Atrazine and it used for both monocot & dicot weeds.
- **Ergot** – cuased by *Claviceps fusiformis*, it is seed born diease.

BARLEY / JAU

Scientific name	:	*Hordeum vulgare* L.
Family	:	Gramineae
Origin	:	Asia and Ethopia

Importance

- It is staple food of Nepal and Bhutan.
- In European countries used as Breakfast food.
- **Malt** is used for manufacturing of **Beer, Whisky and Brandy.**
- **Sattu** is a mixer of barley flour mixed in sugar and water.

- Barley can grow all climates except tropical.
- Barley classification was given by **Aberg & Wiebe (1946).**
- Based on number of row of grains & their arrangement
  - Six row Barley - *Hordeum vulgare*
  - Two row Barley - *Hordeum distichon, Hordeum irregulare*
- Fertile spikelet in six row barley - All three spikes fertile out of six.
- Barley is drought resistant and salt tolerant crop among all cereal.

Spacing and depth of barley

S. No.	Condition	Spacing (cm)	Depth (cm)
1.	Irrigated	22.5	4-5
2.	Rainfed	23-25	6-8

- Varieties - Haritma, Geetanjali, DWR-28, Gopi, Dolma, Munjula and Rekha.

MINOR MILLETS

Finger millet

Scientific name : *Eleusine coracana*

Family : Gramineae

Origin : India (all millet)

Importance

- It's also known as **Mandau/ Ragi**
- Used as Silage
- Medicinal used in Diabetes
- Growing season is kharif (all millet)
- Climate for all millets is tropical

Seed rate

- Under direct sowing - 8-10 kg/ha
- Under Transplanting - 4.0 kg/ ha

PROSO MILLET

Scientific name : Panicum miliaceum

Family : Gramineae

Origin : India

Importance

- It's also known as **Cheena**
- Used in **Kheer**
- Varieties *e.g.* - Sagar, Bhawna, Naga Arjun
- Seed rate - 8-12 kg
- Crop duration is 65-75 days.

BARNYARD MILLET

Scientific name	:	*Echinochloa frumentacea*
Family	:	Gramineae
Origin	:	India

Importance

- It's also known as **Sawan**
- Grain look like as which crop rice
- Used in making of Kheer
- It is resistant to drought
- Seed rate - 8-10 kg /ha

KODO

Scientific name	:	*Paspalum scrobiculatum*
Family	:	Gramineae
Origin	:	India

Importance

- Seed rate - 10-15 kg / ha
- **Highest drought resistance** among all millet

FOX TAIL MILLET

Scientific name	:	*Setaria italica*
Family	:	Poaceae
Origin	:	china

Importance

- It's also known as Fox tail/Italian/German millet.
- It is not good for grazing of Horse.
- Seed rate- 8-10 kg/ha.

BUCK WHEAT

Scientific name	:	*Fagopyrum esculentum*
Family	:	Polygonaceae
Origin	:	Southwest China

Importance

- It also known as **Pseudo cereal.**
- Despite the name, buckwheat is not related to wheat, as it is not a grass.
- Fruit type is Achene.
- Variety - **Tokyo**
- Leave & flower contain (alkaloid) - **Ruti**

PULSE CROPS

- Pulse is a word of **French** language **Pottage/Poridge** means **Soup/dal.**
- Legume is a word of **Latin** language **Legne** means **Collect/Gather.**
- Bacterial nodules found on Secondary root.
- Low production of legume is due to less development of sink.
- Filling of pod acoording to **Hydrodynamic** model.
- Growth pattern of legume is indeterminate type
- **Type of germination**
  - All Kharif legumes have *epigeal germination* (except arhar)
  - All Rabi legumes have *hypogeal germination* (except rajmash)
- **Cowpea** is the rainfall tolerant pulse.
- **Gaur/cluster bean** is the Gum manufacturing pulse
- Chickpea is the oldest pulse.
- **Chromosome number**
  - In all Kharif pulses - 2n=22 except Horse gram/kulthi (2n= 24)
  - In all Rabi pulses - 2n =14 except Kabuli chickpea (2n =14, 16)

CHICKPEA

Scientific name	:	*Cicer arietinum* (roman word)
Family	:	Leguminosceae
Origin	:	South – Eastern Turkey and adjoining Syria (South-Western Asia)

Importance

- Synomous name of chickpea are Bengal gram and king of gram.
- Product made as vegetable - Chhole.
- It is a C_3 and long day plant.
- Seed used to make flour (besan) and sweets
- Medicinal use as blood purifier
- India rank in production first followed by Pakistan

Scientific name and number of chromosomes

- **Desi /brown gram** - *Cicer arietinum* / 2n=14, 16 (most widely grown, short and highly branched, seeds are small and dark brown)
- **Kabuli** - *Cicer kabulium* -2n= 16 (taller plant with few branches, seeds are bold and white)
- Wilt resistance variety - **Avrodhi** and **Sadabahar**
- Suitable variety for late planting - Uday (KP-750)
- **RS -11**is a drought resistance variety.
- **Seed rate** – Desi – 60-80 (kg/ha) and kabuli – 80-100 (kg/ha).
- Seed sowing depth should be kept dipper around 8-10 cm to escape wilt problem.
- Sour test of leaf due to **Mallic acid (90-94%) + Oxalic acid (4-10%).**
- **PUSA -256** is the best var under Rainfed condition.
- Ascochyta blight resistant variety is Gaurav.
- Radhey is a late sowing variety.
- RS-11is a mutant variety.
- **Aparna** (2006) cultivar is having seed remain green color.
- C-235is a wildly adopted var in dry-land condition.
- Early & shallow sowing crops are more liable to wilt.
- RS-10 is a suitable for Kota and Ajmer division.
- Early sowing suffers to wilt and it occurs due to high temperature.
- Mostly chickpea grown as a rainfed crop.

Nipping /Topping:

- It is important practice used for high production.
- It's comes between **45 DAS.**
- Grazing by sheep's flock is general practice for nipping.
- Chemical **TIBA @ 75 ppm** used for nipping.
- Gram is highly susceptible to frost at flowering stages.

FIELD PEA

Scientific name : *Pisum sativum*

Family : Leguminosceae

Origin : Mediterranean region of southern Europe and Western Asia.

Importance

- Scientific name of Field pea - ***Pisum sativum var arvense***

Features of field pea:

- Ripe mature seeds are used as the pulse (DAL)
- Forage or green manure purpose
- Rounded & angular seed
- Coloured flower colour
- Frost resistant
- Scientific name of Garden pea- ***Pisum sativum var hortense***

Features of garden pea:

- Canning purpose
- Wrinkle seed
- White flower colour._
- Germination type is a hypogeal.
- Pod contain number of seeds are 4 to 9.
- Length of the pod is 5-9 cm.
- Optimum temperature for germination is 22^{0}c.

Sowing time and seed rate

S. No.	Sowing time	Spacing (PxR cm)	Seed rate (kg/ha)
1.	Early sowing	20 (closer)	100- 125
2.	Late sowing	30 (wider)	75-80

- Suitable varieties for Rajasthan - Arpana, HUP-2, Rachna
- Aparna – first hybris of field pea
- Salvia – edible pod variety
- Arkel – early maturity and introduced from UK.
- Tendrometer is used to determine pea maturity (reading value – 105-120)
- **Powdery mildew** is the serious disease in pea cultivation as biotic stress. Whereas, **frost** as abiotic stress.

PIGEON PEA

Scientific name : *Cajanus cajan* L.

Family : Leguminoceae

Origin : According to Vavilon (1951)

- India – Pennisular part
- Katjan – Malyalam word

Importance

- It's also known as **Red gram/Arhar**
- **Most drought tolerant** crop among all pulse
- Second most important pulse in India

It is grouped into two types

1. ***Cajanus indicus* var bicolor**

   Features:
   - Late maturing
   - Perennial in nature
   - 4-5 grain/pod and long pod
   - Commonly grown in N-India

2. ***Cajanus indicus* var flavu**

   Features:
   - Annul plant
   - short and early maturing
   - 2-3 grain /pod
   - Seed rate – 12-15 kg/ha
   - Spacing - 60 x15 cm
   - Optimum plant population per hectare is 55-66 thousand.
   - **Prabhat** (110-120 days) is a extra early maturing variety.
   - **ICPH-8** - First hybrid of the world developed by ICRISAT, Hydrabad (1991).
   - Wilt resistant variety - Mukta (150-180 medium).
   - Long duration variety - Pusa-9 (180-300 days).
   - Varieties for Rajasthan - Manak, UPAS-120, Pusa-33, 955.
   - Other varieties - Amar /Azad, Paras, Laxmi, Pusa Gati, Bahar.
   - **Harvest index** is 19% (lowest among pulse crops)

GREEN GRAM

- It is also known as **green gram/ golden gram.**

  Scientific name : *Phaseolus aureus*

  Family : Leguminosceae

  Origin : India

Time of sowing and seed rate

S. No.	Season	Seed rate kg/ha	Sowing time
1.	Kharif	12-15	20 June -10 July
2.	Rabi	20	15 October - 10 November
3.	Summer	20	15 March -15 April

- Yellow mosaic virus is a viral disease and Its resistant varieties are Asha, PDM -11, 54 and ML-337
- Suitable varieties for Rajasthan state- RS-4 ,R-288, RMG-62 and Pant M-2
- **Pusa Vishal** is uniform maturity and suitable variety for both summer/kharif seasons.
- **Kopergoan** is a bold seed & green shining variety.
- RS-4 is a suitable variety for Udaipur / Kota region.

BLACK GRAM

Scientific name : *Vigna mungo*

Family : Leguminosceae

Origin : India

Importance

- Byproducts of Urd bean are **Idli, Dosa, Halwa, Imarti, Paped, Bari (spide ball)**
- Climate - Tropical (Hot +Humid)
- Classification Urd bean was given by **Bose (1932)**
- **Seed rate** -12-15 kg/ha
- **T-9** is first genotype as cultivar in the black gram history.
- **Barkha (T-44)** – variety develop from Rajasthan.

1. ***Vigna mungo var niger***

   Features:

   - Short stature
   - Early maturing
   - Seed bold and black in colour

2. ***Vigna mungo var viridis***

   Features:
   - Large stature and late maturing
   - Seed small size and green in colour

COWPEA

Scientific name : *Vigna sinensis* L.
Family : Leguminoseae
Origin : Central Africa

Importance

- It is also known as **Lobia/Indian bean and vegetable meat**
- Classification of cowpea given by **Piper**

1. ***Vigna sinensis var catjang***
   - Indian cowpea (7-23cm)
2. ***Vigna sinensis var sesquipedalis***
   - 30-60cm
   - Climate - Tropical
   - Seed rate - 20-25 kg/ha for vegetable and grain purpose
   - 35-40 kg/ha for green manuring
   - **Pusa Phalguni**, V-240 and V-16 are suitable varieties for Rajasthan.
   - Important varieties for both seasons are **Pusa Phalguni, Pusa Barsati, Pusa Dofasli and Pusa Rituraj**.
   - **Sirsa -10** is a fodder purpose variety.
   - Hairy caterpillar is the most serious pest of cowpea.
   - Cowpea crop duration is 60 -70 days.

OIL SEED CROPS

SOYBEAN

Scientific name : *Glycin max* L. (Progenitor of this is *Glycine ussuriensis*).
Family : Leguminoseae
Origin : North China

Importance

- It is also known as Wonder **crop/yellow jewel**.
- Soybean byproducts are **Milk, vanspati ghee, sweets and bread making**.

- Soyabean have the first rank in vegatable oil production at world level.
- It provides vegetable protein & oil percent are 40 and 20%, respectively.
- **Lysine** is most abundant amino acid in soybean.
- Vitamin is found 5.0%
- Number of seeds/pod is 2-3.
- Sowing depth of seed is 3-4 cm.
- Optimum number of plants/hactare - 3, 00,000 - 4, 00,000.
- Spacing – Row to row 45-60 cm and plant to plant 15-20 cm.
- Self pollinated and compeletly self fertile.
- Photoperiod – Short Day Plant.
- Seed treatment is done with *Rhizobium japonicum* @ 500 gm/70 kg seed.
- June planting suffers more with Yellow Mosiac Virus due to longer period to maturity.
- **Girdle beetal** and Shoot fly are serious pest.
- 13-14% moisture content is required for threshing.
- Crop duration (days) - 90-120

Varieties

- Varieties introduced from USA - Bragg, Lee, Kent, Clark-63, Black Heart
- **Improved Pelicon** is a suitable for South zone
- Ankur, Alankar & Sheelajit variteis developed from GBPUAT, Pantnagar
- Varieties suitable for Rajasthan - Bragg, Durga, Guarav, Ahiliya -3 and Monetta
- **Pooja and Shivalik** are resistant variety to Yellow Mosiac Virus.
- Ankur is a Intermediate habit variety.

GROUNDNUT

Scientific name : *Arachis hypogeae* (Greek word)
Family : Leguminosceae
Origin : Brazil or South America

Importance

- The biological value of ground nut protein is highest among the vegetable protein & equal to casein protein

Classification of groundnut

- Erect /bunch type - *Achrus Arachis hypogea var **fastigaita***
- Trailing/spreading type - *Achrus Arachis hypogea var **procumbens***

- *Achrus Arachis hypogea **var oleifera*** is a most cultivated species.
- Peg is also known as **Gynophore.**
- Spreading type ground nut is having dormancy.
- **Cotyledon** is an oil containing part.
- Covering of seed /nut by **Testa**
- Spreading type of groundnut is a late maturity.
- **Arachis word means Legume** and it was coined by Greek language
- Hypogeal is a Greek word means **below ground.**
- **Ammonium Sulphate** is the best fertilizer for N_2 application and SSP for Phosphorus supply.
- **Popping/hollow heart** – pod without kernels due to **calcium deficienc** ; it is managed by application gypsum @ 250 kg/ha at **peak flowering stage.**
- **Aflatoxi** caused by ***Aspergillus flavus*** (high moisture containing seed). Therefore, seeds moisture content should be kept < 8%
- Pegging starts after 55 DAS.
- Important variety for bunch type - Jyoti, Khushal, Phule Pragati
- Spreading variety - Chandra, RS-1, Amber and Prakash
- Suitable variety for Rajasthan- RS-1, RSB-103-87, especially for Kota division (heavy soil)
- Semi spreading variety - Chitra, RSB-103-87 and Vikram
- Optimum temperature for nut production is 21-26.5 ^{0}c
- Suitable soil is sandy or sandy loam with p^H range 6.0-6.5.
- Shelling percent in ground nut is 70-75%.

Spacing and seed rate as per type of groundnut species

Type	Spacing	Seed rate kg/ha	No. plant/ha	Yield q/ha
Spreading	45-60x20	60-80	1.11 lakh	20-30
Bunch	30-40x15	80-100	2.22 Lakh	15-20

- Non filling of pod (without kernel)/ popping due to **Calcium** deficiency. It is control with application of Gypsum @ 120-125 kg/ha at pre flowering stage.
- Herbicide used in groundnut is **Nitrofen (TOK-E -25) Banned**
- **Earthing up** is a most common practice in groundnut or pea nut -
- **Tikka** is a most common disease in groundnut.

SESAMUM /TIL

Scientific name : *Sesamum indicum*

Family : Pedaliaceae

Origin : South – West Africa

- It is major substitute for **Poor man's butter (Ghee)**
- Calories/100gm is 592
- NPK content in cake is 6:2:1 per cent
- Sandy is a best soil for cultivation.
- Two type of varieties - **White seeded, black seeded**
- Suitable variety Rajasthan - Pratap (especially for Kota), RT-46, TC-25
- Optimum temperature - 25 -27°c
- It is the best mixed cropping with groundnut
- **Phyllody** is a most destructive disease for sesamum cultivation.

CASTOR/ ARANDI

Scientific name : Ricinus communis

Family : Euphorbiaceae

Origin : Ethiopia

Importance

- It is an important industrial oil seed crops especially used in **Aeroplanes**
- Medicine use as **Purgative**
- **Ricin** is a poisonous chemical makes it unfit for cattle feed.
- Important varieties - Aruna, Bhagya, Saubhagya, Rosy (TN), Dwarf mutant and Girija
- Suitable var for Rajasthan - Aruna, Saubhagya
- Important practice is a **Nipping**
- Jassid is a regular & serious pest of castor.

MUSTARD

Scientific name : *Brassica spp.*

Family : Cruciferae (Brassicaceae)

Origin : China (Asia Minor)

Importance

- Mustard is the major rabi oilseed crop in India
- Rajasthan is a leading state in production.
- Rape seed also called **Sarson /toria/lahi.**
- Mustard is called as **Rai/ Raya /Lohi.**
- It is used as Greases and **Tanning industries to soften the leather**
- Oil is rich in **Sulphur** mineral.

- Pungency in oil is due to **Isothiocynate** and it cause disease **Goiter** (affect the growth & development so its limit should be having **< 0.5 % in edible oil.**

Botanical name and Chromosome Number

S. No.	Common name		Scientific nam	2n
1.	Sarson	:	*Brassica compestris var sarson*	20
2.	Toria	:	*Brassica compestris var toria*	20
3.	Rai	:	*Brassica compestris var juncea*	36
4.	White mustard	:	*Brassica compestris var alba/hirta*	
5.	Tara mira	:	*Eruca sativa*	22
6.	Indian mustard	:	*Brassica compestris var juncea*	
7.	Black mustard	:	*Brassica compestris var* nigra	16

- Climate - Cool and dry
- First var of *Brassica juncea* was developed **Pusa Jai Kisan (Bio - 902)** and it was developed by **Dr. V. L. Chopra.**
- Mixed cropping is taking in UP (40 state.

Varieties

- Toria - Sangam, Bhawani, Panchali
- Rape seed - Pusa Kalyani (brown), Vinod (Yellow)
- Mustard (rai/raya) - Sita, Bhagirathi, Varuna, Shekhar, Krishna, Vaibhav, Vardan, Rohini, Durgamani, Saurabh, Laxmi, Pusa Bahar, Pusa Jai Kisan, Rajat, Pusa Bold and Kranti
- New released varieties_- Laxmi, Vasundhra and Ashirwad ,
- Laxmi is the highest yielder varieties.
- Bio -902 variety was developed by **Somatic hybridization.**

LINSEED/ALSI

Scientific name : *Linum usitatissimum* L.
Family : Linaceae
Origin : Mediterranean region (Turkey, Afghanistan and India).

Importance

- It is an important oilseed and fiber crops
- It is also known as **Flax.**
- **Linen is a** product of linseed.
- **Capsule/seed ball** is a type of fruit.
- It is a cool season crop and Long Day Plant.

- It is having 85 per cent digestible protein.
- Small seeded type grown for oil and bold seeded for fiber production.
- Varieties - Hira, Neelam, Mukta, Padmini, Garima, Kiran, Chambal, Jeevan and Ngarkot
- Varieties suitable for Rajasthan - Chambal, Karan , sheetal , Jeevan, Pusa -2, 3
- Dual purpose (oil and fiber) varieties - Jeevan, Nagarkot and Gaurav
- When it is sown as mixed cropping then it is called **Paira crops** because it is broadcasted in standing paddy
- Its cultivation called in MP and Bihar as **Utera and Paira,** respectively.
- Most serious disease is Rust (*Melampsora lini*).

SAFFLOWER/KAUSUM

Scientific name : *Carthamus tinctorius* L.
Family : Compositeae
Origin : India, Afghanisthan and Ethiopia

Importance

- Fruit type - **Achene**
- Acid - **Linoleic acid (78%)**
- Medicinal used for **heart disease.**
- Industrial use - **Roghan** (Leather & water proof cloths)
- Season - Cool
- Drought resistant crop
- Variety - **Manjira**

SUN FLOWER/SURAJMUKHI

Scientific name : *Helianthus annuus* L. (Greek word)
Family : Compositeae / Asteraceae
Origin : Southern United States and Mexico

Importance

- It is a **familiar and Ornamental** plant in India.
- It is ideal as Catch crops.
- Linoleic (64%).
- Medicinal use for heart patient.
- High temperature is decreased linoleic acid content.
- It is a **photo insensitive** plant so grown every season.

Crop duration (days)

- Kharif - 80-90
- Rabi - 105-130
- Spring - 100-110

- Varieties - Divyamukhi, Badshah , Perodovic, Surya and Sunrise

FIBRE CROPS

COTTON/ KAPAS

Scientific name	:	*Gossypium spp.*
Family	:	Malvaceae
Origin	:	India

Importance

- Most important fiber crops playing a key role in economic & social affair in whole the world
- Genetic traits - Fiber length /fineness
- Fiber maturity is measured by **Arealometer.**
- Fiber thickness is measured by **Nepiness.**
- Fineness of fiber is measured by **Micronaire.**
- Unit of fineness is a **Microgram.**
- Toxic chemical is a **Gossypol.**
- American is a higher oil containing group.
- Cotton classification was given by **Hutchinson (1947).**
- One bale of cotton weight is **170 Kg (1 bale of Jute - 180 Kg)**
- Seed cotton consist = seed **+ lint**, cotton seed = seed only (after removing lint)
- Flower inducing chemical - Planofix (a-NAA), Cycocel @10 ppm Application after 80-90 (DAS)
- Bad opening of ball is called **Tirak** and it is remarked by **Prof. Dastur**.
- Cotton is known as world wide as **White gold /king of appraisal fibe** .
- Growing area is 8.4 mha & 86% covered by Bt. cotton in India.
- Cultivated cotton is annual
- **Monopodial** - Vegetative branches of cotton are called monopodial.
- **Sympodial** - Reproductive branches.
- **Capsule/boll** - Three to five locule fruit is called capsule.
- **Lint** - The long out growth form of fiber is also called staple.
- **Fuzz** - Shorter out growth form of fiber is called fuzz.
- **Topping** - Removal of axillary branches of plant to encourage sympodial branching or diversion of energy flow for ball development.
- **Ginning** - Separation of fiber from seed cotton is known as ginning.
- Ginning % - 30-35%

- True appearance of flower bud in cotton is known as square.
- Optimum temperature for cultivation - 21-27 ºc
- Minimum temp. for boll opening - 16ºc
- Number of seed /ball - **24-50**

Soil requirement

- Rainfed cropping - Black
- Irrigated cropping- Alluvial
- Seed treatment for defuzzing by H_2SO_4.
- Seed/acid ratio for delinting/defuzzing treatment - 10:1

Ultra narrow method of planting

- Spacing - 19x19cm^2
- Number of plant/ha – 2,77,000
- It was developed by USA
- **Diuron is a common horbicide used in cotton.**
- Cotton is highly sensitive to **2,4-D/Glyphosate.**
- It injured by Spray drift
- 2,4-D resistant cotton gene is a **tfd –A.**
- American cotton is susceptible to Red leaf.
- Crinkle leaf of cotton is caused by **Mn toxicity.**
- Usually two or more jointed seeds show affected by pink bollworm.
- Fiber (lint) length (cm) of Sea Island (Sudan) cotton is 3.6 - 5.0.

Cotton group	B. Name	Ploidy level (2n =)	Growing area (%)	Seed rate (kg/ha)	Spacing (cm)	Yield (q/ ha)	Fiber (lint) length (cm)
Desi	*G. arboreaum* *G. herbaceum*	Diploid (26)	16 8	10- 18	60x30	15-20	1.25-2.10
Amrican	*G. hirsutum* *G. barbadense*	Tetraploid (52)	36 0.2	15-25	60x45	20-25	1.8-3.10
Hybrid			40	2.5-3.0	120-150x60 50,000-80,000 (Aug-66000)	25-30	

- Depth of sowing seed is a 4-5cm
- Gap filling should be done within 7-10 days.

Important practices

- **Thinning** - light interpretation and proper plant development.
- **Topping** - 1.0-1.2 m in height, it done 80 - 90 DAS.

Varieties of cotton

S. No.	Cotton species	Varieties
1.	*G. arboreaum*	Lohit
2.	*G. hirsutum*	Ganganagar Agati, Bikaneri Nerma, Laxmi
3.	*G. herbaceum*	Diyvijay, Sanjay
4.	*G. barbadense*	Sujata
5.	Hybrid cotton	H-4, Vira Laxmi, Maru Vikas, Surya, Savita

- **First hybrid variety** of cotton in the world is **H-4 (Intra – hirsutum cross) and** it developed by **Dr. C.T. Patel** at GAU, Surat (1970)
- It's parent - **G-67 X American Neetorless**
- **First interspecifi hybrid** variety is a **Varlaxmi** and it's Parents (Hirsutum x Barbadence) at UAS (Dharward).
- MCO -5 is an extra/early stable variety.
- **Maru Vikas (RAJ HH -16)** is the first hybrid of hirsutum cotton in Rajasthan.

JUTE

Scientific name : *Corchorus spp.*
Family : Tiliaceae
Origin : *C.olitorius* Africa , *C. capsularis* Indo-Burma

Importance

- 70 per cent world production is contributing by Both India & Bangladesh.
- Jute is an important fiber crop next to cotton.
- Fiber source in jute is a **Phloem.**
- Leaf bitter taste is due to **Corchorin**.
  - **Tita pat** - Bitter taste leaf (White leaf)
  - **Mitha pat** - Tasteless leaf (Red /yellow)
- **Dazed** - Over retting fiber
- **Croppy** - Under retting fiber
- **Mesta** is fiber crop which grown where jute could not grow.

Commercial sp. of jute and about their

S. No	Scientific nam	Trade Name	Sowing Time	Seed Rate(kg/ha)
1.	*Corchorus capsularis*	White Jute	March- Aril	6-8
2.	*Corchorus olitorius*	Tossa jute	April –May	4-5

- Harvesting stage - **Small pod stage** (130-140 DAS).
- **Retting** – it is a microbial process in which aerobic and anaerobic bacteria and fungi loosen the fiber by decomposing & dissolving the pectin, hemicelluloses and other cementing agents is called retting.
  - Duration of retting (Days) - 8-30
  - Water quality - Pure and clear water required for best retting
  - Best **retting temperature** - 34°c
  - Retting required relative humidity (%) - 100
- Weak fiber (dazed fiber) occurs due to over retting.
- Practice of retting in China is also known as **ribboning.**
- Steeping means soaking of water.

SUNHEMP

Scientific name . *Crotalaria juncea* L.
Family : Leguminosceae
Origin : India

Importance

- Climate - Warm season
- Harvesting stage at **Pod formation**
- Optimum temperature for retting - 21-27°c

Varieties

- White type – Baldev and Joydev,
- Tossa – Basudev and Mahadev
- JRO-875 is a pre mature/flowering resistant variety.

Fodder Crops

BARSEEM

Scientific name : *Trifolium alexandrium* L.
Family : Leguminosceae
Origin : Egypt

Importance

- Berseem is also known as **Egyption clover**.
- Most important varieties are BL-1, Miscavi, Khadarvi
- **Pusa Gaint** is an **autoteteraploid** var and released from IARI, New Delhi.

Harvesting

- First harvesting at 50-55 DAS
- Second at 25-35 after first cutting

LUCERNE

Scientific name : *Medicago sativa* L.
Family : Leguminosceae
Origin : Iran

Importance

- Most important fodder crops of India.
- It is an **oldest cultivated fodder crops** in the world.
- It is also known as **Alfalfa.**
- **Persia (Iran)** is a founder cultivated country of Lucerne in the world.
- Digestible protein - 72%.
- Leaves are trifoliate.
- ***Rhizobium melilotii*** culture is used to inoculation of lucerne.
- Sowing depth is 1-2 cm
- **Dodder (Cuscuta) is an** associate weed of lucerne.
- Bacterial Wilt is a most serious disease, cuased by *Aplanobactor insidiosum.*

Harvesting

- First cutting (DAS) - 50-60
- Next cutting (interval days) - 20-30

Varieties

- Annual - Sirsa No-8
- Perennial - Sirsa No-09
- Suitable variety for Hilly areas - **Rambler** (grown in Canada)

NAPIER GRASS (Elephant grass)

Scientific name : *Pennisetum purpureum*
Family : Gramine
Origin : S. Africa (Rhodesia)

Importance

- Variety - Pusa Gaint
- Sowing time - March

Cash Crops

SUGARCANE / GANNA

Scientific name : *Saccharum spp* (*officinarum*)
Family : Gramineae
Origin : India

Importance

- Main source of sugar in the India.
- Saccharum is **Sanskrit** word ***Sarkara* means - Sugar**.
- Tillering is reduced by low temperature.
- Scientific name of wild cane is a *Saccharum spontaneum.*
- Nobel cane was developed by **T.S Vankataraman**.
- Inter cropping with Potato and Moong
- Best propagation portion of set is 1/3 to1/2 top portion
- Bottom portion is not used for propagation becuase it is rich in sugar so takes a long time in germination that's why it is not used and therefore, bottom portion is used to make Jaggery.
- Combination of **Gasohol product** - Petrol & Alcohol (**Ratio 80:20**), used in Auto mobile
- Number of sugar mills in India - 400 {UP (109), MH (90)}

Sugarcane used as in (%)

- White sugar - 50%
- Jaggery - 30%
- Alcohol - 20 %

Product and it's by products

S. No.	Product	It's by products
1.	Juice	White sugar, Brown sugar (khandsari), Jaggery (Gur)
2.	Bagasse	Fuel, Board paper, Plastic
3.	Molasses	Alcohol, Rum

Classification of sugarcane

S. No.	Group	B. Name	Chromosome No.	Remark
1.	Tropical cane/ Noble cane/ Thick cane	*Saccharum officinarum*	2n =80	Good for chewing, high sugar content
2.	Indian cane	*Saccharum barberi*	2n =82-124	Dwarf and early maturity cane. Developed by **Dr. C. A. Barbar**
3.	Chinese cane	*Saccharum sinense*	2n =118	Medium sucrose, high fiber content

Ideal condition for tillering:

- Temperature - 25-30°c
- Relative Humidity - 75-90 %

Institutes related to sugarcane in India

S. No.	Institute	Full Name	Suited at
1.	SBI	Sugarcane Breeding Institute	Coimbatore, TN (1912)
2.	ISI	Indian Sugar Institute	Kanpur, UP
3.	IISR	Indian Institute Sugarcane Research	Lucknow, UP

Growth phase of Sugarcane (4)

S. No.	Growth phase	DAP	Remarks
1.	Germination phase	Up to 60	Optimum temp.26-32°c
2.	Formative phase	60-120	Max. water requirement phase, opt. temp. 21°c (70°f)
3.	Grand phase	120-250	Max. growth
4.	Mature phase	250-365	

Method of planting setts

S. No.	Type of planting	Remarks
1.	End to end	Low Sett rate is required
2.	Eye to eye	High sett rate is required
3.	Double row system	For thicker and Off season planting

Use and their purpose of chemicals

S.No.	Chemicals	Purpose
1.	Balsario	@ 4.5 kg /1000 setts for early maturing
2.	Polaris/ Na Metasilicate	Improve juice & sucrose content
3.	Glyphosate/ Round -Up	Sugarcane ripener

Maturity indices judged by:

- Cane produces sound - Metallic
- Brix reading (°) - 16-18
- Glucose content - 0.5%
- Glucose content is < 0.5 % tested by **Fehling solution.**
- Sucrose get converted into glucose and quality of the product become poor due to high temperature

Recovery percentage of sugarcane products

S.No.	Products	Recovery percentage
1.	Jaggery from juice	9-10
2.	Crystallized sugar from Gur	62.5
3.	Sucrose content in cane	13-24
4.	Sugar from juice	6-10
5.	Molasses	3.5
6.	Rab	18-20 % of juice
7.	Juice % by ordinary crusher	50-60
8.	Juice % by Power crusher	60-70
9.	Juice % by Vacuum crusher	70-75

- Flowering comes (DAP) - 300-350
- Most popular planting method in N. India is flat bed method.
- Spacing (RxR) - 90cm

Sowing time and cropping duration

S. No.	Cropping	Sowing time	Crop duration	Yield (q/ha)	Remark
1.	Eksali	Jan - Feb	12-13	400-500	It take in South India
2.	Adsali	Jun –July	18 month	800-1200	
3.	Autumn	Oct- Nov.		500-600	

Depth of bed

- Flat bed- 8-10 cm
- Furrow (South India) - 20 cm
- Furrow (North India) - 10-15 cm
- Trench method of planting is adopted to avoid logging.
- Crop logging concept was given by **H.F. Clements.**
- Unstable sugar content is found in saline soil.

Best N_2 fertilizer use under specific condition

- Normal soil - Ammonium Sulphate
- Saline - Urea
- Acidic- CAN
- **Furrow method** is a most common irrigation method in India.
- Ratoon crop required 20 % more N_2.

Important practices follow in sugarcane cultivation

S. No.	Practice	Objects
1.	Earthing up	Better soil aeration
2.	Tying	To avoid crop lodging
3.	Blind hoeing	Weed control
4.	Burning	Improving sucrose content & juice quality
5.	Wrapping and Propping	To avoid crop lodging

TOBACCO/ TAMBACCAU

Scientific name : *Nicotiana spp.* L.
Family : Solanaceae
Origin : Central America and Mexico.

Importance

- Central Tobacco Research Institute is suited at Rajahmundry (Andra Pradesh)
- Bidi Tobacco Research Institute situated at Anand (Gujarat)

Indian tobacco has two species:

S. No.	Species	Nicotine %	Uses
1.	*Nicotiana tabacum*	0.5- 5.5	Smoking /chewing purpose
2.	*Nicotiana rustica*	3.5- 8.0	Hookah / snuff

- Mild acidic (5-6) soil is suitable for tobacco cultivation.
- **Sodic soil** is not suitable for tobacco because tobacco absorbs a lot of chloride ions in leaves which cause poor burning quality.
- **Root** is the nicotine **synthesis site** in tobacco plant. Whereas, **leaves** are nicotine **stored site** in plant.
- In Gujarat, tobacco mainly grown in sandy to sandy loamy soil under irrigated condition is called **goradu.**
- **Pikka** - its same condition of cultivation in Orissa is called pikka
- **FCV stand for** - Flue Cured Virginia, it used for cigarette
- **Non Virginia** - used in bidi, natu, chewing, hookah, cheroot, snuff and it cover 70% area and production.

- Bidi tobacco mainly grown in Gujarat state.
- Optimum temperature for germination & growth is 27-32 °C
- **Flue curing** is used for cigarette.
- **Potassium and phosphorus** nutrients are required in **more and less** in quantity for quality, respectively
- **Orbanche** is an associated weed of tobacco.
- 1.0 gram seed contain - 10,000-12,000

Tobacco nursery

- Nursery time - Mid August
- Transplanting stage of plant - 4-5 leaves
- Seedling ready in tabaccum 7-9 week and in rustica 5-6 week
- Transplanting time in summer - End of March to first week of April
- In winter October -November

Important practices of tobacco

S. No.	Practices	Objects
1.	Topping	• Removal of flower buds either alone or with few upper or top leaves to improve the size, body and quality of leave in all tobacco except wrapper tobacco • No. of leaves left on plant for specific purpose • Cigarette and cheroot - 8-9 • Wrapper (cigar) - 10-12 • Hookah, Bidi - 10-14
2.	Desuckering	• Divert energy and transfer of nutrient from flower bud to leaves
3.	Priming	• Removing of lower leaves which come in contact with soil and loses their commercial value. • Such type of leaves are called sand leaves
4.	Curing	• To improve the colour, texture and aroma(for quality product) • 32-35 °c (160 °F) temp. and 80-85 % RH req. for drying **Type of curing** **1. Air curing** - For wrapper tobacco **2. Sun curing** - For bidi, cigar and chewing **3. Fire curing** - Chewing **4. Flue curing** - Cigarette

Problem in tobacco cultivation

S. No.	Melody	Symptoms
1.	Splongles	when top leave developed into red rusty spot
2.	Sponging	development of brownish tinges on leaf due to oxidation of tannin like substances
3.	Scalding	A sudden rise in temperature when leaves are still wet, and result in a bluish black discoloration spot appear on leave is called scalding
4.	Leaf curl	Vector - White fly
5.	Cigarette beetle	Storage pest of tobacco

IMPORTANT FACTS OF AGRONOMY

- Water absorption through xylem due to : Transpiration Pull
- 1st photosynthetic product in maize : **Malic acid**
- ABO blood group is the example of : Multiple allele
- Phosphorus mobilize in soil by : **VAM**
- pH sewage : 7-7.5
- Max. albedo in : Dry land dunes
- Who was given "The Agroclimatic law" : **Hopkins**
- PPDA is : Urease inhibitors
- Mostly herbicides decomposed in soil by : Microbial
- Weed index is : [{x-y/x}x100]
- **Biuret content % max. limit in urea** : 1.5 per cent
- Nicotine synthesize in tobacco : Roots
- Three kind dormancy is found in : ***Avena fatua***
- Maximum seed produced by which weed : Orobanche
- In Tobacco use sequence of practices : Topping>Desuckering> priming>curing
- Test weight of Pusa Basmati rice : 21.0 g
- IFOAM head quarter in : Germany
- Photosynthesis is a process : Oxidation- Reduction
- A lady Bird Beetle is a : A predator
- Weight of 1 ha furrow slice up to depth of 0-15 cm : 2.25×10^{-6} Kg
- Power tiller is most suitable for the cultivation of : Rice

- Pahala blight diseases in sugarcane due to deficiency of : Magnease
- Bragg and Clark-63 are varieties of : Soybean
- Delinting of cotton seed by which chemical : H_2SO_4
- Lysine is deficient in : **Cereals**
- Typhoon system ` of irrigation is adopt in : Sugarcane
- **Double Zero** var of : Mustard
- Clipper var of : Barley
- Bamboo irrigation followed in which state : Meghalaya
- Dapog method of rice nursery was introduced from : Philippines
- **Oryazein protein** is found in : Rice
- which plant nutrient is responsible for pollen viability : Boron
- Most limiting factor in acidic soil is : Al toxicity
- The "Law of Agrobiology" was given by : Wilcox
- Which acid is used for preservation of food : Benzoic acid
- 2,4-D susceptible variety of wheat is : HD-2009 (Arjun)
- What is the working range of Tensiometer : 0.85 bar
- What is the water potential of pure water : Zero
- Most of the weather phenomenon like rainfall, dew, storm are happened in : Troposphere
- **"Inverse yield nitrogen law"** was given by : Wilcox
- **"Law of limiting factor"** was given by : Black man
- RGR proposed by : Fisher
- **"Law of tolerance"** was given by : Schofield
- The unit of measurement of Eh is : Volt/milli volt
- The elastic substance found in wheat flour that gives adhesiveness : Gluten
- **Nutrient mobility concept** was propounded by : Bray
- Measurement of sunlight intensity is expressed in : Lux unit
- Instrument used for measuring of depth of water is called : Piezometer
- Condensation of two molecule of urea with loss of one molecule of ammonia leads to the formation of : Biuret

- ....method of sowing is more energy and time consuming but reduces seed rate considerably : Dibbling
- Water hyacinth accumulates ....and thus can be used for water pollution control : **Pb, Ni, Cd**
- What do you understand by silage crops : Crops cut and preserved in green condition used for feeding of cattles
- Mycorrhizal fungi helps to increase the availability of : Phosphorus
- Most suitable range of temperature for the action of most of the herbicides are : 20-25^{0}c
- The herbicide recommended to control *Cynodon dactylon* in fallow field is : Glyphosate
- **Silking** occurs in which crop : Maize
- Utera cultivation is related to which crop : Lentil
- Nitrogen use efficiency in low land rice is around : 30%
- A substance intended to counteract the effect of a poison is called : **Antidote**
- A substance that retard or inhibit transpiration is called : Antitranspirant
- Allowing of weeds to germinate by rain or wetting and tilling to kill them before sowing seed is called : Stale seed bed
- An abnormal warming of surface ocean waters in the eastern tropical pacific : El–Nino
- Control of Isoproturon resistance *Phalaris minor* : **Sulfosufuron**
- Enzymes involved in plant defense against herbivores : Thioglucoside glucohydrolase
- LER formula is : yab/yaa+yba/ybb
- Linear equation is : y=a+bx
- Lint index of cotton is : 33 per cent
- Man-made cereal is : Triticale
- Maximum range weather forecasting is done for : < 2 per cent
- **Molasses is** the byproduct of : Sugarcane industry
- The amount of dry matter that can be produced from an given quantity of water is called : WUE
- The cultivation of an additional crop from the regrowth of stubbles of previous main crop after its harvest is called : Ratooning

- The effect if short wave radiation can easily pass through the atmosphere to reach the earth surface while the outgoing radiation from the earth are not permitted to pass through the atmosphere is called : Green house effect
- The poisonous compound found in sorghum is : **HCN**
- The tree species having root and stem nitrogen fixing nodules is called : Legume forage tree
- The unit for measuring radiation is : Langlay
- **Trypsin** is produced from : Pancreas
- Wave length measuring unit is : Angstrom
- …………… is fiber and oil seed crop : Linseed
- Most common using herbicide in paddy rice : Butachlor
- Type of mulching in dry land areas : Stubble mulch
- Contour farming is practiced where slop is : 1-6 per cent
- Which micronutrient is deficient in dry soil : Zinc
- Soil temperature mainly influenced by : Organic matter
- In acid soil which fertilizer efficiently used : Rock phosphate
- Volatilization losses of N occurs in soil : Saline
- North India receive rainfall mostly from : S.W. Monsoon
- Highest Sulphur uptake by which crops : Oil seed crops
- Guntur district is famous for : Tobacco cultivation
- An area from where entire rain water is collected and then conducted to a reservoir is called : Catchment area
- Water flow measured by : **Parshall Flume**
- Highest staple length cotton sp. : *G. barbadense*
- Casperian strips found in : Phloem
- Plant that grow in saline soil is called : Halophytes
- Rice variety not matured in 130-135 DAS : Sona
- ………Is done when the slope is more than 15% : Bench terracing
- TZ (Tetrazolium) test is done for : Seed viability
- Cereals are deficient of which protein : Lysine
- Bordeaux Mixture was development : PMA Millardet
- Father of microbiology : Louis Pasteur
- Bacterium was discovered by : Antonie van Leeuwenhoek
- Bacterial disease is control by : Streptocyclin

- Pest population of gram pod borer can be minimize by using : NPV
- Rice (photoperiod) : Quantitative SDP
- Dried nuts are important source for : Fats
- Density of water is maximum at…….ºC : 4
- **Diara method** of cultivation is adopt in : Cucurbit crops
- In saline soils method of irrigation is used : Ridges and furrow
- …….is the oldest method of selection : Mass selection
- The limit of regression coefficient is : -∞ to +∞
- The value of mean , mode, median are same in : Normal distribution
- Cytokinin is mostly synthesized in : Root tips
- Which growth regulator mostly affects the mobilization of nutrients : Auxin
- The most abundant enzyme on the earth is : Rubisco
- Most of the carbohydrate are trans located in the from : Sucrose
- Dr. B. P. Pal is associated with : Rose
- The fruit crop which require the highest number of irrigation : Banana
- Conversation of fat in to carbohydrate is : Glyoxylate
- The site of protein synthesis : Ribosomes
- Machete trade name : Butachlor
- Bioassay of atrazine : Soybean and cucumber
- Index expressing the reduction in the yield due to presence of weeds in comparison with weed free situation : Weed index
- Collego is : Mycoherbicide
- Alternate host of gram caterpillar : Amaranthus
- Salt tolerant weed : Chenopodium
- PET is given by : **Thornthwaite**
- Cultivation of woody plants, particularly those used for decoration and shade is called : Arboriculture
- Downward entry of water in to the soil is : Infiltration
- Albedo of dry lands : 15-35%
- **Mar** is : Raw humus
- First dry land research station in India at : Manjeri (1923)

- The main objective of Puddling : To reduce percolation losses of water
- First cotton variety with terminator gene was developed by : Monsanto
- The first interspecific variety of cotton : Veralaxmi
- Rabi, Kharif and zaid are derived from : Arabic words
- The quality of beetle vine is most affected by : Light intensity
- Why nitrogenous fertilizer applied in reduced zone : To control denitrification and leaching loss
- **Natural farming concept** was given by : Masanobu Fukuoka
- Luxury consumption nutrient e.g : Potassium
- Highest ground water polluting fertilizer : **NO_3**
- The Phyllody is a problem in : Sesame
- The journal Agroforestry Today" is publishes from : ICRAF
- Karnal bunt of wheat first time reported at : Karnal (1930)
- The rate of Application of BGA is : 10.0 kg/ha
- The area required for nursery in Dapog method is : 30-40 m^2
- Termites does not affect : *Tectona grandis*
- Haveli system" mean : Weed control through water ponding
- **Chelate** is an : Organic compound
- Khejari (*P.* cineraria) is commonly grown in : Rajasthan
- **White bud** of maize is caused by : Zinc deficiency
- Sunflower oil contains : PUFA
- Amongst the other things the wood quality of eucalyptus depends on its : Lignin content
- Relationship between density and yield is : Parabolic
- Azolla is a : Fern
- **Queen of cereal** is : Maize
- In INM, most important fertilizer is : Organic
- P and B is required for : Energy transfer
- For export, which type of onions variety are prefer : Red colour
- The fertilizer element which improves quality of the fruit is : Potash
- Which of the following is nitrification inhibitor : N-serve

- Which one of under mentioned weeds is a mimicry weed of cumin : *Rumex acetosella*
- Single seeded cucurbit is : Chow- Chow
- Which cultivar of moth bean is popular as a *wonder variety* : **RMO-40**
- Highest oil content crop is : **Coconut**
- Name of the crop which gives synergistic response with sugarcane : Potato
- The source of boron in the soil is : **Tourmaline**
- The "farmers day" celebrate on : 23rd December
- The "world food day" celebrate on : 16th October
- **Aflatoxi** developed by : Fungi
- Pusa Bold is the variety of : Mustard
- CO_2 compensation point for C_4 plants : 0-10 ppm(C_3=25-100ppm, CAM=0-5 ppm)
- Denitrification losses of nitrogen in paddy field can be prevented by intercropping ammonical nitrogen into : **Reduce Zone**
- "Dropsy" diseases caused by oil of : *Argemone maxicana*
- Highest albedo of : **Fresh snow**
- Highest CV for SW monsoon in : Rajasthan
- Highest irrigation area covering canal in the world : Indira Gandhi Canal
- Stem nodulating green plant is : Sesbania
- The Bio-drainage plant is : *Eucalyptus tereticornis*
- The pF value at permanent wilting point is : 4.2
- Which crop is known as gold of America : Soybean
- Basic tool for weather forecasting is : Synoptic chart
- Mixture of 2,4-D+2,4,5-T are known as : Brush killer
- *Avena fatua* and *Phalaris minor* are : **Satellite weeds**
- Rockphosphate+Sulphate+Thiobacillus oxidant are : Bio super
- Which micronutrient is deficient in Indian soil : Zinc
- Which nutrient help in oil synthesis : Sulphur
- Highly tolerant to Na (crop) : Sugarbeet
- Saline soil (Russian words) : **Solan chak**

- Alkaline soil (Russian words) : **Solanetz**
- Maize variety for dryland farming : Ageti-76
- Chikori weed associated with : Barseem
- :
- **Meskawi** is variety of : Barseem
- Double dwarf wheat should be sown at depth : 4.0 cm
- Triple dwarf wheat should be sown at depth : Less than 4 cm
- Root hairs are characteristically absent both on main and lateral roots in which crop : Groundnut
- **Brace root** found in : Sorghum
- Butachlor application time water require in the rice field : About 1 cm standing water
- Cotton is very susceptible to which herbicide : 2,4-D
- Oil seed cum pulse crop : Soybean
- Oil cum fiber crop : Linseed
- Legume cum pasture crop : Stylosanthes hamata
- Condensed water vapour within the atmosphere : Fog
- Madhaya Pradesh is known as : Soybean state
- Triticale gives more yield than wheat in : **Dry land**
- Green manure of dhaincha reduce alkalinity of soil by production of : Organic acid
- Who was given **"The concept of residual carbonate"** : Eaton (1950)
- Test weight of hybrid maize : **200 g**
- Loomis Shull (1937) classified the nutrient on the basis of requirement : Macro>IPPM Micro<IPPM
- Per cent organic matter : % Organic carbon x 1.724
- % O.M. : N in soil x 20
- National Biofertilizer development centre situated at : Ghaziabad
- N-E monsoon period is : Nov. to Dec.
- Crop water calendar prepared by : Division of Agril. Meteorology, Pune
- Albedo of crop plants is : **15-25 per cent**
- Albedo of water surface is : **6-9 per cent**

- **Study of clouds** is called : Nepthology
- Mycorrhiza was discovered by : **Frank (1855)**
- One factor increase another is decrease is : **Antagonistic effect**
- One factor increase another is increase is called : **Synergistic effect**
- Study of crop science is called : Agriculture
- Pesticides used in agriculture at Global level : Herbicide > Insecticide > Fungicide
- Pesticides used in agriculture in India : Insecticide > fungicide > Herbicide
- **Urea synthesis** by : Wholer (1928)
- Most volatile form of 2,4-D is : Ester
- Sterile spiklet in rice due to : Low Temp
- Soil colour measure by : Munsell colour chart
- **Principal mineral** in earth crust are : Feldspars (48%), Quartz (36%) and Mica (10%)
- **Healthful herb** is : Tea
- Metabolic transformation of inorganic Plant nutrients into organic plant constituents is known as : Assimilation
- In C_4plants CO_2 reduction taken place in : **Bundle sheath cells**
- In C_3 plants CO_2 and NO_2 reduction takes place in : **Mesophyll cells**
- Harvest index (1962) Competition index(1963)and Ideotype (1968) given by : **Donald**
- Fast growing plants and trees are known as : **Energy faring**
- The process of breakdown of H-clay under alkali condition is known as : Solodisation
- First irrigation before sowing is called : **Palco**
- Super rice idea given by : G.S. Khus
- Super wheat idea given by : S. Nagarjan
- Husk less variety of barley is : Dolma
- Rancidity in sunflower oil is due to : Oxidation (90DA Threshing
- Urea is first organic compound are given by : Wohler in 1828
- Narrow windo herbicide is : Pendimethalin
- Wider windo herbicide : Atrazine, 2,4-D, Butachlor
- Solar energy efficiency in ecosystem is : **0.1 Percent**

• Moisture sensitive crop during entire growth period	:	Tobacco
• Very acid soil with low base saturation is called	:	Acrosol
• Residue of sugarcane factories is	:	**Baggase**
• **Cereals (Ceres) world** of which language	:	Roman
• Rhizobium	:	Frank (1889)
• Mycorrhiza association	:	Frank (1885)
• Concept of soil catena	:	Milne (1935)
• Organic farming	:	Howard
• Western disturbance cause due to the creation of low pressure zones. They cause severe damage to the coastal zones. On an average 2-3 storms may be expected in	:	Fall in the hills
• In India areas of high annual PET are extreme in	:	Rajasthan (Jaisalmer)
• And extreme in south part of country	:	Tuticorin, TN
• Study of ancient geological fossils is known as	:	Paleontology
• The plant having the biggest flower in the world is	:	Rafflesia
• **Oldest tree** was found in	:	Taiwan
• **"worlds food basket"** is	:	Prairies of north America
• **"Sugar bowl of the world"** is	:	Cuba
• Penicillin is obtained from	:	Fungi
• **Boneless meat**	:	Soyabean
• **Golden fibe**	:	Jute
• Tractor has air cooled engine	:	Eicher
• First tractor plant in India was set up by	:	Eicher
• Which gas is generated in Gobar gas plant	:	**Methane**
• All legumes may or may not be	:	Pulses
• All pulses are	:	Legume
• Pore space is highest in	:	Sandy soil
• Vitamin mainly act as "........" for enzymatic activity	:	Cofactor
• Eukaryotic *e.g.*	:	Fungi, Protozoa
• Prokaryotic e.g	:	Bacteria Actinomycetes, BGA
• Largest leaf *e.g.*	:	Water lily

- Largest earthworm : Gaint earthworm (4.6 feet)
- **Fibroin** structural protein is found in : Silk
- **Elastin** structural protein found in : Insect wings
- Effect of pollen on endosperm expression is known as : Xenia
- Effect of pollen on seed coat colour is known as : Metaxenia
- **Solum** is a : A+ B horizons
- **Regolith** is a : A+ B + C horizons
- Laterite soil is used for preparing : Bricks
- Stoke's law is applied in the determination of : Soil texture
- Desert gold known as : Jatropha
- Hull less variety of barley : Karan-18
- Which type structure is best for cultivation : Crumby type
- Soil pH increase in winter and decrease in : Summer season
- **Leg hemoglobin** is a carrier of : Oxygen
- Optimum population of BGA in rice field : 43000 per gm of soil
- Wonder crop known as : Soybean
- Wonder cane is : **CO- 419**
- **Nitro –positive crop** is : Bajra
- Area irrigated by1 cusec of water is called : **Duty**
- Total depth of water required by a crop : **Delta**
- Which state have highest area of drip irrigation in India : Maharashtra
- Which mineral have highest content in earth 'crust : **Feldspars (48 Per cent)**
- Judged by soil physical, chemical and biological properties is known as : Soil health
- Symbiotic Actinomycetes is : Frankia
- Direct sunlight is injurious to most of MO except : **Algae**
- Starch is the polymer of : Glucose
- The basic component of lipid is : Glycol
- Humus is a : Acidic
- A serious weed in tea garden is : Loranthus
- Biological farming / Rishi Kheti is also called : Organic farming

• Study of fresh water is called	: Hydrology
• **Lowest category of soil classificatio**	: Series
• A process of Fe remove is called	: Gelization
• Which group of microorganism have add highest biomass in soil	: Fungi
• 95% constitute of total dry matter of plant is	: CHO
• Best methods of application for Fe and Zn	: Foliar application
• Application of Ca, B, S and Mo is suitable methods	: Soil application
• Least resistant mineral to weather ability is	: **Olivine**
• Most resistant mineral to weathering is	: **Quartz**
• Equation of universal soil loss	: A = RKLSCP
• Nutrient mobility was given by	: Bray s
• First product of urea if apply in soil	: Ammonium Carbonate
• Blast elements e.g	: Al and Si
• The use of synthetic fertilizers is combined with organic manure rich in humus and soil microbes	: IPNS
• Which sugarcane variety covered highest area in north India_	: CO-0238 (Karan-4)
• Which variety recorded highest sugar recovery (12.1%) ever in sub-tropical India	: CO-0238
• Basmati rice export affected due to high level of ……..in Indian basmati rice	: Tricyclazol
• Recently government of India ban on	: Tricyclazol and Buprophazine
• CO 0238 (Karan 4) developed by	: Sugarcane Breeding Institute, Regional Centre, Karnal cross Co LK 8102 x Co 775
• Most successful salt tolerant rice variety	: CSR-30 and KRL-210
• CO 3, CO (CN) 4 and CO (BN) 5 BNH developed by	: Tamil Nadu Agricultural University
• Pusa Basmati-1637 improved of Pusa Basmati-1 against…………..	: **Neck blast**
• Pusa Basmati-1718 improved of Pusa Basmati-1121 and Pusa Basmati-1728 improved of Pusa Basmati-1401, both of them incorporating the Xa21 and xa13 genes that confer resistance……………….against	: bacterial blight pathogen

AGRONOMY OBJECTIVE QUESTIONS (MCQ)

1. Current Science is published from
   A. Bangalore B. USA C. China D. Africa
2. Who is known as milk man of India?
   A. Rajendra Prasad B. Govind Singh Khurana
   C. Swaminathan D. Dr. Vergeese Kurien
3. Difference between seed and grain is
   A. Size B. Genetic purity C. Germination % D. Purity %
4. Jalna city of Maharastra is known as
   A. Seed city B. Lake city C. Gold city D. Silver city
5. The upper limit of water potential is
   A. 2 B. 0 C. 1 D. 3
6. Soil water potential always has
   A. Positive sign B. Only positive C. Negative sign D. None
7. Haryali yojna is related to
   A. Crop management B. Soil management
   C. Insect pest management D. Water management
8. In splash erosion the raindrop energy acts as in the form of
   A. Atomic energy B. Gibbs free energy
   C. Kinetic energy D. Potential energy
9. Dust mulch is effective in
   A. Heavy textured soil B. Light soils
   C. Organic soils D. Saline soils
10. Cotton herbicide is .....
    A. Atrazine B. Diuron C. Metribuzin D. Clodinofop
11. Butachlor belongs to which group of herbicide
    A. Allo group B. Methyl group C. Fop group D. Amide group
12. Who is the father of soil erosion?
    A. V.V. Dokuchav B. Swaminathan C. H.H.Bennet D. Milne
13. Agronomy journal is published from
    A. Bangalore B. USA C. China D. Africa
14. An agricultural implement used for leveling paddy fields was
    A. Star weedar B. Leveller C. Country plough D. Parambu
15. High dose of nitrogen in sugarcane cause
    A. Decrease sugar quality B. Increase sugar quality
    C. Constant sugar quality D. All

16. Harmful substance in urea is
    A. Methyl B. Cyanide C. Ethyl D. None
17. Kharif, Rabi and Zaid words belong to
    A. Latin B. Greek C. Arabic D. Sanskrit
18. One sq. km. consists of
    A. 1000 ha B. 10 ha C. 500 ha D. 100 ha
19. Relative water content (RWC) was proposed by
    A. Briggs B. Tuerners C. Sorenson D. Milne
20. The direct sown rice is
    A. Virula B. Balua C. Aus D. Aman
21. Greening of potato results in
    A. Increase in nutritional quality B. Constant in nutritional quality
    C. Decrease in nutritional quality D. None
22. The catchment area where ground water reaches the surface and runoff is termed as
    A. Basin area B. Springs C. Drainage area D. All
23. The crop with yarns of sunhemp is prepared in
    A. 12-15 weeks B. 8-10 weeks C. 2-4 weeks D. 15-20 weeks
24. Actual layout of any meteorological observatory is
    A. 30 × 55 m B. 36 × 50 m C. 36 × 55 m D. 32 × 56 m
25. Ozone layer discovered in
    A. 1920 B. 1913 C. 1925 D. 1915
26. Maximum rainfall occurs in the
    A. Windward side B. Leeward side C. Both D. None
27. Carbon di oxide fixation in crop plants is
    A. 10% B. 30% C. 40% D. 20 %
28. CWSI stands for
    A. Crop weed stress index B. Crop water saturation indices
    C. Crop water stress indices D. None
29. Lichens indicate pollution by
    A. NO_2 B. SO_2 C. CO_2 D. All
30. Saffron revolution related to
    A. Solar energy B. Cattle walfare C. Fishries D. Food grain
31. Crops to have its genome decoded
    A. Soybean B. Rice C. Finger millet D. Wheat
32. Weed to have its genome decoded
    A. Phylanthus niruri B. Chenopodium murale
    C. Melilotus alba D. Arabidopsis thaliana

33. Toxin in pearlmillet
    A. Oxalic acid B. Resins C. Tripsin D. Glucosides
34. Golden rice produced by
    A. Daffodils B. Ervinia vredivora
    C. Agrobacterium tumifacience D. All
35. Toxin substance in subabul
    A. Oxalic acid B. Mimosine C. Tripsin D. Glucosides
36. Dormancy resistant rice variety
    A. Masuri B. Jaya C. TN -1 D. IR-8
37. Highest fat in cereals
    A. Rice B. Wheat C. Pearlmillet(5%) D. Sorghum
38. Highest water use efficiency
    A. Barley B. Rice C. Sorghum D. Finger millet
39. Crop works as mini fertilizer crops
    A. Arhar B. Gram C. Green gram D. Black gram
40. Blind or light hoeing is done at
    A. 4-5 weeks after planting
    B. 1-2 weeks after planting
    C. 5-7 weeks after planting
    D. 7-10 weeks after planting
41. The sequence of different operations in tobacco crop is
    A. Topping-Desuckering-Priming-Curing
    B. Topping- -Priming-Curing-Desuckering
    C. Desuckering-Topping-Priming-Curing
    D. Topping-Curing-Desuckering-Priming
42. Measuring uniform flow of water
    A. Rectangular weir B. Trapezoidal weir
    C. Both D. None
43. Weed used as pollution indicator
    A. Brassica kaber (NO_2) and Stellaria redia (SO_2)
    B. Crotolaria
    C. Lantana camera
    D. All
44. The polish percentage in paddy
    A. 1% B. 2% C. 3% D. 4%
45. What is the moisture percent of hay?
    A. 25-30% B. 30-35% C. 25-40% D. 15-20%

46. The component herbicides of the herbicide mixture 'Odyssey' used to control weeds in soybean are
A. Imazethapyr and Imazamox B. Imazethapyr and Quizalafop
C. Imazethapyr and Alachlor D. Imazethapyr and Pendimethalin
47. The husk : seed ratio of isabgol by weight is
A. 50:50 B. 25:75 C. 40:60 D. 75:25
48. Regulator in K/Ca ratio
A. B B. C C. Mo D. Si
49. FACE stand for
A. Free Air C Enrichment B. Free Aqua CO_2 Efficiency
C. Free Air CO_2 Enrichment D. Free Air C Energy
50. Methane in one ton rice residue burn
A. 2.37 kg B. 3.15 kg C. 4.50 kg D. 5.30 kg
51. NO_2 and H_2S pollution weed
A. Lantana camera B. Chenopodium album
C. Sorghum halpense D. Melilotus indica
52. Deficiency of surface and sub surface water is known as which drought?
A. Hydrological drought B. Meteorological drought
C. Soil drought D. Atmospheric drought
53. Gurubacchan singh related to which field
A. Horticulture B. Bee-keeping C. Agronomy D. Agroforestry
54. 5 ha mm equals to
A. 50 m^3 B. 100 m^3 C. 500 m^3 D. 5000 m^3
55. Legume require fertilizer as well as cereals
A. Gram B. Rajma C. Mothbean D. Green gram
56. NDVI stands for
A. Normal Difference Vegetation Indices
B. Normalized Diagnosis Vegetative Index
C. Normalized Differential Vegetation Index
D. Normally Division Vegetation Indices
57. 'One Straw Revoluation' book written by
A. M.Fukouka B. Beaufils
C. Rajendra Prasad D. Swaminathan
58. Low dose herbicide group
A. Chlor group B. Sulfonyureas group
C. Allo group D. Fop group
59. Metric potential always
A. 0 B. One C. Positive D. Negative

60. Unpredictable legume
    A. Groundnut B. Chick pea C. Soybean D. Guar
61. Transplanting of pearlmillet
    A. After 2 weeks B. After 3-4 weeks or 1 month
    C. After 5-7 weeks D. After 7-10 weeks
62. Moisture and drought sensitive crop
    A. Sorghum B. Pearlmillet C. Maize D. Wheat
63. Crop weed compition period of soybean
    A. 30-40 DAS B. 10-20 DAS C. 50-60 DAS D. 15-20 DAS
64. Consumptive use of water
    A. E + IR
    B. ET + IW
    C. T + water used in metabolic purpose + deep percolation
    D. ET + water used in metabolic purpose
65. Conjuctive use of water
    A. Surface water + Ground water
    B. ET + IW
    C. ET + water used in metabolic purpose
    D. ET + deep percolation
66. Crop used in phytoremediation
    A. Maize B. Rice C. Mustard D. Barley
67. Tilth index range
    A. 0-1 B. 1-5 C. 5-10 D. 10-15
68. Concept used to evaluate the sprinkler irrigation system is
    A. Correlation coefficient B. Uniformity coefficient
    C. Regression coefficient D. Crop coefficient
69. Major driving force for hydrological cycle
    A. Rainfall B. Wind C. Pressure D. Solar radiation
70. Factor does not have direct effect on distribution of crop plants
    A. B B. N C. Mo D. S
71. Limiting factor range from
    A. 1-3 B. 3-5 C. 5-8 D. 8-10
72. Silica percent in paddy straw
    A. 1-3 % B. 3-6 % C. 4.6-7 % D. 7-10 %
73. Rain water contains CO_2
    A. 0.45% B. 0.60 % C. 0.75 % D. 0.85 %
74. Narrow window herbicide
    A. Atrazine B. Pendimethalin C. Butachlor D. 2,4-D

75. Baroda, Babapuri and Pusa moti are varieties of
    A. Pearl millet B. Maize C. Sorghum D. Rice
76. Instrument used for nitrogenase activity
    A. Anemometer B. Gasometer
    C. Gaschromatogrphy (GH) D. Arealometer
77. Incidence of blast disease in rice can be reduced by
    A. Silicon application B. Herbicide spray
    C. Insecticide spray D. None
78. Cancer of wheat
    A. Rust B. Karnal bunt
    C. Smut D. Nemetode caused disease
79. Bioassay term was given by
    A. J.C.Boss B. Linnaeus C. Ferney D. Burgiston
80. International size of herbarium
    A. 16.5×10.5 square inch B. 19.5×12.5 square inch
    C. 16.5×19.5 square inch D. 10.5×15.5 square inch
81. Zinc accumulator weed
    A. Chenopodium album B. Sorghum halpense
    C. Setaria lutescence D. Prosopis juliflora
82. Present in walnut
    A. Juglone B. Patulin C. Phloridzin D. Camphore
83. 'Isoguard' combination of herbicides
    A. Anilophos + 2,4-D B. Isoprturon + 2,4-D
    C. Anilophos + Pendimethalin D. Anilophos + Clodinofop
84. Who coined the term Control Drop Application
    A. E.J.Bals B. J.C.Boss C. Norfork D. Donald
85. National Agricultural Education Day celebrate on 3rd December on birthday of
    A. Choudhary Charan Singh B. Dr. Rajendra Prasad
    C. Swaminathan D. Borloug
86. Test weight of sunflower seed ranges from
    A. 30-40 gm B. 40-60 gm C. 60-70 gm D. 70-80 gm
87. Udbatta disease is associated with which of the following crop
    A. Sorghum B. Barley C. Cotton D. Rice
88. The base temperature of rice for calculating GDD is
    A. 4.5^{0}C B. 5^{0}C C. 8^{0}C D. 10^{0}C
89. Agronomic efficiency is expressed as
    A. kg/kg B. kg/ha C. kg/ha/day D. kg/day

90. Pendimethalin belongs to which group of herbicide
    A. 2,4-dinitrophenols B. 2,4-dinitroanilines
    C. Nitrophenyl ethers D. Phenylureas
91. Which part of leaf is present in wheat but not in the grassy weeds such as Phalaris minor, wild oat?
    A. Ligule B. Auricle C. Both D. None
92. Which is called "Green Gold"?
    A. Green gram B. Sisal C. Bamboo D. None
93. Growing degree day concept is useful in forecasting of
    A. Rainfall B. Drought
    C. Frost D. Crop harvest dates
94. The optimum temperature for wheat at flowering is
    A. 15°C B. 20°C C. 25°C D. 30°C
95. A strain resulting from exposure to a particular environment is termed as
    A. Ecotype B. Ideotype C. Ecophone D. Ecotone
96. Brown manuring is generally recommended in which crop
    A. Wheat B. Mustard C. Rice D. Potato
97. Protein content of milled rice is about
    A. 4-5% B. 6-7% C. 9-10% D. 12-13%
98. Nozzle type best suited for herbicide spray is
    A. Tee jet B. Hallow jet C. Flat jet D. Turbo jet
99. Computation of weed index is based on
    A. Weed dry weight B. Crop yield
    C. Crop population D. Weed population
100. The basic objective of surge irrigation in ridges and furrow methods to
    A. Advance wetting length B. Increase lateral movement
    C. Reduce evaporation loss D. Reduce irrigation requirements
101. A loose, finely granular or powdery condition on the surface of the soil, usually obtained by shallow cultivation is
    A. Surface crust B. Soil crust C. Soil powder D. Dust mulch
102. The upper surface of ground water or that level below which the soil is saturated with water, is known as
    A. Water level B. Water table C. Water depth D. Water height
103. The oil content in niger seed
    A. 20-22% B. 25-26% C. 28-30% D. 35-40%
104. Which fungus sp. Produces aflatoxins in groundnut during storage
    A. Aspergillus awamori B. Aspergillus niger
    C. Aspergillus flavus D. Aspergillus funiculosum

105. The number of tobacco seeds in 1 gram seed weight ranges between
A. 1100-1200 B. 11000-12000 C. 500-600 D. 5000-6000
106. Movement of methane through rice plant aerenchyma and out from the leaves is known as
A. Emission B. Ebullition C. Osmosis D. Diffusion
107. Most widely used classification for weeds is based on
A. Duration B. Cotyledons
C. Morphological character D. None
108. Optimum time of harvesting of maize for hay making is
A. Early bloom stage B. Mature grain stage
C. Any time D. Soft dough to medium dough stage
109. Khaira disease of rice is diagnosed as
A. Necrolic lesions B. Dark reddish brown pigmentation
C. Yellowish spots on leaves D. Black spots on leaves
110. The sugar percent in sugarcane juice is
A. 5-6% B. 7-8% C. 8-10% D. 11-12%
111. Urea can be safely used as foliar spray in wheat crop at concentration of
A. 2-3% B. 4-5% C. 6-7% D. 9-10%
112. Pre mature leaf fall due to deficiency of
A. P B. N C. Fe D. Ca
113. The average yield of latex obtained from opium poppy is around
A. 25 kg/ha B. 55 kg/ha C. 85 kg/ha D. 110 kg/ha
114. Seed rate of rice required in wet nursery is
A. 20-30 kg/ha B. 30-40 kg/ha C. 40-50 kg/ha D. 60-80 kg/ha
115. The lowest temperature below which there is no growth is called
A. Optimum temp. B. Base temp.
C. Minimum temp. D. Maximum temp.
116. Gum content in Guar seed varies between
A. 10-15% B. 15-20% C. 20-25% D. 28-32%
117. Which sugarcane variety is most suitable for ratooning?
A. CO-1148 B. CO-997 C. CO-527 D. None
118. Wind breaker and avenue crop
A. Green gram B. Red gram C. Black gram D. Mothbean
119. Soil destructive legume
A. Soybean B. Rajma C. Groundnut D. Guar
120. Nitrification bacteria was isolated by
A. Benjerik B. Winogradsky C. Flamming D. Aurther young

121. Percent depletion of available moisture is optimum for ploughing
A. 0-25 B. 25-50 C. 50-75 D. All

122. Spraying of which of the following conserve moisture in chickpea
A. KCL B. Urea C. $FeSO_4$ D. All

123. Most critical stage for nitrogen requirement of soybean is
A. Pod maturity B. Branching stage
C. Bud initiation to flowering D. None

124. When AET=PET
A. 2 days after irrigation or at field capacity
B. 5 days after irrigation
C. 7 days after irrigation
D. 10 days after irrigation

125. 1 PF equals to
A. 5 cm water column B. 10 cm water column
C. 15 cm water column D. 20 cm water column

126. SALT stands for
A. Slope Agricultural Land Technology
B. Soil Agriculture Land Technically
C. Salt Agricultural Lawn Technology
D. Sale Agronomic Law Technique

127. Resistance block not used for
A. Black soils B. Saline soils C. Alkali soils D. Sandy soils

128. Mycoherbicide for puccinia spp.
A. Devine B. Collego C. Biosedge D. Biophos

129. Strigol produced by
A. Cotton roots B. Sorghum roots C. Barley roots D. Wheat roots

130. A unit volume of water used for produce
A. Green water B. Virtual water C. Blue water D. Gray water

131. Water used in crop root zone
A. Green water B. Virtual water C. Gray water D. Blue water

132. In dryland crop reduce transpiration by rolling leaf
A. 20-30 % B. 30-40 % C. 40-50 % D. 60-70 %

133. Green seeker used for
A. N requirement B. P requirement C. S requirement D. K requirement

134. Drosometer is used to measure
A. Humidity B. Radiation C. Field dew D. Cloudiness

135. With increase in crop canopy
A. LAI decrease B. LAI increase C. LAI constant D. None

136. Generally pudding is done at
   A. Upper plastic limit B. Lower plastic limit
   C. Ultimate wilting point D. Any time
137. IIVS is used for
   A. Crop canopy B. Crop breath C. Crop width D. All
138. The application of nitrogen at the time of flowering will
   A. Increase growth B. Increase root development
   C. Increase panicle density D. None
139. Parboiling of rice is a
   A. Thermal treatment B. Chemical treatment
   C. Biochemical treatment D. None
140. The general rate of application of bioherbicide for one ha of main field is
   A. 2 kg B. 5 kg C. 7 kg D. 10 kg
141. Delta in irrigation is measured in which unit
   A. Hours B. cm C. Cusec D. Liter/Second/ha
142. What is the effect of nitrification on soil Ph?
   A. Increase B. Decrease C. Remains same D. All
143. The result of soil moisture stress due to imbalance between available soil moisture and evaporation of a crop is
   A. Agricultural drought B. Meteorological drought
   C. Atmospheric drought D. Hydrological drought
144. Which of these herbicides produces H_2O_2 in plants?
   A. Metsulfuron B. Paraquat C. Glyphosate D. Imazethapyr
145. What is the meaning of sod seeding?
   A. Zero tillage B. Minimum tillage
   C. Soil compaction D. Soil pudding
146. Celiec disease in human being is caused by
   A. Lysine allergy B. Zein allergy
   C. Gluten allergy D. Tryptophane allergy
147. In which crop auricles are completely absent?
   A. Oat B. Wheat C. Barley D. Triticale
148. The drainage coefficient helps to determine
   A. Drain spacing B. Drain size C. Drain depth D. Drain design
149. The boll weight of cotton ranges between
   A. 2-5 gm B. 5-8 gm C. 9-12 gm D. 13-16 gm
150. Biuret used for foliar spray
   A. 0.25 % B. 0.50 % C. 0.75 % D. 1 %

ANSWERS

1. A	2. D	3. B	4. A	5. B	6. C	7. D	8. C
9. A	10. B	11. D	12. C	13. B	14. D	15. A	16. B
17. C	18. D	19. B	20. A	21. C	22. B	23. A	24. C
25. B	26. A	27. D	28. C	29. B	30. A	31. B	32. D
33. A	34. D	35. B	36. A	37. C	38. D	39. A	40. B
41. A	42. C	43. A	44. B	45. D	46. D	47. B	48. A
49. C	50. A	51. B	52. A	53. D	54. A	55. B	56. C
57. A	58. B	59. D	60. A	61. B	62. C	63. A	64. D
65. A	66. C	67. A	68. B	69. D	70. B	71. A	72. C
73. A	74. B	75. A	76. C	77. A	78. B	79. C	80. A
81. C	82. A	83. B	84. A	85. B	86. B	87. D	88. D
89. A	90. B	91. B	92. C	93. D	94. C	95. A	96. C
97. B	98. C	99. B	100. A	101. D	102. B	103. D	104. C
105. B	106. A	107. C	108. D	109. B	110. C	111. A	112. B
113. B	114. C	115. B	116. D	117. A	118. B	119. A	120. B
121. B	122. A	123. C	124. A	125. B	126. A	127. B	128. C
129. A	130. B	131. A	132. D	133. A	134. C	135. B	136. A
137. D	138. C	139. C	140. A	141. B	142. B	143. A	144. B
145. A	146. C	147. A	148. B	149. A	150. A		

Chapter

5

AGRICULTURAL ENTOMOLOGY

INTRODUCTION

- Insects are members of the class Insecta in the invertebrate phylum Arthropoda and the **largest in the Animal kingdom.**
- Out of 1.35 million living species of animals, in which **9 lakh species are insects.**
- **Arthropoda** comprises some **85% of all living animals** and of these the vast majority is insects.
- Arthropods are members of the Phylum Arthropoda. The word Arthropoda comes from two **Greek words arthron and podos, where arthron means "joint", and podos means "foot",** which together mean "**jointed feet**". It divided into two subphylum Mandibulata and Chelicerata.
- **Class: Arachnida:** It includes spiders, mites, ticks, scorpions, and their relatives and are characterized by presence of six pairs of jointed appendages: Anterior pair is chelicerae-jaw like, second pair is pedipalp and four pairs of legs. Two body regions: a fused head and thorax (cephalothorax) and an abdomen. The appendages are borne by the cephalothorax. *e.g.* **spiders, mites, ticks** *etc.*
- **Class**: **Crustaceans** include crayfish, lobsters, crabs, shrimp, copepods and their relatives and are characterized by presence of two pairs of antennae and one pair of mandibles, two pairs of maxillae, 3 pairs of maxillipedes, 2 body region-cephalothorax and abdomen, excretion through green glands and respiration by gills. *e.g.* **Crayfish, crabs, lobsters, shrimps, prawns, barnacles, woodlice,** *etc.*
- **Class**: **Myriapoda** has characterized by single pair of antennae and body is trunk like without differentiation. **It is ancestor of Class insect.**
- **Class**: **Diplopoda** has cylindrical greater number of body segments and each segment carries two pairs of legs and two pairs of spiracles. *e.g.* **Millipedes**
- **Class**: **Chilopoda** are dorso-ventally flattened and has single pair of appendages and a pair of spiracles. The first pair of legs is modified to form poison claws. *e.g.* **Centipedes**

- **Class**: **Onychophora** include peripatus and its relatives. Small group of terrestrial animals and they show morphological similarities with both annelids and arthropods. It is a elongate, worm like bodies with several bilateral pairs of lobe like legs. Paired nephridia (excretory organs) along the body that open near the leg bases. *e.g.* **Peripatus**
- **Class**: Hexapoda or Insecta (Subphylum: Mandibulata or Antennata: Antennae present, mandibles present). The body is divided into three distinct regions: head, thorax and abdomen. **Tracheal system of respiration** well developed. *e.g.* **Insect.**
- Entomology is the branch of zoology which study of insects.
- Entomology has been derived from the Greek words, ***entomon*-insects and *logos*-study**.
- The insects are the largest class not only of the animal kingdom but also of the whole living world.
- The largest order in class Insecta is **Coleoptera (3 lakh).**
- The largest family in class Insecta is **Cucurlionidae**.
- The smallest order in class Insecta is **Mantophasmatoidea**.
- The phylum arthropoda evolved (ancestor) **from phylum Annelida**.
- The class insecta evolved (ancestor) from **class Myriapoda.**
- The insect evolved in **Devonian period (350 million years ago)**.
- The earliest known **fossil insect is *Rhyniella praecursor***, a collembolan from Devonian period of Scotland.
- Lepidopterans are evolved during the period **Crestaceous.**
- Highest number of biotypes was observed in the insect **Hessian fl .**
- Insect with wings similar in structure and texture that are shed along a definite suture and have established social life belong to **Isoptera.**
- The both antennae and wings are absent in which order **Protura.**
- Gamma taxonomy is defined to **Study of speciation.**
- The recapitualation theory was proposed by **Hackel.**
- Silk spinning legs are found in which order **Embioptera.**
- The insect originated Paleozoic era of **Devonian period.**
- Pusa bin was developed by **S. Pradhan.**
- The study of the kinds and diversity of organism and the relationships among them called **Systematics.**
- When one name has been given to two or different taxa is **Homonym.**
- The smallest insect order (having 3 members) that reported in 2002 is **Mantophasmatodea.**
- The biotic factors plays an important role in locust life cycle was given by **S. Pradhan.**

- Locust warning organization was established in which year **1939.**
- Tympanum organs on the fore tibia and stridulatory organs on tegmina are present in sub order **Encifera.**
- Triungulin larvae are found in family **Meloidae.**
- National Centre for Integrated Pest Management located at which place **New Delhi.**
- Indian Lac Research Institute, Ranchi was established in which year **1925.**
- Insect are **Incapable of moving the eyes.**

HISTORY OF ENTOMOLOGY

- 1745 -1808: J.C. Fabricius made an extensive study of Indian insects for the first time and classified insects into 3 orders based on the type of mouth parts.
- 1758: **10th edition of 'Systema Naturae'** written by **Linnaeus** includes only 12 Indian insects.
- 1800: Denovan published Natural History of Insects which was the first contribution on the insects of Asia and was revised in 1842 by West Wood.
- 1897: **Sir Ronald Ross** made the discovery of the malaria parasite in *Anopheles* mosquito.
- 1901: **Lionel de Nicevelle** was the first entomologist to the Government of India.
- 1905: Establishment of **"Agricultural Research Institute" at Pusa (Bihar) by Lord Curzon**. The land was donated by **Mr. Phiffs of USA** after whom the place was named as Pusa. In 1911 renamed as Imperial Agricultural Research Institute. Subsequently this Institute was shifted to New Delhi in 1934.
- 1906: Maxwell Lefroy published **"Indian Insect Pests" in 1906** and **"Indian Insect Life' in 1909**. Subsequently Madras, Punjab and Uttar Pradesh appointed their first Government Entomologists in 1912, 1919 and 1922 respectively.
- 1914: T.B. Fl*etc.*,her was the first Government Entomologist of Madras State, published his book **'Some South Indian Insects'**.
- 1925: Indian Institute of Natural Resins and Gums (IINRG) earlier known as **Indian Lac Research Institute at Ranchi (Jharkhand)**.
- 1937: Entomology division was **started in IARI, New Delhi**.
- 1939: Started the "**Entomological Society of India**" at IARI, New Delhi.
- 1939: **Locust warning organization** was established after the locust plague during 1926-32.
- 1939: Insecticidal property of **DDT** was discovered by **Paul Muller of Switzerland**. Paul Muller was awarded Nobel Prize in 1948 for discovering insecticidal property of DDT.

- 1946: Government of India started the **"Directorate of Plant Protection, Quarantine and Storage" at Faridabad.**
- 1946: **National Institute of Plant Health Management (NIPHM)** formerly known as Central Plant Protection Training Institute, later National Plant Protection Training Institute at Hyderabad.
- 1962: Rachel Carson's written the book **'Silent Spring appears'**. The book 'Silent Spring' created awareness about ill effects of pesticides.
- 1963: The book **'Text book of Agricultural Entomology'** by H.S. Pruthi.
- 1975: **'Elements of Economic Entomology'** written by B. Vasantharaj David and T. Kumara Swami.
- 1988: **National Centre for Integrated Pest Management** was established at New Delhi. Now it is National Research Centre for Integrated Pest Management.
- **National Bureau of Agricultural Insect Resources** (formerly known as National Bureau of Agriculturally Important Insects, 2009) (earlier known as Project Directorate on Biological Control started in 1993) at Bangalore.

FACTORS FOR INSECT DOMINANCE

- Capacity to flight: Flight for the purpose like food, shelter, oviposition site, migrating (*i.e.* for long distance travel *e.g.* Locusts) from one place to another and protect from natural enemies.
- More adaptability: Insect has a wide range of climate from -50°C to 52°C. ***Psilopa petroli*** found in crude petroleum oil**. Ephydra fly** living in Great Salt Lake.
- Higher reproductive potential: **Queen termites lays 6000-7000 eggs per day** and development period is short *e.g.* corn aphid produces 16 nymphs per female which reaches the adulthood within 16 days.
- **Polyembryony:** development of many individuals from single egg *e.g.* parasitic wasps.
- **Parthenogenesis:** reproduction without male or without fertilization, *e.g.* aphids.
- **Paedogenesis:** reproduction by immature stages. *e.g.* certain flies.
- Thanatosis-: Insects pretends as if dead. *e.g.* **some beetle**. It is a defense strategy adopted by some insects through feigning death or imitating the voice of dangerous insects or mimicry. *e.g.* **colarado potato beetles** when disturbed draw their legs beneath and drop to the ground and pretend as if dead.
- Colorational: Presence of protective colors. *e.g.* **stick insects**.
- **Physiological adaptations**: Some insects produce or release poisonous or unpleasant odors from their body or possess warning coloration by imitating certain distasteful insects. *e.g.* : Stink bugs have specialized exocrine glands

located in the thorax or abdomen that produce foul smelling hydrocarbons. Larvae of swallow tail butterflies have eversible glands called osmeteria, located just behind the head when disturbed they release repellent volatile and waves their body back and forth to ward of intruders. Some blister beetles (Meloidae) produce cantharidin, a strong irritant and blistering agent.

INSECT TAXONOMY

- Similarly some more codes of nomenclature were evolved in France and Germany. In order to promote stability and universality in the scientific names of animals, an **international code of zoological nomenclature was evolved in 1901 at Berlin**.
- At the 16th session of International congress of Zoology in Washington, **the latest international code of Zoological nomenclature was approved and the same was revised and published in 1964**.
- According to it, the descriptions of unrecorded species should be based on a single specimen whether it is a male or female and the specimen is to be referred as **'holotype'**.
- The opposite sex specimen which is described along with the holotype is termed as **allotype**.
- The other specimens of the species kept along with the holotype and allotype are called **paratypes.**
- The names are to be given in **Latin**. The authors name in full should be written at the end of species name without any punctuation.
- The generic name has to be a simple or compound word. The supraspecific categories like tribe, sub family, family and superfamily are denoted by the **endings -ni, -inae, -idea and -oides** respectively.
- The underlying principle of International code of Zoological nomenclature is **'law of priority'** *i.e.* **the first published name remains in official records** while all the subsequently published names go as synonyms.
- If the same name is given by the different scientists to the different organisms, it is called **Homonymy**.
- **Systematics:** It is defined as the scientific study of kinds and diversity of organism and of any or all relationship among them (Simpson, 1961). Systematics includes **taxonomy and evolution.**
- **Taxonomy:** Simpson (1961) has defined taxonomy as the theoretical study of classification including basis, principles, procedures and rules. Taxonomy includes nomenclature and classification.
- **Biosystematics:** It is a branch of biology incorporating ecological, biological, morphological, cytological and biochemical aspects in the identification and classification of organisms, especially within an evolutionary frame work.

- **Species:** These are a group of individuals which are similar in their structure, capable of interbreeding and producing fertile off spring, but at the same time reproductively isolated from other groups.
- **Subspecies:** is an aggregate of phenotypically similar populations of a species, inhabiting a geographic subdivision of the range of a species and differing taxonomically from other populations of the species.
- **Biotype:** A population capable of damaging and surviving on plants previously known to be resistant to other populations of the same species.
- **Sibling species** (also known as **aphanic species**) are species that are very similar in appearance, in behavior and in other characteristics, but they are reproductively isolated. Sibling species may arise as a result of allopatric speciation through geographic isolation, parapatric speciation, or sympatric speciation.
- Order Thysanura: The name Thysanura is derived from the Greek Thysanos means fringe, tassel, bristle and oura means tail. *e.g.* Bristle tails, silverfish
- Order Diplura: Compound eyes absent. **Cerci present with long and slender, or forceps-like in appearance**. *e.g.* Two pronged bristle tails
- Order Protura: **Antennae and compound eyes absent. Front legs directed forward (probably sensory in function).** *e.g.* Proturans
- Order Collembola: It is normally less than 6 mm long, have six abdominal segments and possess a tubular appendage (the collophore or ventral tube) with eversible vesicles, projecting ventrally from the first abdominal segment. Furcula (springtail) attached ventrally to fourth abdominal segment. *e.g.* Spring tails
- Order Ephemeroptera: Ephemeroptera is derived from the Greek '*ephemera*' meaning short-lived and '*ptera*' meaning wings. **Abdomen slender, bearing two (or sometimes three) long terminal filament** . *e.g.* May flies
- Order Odonata: The name Odonata is derived from the Greek '*odonto*' meaning tooth refers to the strong teeth found on the mandibles of most adults. Two pairs of membranous **wings having similar shape and size.** One distinctively pigmented cell (stigma) on leading edge of wing. *e.g.* Dragonflies, damselflies
- Order Plecoptera: The name Plecoptera is derived from the Greek '*pleco*' meaning folded and '*ptera*' meaning wing and means pleated hind wings which fold under the front wings when the insect is at rest. **Two pairs of membranous wings that are held curved around the body at rest.** *e.g.* Stoneflies
- Order Phasmida: The name Phasmida is derived from the Greek '*phasm*' meaning phantom and **means the cryptic appearance and behaviour of these insects**. *e.g.* Stick insects, leaf insects.
- Order Dermaptera: The name Dermaptera is derived from the Greek '*derma*' means skin and '*ptera*' meaning wings and means **thickened forewings that**

cover and protect the hind wings. Cerci enlarged to form pincers (forceps). *e.g.* Earwigs.

- Order Embioptera: The name Embioptera is derived from the Greek '*embio*' meaning lively and '*ptera*' meaning wings and means the fluttery movement of wings that was observed in the first male. **The front tarsi is enlarged which contains silk glands used to create the galleries.** *e.g.* Web-spinners.
- Order Zoraptera: The name Zoraptera is derived from the Greek word '*zor*' meaning pure and '*aptera*' meaning wingless. *e.g.* Zorapterans.
- Order Psocoptera: The name Psocoptera is derived from the Greek '*psokos*' meaning rubbed or gnawed and '*ptera*' meaning wings. **Wings held tent like over the body.** *e.g.* Booklice, psocids.
- Order Mallophaga: The name Mallophaga id derived from the Greek word '*mallos*' meaning hair and '*phagein*' meaning eat. **Dorsoventrally flattene** means that is flattened from top to bottom. *e.g.* Biting lice and bird lice.
- Order Siphunculata: **Sucking lice (Anoplura** formerly known as **Siphunculata**) have around 500 species. **Anoplura are all blood-feeding ectoparasites of mammals**. *e.g.* Sucking lice.
- Order Neuroptera: The name Neuroptera is derived from the Greek word '*neuron*' meaning sinew and 'ptera' *meaning* wings. The nerve-wings is appropriate because it extensive branching found in the wing veins. **Forewings and hindwings membranous**, equal or subequal in size and narrowed at base, and held roof-like over the abdomen when at rest. **Larvae are predaceous,** usually with piercing and sucking mouthparts. *e.g.* Lacewings, antlions.
- Order Strepsiptera: The name Strepsiptera is derived from the Greek word '*strepsi*' meaning turned or twisted and '*ptera*' meaning wings and refers to the resting position of the male's large hind wings. **These insects are very small parasites** of other insects and are rarely seen as they spend most of their life within the host. *e.g.* Stylopids.
- Order Mecoptera: The name Mecoptera is derived from the Greek word words '*meco*' meaning long and *"ptera"* meaning wings and means the shape of both the front and hind wings. Two pairs of membranous **wings of almost equal size**. Mandibulate **mouthparts at the end of a beak-like projection on the head**. *e.g.* Scorpion flies.
- Order Siphonaptera: The name Siphonaptera is derived from the Greek words '*siphon*' meaning a tube or pipe and '*aptera*' meaning wingless. **Backward pointing hairs and bristles** for ease of movement through the hair of a host. **Body bilaterally flattene** . *e.g.* Fleas
- Order Trichoptera: The name Trichoptera is derived from the Greek words '*trichos*' meaning hair and '*ptera*' meaning wings and refers to the long, silky hairs that cover most of the body and wings. **Wings held tent-like over the abdomen**. *e.g.* Caddis flies

- Order Orthoptera: The Orthos means straight; Pteron mean wings; common names: grasshoppers, locust, katydid, cricket, mole cricket. The forewings are narrower than the hind wings and hardened or leathery at the base is known as **tegmina**. The hind wing is membranous and held folded fan-like under the forewings when at rest. Fore legs are modified for digging (fossorial) as in molecrickets or hind legs modified for jumping (saltatorial) as in grasshopper. Auditory or tympanal organs are also well developed and are located on either side of the first abdominal segment or at the base of fore tibiae.
- Family Acrididae (Short horned grasshopper): **Auditory or tympanal organs situated on either side of 1st abdominal segment. Stridulation femoro-alary type**. *e.g.*Rice grasshopper, *Hieroglyphus banian;* rice small grasshopper, *Oxya chinensis*; desert locust, *Schistocerca grageria.*
- Family Tettigonidae (Long horned grasshopper and Katydids): **Auditory or tympanal organs situated on the fore tibia**. Sound production is alary type. A thick region on the hind margin of the forewing (scraper) is rubbed against a row of teeth on the stridulatory vein (file) present on the ventral side of another forewing which throws the resonant area on the wing (mirrors) into vibrations to produce sound. *e.g.* Surface grasshopper, *Conocephalus indicus*

Order Isoptera

- The Iso means similar; Pteron means wings; common names: termites or white ants. Wings are membranous with somewhat reduced venation and **are capable of being shed by means of basal structures**.
- Reproductives or Primary reproductives of termite (King and Queen): They leave the colony in a swarm, mate and individual pairs establish new colonies, and **wings are shed after máting leaving remnants called stubs**.
- **Soldiers of termites** are also **sterile wingless adults with greatly enlarged head and mandibles**.
- Termites frequently groom each other with their mouth parts as a result of the attraction of some secretions available on the body (**trophallaxis: Mutual exchange of food *i.e.* secretions on the body**).
- **Family Termitidae:** This is the largest family of termites worldwide included about 2/3rd of recent isopterans. *e.g. Odontotermes obesus*, *Microtermes obesi.*

Order Thysanoptera

- Thysano means fringe; ptera means wings; common name: Thrips. **Mouthparts are rasping and sucking type**. Mouth cone is formed by the labrum and labium together with basal segments of maxillae. There are three stylets derived from two maxillae and left mandibles. **Right mandible is absent.** The mouthparts are asymmetrical. Nymphal stage is followed by prepupal and pupal stages which are analogous to the pupae of endopterygote insects.

- **Family Thripidae:** This is the **largest and most important injurious family of Isoptera**. *e.g.* Onion thrips, *Thrips tabaci*; chilli thrips, *Scirtothrips dorsalis*

Order Hemiptera

- Hemi means half; ptera means wing; common name: true bugs. **Head is opisthognathous. Extra-oral digestion is apparently widespread**. There are two sub orders *viz.*, Heteroptera and Homoptera.
- **Family Pentatomidae (Stink bugs or shield bugs): Scutellum large triangular** sometimes extends posteriorly to the extent of covering the wings entirely. *e.g.* Green (stink) bug, *Nezara viridula*; red pumpkin bug, *Aspongopus janus*; cabbage painted bug, *Bagrada cruciferarum*.
- **Family Coreidae (Squash bugs or leaf footed bugs):** Stink glands are found inside the metathorax and glands openings are found on the sides of the thorax between middle and hind coxae. They emit a bad colour. The hind femora and tibiae may have conspicuous enlargements or leaf like **dilations** and hence the name leaf footed bugs. **Metathoracic gland openings present.** *e.g.* Rice gundhi bug, *Leptocorisa varicornis*; pod bugs, *Clavigralla gibbosa*, *Riptortus pedestris* on pulses.
- **Family Pyrrhocoridae (Red bugs or Stainers): Feeding injury** caused by these bugs leads to the contamination by the fungus ***Nematospora gossypii*** resulting in yellowish brown discolouration of the lint. *e.g.* Cotton stainer, *Dysdercus cingulatus*.
- **Family Lygaeidae (Seed bugs or Chinch bugs): Cuneus is absent and clavus is elongate in hemelytra**. *e.g.* Dusky cotton bug, *Oxycarenus hyalinipennis*; groundnut pod bug, *Aphanus sorididus*.
- **Family Cicadellidae or Jassidae (Jassids or leaf hoppers): One or two rows of small spines are present on hind tibia.** Elongate insects with a **wedge shaped body** usually rest in a position ready for jumping. **It transmits tungro disease in rice.** *e.g.* Cotton leafhopper, *Amrasca biguttula biguttula*; paddy leafhopper, *Nephotettix virescens*; mango hoppers, *Amritodes atkinsoni*.
- **Family Delphacidae (Plant hoppers):** The characteristic feature of this family is the **presence of large mobile apical spur on hind tibiae.** e.g. Brown plant hopper of paddy (BPH), *Nilaparvata lugens*; white backed plant hopper on rice (WBPH), *Sogatella furcifera*.
- **Family Aphididae (Aphids or plant lice):** A pair of cornicles or siphonculi or wax tubes is present in the dorsum of fifth or sixth abdominal segments which secretes wax like substance. *e.g.* Tobacco aphid, *Myzus persicae*; cotton aphid, *Aphis gossypi*; groundnut aphid, *Aphis craccivora*.
- **Family Coccidae (scale insects or soft scales):** Metamorphosis complex. 1st instar nymph has legs and antennae and active known as crawlers after 1st

moult, become sessile a waxy or scale like covering is secreted. In males last instar preceding adult is quiescent and called pupa. Females have one less instar than males. *e.g.* Guava scale, *Pulvinaria psidi*; cottony cushion scale, *Icerya purchasi*; coffee green scale, *Coccus viridis*.

- **Family Aleurodidae (white flies):** The presence of vasiform orifice which opens on the basal surface of the last abdominal segment in both nymphs and adults. The opening is provided with an operculum and beneath it a tongue shaped organ known as **linguae** is present. The anus opens within the orifice at the base of the linguae. Honey dew is excreted through this orifice. The eggs are very characteristic being provided **with a pedicel**, which sometimes exceeds the length of the egg. *e.g.* Sugarcane whitefly, *Aleurolobus barodensis*; cotton whitefly, *Bemisia tabaci*; castor whitefly, *Trialeurodes ricini*.
- **Family Pseudococcidae (mealy bugs): Wings are absent in female but present in adult male.** *e.g.* Coconut mealy bug, *Pseudococcus longispinus*; brinjal mealy bug, *Planococcus insolitus*, citrus mealy bug; *Planococcus citri*; sugarcane mealy bug, *Saccharicoccus sacchari*.

Order Lepidoptera

- Lepido means scale; ptera means wings; common name: moth, butterflies and skippers. Lepidoptera with some **140,000 described species in 70 families is** one of the major orders of Holometabola. Large labial palps usually are present, but other mouthparts are absent, except that mandibles are present primitively in some groups. Larvae are polypod-eruciform type. Two to five pairs of fleshy unsegmented prolegs are found in the abdomen. At the bottom of the proleg crochets are present. **Pupa is generally obtect**. It is either naked or enclosed in a cocoon made out of soil, frass, silk or larval hairs. Majority of **Lepidopteran insects (97%)** are grouped under the **suborder Ditrysia** in which the female insects have two pores.
- **Family Noctuidae (Noctuid moths, army worms, cutworms):** Crochets on the larval prolegs are all of one size and arranged in semi-circle. Pupation is in an earth cell in the soil and some pupae are characterized by the presence of labial palpi and maxillae extending to the caudal margin of the wings. *e.g.* Redgram podborer, *Helicoverpa armigera*; tobacco cutworm, *Spodoptera litura*; fruit sucking moth of citrus, *Othereis fullonica*; climbing cutworm of paddy, *Mythimna separata*.
- **Family Sphingidae (Sphingid or hawk moths or horn worms):** Medium to large sized, heavy bodied powerfully flying moths with **spindle shaped body tapering and pointed both anteriorly and posteriorly**. Antennae are thickened in the middle or **towards the tip and hooked at tips**. Markings present on the thorax of the adult moth resemble human present on the thorax of the adult moth resemble **human skull**. *e.g.* Gingelly deaths head (robs honey from honey comb), *Acherontia styx*. It is a defoliator on gingelly.

- **Family Pyralidae (Snout moths): Labial palpi well developed and projected forward appearing as a snout in front** of the head and hence are called snout moths. **Forewings elongate or triangular with cubitus** appearing four branched. *e.g.* Paddy stem borer, *Scirpophaga incertulas*; jowar stem borer, *Chilo partellus*; brinjal shoot and fruit borer, *Leucinodes orbonalis*.
- **Family: Gelechiidae (Small moth): Forewings trapezoidal** and narrower than hindwings. *e.g.* Angoumois grain moth, *Sitotroga cerealella*; cotton pink boll worm, *Pectinophora gossypiella*; groundnut leaf miner, *Aproaerema modicella*.
- **Family Arctiidae (Tiger moths): Larva is either sparsely hairy or densely hairy (wooly bear)**. *e.g.* Red hairy caterpillar, *Amsacta moori*; black hairy caterpillar, *Pericallia ricini*; sunhemp hairy caterpillar, *Utetheisia pulchella*.
- **Order Coleoptera:** Coleo means sheath, pteron means wing, common name: beetles and weevils.The major synapomorphic feature is the development of the forewings as **sclerotized rigid elytra. Mesothorax and metathorax are fused with the first abdominal** segment. The **larvae are known as grubs** and generally thoracic legs are present. **Polyphaga** contains the majority (**> 90% of species**) of beetles, with about 300,000 described species.
- **Family Coccinellidae (Lady bird beetles or Coccinellid beetles):** The **brownish hemispherical** beetle has 12 - 28 black spots on the elytra. **Grubs are compodeiform and spiny**. The female lays elongate, spindle-shaped yellowish eggs in groups of 10 - 20 on the under surface of leaves. Spotted leaf beetle (or Hadda Beetle), *Henosepilachna vigintioctopunctata*; *Menochilus sexmaculata* predator on aphids; *Rodalia cardinalis* predator on cottony cushion scale.
- **Family Galerucidae or Chrysomelidae (Flea beetles, leaf beetles):** Usually small to medium, oval or spherical beetles with the upper surface of the body generally bare and shining, frequently with metallic colouration. **Larvae are root feeders while adult bite holes on leaves**. *e.g.* Sweet potato tortoise beetle, *Metriona circumdata*; rice hispa, *Dicladispa armigera*; red pumpkin beetles, *Raphidopalpa foveicollis*
- **Family Cerambycidae (Longicorn beetles): Antenna is long as or longer than the beetle itself. Grubs are called round headed borers**. They are apodous but have psuedopods both on dorsal and ventral side. **Grubs are wood borers**. They develop beneath the bark and tunnel into the branches or main stem. Grapevines stem gridler, *Sthenias grisator*; longicorn beetles on cucurbits, *Apomycyna pertigera*; mango stem borer, *Batocera rufomaculata*.
- **Family Curculionidae (Weevils, snout beetles):** Head prolonged to form a **pronounced snout or rostrum** of variable length and shape. **Palpi reduced and rigid, mouth parts small and arranged at the end of the snout**. *e.g.* Red palm weevil, *Rhynchophorus ferrugenius*; sweet potato weevil, *Cylas formicarius*; rice weevil, *Sitophilus oryzae*; Grey weevil, *Myllocerus subfasciatus* or *M. maculosus*.

- **Family Bruchidae (Pulse beetles, seed beetles): Elytra are short and do not cover the abdomen fully**. Larvae usually undergo **hypermetamorphosis**. First instar with well developed legs and possess spined or toothed thoracic plates to aid in entering smooth and hard seeds. After first moult it becomes partially or wholly apodous. *e.g.* Pulse beetles, C*allsobruchus chinensis*, *C. maculates*. It is serious pest on stored pulses.
- **Family Melonthidae or Scarabaeidae (Chafer beetles, dung rollers, root grubs):** Antennae 8-11 segmented and lamellate type. Grubs typically scraboid and feed on roots known as root grubs which are very serious pests of most of the crops, while some feed on organic matter living in manure pits. *e.g.* Groundnut white grub, *Holotrichia consanguinea*, it is a serious pest on groundnut under rainfed condition. Rhinoceros beetle on coconut, *Oryctes rhinoceros*; chafer beetles on fruit trees, *Anomala varians*.

Order Dictyoptera

- Dictyoptera is divide in to Dictyon means net work; ptera means wings; common name: cockroaches and preying mantids. **Head is usually hypognathous. Eggs are laid in ootheca**. It is divided into two suborders **Blattaria** (cockroaches) and **Mantodea** (preying mantids).
- **Family Mantidae (Preying mantids): Head is triangular and deflexed with filiform antennae**. The eggs are laid in water tight egg cases which are fixed to the plants. Head moves upto 180° into backside of mantid.
- **Famil: Blattidae (Cockroach):** A pronotum is a plate-like structure that covers all part of the dorsal surface. Leathery forewings with delicate hind wings. It has a pair of large compound eyes, each having over 2000 individual ommotidia and is a very active at night.

Order Hymenoptera

- Hymenoptera is derived from Greek word. Hymen means membranous; pteron means wing; common name: wasp, bees, ants, sawflies. Abdomen is basally constricted. The first abdominal segment is called **propodeum**. It is fused with metathorax. The first pair of abdominal spiracles is located in the propodeum. The second segment is known as **pedicel** which connects the thorax and abdomen. Abdomen beyond the pedicel is called **gaster or metasoma**. Ovipositor is always present in females. It is variously modified for oviposition or stinging or sawing or piercing plant tissue.
- **Family Tenthridinidae (Mustard saw fly):** Stout wasp like insects without abdominal pedicel. Resemble to those of lepidopteran larvae but are provided with 6 to 8 pairs of abdominal legs which are without crochets called **pseudocaterpillar**. Pupation takes place in an elongated oval silken cocoon or in an earthern cell. *e.g.* Mustard sawfly, *Athalia lugens proxima*

- **Family Apidae (Honey bees):** Body is covered with **branching or plumose hairs**. Mouthparts are **chewing and lapping** type. Legs are specialized for pollen collection. **Scopa (pollen basket) is present on hind tibia**. *e.g.* Indian honey bee, *Apis indica*; rock bee, *Apis dorsata*; little bee, *Apis floria* and Italian bee, *Apis mellifera*.
- **Family Trichogrammatidae (Egg parasitoids):** They are very tiny insects (**0.3 to 1.0 mm long**). Tarsus is three segmented. **Forewing is broad with pubescence** (microscopic hairs) in rows. Hind wing is reduced and **fringed with hairs along** with margins. *e.g. Trichogramma* sp. is extensively used in the biological control of sugarcane borers.
- **Family Ichneumonidae (Ichneumon flies** : Trochanter is two segmented. **Hind femur** is with trochantellus. Sternites of the gaster are membranous. *e.g. Eriborus trochanteratus* is an exotic larval parasitoid on coconut black headed caterpillar.
- **Family Braconidae (Braconid wasps):** Forewings have one recurrent vein. Sternites of the gaster are partly membranous. They are **gregarious parasitoid**. *e.g. Bracon brevicornis* is mass multiplied and released for the control of coconut black headed caterpillar.

Order Diptera

- Diptera is derived from Greek word. Di means two; ptera means wing; common name: True flies. Hind wings are highly reduced, knobbed at the end and are called **halter**. They are rapidly vibrated during flight. It is known as **balancing organ**. Larvae are **known as maggots**. They are apodous and acephalous. Pupa is generally with free appendages, often enclosed in the hardened last larval skin called **puparium**. Pupa is coarctate type.
- **Family Cecidomyiidae (Gall midges):** A dark sclerotised area is present midventrally on the pro-thorax in the larva **called 'chestbone'** *e.g.* Rice gall midge, *Orsella oryzae*. Larval feeding on plant and produces galls.
- **FamilyTephritidae (Fruit flies):** Sub costa bends apically. Wings are spotted or banded. *E.g.* Cucurbit fruit fly, *Dacus curcurbitae*
- **Family Tachinidae (Tachinid flies** : Head is large and free. *e.g. Stomatomia bezziana* on parasitoid on black heade caterpillar on coconut; *Exorista civiloides* parasitoid on many caterpillars.
- **Family Agromyzidae (Leaf miner flies):** Femora of the legs **bristled**. Most of the larvae are phytophagous, mine in the leaves producing characteristic blotches. *e.g.* Redgram pod fly, *Melanagromyza obtuse* and pea leaf miner, *Phytomyza atricornis*.

INSECT EXTERNAL STRUCTURES

Integument

- The **body wall or integument** of insects forms an exoskeleton covering the insect body.
- The body wall of an insect is consisting of three layers *i.e.* **cuticle, epidermis or hypodermis and basement membrane.**
- **Cuticle**: It is outermost thick layer of integument secreted by epidermis.
- **Epicuticle**: It is a thin outermost layer and **contains no chitin**. It is about 1-4μ thick. It consists of the following 4 layers.
- **Cement layer**: It is secreted by dermal glands and is composed of lipoprotein. It protects the body from external damage.
- **Wax layer**: It serves as water proof layer preventing water loss from the body.
- **Polyphenol layer**: It is **resistant to acids and organic solvents.**
- **Cuticulin layer**: It is composed of polymerized lipoprotein and serves the purpose of permeability and also acts as **growth barrier**.
- **Exocuticle** is much thicker layer consisting mainly of chitin and a hard protein called sclerotin which gives rigidity to the cuticle.
- **Endocuticle** is the inner most and thickest layer of cuticle which contains chitin and protein. It is soft, flexible, light coloured and unsclerotized. During moulting endo cuticle is dissolved.
- **Basement membrane**: It is the basal part of the body wall formed from degenerated epidermal cells. It is non cellular measuring about 0.5μ in thickness and consists of fibrous protein, glycosaminoglycans which are polymers of disaccharides.
- **Chitin**: It is the nitrogenous polysaccharide and long chain polymer of **N-acetylglucosamine**. The chitin content in cuticle is about 25 to 60 percent on dry weight basis.
- **Arthropodin**: An untanned cuticular protein, which is water soluble.
- **Sclerotin**: Tanned cuticular protein, which is water insoluble.
- **Resilin**: An elastic cuticular protein responsible for the flexibility of sclerites, *e.g.* wing articulatory sclerites.
- **Endoskeleton**: Cuticular inner growth of body wall providing space for muscle attachment is known as endoskeleton.
- Endoskeleton is two types *i.e.* **apodeme** is hollow invagination of body wall and **apophysis** is solid invagination of body wall.
- **Moulting (Ecdysis):** Periodical removal of remnants of old cuticle accompanied by the formation of new cuticle is known as **moulting or ecdysis**.
- The remnants of the old, empty exoskeleton are called **exuviae**.

- The time interval between the two moulting is called as **stadium** or stage and the form assumed by the insect during a particular stadium is called as **instar**.
- An insect that is actively constructing new exoskeleton is said to be in a **pharate** condition.
- Separation of old exoskeleton from epidermis is known as **apolysis.**
- Insect body is divided in to a series of segments, which in primitive arthropods are known as **"somites" or "metameres"**.
- Insect body is divided in to three regions or tagmata namely head, thorax and abdomen. This grouping of body segments in to regions is known as **tagmosis.**

Head

- It is the anterior part of the body formed by the fusion of **six segments** (preantennary, antennary, intercalary, mandibular, maxillary and labial segments) during embryonic stage. It bears a pair of antennae, a pair of compound eyes and three dorsal ocelli.
- **Compound eyes** are paired structure, dorso-lateral in position and are located on either side of the head.
- **Ocelli** are generally three in number and situated in between the compound eyes.
- **Tentorium:** Endoskeleton of head is known as tentorium.
- Head is connected to thorax by a membranous region is known as **neck or cervix**. The cervical membrane is quite **flexible and allows movement** of head.
- Epicranial suture: It is inverted **`Y'** shaped suture located at posterior between the eyes.
- Epistomal suture: It is located between frons and clypeus.
- **Post occipital suture:** It is the line indicating the fusion of maxillary and labial segment.
- Hypognathous (Hypo means below and gnathous means jaw) or orthopteroid type: The head is vertical; **mouth parts are ventrally** and project downwards. *e.g.* **Cricket, grasshopper, cockroach**.

Thorax

- Prognathous (Pro means infront and ganthous means jaw) or coleopteroid type: The head is horizontal and **mouth parts are directly forward**. *e.g.* **Tiger beetle, dung beetle, ground beetle**.
- Opisthognathous (Opistho means behind and gnathuos means jaw) or hemipteroid type: Head is deflexed and **mouthparts are directed backwards** held in between the forelegs. *e.g.* **Coried bug, stink bug.**

- The meso and metathorax are called as **Pterothorax**. The thorax also known as centre of the locomotory organs because the presence of legs and wings.
- The thorax is made up of three scleritic *i.e.* dorsal body plates (**tergum or nota**); ventral body plate (**sterna**) and lateral plate (**pleura**).
- Each segments of thorax bears a pair of legs; first pair of wing attached with mesothorax and second pair of wing found in mesothorax.
- Pterothoracic pleuron provides space for articulation of wing and leg. **Thoracic appendages are three pairs of legs and two pairs of wings.**

Abdomen

- In many Hymenoptera and other orders, the number of abdominal segments is reduced and in the honey bee only seven segments are visible. In the Collembola the abdomen has **only six segments**.
- Spiracles usually found in the conjunctive tissue between the **terga and sterna of abdomina l 1-8 segments**.
- Reproductive structures are located on the **9th segment** in males and on the **8th and 9th abdominal segments** in females.
- Male genitalia consist of a median intromittent organ or adeagus and lateral clasping organs called **parameres**.

Antennae

- The antennae are a pair of sense organs located between or behind the compound eyes. It is also known as **feelers**.
- The three basic segments of insect antenna are the scape (base), the pedicel (middle) and finally the flagellum, which often comprises many units known as flagellomeres.
- Both scape and pedicel are provided with intrinsic muscles. The remaining annuli or flagellomeres are known as flagellum or clavola which lack individual muscle.
- The pedicel (the second segment) contains the **Johnstons organ** which is a collection of sensory cells.
- Surface of the flagellum is supplied with many sensory receptors that are innervated by the **duetocerebrum of brain**.
- **Setaceous (bristle like):** The size of the segments decreases from base to apex of antenna. *e.g.* **Leafhopper, dragonfl , damselfl** .
- **Filiform (thread like):** The segments are usually cylindrical and same thickness of all the segments if antenna. *e.g.* **Ground beetle, cockroach, grasshopper**.
- **Moniliform (bead like):** Segments are globular or spherical with prominent constriction in between segments. *e.g.* **Termite**.

- **Serrate (saw like):** The antennae segments are short triangular projections on one side. *e.g.* **Longicorn bettle, pulse beetle.**
- **Bipectinate (double comb like):** The antenna segment has long slender projection on both the sides. *e.g.* **Silkworm moth.**
- **Clavate (clubbed):** The segments gradually broader towards the tip and last segment is round core. *e.g.* **Blister beetle.**
- **Capitate (knobbed):** Terminal segments become enlarged and make large cap. *e.g.* **Butterfl .**
- **Lamellate (plate like):** Antennal tip is expanded laterally on one side to form broad leaf like plates. *e.g.* **Scarabaeid beetle.**
- **Aristate:** The terminal segment is enlarged. It bears a conspicuous dorsal bristle called arista *e.g.* **House fl .**
- **Stylate:** Terminal segment bear a style like process *e.g.* **Horse fl , robber fl .**
- **Plumose (Feathery):** antenna with long whorls of hairs arises from the joint of each segment. *e.g.* **Male mosquito.**
- **Geniculate (Elbowed):** The first segment is long, second is short and flagellum is made of small segments with bent on the scape. *e.g.* **Ant, weevil and honey bee.**
- Antennae are useful for **hearing in mosquitoes** and communication in ants. Rarely is it also useful to **clasp the mate (flea** and grasp the prey.

Mouth Parts

- Mouth parts of insects are best examples to homologous organs. Organs which are similar in their origin and structure are called **homologous organs.**
- **Biting and chewing type:** The basic and most primitive type of mouthparts present in grasshopper, cockroach and beetles. In order of appearance, from anterior to posterior, chewing mouthparts consists of a single labrum (upper lip). **Epipharynx** is identified as a swollen area of the ventral surface of the labrum, which is an organ of taste. Distal teeth are sharply pointed and area called **incisors or cutting teeth**. The proximal teeth are called molar or grinding teeth. This is joined to the central body of maxilla, the stipes. On the outer side of the stipes is a more or less distance sclerite known as the **palpifer** to which the palpus is attached. The outer lobe is called galea and the inner lobe lacinia which is toothed. Whenever these two distinct lobes are fused together to form a single lobe (*e.g.* Coleopterous larvae) it is termed as **mala**. Distally prementum bears two pairs of lobes. The outer pair is called paraglossae and inner pair glossae. Both pairs when fused together to form a median lobe (*e.g.* Grasshopper), it is termed as **ligula.**

- **Piercing and sucking or hemipterous or bug type:** The labium is a long, four-segmented beak like hollow structure called as **rostrum or proboscis.** This enclosed four needle like structures called as stylets, in the dorsal groove. Each mandibular and maxillary stylets alternately have downward and upward movement for the purpose of piercing due to retractor and protractor muscles.
- **Piercing and sucking or dipterous or mosquito type:** Mouthpart of female mosquito consists of **six stylets** are composed of labrum-epipharynx (enclosing the food canal), the hyphophrynx (containing the salivary canal), two maxillae and two mandibles. The sensory lobes present at the tip of proboscis are known as **labellae** and they are connected by **Dutton's membrane**. *e.g.* Mosquitoes
- **Rasping and sucking: It is asymmetrical type of mouth parts because the right mandible absent**. *e.g.* Thrips
- **Chewing and lapping type: The labrum and mandibles are just like chewing type, while the labium and maxillae are modified into proboscis**. The glossa terminates into a small circular spoon shaped lobe called **spoon or flabellu** which is useful to lick the nectar. *e.g.* honey bee
- **Mandibulosuctorial type:** Mandibles are elongate sickle shaped and grooved on the inner surface. Each maxilla is elongated and fits against the mandibular groove to from a closed food canal. *e.g.* grub of antlion
- **Sponging type:** The labellum consists of two lobes known as **labellae.** Minute channels present on labellae are known as **pseudotrachea**. Labellum has sense organs of **taste and smell**. *e.g.* house fly
- **Siphoning type:** They are modified for sucking the juices from flowers or nectar from flowers. The proboscis which helps in sucking the food is formed by apposition of two galea of 1st maxillae. Maxillary palps and labial palps are present in reduced condition.During the resting phase proboscis is highly coiled and is situated beneath the head, it is also called **watch spring like proboscis**. *e.g.* moths and butterflies

Insect legs

- Each leg consists of **five segments *viz.*, coxa, trochanter, femur, tibia and tarsus**.
- Two segmented trochanter (only one is musculated) occurs **in the Odonata** and in some **Hymenoptera**, but the second trochanter actually appears to be a part of the femur.
- In some groups, the hind coxae are fused to the body wall (**Coleoptera: Adephaga**) or both the middle and hind coxae are fused to the body wall (**Lepidoptera**).
- Trochanter is usually a small segment; it has a dicondylic articulation with the coxa which limits movement to the vertical plane. *e.g.* Double trochanters in

Hymenoptera and Odonata.

- Femur is small in larval insects, but usually the largest and strongest leg segment in adult insects. **Femur is the largest and stoutest part of the leg**.
- A pad-like ventral surface or **euplantula (sometimes called pulvilli)** on the tarsal segments may enhance the traction of the insect.
- **Pretarsus of most insects** consists of a membranous base supporting a median lobe, the arolium and a pair of tarsal`claws.
- A median pad between the claws is usually known as **arolium** and a pair of pads, at their base is called **pulvilli**.
- There is no arolium in most Diptera, but rather a membranous pulvillus arises from the base of each auxillia, while a median empodium arises from the unguitractor (may be **spine-like or lobe-like).**
- **Ambulatorial (to walk):** Leg which is usually adapted for walking. Femur and tibia are long. Legs are suited for walking. *e.g.* Fore leg and middle leg of grasshopper, bugs (order Hemiptera), leaf beetles (order Coleoptera).
- **Cursorial (running leg):** Used leg for walking or running. Femur is not swollen. *e.g.* **Cockroach**.
- **Saltatorial (jumping leg)**: Hind legs adapted for jumping; characterized by an elongated femur and tibia Trochanter is fused with femur. Tibia is long and slender. *e.g.* **Hind leg of grasshopper**.
- **Fossorial (digging leg):** Fore legs and tibiae specialized for digging. The tibia is provided with four strong scrapers like prolongations at terminal end which are used for digging the ground and cutting the roots of the plants. Tympanum is present in fore tibia. *e.g.* **Fore legs of mole cricket, cicada.**
- **Scansorial (climbing leg):** Legs are modified for clinging to the hairs of its host. The coxa is widely separated and one or two segmented tarsi bear a single claw. Tibia is stout and at one side bears a thumb like process. *e.g.* **Head louse.**
- **Foragial leg (Pollens collecting and carrying leg):** The foreleg has three important structures. **Eye brush:** Hairs on tibia useful to clean the compound eyes. Antenna cleaner: Velum is a movable clasp present at distal end tibia. Antenna comb semicircular notch lined with small spines. **Pollen brush:** Bristles on basitarsus from the pollen brush which is useful to collect pollen from the head and mouth parts. *e.g.* **Honey bee.**
- **Sticking leg:** Pretarsus consists of a pair of lateral adhesive pads under spine like structure empodium is present. The pulvilli are covered with dense mats of tiny glandular hairs called tenent hairs. Secretions of these glandular hairs are helpful in clinging to smooth surface and to walk upside down on the ceiling. *e.g.* **House fl** .
- **Raptorial (grasping leg):** Fore legs modified for grasping. Coax is long and mobile. The femur is large and grooved along the ventral surface with spines on the two opposing surfaces of femur. *e.g.* **Forelegs of preying mantis**.

- **Natatorial (swimming leg):** Fore or hind legs adapted for swimming; charachterized by elongated setae on tarsi. Femur, tibia, and the first four tarsomeres are all broad and flattened. *e.g.* **Hind legs of water bug.**

Insect wings

- Wings not only organ of flight but also adapted as protective covers (Coleoptera and Dermaptera), thermal collectors (Lepidoptera), gyroscopic stabilizers (Diptera), sound producers (Orthoptera) and visual cues for species recognition and sexual contact (Lepidoptera).
- Wing venation and the lines of folding and flexion all contribute to patterns that can assist in identification. The lines of venation have been 'mapped' with a common terminology called the **Comstock-Needham system,** which recognises the homology of wing veins across the insect orders.
- **The Comstock-Needham** system was developed by **John Comstock and George Needham in 1898.**
- Arrangement of veins on wing surface is known as **wing venation,** which consists of two types of veins *i.e.* longitudinal veins extend from base of the wing to the margin and cross veins that interlink the longitudinal veins.
- The costa (C) is the vein that runs along the leading edge of the wing and it stems from a small sclerite, called the **humeral plate at the wings base.**
- **The jugal region** is found in some **Lepidoptera and Trichoptera**. It the posterior basal lobe below the jugal fold.
- The anterior (upper) part of the wing towards coastal margin where more no of longitudinal veins are present is called remigium. The posterior part of the wing where veins are sparsely distributed is known as **vannal area,** which is called as **clavus in forewings and vanus in hindwings.**
- **Jugum** is the inner most portion of the wing that is cutoff from the main wing by jugal fold. Wings are very thin broad leaf like structures strengthened by a number of hollow narrow tubular structures called veins.
- **Tegmina: Forewings** are leathery and tough. They protect the membranous hind wings. *e.g.* **Forewings wings of cockroach, grasshopper.**
- **Elytra:** Hard, shell like without clear venation. They form horny sheet and protect the membranous hind wings and abdomen. *e.g.* **Forewings beetles.**
- **Hemelytra:** The base of the wing is thick like elytra and the remaining half is membranous. This thickened portion is divided in to **corium, clavus cuneus and embolium**. They are useful of protection and flight *e.g.* **Forewings bugs.**
- **Membranous:** Naked thin with clear venation. Always useful of flight *e.g.* **Both the wings of dragonflies, bees and wasps, hind wings of grasshopper, beetles** *etc.*

- **Scaly wings:** Wings thin, membranous but covered with unicellular scales all over the surface .They are useful for flight *e.g.* both the wings of moths and butterflies.
- **Fringed wings:** Wings are highly reduced with reduced venation. The wings are fringed with long marginal hairs giving a feather like appearance. *e.g.* both the wings of thrips.
- **Fissured wings:** Forewings are longitudinally divided twice forming a fork like structure whereas hind wings are divided twice in to three arms. All the forks possess small marginal hairs. They are useful for flight. *e.g.* both the wings of plume moth.
- **Halteres:** The **hind wings** of houseflies are modified in to small microscopic structures called halteres and are divided in to three regions namely scabellum, pedicel and capitellum. They act as balancers. *e.g.* hind wings of housefly and front wings of male stylopids.
- **Pseudohalteres:** They are short and modified in to pseudo halters which are dumbbell shaped. *e.g.* **Front wings of Strepsiptera**.
- **Jugate type or jugum type:** The costal margin of the front wing possesses a small lobe at base called fibula. Which rest on the surface of the hind wing or sometimes engages with spines present on the upper surface of hind wings. *e.g.* **Primitive Lepidoptera of the family Hepialidae**.
- **Frenulum and retinaculum type:** The hind wings posses bristle or spine like structure or group of hairs known as frenulum. The forewings possess hook like retinaculum on anal side. During flight the frenulum passes beneath the retinaculum and thus the both the wings are kept together. *e.g.* **Moths**.
- **Amplexiform:** Costal margin of hind wing and anal margin of forewing overlap one above the other *e.g.* **Butterfl**
- **Hamuli:** Small curved hook like structures present on the costal margin of the hind wing known as hamuli that fit into the upward fold of the anal margin of the forewing. *e.g.* **Hymenopterans (wasps and bees).**

Insect sensory organs

- **The campaniform sensilla:** The campaniform sensilla cannot be seen externally but recognized from the dome-shaped cuticular areas. They elevated above or depressed below the general body surface. They occur in various parts of the body, wing-base, halters, cerci, palps and on the base of trochanter, femora, tibia and tarsal segments.
- **Chordotonal organ:** Chordotonal organ consists of single unit or group of similar unit is called scolopidia. They are sub-cuticular and are attached to the cuticle at one or both end often no sign of their presence. The chordotonal organs occur in legs at femoral, distal tibial and tibio-tarsal regions, in abdomen and wing base.

- **Mechanism of chordotonal organ:** They are stimulated by passive movement of segments, tension of muscles and internal pressure changed due to blood and tracheae. Jonstons organ is a specialized **chordotonal organ in the 2nd antennal segment,** occurs in all adult insect except **Collembola and Diplura**. It consists of single mass or several groups of scolopidia and is highly developed in Culicidae, where the pedicel is enlarged.
- **Location of chemoreceptor**: Organs of taste are common on the mouthparts, especially in palps, antennae (Hymenoptera), tarsi (many Lepidoptera, Diptera, and the honeybee), ovipositor (parasitic Hymenoptera and some Diptera) and general body surface.
- **Photoreceptors:** The ability to perceive light in visible or near visible range of the electromagnetic spectrum. Three types of photoreceptive structures found in insects: compound eyes, dorsal ocelli and lateral ocelli (stemmata).
- **The apposition mechanism:** It is a mechanism of image formation in the bright day light by most of the diurnal insects. The pigment cells envelop the ommatidia completely and thus the rays of light can enter from the central point of the dioptric apparatus. Different ommatidia finally produced a single complete, erect image of an object.
- **The super position mechanism:** It is a mechanism of image formation in poor light generally during night, mostly the nocturnal insects. At night time the pigment cells become contracts so that the light from the wide visual field can enter to the rhabdom of retinula cells obliquely. So light can enter centrally and obliquely.

Insect metamorphosis

- Series of changes that takes place during the development of an insect from egg to adult are collectively known as **metamorphosis**. Metamorphosis is derived from Greek word Meta=change, morph=form or structure.
- Stadium is the interval or time period between two moults is known **as stadium**.
- **Exuviae** is the skin shed during moulting process is known as exuviae. **Imago** (or) adult is the final stage of insect with well developed organs for reproduction, which emerges out from pupal body.
- **Sub-imago** is a pre adult stage with fully developed wings but without reproductive organs. *e.g.* **mayflies (Epheme optera).**
- **Ametabola (No metamorphosis):** These insects have only three stages in their life namely egg, young ones and adult. It is most primitive type of metamorphosis. The hatching insect resembles the adult in all respects except for the size and called as **juveniles**. Moulting continues throughout the life. *e.g.* **Silver fis** .
- **Paurometabola (Gradual, simple, directs metamorphosis):** The young ones are called nymphs. They are terrestrial and resemble the adults in general body

form except the wings and external genitalia. Their compound eyes and mouth parts are similar to that of adults. Both nymphs and adults share the same habitat. Wing buds externally appear in later instars. Later instars nymphs closely resemble the adult with successive moults. *e.g.* **Cockroach, grasshopper, termite, bugs.**

- **Hemimetabola (Incomplete metamorphosis):** In which, accessory organs known as **gills** are found in nymphal stages. These insects also have three stages in their life namely egg, young one and adult. The young ones are aquatic and are called as naiads. *e.g.* **Dragonfl , damselfly and may fly.**
- **Holometabola (Indirect, complete, complex metamorphosis):** These insects have four life stages namely egg, larva, pupa and adult. Compound eyes are absent in larva. Lateral ocelli or stemmata are the visual organs. *e.g.* **Butterfl , moth, fly and bee** .
- **Hypermetabola:** This is more complex type of metamorphosis in which there are several types of larva differing from each other after each instar. *e.g.* **Neuroptera, coleoptera (meloidae, carabidae *etc.*), strepsiptera and parasitic hymenoptera.**
- **Polymetabola:** In some insects from single egg more than one larvae have been seen to hatch out and this process in known as poly embryony or polymetabola. *e.g.* **Hymenoptera (Itonididae and Chalcididae).**
- **Intermediate metamorphosis:** In this case, insects may undergo either hemi or holometamorphosis. If they undergo holometamorphosis, there is a short pupal stage. *e.g.* **Coccids, thrips *etc.***

Insect larvae

- Immature stages of exopterygote insects are known as **nymphs** and endopterygote insects are known as **larvae**.
- **Protopod larva:** The larva emerge (hatch) from egg which is still in an early embryonic phase as the egg contain less yolk material. The larvae are partially developed. They possess well developed head and thoracic segments but lack segmentation in the abdomen. *e.g.* **endoparasitic Hymenoptera.**
- **Oligopod larva:** These are characterized by the presence of well developed thoracic legs, head capsule and without any abdominal legs. These are classified in to two types' *viz.*, **campodeiform and scarabaeiform.**
- **Scarabaeiform larva**: The body of larva is 'C' shaped, stout, sluggish and sub-cylindrical. Head is well developed and thoracic legs are short. *e.g.* **Grub of *Holotrichea* and rhinoceros beetle.**
- **Campodeiform larva**: The body is elongate, depressed dorsoventrally, well sclerotized and head is prognathus type. Larvae are mostly predators and are very active. *e.g.* **Grub of lady bird beetle and antlion.**

- **Polypod larva (Eruciform larva):** The larva possesses well defined segmentation of the body with three pairs of thoracic legs, 2-5 pairs of abdominal legs (3rd, 4th, 5th, 6th and 10th abdominal segment). They are phytophagous and destructive. *e.g.* **Caterpillar (larvae of moth and butterfly** .
  - **Hairy caterpillar:** Larval body is fully covered with hairs *e.g.* **Red hairy caterpillar (*Amsacta albistriga*), castor hairy caterpillar (*Pericalia ricini*).**
  - **Sphingid caterpillar or larva:** The larva consists of a horn (or) hook on the dorsal surface of 8th abdominal segment. *e.g.* ***Acherontia styx* (gingelly death's head moth).**
  - **Looper:** Only two pairs of abdominal legs present on 6th and last abdominal segment. During walking the insect body forms a complete loop like structure hence, the name looper. *e.g.* **mango looper, *Thalassodes quadraria*.**
  - **Semilooper:** First two pairs of abdominal legs (on 3rd and 4th segments) are reduced; hence a part of the insect body forms a small loop during its movement. *e.g.* **Castor semilooper, *Achoea janata*.**
- **Apodous larva:** These are characterized by the absence of trunk appendages or legs. They are usually derived from oligopod type. *e.g.* **Dipterans**
  - **Eucephalous:** The larva consists of a well sclerotized head capsule. *e.g.* **Sub order Nematocera, mosquito (Culcidae).**
  - **Hemicephalous:** Larva possess partially developed head capsule. *e.g.* **Sub order Brachycera, robberflies (Asilidae** .
  - **Acephalous:** The larva is characterized by the absence of head capsule and mouth parts are represented by mouth hooks. *e.g.* **Sub order Cyclorrhapa, muscidae (houseflies)**

Pupa

- It is resting, inactive stage of the holometabolous insects and transitional phase during which the **wings are developed and the insect attain matured sexual organs.**
- The pupa is incapable of feeding, locomotion except in some cases where they crawl *e.g.* **Neuroptera, mosquitoes.**
- **Decticous pupae** are possessing relatively powerful mandibles which are used for escaping of the adult from the cocoon *i.e.* to break the cocoon. *e.g.* **Neuroptera.**
- **Adecticous** pupae do not possess the mandibles but with the help of other appendages, adults escape from the cocoon *e.g.* **Lepidoptera, Diptera.**
- **Exarate pupa:** The pupae have appendages which are free without any secondary attachment to the body. *e.g.* **most of the Coleoptera.**
- **Obtect pupa:** The pupae have appendages which are firmly pressed against the body and the pupa is highly chitinized. *e.g.* **Lepidoptera (moths).**

- **Coarctate pupa:** The pupa remains enclosed in a puparium formed by the last larval skin and the pupa looks like a capsule or barrel. *e.g.* **Cyclorrhapha of Diptera (housefly** .
- **Chrysalis pupa:** It is an obtect type of pupa which has golden colouration and a stalk *e.g.* **butterflie** .

INSECT INTERNAL STRUCTURE

Digestive system

- The alimentary canal of insects is a long, muscular and tubular structure extending from mouth to anus. Insects having the habit of feeding on the solid food material, possess the biting and chewing type of mouth parts where as sap feeding ones have sucking type. Sap suckers possess a filter chamber and solid feeders have well developed gizzard.
- The alimentary canal in insects extends from mouth to anus which is divided in to an anterior stomodaeum (**foregut**), middle midgut (**mesenteron** or stomach or ventriculus) and posterior hindgut (**proctodaeum**).
- The **foregut and hindgut** are ectodermal in origin where as the midgut is endodermal in origin. Compared to the carnivores or sap suckers, the **phytophagous solid feeders possess longer alimentary canal.**
- **Foregut:** It is the anterior part of the alimentary canal which starts with the mouth cavity and ends with the gizzard or proventriculus. It is divided in to pharynx, oesophagus, crop and gizzard. It is ectodermal in origin.
- Crop is the dilated distal part of oesophagus acting as food reservoir. **In bees crop is called as honey stomach where nectar conversion occurs.**
- **Proventriculus or gizzard** is the posterior part of foregut and is musculated. Food flow from foregut to midgut is regulated through cardial or oesophageal valve.
- **The mosquitoes (Diptera)** are a **blood-feeding insect** that is released **anticoagulants and blood thinners** from the **foregut.**
- **Midgut:** It is endodermal in origin. The mesenteron is a simple tube bearing **six large enteric caecae** at the anterior end and numerous malpighian tubes at the posterior end of midgut.
- **Peritrophic membrane** is the internal lining of midgut, secreted by anterior layer of midgut epithelial cells. The peritrophic membrane is separates food from midgut tissue, composed of chitin and proteins.
- **Pyloric valve** is used for mid open in to hindgut and regulate food flow.
- **Filter chamber** also helps in osmo-regulation by preventing dilution of haemolymph. In aphids, the honey dew (rich in sugars) is the substance that is being excreted after passing through the filter chamber.

- **Hindgut:** It is ectodermal in origin and produced by the posterior invagination of ectoderm. The hindgut has three portions **ileum, colon and rectum.**
- The **ileum** is a short tube which tapers posteriorly and on the inside has folds to increase the absorptive capacity.
- The colon is not easily differentiated externally and the rectum is the swollen terminal portion which **bears six papillae**.
- **Rectum** contains rectal pads helping in dehydration of faeces and it opens out through anus.
- **Digestion in insects**: The food ingested by the insects through the mouth cavity enters into the alimentary canal, get digested and the undigested waste material is excreted through anus.
- Ingestion: Food is partially digested in the oral cavity with the help of salivary enzymes. In insects like fluid feeders, carnivorous hemipteran, blowfly larvae *etc*, digestion occurs outside the intestine by a process called **extra intestinal or extra oral digestion.**
- Transportation: The food material entered in to the oesophagus is transported in to the crop by muscular activity that is by **peristatic movements**.
- In termites, digestion takes place in **colon of hindgut where mycetomes** (group of cells which harbour the micro organisms like protozoans) secrete the enzyme cellulase which can digest the wood material rich in cellulose.
- In scarabid beetle larvae, **bacteria** are involved in digestion. In wood feeders, keratin digestion is facilitated by alkaline pH of midgut. In *Tineola* (cloths moth), keratinase secreted by protozoans.
- Absorption: **Midgut epithelial cells absorb** the nutrients from the digested food and pass on the faecal matter and undigested food material in to the hindgut. The Malpighian tubules maintain ionic balance by absorbing Na and K salts from the blood.
- **Egestion:** The waste food material is discharged through the anus due to the action of the anal muscles.

Insect circulatory system

- There are two types of circulatory systems in the animal kingdom. In many animals, the blood travels through vessels like arteries, capillaries and veins. This is known as **closed type of circulatory system**.
- In insects the blood flows through the body cavity (*i.e.* haemocoel) irrigating various tissues and organs. It is known as open type of circulatory system.
- **Dorsal or pericardial sinus**: The area lying in between the tergum and dorsal diaphragm. It contains heart.
- **Ventral or perineural sinus**: The area lying in between the sternum and ventral diaphragm. It contains nerve cord.

- **Visceral sinus**: The area in between dorsal and ventral diaphragms. It harbours the visceral organs like alimentary canal and gonads.
- Aorta: It is the anterior part of the dorsal blood vessel and functions as principal artery. **It is present in the thoracic region** and opens in to the head near the brain. Its attachment with the heart posteriorly is marked by an aortic valve.
- **These alary muscles appear to be distributed fan like over the heart**. Heart consists of number of chambers marked by constrictions and the presence of the opening called the incurrent ostia which allow the entry of blood from pericardial sinus in to the heart.
- Accessory pulsatile organs: Insects consists of sac like structures called accessory pulsatile organs, which are present at the base of the appendages such **as wings, legs and antenna.** They pulsate independently and supply adequate blood to the appendages.
- Circulation of blood takes place in two phases due to the action of the alary muscles as well as the muscles of the walls of the heart.
- First is diastole means during which expansion of heart takes place and second is **systole means contraction of heart** takes place.
- **Diastole:** Expansion of heart (diastole) occurs, when the alary muscles that are spread fan like over the heart and connected to the tergum get contracted. It results in increase of volume of heart and decrease in the area of pericardial sinus.
- **Systole:** Contraction of heart (systole) is brought about by the expansion of the alary muscles as well as contraction of the muscles of the heart wall.
- In between diastole and systole there will be a short period of rest which is known as **diastasis.**
- Blood is colourless (or) green (or) yellowish with different types of haemocytes and plasma. Green colour is due to chlorophyll dissolved in the plasma and red colour is due to **haemoglobin in Chironomus midge**.
- Blood covers up 5-40% of the total body weight that vary with the sex and stage (or) age of the insect.
- pH of the blood generally varies from **6-7**.
- The blood sugar of insects is **trehalose**.
- Blood lacks vitamin 'K'.
- **Phagocytosis:** To get protection from micro organisms like bacteria, viruses and fungi, the haemocytes completely engulf the foreign body and gets autolysed (this is the principal function of haemocytes).

Insect Excretory system

- The organs of insect body involved in the elimination of excess or unwanted materials either toxic or not useful are together known as **excretory system**.

- The organs that are involved in the process of excretion are malpighian tubules, integument or body wall, tracheal system, alimentary canal, nephrocytes, urate cells, oenocytes, labial glands and chloride cells.
- **Malpighian tubules:** These are discovered by an **Italian scientist, Marcello Malpighi** in the year 1669, which were named after him by **Heckel** in 1820.
- The malpighian tubules long, tubular structures which open proximally in between midgut and hindgut and closed distally, floating freely in the haemolymph.
- Their number varies from **2-250 (in coccids -2; in locust-250)**. The shape of tubules may be sac like or papillae like or branched. **Malpighian tubules are absent in aphids and Collembola**.
- In some of the insects such as caterpillars and coleopterans, the distal ends of the malpighian tubules get reattached to the alimentary canal by opening in to the rectum of hindgut. This condition is called **'cryptonephridial condition'**.
- The cryptonephridial arrangement is concerned with **re-absorption of water from rectum**.
- **Malpighian tubules:** Helps in the process of excretion or removal of waste products in order to regulate the internal body environment by maintaining ionic and water balance. In case of glow worms, the distal ends of tubules produce light energy. In case of aphid lion (*Chrysoperla*), the secretions of the tubules produce stalked of eggs. In case of spittle bugs spittle around the immature stages is also a malpighian secretion.
- In some insects, where respiration occurs through body wall, CO_2 is removed through integument as waste product (**cutaneous respiration**).
- **Nephrocytes:** These are the special cells that are distributed in the body cavity and scattered. Nephrocytes are cells that take up foreign chemicals of relatively high molecular weight which Malpighian tubules are incapable.
- **Oenocytes:** These are large cells and are usually present near the abdominal spiracles. **They arise from the ectoderm or hypodermis**.
- **Urate cells:** Some of the fat body cells which store urea or uric acid in the form of granules are known as **urate cells**. Preserved uric acid can be utilized subsequently.
- **Labial glands:** These are seen in **Collembola, Diplura, Thysanura**. They consists of a sac like structures called ampulla that leads to a long coiled labyrinth that open at the base of labium in the head. These glands help to remove ammonia.

Insect respiration

- In insects, exchange of gases takes place through tubular **structures called trachea**. They are distributed throughout the body collectively forming tracheal system. These trachea open outside on the body wall through small **openings called spiracles.**

- The trachea is divided into very fine branches **known as tracheoles.** They supply oxygen to the body tissues.
- The tracheal system with functional spiracles is called the **open tracheal system** and with non-functional spiracles is called **closed tracheal system.**
- **Trachea is fine elastic tubular structures which are ectodermal in origin.** They consist of cuticle, epidermis, basement membrane as in case of general body wall but arranged in reverse manner *i.e.* basement membrane forms the outermost coat of trachea.
- **These spiral thickenings are known as 'taenidia'** which give support to the trachea without being collapsed when there is no air. It consists of chitin, resilin in protein-chitin matrix.
- The trachea ramifies into **very fine branches known as tracheoles,** which are about 0.1-1 μm in diameter. These tracheoles are formed in to cells called **tracheoblast or tracheolar** end cell, which are derived from epidermal cells and **lining the trachea.**
- Tracheoles form a network over the visceral organs including the alimentary canal as well as the **gonads ovaries, testis and penetrate in to the tissues of the organs** and become intracellular and supply oxygen directly to the tissues.
- As the head do not contain spiracles, air is supplied through the first pair spiracles by means of two main branches of the dorsal longitudinal trunk, where one branch supply O_2 to eyes, antenna, brain; other branch to mouthparts and muscles of the head.
- Spiracles: They are the openings of the internal tubular trachea. **Except in Diplura,** in all the orders, spiracles are absent in prothorax and distributed in meso, metathorax and abdomen. A total of ten pairs are present in general in which two pairs in thorax and eight pairs in abdomen.
- Spiracles are situated on pleural surface. They consist of a small ring like sclerite at **opening called peritreme leading to a cavity known as atrium.**
- The hydrophobic nature of spiracles is also due to the presence of modified epidermal glands known as **peristigmatic glands** which secrete a hydrophobe material preventing the wetting of these organs.
- **Holopneustic:** These are primitive type with two pairs of spiracles on thorax and eight pairs on abdomen. All the spiracles are functional (1+1+8). *e.g.* **Dragonflies, grasshoppers an cockroach.**
- **Hemipneustic:** One or more pairs of spiracles become non-functional.
- **Peripneustic:** Metathoracic spiracle is closed (1+0+8). *e.g.* **Larvae of Lepidoptera, Hymenoptera and Coleoptera.**

- **Amphipneustic:** Only mesothoracic and last pair of abdominal spiracles is open (1+0+1). *e.g.* **Larva of cyclorrhapha (Diptera).**
- **Propneustic:** Only one pair *i.e.* mesothoracic spiracles are open (1+0+0). *e.g.* **Mosquito pupa.**
- **Metapneustic:** Only last pair of abdominal spiracles are open (0+0+1). *e.g.* **Mosquito larvae.**
- **Apneustic:** No functional spiracles. *e.g.* **Mayfly larva and nymph of Odonata.**
- **Hypopneustic:** One or two pairs of spiracles may completely disappear or absent *e.g.* **Siphunculata and Mallophaga.**
- **Hyperpneustic:** More than 10 pairs of spiracles are present *e.g.* ***Japyx* sp. (dipluran).**
- **Cutaneous respiration**: When the spiracles are absent, respiration occurs through body wall which forms main source for gaseous exchange. *e.g.* **Protura, Collembola and endoparasitic insects.**
- **Tracheal gills:** Also called as **abdominal gills** which occur as the outgrowths of the trachea in the form of gills distributed on the lateral sides of the body. They are useful for absorption of dissolved oxygen. They may vary in shape as lamellate or filamentous. *e.g.* **Larva of Trichoptera and nymphs of Ephemeroptera.**
- **Spiracular gills:** Peritreme or atrium of spiracles is drawn out in to a long filament like structure known as spiracular gills. *e.g.* **Some aquatic pupae.**
- **Blood gills:** These are tubular or digit form or eversible structures present at the anal end of body ranging from 4-6 in **larva of Trichoptera**. In **chironomid larva of Diptera, 2 pairs of blood gills** are present on penultimate segment and a group of 4 shorter anal gills are present.
- **Rectal gills:** The rectum modifies in to a barrel like chamber where the rectal wall forms in to basal thick pads and distal gill filaments which are richly supplied with tracheoles. They help in respiration. *e.g.* **Dragonfly nymphs (naids)**
- **Air sacs:** In many winged insects, the trachea gets dilated at some points to form thin walled **air sacs which do not contain the taenidia**. These can be seen as glistening sac like structures `mainly function as storage structures of air which change their volume with respiratory movement.
- **Plastron respiration:** The plastron is a special type of air store in the form of a thin film held by a system of hydrofuge hairs, scales or other cuticular processes whose volume remains constant. If there is adequate oxygen dissolved in water, the plastron can act as a permanent physical gill. The trachea opens in to plastron. *e.g.* **Aquatic beetles.**

Insect nerve system

- The nerve cells are called neurons which are derived from ectoderm. Each neuron consists of a prominent nucleated cell body known as perikaryon or soma and an elongated cytoplasmic thin fibre called the **axon** and group of small branches called the **dendrites**.

- The axon gives lateral branches called collaterals. Both the axon and collaterals end in fine fibrils known as **terminal arborizations**.
- **Unipolar or monopolar:** These have a single axon without collaterals and dendrites.
- **Bipolar:** These have either collaterals or dendrites in addition to axon.
- **Multipolar:** Neurons have an axon with several collaterals and dendrites.
- **Sensory or afferent:** Present just beneath the integument and associated with sensory organs. Carry impulses from sense organs to the central nervous system.
- **Motor or efferent neurons**: Always unipolar or monopolar carry impulses from central nervous system to the organs.
- **Association or internuntial neurons:** Associated in between sensory and motor neurons, usually present in ganglia, consists of axons of sensory neurons and soma of motor neurons. The transverse commissures are also formed with these neurons.
- The point at which neurons receive information from or convey to another neuron is known as synapse. **Synaptic gap is approximately 1000A.**
- Nervous system can be grouped in to three parts: **Central nervous system** (CNS), visceral or sympathetic nervous system and peripheral nervous system.
- Protocerebrum: It is formed by the union of the ganglia of pre-antennary segment and forms the greater part of the brain. It gives nerve connection to the **compound eyes and ocelli**.
- Deutocerebrum: It is formed from the ganglia of antennary segment and **innervates antenna**.
- Tritocerebrum: It is formed by the union of ganglia of third or intercalary segment and is relatively small. The lobes of it are attached anteriorly to deutocerebrum and **posteriorly to suboesophageal ganglion**.
- **Sub-oesophageal ganglion** is the ventral ganglionic centre of the head formed by the union of ganglia of the gnathocephalic segments. It gives nerves to **mandibular, maxillary, labial segment, labrum, salivary ducts, part of cervical muscles in the neck region and corpora allata**.
- Oesophageal sympathetic or stomatogastric nervous system: It is directly connected with the brain which supplies nerves for the anterior part of the **alimentary canal (foregut and midgut), heart and certain other parts**. It is dorsal in position.
- Ventral sympathetic nervous system: Consist of a pair of transverse nerves that are connected with each ganglia of VNC. The transverse nerves **pass to the spiracles** of that particular segment.

- Caudal sympathetic nervous system: Some additional nerves arise from posterior compound ganglion of VNC which supply nerves for the posterior **part of the gut and the reproductive system**.
- Peripheral nervous system: It includes all the nerves coming from the ganglia of CNS and that of the **visceral nervous system**.
- Neurosecretory cells: A pair of median and lateral neurosecretory cells is present. The axons of these neurosecretory cells form two pairs of nerve **corpora cardiac ending in carpora cardiaca**.
- Corpora cardiaca: It consists of paired bodies composed of two parts *i.e.,* nervous tissues and glandular tissues. It acts as a storage and release organ for neurosecretory cells. It controls **heart beat and regulate trehalose level in haemolymph**.
- Corpora allata: Corpora alluta are glandular organ closely associated to corpora cardiaca and **secretes juvenile hormone (JH)** there by inhibit metamorphosis. It is needed for egg maturation and functioning of male accessory glands.
- **Precocene** is an anti JH which induce precocious metamorphosis and death in insects.
- Prothoracic glands: These are paired gland present in ventrolateral part of prothorax of larva and are degenerated in adults. It secretes the **moulting hormone ecdysone**. These glands are activated by neurosecretory cells activate prothoracic glands to secrete ecdysone.

Insect reproductive system

- Oviparity that insects reproduce by laying eggs by the female on any substrate either singly or in mass or in groups which later hatch and produce the young ones. *e.g.* **Moths and butterflie** .
- **Viviparity** is the phenomenon of reproduction where the female gives birth to the young ones instead of laying eggs. Embryonic development is completed within the female and the embryo gets nourishment from the mother and produces the larva or young ones within the haemocoel.
- **Ovo-viviparity:** Insects retain the eggs within the genital track until the eggs are ready to hatch or giving birth to young ones. Hence, no special structures are developed for nourishment. *e.g.* **Thysanoptera**.
- **Adenoparous viviparity:** It is a type of viviparity where the eggs have sufficient yolk, complete their embryonic development and retain in the uterus. Eggs hatch and the young ones get nourishment from special nourishment glands called milk glands which contains milk that have lipids and proteins. *e.g.* ***Glossina pupipara*** **of Diptera**.
- **Pseudoplacental viviparity:** It is a phenomenon where insect have eggs with little or no yolk. Hatching takes place within the mother and the nourishment

for the young one is received through embryonic maternal structure called pseudoplacenta *e.g.* **Psocoptera, Dermaptera, aphids *etc.***

- **Gynandromorph (sexual mosaic):** Abnormal individual with secondary sexual characters of both male and female. *e.g.* **Mutant *Drosophila*.**
- **Haemocoelous viviparity:** It is a type of reproduction where the eggs are retained within the haemocoel and the embryonic development as well as the nourishment of young one takes place through the transfer of nutrients from the haemolymph of mother. *e.g.* **Strepsiptera and some larvae of Cecidomyids (Diptera).**
- **Parthenogenesis:** It is the ability of the females to reproduce without fertilization or copulation with males. This usually occurs due to the genetic characters, due to heredity, failure in finding a mate, hormonal changes within the body and weather factors. This parthenogenesis is classified as sporadic parthenogenesis: occurs occasionally *e.g.* **Silkworm.**
- Constant parthenogenesis occurs regularly. *e.g.* **Thrips.**
- **Cyclic parthenogenesis:** It is nothing but the alternation of generations where parthenogenesis occurs in alternation with the sexual reproduction. *e.g.* **Aphids.**
- Arrhenotoky: Only males are produced. *e.g.* **Hymenoptera.**
- Thelytoky: Only females are produced. *e.g.* **Acridids.**
- Amphytoky: Both females and males are produced. *e.g.* **Hymenopterans.**
- Paedogenesis or Neoteny: It is a phenomenon where the immature insects or stages give birth to young ones. This usually occurs due to the hormonal imbalance. *e.g.* **Cecidomyids.**
- **Polyembryony:** It is a type of reproduction where insects reproduce by giving birth to two or more young ones instead of a single one as two or more embryos are produced from a single egg. *e.g.* **Endo parasitic Hymenoptera like *Platygaster*.**
- **Hermaphroditism:** It is a type of reproduction where both male and female gonads are present in the same individual. **It may be a functional hermaphroditism as in case of *Icerya purchsi* and non functional as in case of stonefl , *Perla marginata*.**
- **Castration:** It is a type of reproduction where the separation of the individuals occurs mainly due to the development of the reproductive organs. The insects with well developed ovaries develop in to females (queens), the insects with well developed testis develop in to males (drones) and insects with underdeveloped ovaries develop in to workers. *e.g.* **Social insects such as honey bees.**
- **Alternation of generation:** Insects reproduce by parthenogenesis and bisexual reproduction by alteration of generations. *e.g.* **Aphids which reproduce by parthenogenesis in summer and undergo sexual reproduction in winter.**

- **Segments 8 and 9 bear the genitalia**; segment 10 is visible as a complete segment in many lower insects but always lacks appendages and the small segment 11 is represented by a dorsal **epiproct and pair of ventral paraprocts** derived from the sternum.
- A pair of appendages and cerci articulates laterally on segment 11 typically these are annulated and filamentous but have been modified (*e.g.* **the forceps of earwigs**) or reduced in different insect orders.
- An annulated caudal filament and median appendix dorsalis are arises from the tip of the epiproct in apterygotes like **mayflies (Ephemeroptera)** and a few fossil insects.
- Spermathecae store sperm for varying periods of time and along with portions of the oviducts can control sperm use. **The ducts and spermathecae are lined with a cuticle.**
- The ovaries are made up of a number of egg tubes **called ovarioles,** which vary in size and number by species.
- Accessory glands or glandular parts of the oviducts produce a variety of substances for sperm maintenance, transport, and fertilization, as well as for protection of eggs. They can produce glue and protective substances for coating eggs or tough coverings for a batch of **eggs called oothecae**.
- **Spermathecae** are tubes or sacs in which sperm can be stored between the time of mating and the time of egg is fertilized.
- **Panoistic ovarioles:** The nutritive cells are absent and the development of oocytes takes place with the help of follicular epithelial cells *e.g.* **Odonata, Dictyoptera, Orthoptera and Ephemeroptera.**
- **Meriostic ovarioles:** They contain trophocytes or nutritive cells which vary in their position.
- Based on the position of trophocytes meriostic ovarioles are classified into polytrophic ovarioles where developing oocyte and trophocytes arranged alternatively within the vitellarium. *e.g.* **Mecoptera, Dermaptera and Psocoptera**.
- **Acrotrophic ovarioles:** Also called **teletrophic ovarioles** where the trophocytes are present in the germarium (apex) and are connected with the growing or developing oocytes by cytoplasmic strands. *e.g.* **Hemiptera and Coleoptera**.
- The main component of the male reproductive system is the testis suspended in the body cavity by tracheae and the fat body.
- The more primitive apterygote insects have a single testis, and in some lepidopterans the two maturing testes are secondarily fused into one structure during the later stages of larval development although the ducts leading from them remain separate.

- A portion of the vas deferens is often enlarged to form the seminal vesicle which stores the sperm before they are discharged into the female.
- The ejaculatory duct is derived from an invagination of the epidermal cells during development and as a result has a cuticular lining.
- The terminal portion of the ejaculatory duct may be sclerotized to form the intromittent organ the aedeagus.
- The remainder of the male reproductive system is derived from embryonic mesoderm, except for the germ cells or spermatogonia which descend from the primordial pole cells very early during embryogenesis.
- Which type of ovarioles, where nutritive cells are present and alternate with oocytes **Polytrophic.**
- During embryonic development germ band of many insect undergo sequence of movement is known as **Blastokinesis.**
- The nutritive cells absent in which ovariole **Panoistic.**

BENEFICAL INSECTS

Honey bee

- **Apiculture or bee-keeping** is the art of caring for and manipulating colonies of honeybee in large quantity over and above their own requirement.
- The rearing of bees is called **Apiculture or Bee keeping**. The location where the honey bees are reared is **called apiary**.
- The wild and cultivated flora which are source of pollen and nectar to the honey bees are collectively called as **bee pasturage or bee forage.**
- Honey bee is originated from **Indian and South East Asia.**
- Rev. L. L. Langstroth found **Bee space** which helped to improve bee hive in 1851.
- **Bee space or passage** way is the space required between the any two frames for the bees to move about conveniently between the combs and is too small to encourage bees building combs and too large to induce their depositing propolis.
- Scientific way of beekeeping was started in **1907 at Pusa Institute on Indian and Italian bees**.
- Rev. Fr. Newton of Tiruchirappalli developed a hive in 1919 with movable frames (**Newton's Bee hive**) and successfully maintained a few hives at Shenbaganur in Kodaikanal hills and at Marthandam in Kanyakumari district.
- **Rock bee, *A. dorsata*:** Rock bees' also known as giant bee is common in plains and hills up to 1200 m altitude in the country. The annual **yield is 37 kg per year per colony**.

- **Little bee, *A. florea*:** It is known as the little bee since it is the smallest. Builds single small combs in bushes, hedges, *etc.* It yields very little honey, about **0.5 to 1 kg per year per colony.**
- **Indian bee, *A. indica*:** The Indian hive bee is also considered as sub-species of the Asian or Oriental or Eastern hive bee, *Apis cerana.* **Honey yield is 2 to 5 kg per colony per year**. A race of this species is found in hilly tracts and they are larger in size with darker in colour.
- **Dammer bee or mosquito bee, *M. iridipennis*:** The dammer bee is tiny with a vestigial sting. It inhabits crevices in walls and hollow trunks of trees. The comb is made up of a dark material called **'Cerumen'** which is a mixture of wax and earth or resin. Honey yield is **60 to 180 ml per colony per year.**
- **Italian bee, *A. mellifera*:** European bees originated in Italy and have been introduced into all the countries of the world where they formed the well recognized races. It has a prolific queen, swarms less and has good honey gathering qualities, yielding about **45 to 180 kg per year per colony**.
- The queen is a **fertile female** larger than the other two castes. Later, it goes out of the colony on mating or nuptial flights followed by several drones. During the mating process, the stock of sperms of male bee goes into the **spermatheca of the queen, where they can survive for over 3 years**. A queen can lay up to **15,000 eggs per day**. It is fed on royal jelly, **a highly proteinous, enzyme and hormone rich** food produced by the worker bees. **The worker brood cells receive fertilized eggs, while the larger drone cells receive unfertilized egg.**
- Worker bees are **infertile female** developed from fertilized eggs. A bee colony of *A. cerana indica* has 15,000 workers and a strong one has 80,000. The **workers (sterile females)** are smaller than drones and queen. These have underdeveloped reproductive organs due to the influence of the pheromones produced by the queen. At any given time, a colony may have 2 or more batches of worker bees, having different male parentage. A **worker bee has a life span of 6 weeks.**
- The drones are male developed from **unfertilized eggs** parthenogenetically. The male bees are produced in the colony for the purpose of **mating and fertilizing the queen bees**. The larvae that develop from **unfertilized eggs are fed royal jelly and give rise to drones**.
- The transformation of 3 castes depends upon the amount of **brood food or royal jelly produced by pharyngeal salivary glands of worker** fed to larvae.
- Swarming is a method of **reproduction**. A part of the colony migrates to a new site to reproduce a new colony.
- **Floral fidelity** A bee visits same species of plant for pollen or nectar collection until exhausted. Bees travel 2-3 km distance to collect pollen or nectar.
- **Queens retinue:** A group of house bees form a circle around the queen and attends its feeding, cleaning and assists her in egg laying. Bees in this retinue

constantly lick the queen's pheromones and share this substance with other house bees.

- Supersedure is queen replacement without colony division. A failing queen who is unable to lay as many eggs as required, or who begins to run out of spermatozoa and lay a high proportion of male eggs will need to be superseded by supersedure queen. **The worker bees are induced to build supersedure cells which are 2 or 3 large queen cells** in the middle of the nest, unlike a dozen of comparatively small queen cells build at edges of nest during the swarming period.
- **Emergency queen:** In absence of queen substance due to death of queen, the workers are stimulated to get set for producing an emergency queen.
- **Bee dance**: The bees that go out first to find out new sources of food materials are called **searchers or scout bees.** They return to the hive and communicate the message to young foraging bees by means of definite patterns of **dancing is called bee dance.**
- The **scientist Karl Von Frisch,** who won **Nobel Prize in 1946**, accomplished the decoding of the **dance language of the honey bees.**
- The foraging source is rich the scout bees open the scent glands **situated in 5th and 6th abdominal tergites and the odour emitted** attracts the other forages to the vicinity.
- **Round dance:** When a scout bee finds a plant with flowers **within 50 m of the hive**, the round dance is performed. The scout bee runs one to two circles clockwise and then anti clockwise.
- **Wagtail dance:** The food source is far away from the **colony about 100 meters or more**, this dance is performed. Here the bees make a half circle on one side and then run in a straight line to starting point and make a half circle in the opposite direction, thus completing a full circle.
- **American foulbrood disease (AFB):** Beekeepers in temperate and sub-tropical regions around the world generally regard **American foulbrood (AFB)** as possibly the most destructive bacterial disease affecting bee brood. It is widely distributed wherever colonies of *Apis mellifera* are kept. American foulbrood disease is caused by a **spore forming bacterium, *Paenibacillus***, which only affects bee brood; adult bees are safe from infection. **Antibiotics and sulfonamides** prevent multiplication of the agent though it will not kill the spores.
- **European foulbrood disease (EFB):** Reports from India indicate that *A. cerana* colonies are also subject to EFB infection. The pathogenic bacterium of EFB is *Mellissococcus pluton*. **Honey bee larvae killed by EFB are younger than those killed by AFB.**

- **Chalkbrood disease (*Ascosphaerosis*):** In Asia, chalkbrood is rarely considered to be a serious honey bee disease and **Chalkbrood is a disease caused by the fungus *Ascosphaera apis*.**
- Stone brood: Stonebrood is a fungal disease caused by *Aspergillus fumigates, Aspergillus flavus* and *Aspergillus niger.* It causes mummification of the brood of a honey bee colony.
- **Sacbrood disease:** Sacbrood disease (caused by *Morator aetotulas*) is perhaps the most common viral disease of honey bees. In Asia, at least two major types have been recorded. Sacbrood disease that affects the common honey bee ***Apis mellifera*** and the sacbrood disease of the Asian hive bee ***A. cerana.***
- **Acute bee paralysis virus (ABPV) or (APV):** It is considered to be a common infective agent of bees. It is frequently detected in apparently healthy colonies. Apparently, this virus plays a role in cases of sudden collapse of honey bee **colonies infested with the parasitic mite, *Varroa destructor*.**
- **Israeli acute paralysis virus (IAPV):** A related virus described in 2004 is known as the "Israeli acute paralysis virus". The virus is named after the place where it was first identified its place of origin is unknown. IAPV has been suggested in September 2007 as a marker associated with colony collapse disorder.
- **Kakugo virus (KV):** This is an iflavirus infecting bees. **Research suggests that *Varroa* mites mediate KV prevalence.**
- Nosema disease (*Nosemosis*): Nosema disease is generally regarded as one of the most destructive diseases of adult bees, affecting workers, queens and drones alike. The disease is **caused by the protozoan *Nosema apis*, whose 5 to 7 mm** spores infest the bees, are absorbed with the food and germinate in the midgut.
- **Colony Collapse Disorder (or CCD)** is a poorly understood phenomenon in which worker bees from a beehive or Western honey bee colony abruptly disappear. CCD was originally found in Western honey bee colonies in North America in late 2006.
- **Varroa mite (*Varroasis*):** This mite is a native parasite of *A. cerana* throughout Asia. Occasionally in *A. Melllfera* and more frequently in *A. Cerana* heavy infestation may cause absconding. Today this parasite is found throughout the world, except for Australia and New Zealand South Island.
- *Varroa destructor* (previously confused with *Varroa jacobsonii*) is quite large as compared with other mite species and can be seen with the unaided eye.
- ***Tropilaelaps* mite**: The mite is a native parasite of the giant honey bee *A. dorsata* widely distributed throughout tropical Asia.
- **Tracheal mite:** *Acarapis woodi* infests the tracheal system of adult bees, queens, workers and drones, which are all equally susceptible to its attacks. Since it was **first reported in *Apis mellifera* colonies in Europe in 1921**, opinions regarding the extent of the damage it can cause to honey bee colonies have varied. **Reports**

from India and Pakistan indicate that the tracheal mite caused severe losses of *A. cerana* colonies.

- The **greater wax moth (*Galleria mellonella*)**: The greater wax moth is often reported to cause damage both to honey bee colonies and to bee products in tropical and sub-tropical Asia.
- The **lesser wax moth (*Achroia grisella*)**: As its name implies, the lesser wax moth is generally smaller than the greater wax moth except when the latter is dwarfed owing to poor diet during its larval stage.
- **Honey:** Honey prepared from the nectar of a single flower source is called **unifloral or monofloral** honeys while honeys prepared from nectars of several sources is called **multiflora** .
- **Beeswax:** Beeswax is a **metabolic product of honeybees. Worker honeybees** of less than two weeks of age produce beeswax in a process of synthesizing esters in the fat body behind the wax glands.
- Beeswax is a highly complex **mixture of hydrocarbons and esters of monohydric alcohols combined with straight chain fatty acids and hydroxy-acids.**
- **Royal jelly:** Royal jelly is a special liquid food, rich in proteins, hormones, vitamins, organic acids and minerals secreted by the **hypopharyngeal gland of young worker** (nurse) bees.
- **Propolis:** Propolis is a mixture of the beeswax and the resins collected by honeybee from plants. In the process of collection of resins, it is mixed with some saliva as well as with wax.
- **Venom:** Honeybee venom is a clear, odourless and watery liquid. Honeybee venom is produced by two glands associated with the sting apparatus of worker bees.

Silkworm

- According to the Chinese records, the discovery of silk production from *B. mori* occurred about 2700 BC. It is believed that **empress Si-lung-Chi was asked by emperor Huang-ti to fin** the cause of damaged mulberry leaves on trees in their garden.
- Silk industry began in China where the source of silk was kept a secret for more than 2000 years. After some time, **China lost their monopoly in silk production, sericulture reached Japan through Korea and then to other countries.**
- India has the unique distinction of being the only country producing all the five kinds of silk **Mulberry, Eri, Muga, Tropical Tasar and Temperate Tasar**.

- In India, because of the prevalence of favourable climatic conditions, mulberry is cultivated mainly in five states *viz.*, **Karnataka, Andhra Pradesh, Tamil Nadu, West Bengal and Jammu and Kashmir**.
- **Karnataka occupies firs** place in India.
- These five states collectively account for **97% of the total area under mulberry cultivation and 95% of raw silk production in the country**.
- Total silk produced in 2012 (23060 MT) is 13% more than produced in 2010-11(16360MT). Among the four varieties of silk produced during 2010-11, **mulberry accounts for 80.73% (16957MT), eri 13.14% (2760MT), tasar 5.55% (1166 MT) and munga negligible (122 MT)** of the total raw silk production in the country.
- More than 80 % of mulberry silk is produced in only two states namely Andhra Pradesh (32%) and Karnataka (43%).
- **Spun silk yarn:** Yarn composed of silk filaments of **lengths ranging from 1 to 8 inches** produced by bleaching, dressing and spinning the silk waste which is the by-product of the raw silk reeling industry.
- **Noil yarn:** It is **short-staple residue obtained** during dressing operations in silk spinning from silk waste. It is a by-product of spun silk industry. This can be spun into noil yarn of coarse counts.
- **The process of killing the pupae by keeping the cocoons in hot water is called stiflin** .
- About **50,000 cocoons are required to produce one kg of silk.**
- **The process of removal of silk thread from the cocoon is called reeling.**
- Mulberry silk, *Bombyx mori* is the most common among them contributing to **nearly 95% of world's silk** production. **It is domesticated silk worm.** Mulberry silk is produced mainly in **Karnataka, West Bengal, Jammu & Kashmir, Tamil Nadu and Andhra Pradesh** although some other states have made some progress in this direction under their development plans.
- Eri silk worm, *Philosamia* (=*Attacus*) *ricini*: It is spun from cocoons of silkworms belonging to **Saturniidae family, which are fed on castor** leaves. Eri yarn is produced in **Assam, Bihar, Manipur, Meghalaya, Nagaland and West Bengal**.
- Munga silk worm, *Antheraea assama*: Munga silkworm is found **only in Assam. It is semi-domesticated silk worm.** It feeds on two local species of shrubs ***Machilus bombycina* and *Litsae polyantha* producing a strong golden yellow thread.**
- Tasar silk worm, *Antheraea mylitta*: It is silk reeled from cocoons of silkworms belonging to saturniidae family which are fed on **leaves of oak, asan and arjuna trees**. Tasar silk is mainly produced in **Jharkhand, Bihar, Chhattisgarh, Madhya Pradesh, Orissa, West Bengal and Andhra Pradesh.**

Tasar silkworms **are wild**. The Indian tasar worm feeds on trees **of *Terminalia*** species and other minor host plants, while the Japanese and Chinese worms feed on oak and other allied species.

- Spinning of silk thread is from outer to inside. The larva with silk thread is called cocoon.
- The silk thread is made **of insoluble protein fib oin**.
- Worms are introduced through DFLs (Disease Free Layings, *i.e.* eggs) procured from a quality centre (called **grainage**).
- **Uzi fl , *Tricholyga bombycis* (Diptera):** These flies lay eggs on silkworm and the maggots on hatching eat away the body of the caterpillars.

Lac insect

- Lac is the scarlet resinous secretion of a number of species of insects, namely some of the species of the genera *Metatachardia*, *Laccifer*, *Tachordiella*, *Austrotacharidia*, *Afrotachardina* and *Tachardina* of the superfamily *Coccoidea*, of which the most commonly cultivated species is ***Kerria lacca***.
- **Our country is the largest producer of lac** (a resinous compound secreted by lac insect while feeding on phloem sap of certain plants called lac hosts) in the world, accounting for about **50–60% of the total world lac production**.
- The major lac-producing states are **Jharkhand (57% of the country's production), Chhattisgarh (23%), West Bengal (12%)**, while Orissa, Gujarat, Maharashtra, Uttar Pradesh, Andhra Pradesh and Assam are minor producers.
- **India** is the leading country in the production of lac.
- On the suggestions of this Committee, lac merchants organized themselves into the Indian Lac Association for Research under the aegis of which, the foundation stone of **Indian Lac Research Institute (ILRI) was laid at Ranchi on September 20, 1924**.
- Lac research centre is located at **Namkum near Ranchi, in Jharkhand**.
- **Uses of lac** in medicine as **aphepatoprotective and anti-obesity drug.**
- Lac is used in the **manufacture of toys, gramophone records, varnishes, paints, buttons and bangles. Lac is used as a sealing material.**
- The raw lac collected directly from branches of trees is **called stick lac**.
- The harvested stick lac is crushed and sieved to remove impurities. The sieved material is then repeatedly washed to remove insect parts and other soluble material. The resulting product is known as **seed lac**. Seed lac which still contains 3-5% impurities is processed into shellac by heat treatment or solvent extraction.
- Purified and commercial lac is called **shellac**.

- **Shellac is a natural bioadhesive polymer and is chemically similar to synthetic polymers and thus can be considered a natural form of plastic.** It can be turned into a moulding compound when mixed with wood flour and moulded under heat and pressure methods, so it can also be classified as thermoplastic.
- In India the most common host trees are **dhak (*Butea monosperma*), ber (*Ziziphus mauritiana*) and kusum (*Schleichera oleosa*).**
- In Thailand the most common host trees are rain tree (*Samanea saman*) and pigeon pea (*Cajanus cajan*).
- In China the common host trees include pigeon pea (*Cajanus cajan*) and *Hibiscus* species.
- In India two types of lac crops *viz.*, **Rangini and Kusumi are grown in a year.**
- The Rangini crop is of two types called as **Kartiki and Baisakhi crop, which produce Kartiki and Baisakhi lac respectively**.
- The Kusumi crop is also of two types *viz.*, **Agahani and Jethi** which produce **Agahani and Jethi** lac respectively.
- The inoculations of Kartiki, Baisakhi, Agahani and Jethi crops are recommended in the months of June and July, October to November, July to August and January to February respectively.
- Two important predators of lac insect *viz.*, the **black moth *Pseudohypatopa* (*Holcocera*) *pulverea* and the white moth, *Eublemma amabilis*** attacked lac insect of which the former was more prevalent.

PESTS OF CROPS

Polyphagous pests

- Insects feed on a variety of cultivated crop plants is known as polyphagous pests. *e.g.* Red hairy caterpillar, White grub, Termite, Locust, Grasshopper.
- **Red hairy caterpillars *Amsacta albistriga* and *Amsacta moorei* (Arctiidae: Lepidoptera):** The larvae that hatch out from the eggs feed gregariously by **scraping the under surface of tender leaflet** leaving the upper epidermal layers. The **creamy or bright yellow eggs** are laid in groups mostly on the under surface of cowpea leaves usually sown along with groundnut as an inter crop and eggs laying last for 2-6 days. **Digging trenches around the fiel** and dusting them with malathion dust 25 kg per hectare.
- **White grub, *Holotrichia serrata, H. consanguinea* (Scarabaeidae: Coleoptera):** The beetles are polyphagous and may feed on the foliage of a wide variety of host trees and bushes found in the close vicinity. However, they do exercise some preference for certain hosts like **jujube (ber), *Prosopis cineraria* (khejri), neem, cluster bean, jambolana (Jamun) and drumstick (sainjana)**. The **eggs are laid at the depth of 6 to 10 cm inside the soil**. Freshly laid eggs are white, elliptic, about 3.2 mm in length and 1.9 mm in width.

- **Termite, *Odontotermes obesus* Rhamb. (Termitidae: Isoptera):** They are **polyphagous** and have the widest range of host plants. The rest of the colony consists of the king, the functional queen, **three sub-castes of fertile reproductives and two sterile castes of workers and soldiers**. In some species of termites, there is a special rostrum projecting out from the head region and an offensive secretion from a special gland is ejected out from the top of the rostrum; this represent the evolution of chemical warfare even in such structurally primitive insects; soldiers possessing this organ are called **Nasute soldiers**. Queen after fertilization enlarges in size to a length of 11 cm. Eggs are laid @ 30,000 per day and the longevity of the queen is 5-15 years, even up to 50 years. **It is also known as egg laying machine**. Soil application of chlorpyriphos 50 EC @ 10 ml/l as a soil drench at sowing time in termite prone soils.
- **Locusts** are the major agricultural pests of the world. Out of 5000 different species of grasshoppers, those which can live in two different phases *viz*. **solitary and gregarious phase are called locusts**. Of these, only **nine species** have been recognized as locusts, three species belonging to Indian subcontinent. Species of locust: **desert locust, *Schistocerca gregaria*; migratory locust, *Locusta migratoria* (important only in Maharashtra, Gujarat, and Rajasthan) and bombay locust, *Patanga succineta*.** The desert locust, *Schistocerca gregaria* is the **most destructive of them**. The swarms that form due to breeding in West Iran and Eastern Arabia during winter-spring migrate during May-July to Indo-Pakistan region.
- The Bombay locust, *Patanga succincta* is known from **India, Sri Lanka and Malaysia.**
- The **migratory locust, *Locusta migratoria*** breeds during the spring in Baluchistan and the adults migrate to desert areas of India and breed there during the summer.
- The desert locust exists in two distinct phases namely *solitary* phase during the periods of recession and *gregarious phase* during the periods of outbreak and that it gets transformed from one into the other through the *transiens* phase.
- The eggs are laid in batches of up to 150 or more deposited in soil at depths of 8 to 15 cms. The female locust is able to deposit its eggs deep in the soil by thrusting and extending the hind end of its abdomen which allows much elongation for this purpose.
- Loose sandy soil is suitable for egg-laying. An individual egg looks like a **grain of rice** which is yellowish in colour, 7- 8 mm in length and about one millimeter in girth.
- In the *solitary* phase, the hoppers are either **green or acquire the colour** of the vegetation; they behave like ordinary grasshoppers, remain scattered and as adults they stay grey and fly mainly during the night as individual.

- In the gregarious phase the hoppers exhibit a **black colour pattern** and march for long distances in definitive bands.
- **Phadka grasshopper, *Hieroglyphus nigrorepletus* (Acrididae: Orthoptera):** *H. nigrorepletus* is the *kharif* grasshopper popularly known as "*Phadka*" distributed throughout India, Pakistan and Bangladesh.
- **Rice grasshopper, *Hieroglyphus banian* (Acrididae: Orthoptera):** The nymphs and adults cause enormous loss to the crop by chewing and cutting various plant portion *viz.,* leaves, flowers and grains.
- Which insect has only one generation in a year **White grub.**
- *Hieroglyphus banian* has ***One generation per year.***

Insect pests of rice (*Oryza sativa*)

- **Yellow stem borer,** leaf folder, gall midge and other defoliators are important and cause significant reduction in yield in rice growing areas.
- **Rice stem borer or yellow stem borer, *Scirpophaga incertulas* (Pyralidae: Lepidoptera):** The pest is widely distributed in all Asian countries, its **monophagous** pest and is a major pest on rice in India. Larva feeds inside the stem causing drying of the central shoot called **'dead heart'** in young plant or drying of the panicle called **'white ear'** in older plants. ETL are 2 egg masses per meter^2; **10% dead hearts in vegetative stage**; 2% white ear in flowering stage. Release the egg parasitoid, ***Trichogramma japonicum*** @ 50,000/ha twice on 30 and 37.
- **Rice gall midge, *Orseolia oryzae* (Cecidomyiidae: Diptera):** The gall formed by this fly is popularly known as **'silver shoot' or 'onion shoot'** because of the formation of hollow pink or purple, dirty white cylindrical tubes bearing at their tips and green reduced leaf blade complete with ligules and auricles. ETL is **10% silver shoots. Gall is the modified leaf sheath**. Release larval parasitoid, ***Platygaster oryzae*** through parasitized galls @ 1 per 10 m^2 in the main field at 10 days after transplanting (DAT).
- **Green rice leaf hoppers, *Nephotettix nigropictus* and *N. virescens* (Cicadellidae: Homoptera):** Both nymphs and adults suck the plant sap and cause browning of leaves and cause **"hopper burn"** due to heavy infestation. Both the species are known to be vectors of virus diseases such as **rice transitory yellowing and rice yellow dwarf**. The female of *N. nigropictus* is green and the male has two black spots extending upto the black distal portion on the fore wings.
- **Brown plant hopper, *Nilaparvata lugens* (Delphacidae: Homoptera):** The affected plant dries up and gives a scorched appearance called **"hopper burn"**. ***N. lugens* is a phloem feeder**. It is the **vector of grassy stunt, ragged stunt and wilted stunt diseases**. Conservation of natural enemies like **spider, *Lycosa pseudoannulata*; mirid bug, *Cyrtorhinus lividipennis*.**

- **Rice hispa, *Dicladispa armigera* (Hispidae: Coleoptera): Whitish leaf tips of young leaves** giving dried up appearance. **Clipping of leaf tips of** seedlings while transplanting eliminates eggs laid towards the tip.
- **Rice leaf folder, *Cnaphalocrocis medinalis* (Pyraustidae: Lepidoptera):** Severely infested field the whole crop gives a **sickly appearance with white patches**. The Ichneumonid, ***Xanthopimpla emaculata*** is parasitic on the pest larvae. **Passing a rope 2-3 times over the crop at tillering stage** mechanically to dislodge caterpillars.
- **Rice earhead bug, *Leptocorisa acuta* (Alydidae or Coreidae: Hemiptera):** Both nymphs and adults feed on the sap of peduncle, tender stem and **milky grains making them turn chaffy**. The bugs emit characteristic **unpleasant odour indicative of their presence in the field, hence the name is gundhi bug**.
- **Rice case worm, *Nymphula depunctalis* (Pyraustidae: Lepidopera):** The caterpillars feed on green tissues of the leaves and form tubular cases around them by cutting the apical portion of leaves which float on water. Larva has **filamentous gill** on the sides of the body that helps to lead a semi aquatic life.

Insect pests of sugarcane (*Saccharum officinarum*)

- **Early shoot borer, *Chilo infuscatellus* (Crambidae: Lepidoptera):** The borer enters into young shoots and tunnels downwards. The upper portion of the central leaf whorl is thus cut off and dries up causing dead hearts in shoots from about a month old to 2-3 months crop. Release of 125 gravid females of ***Sturmiopsis inferens*** (Tachinidae) per ha at 45th day of crop growth.
- **Internode borer, *Chilo sacchariphagus indicus* (Crambidae: Lepidoptera):** It infests the crop after the third month when internodes begin to form and continues till the time of harvest. Symptoms will be visible only on close examination.
- **Top borer, *Scirpophaga excerptalis* (Pyraustidae: Lepidoptera):** Tunnelling of midrib in a leaf, small holes in a parallel line in the freshly appeared leaves, dead heart which is reddish brown in the young crop (2-4 months old) and a bunchy top in a grown up crop. Moth is uniformly creamy white in colour. **The first pair of wings of certain moths bears a black spot on each wing.** Females have a **crimson hairy tuft at the anal end**. There are **5 broods** reported in a year. Release of egg parasitoid, *Trichogramma minutum* and **larval parasitoid, *Isotima javensis*** during November and December in Adsali sugarcane.
- **Sugarcane leaf hopper, *Pyrilla perpusilla* (Lophopidae: Homoptera):** Adults are straw coloured with **two pairs of wings folded like a roof on the back** and the head prominently drawn forward as a sort of rostrum. *Tetrastichus pyrillae* and **lepidopteran parasite, *Epiricania melanoleuca*** naturally suppress the population.

- **Sugarcane woolly aphid, *Ceratovacuna lanigera* (Aphididae: Homoptera):** Nymphs and adults suck sap from under surface of leaves resulting in white and yellow spot on leaves, drying of leaves, sooty mold on honeydew and activity of ants. Release of the natural enemies, ***Dipha aphidivora.*** **or *Micromus igorotus*** in shade net cages @ 50/cage and allow them to develop and disperse by removing the cages.

Insect-pests of cotton (*Gossypium hirsutum, G. herbaceum, G. barbadense* and *G. arboreum*)

- **Cotton mealy bug, *Phenacoccus solenopsis* or *Maconellicoccus hirsutus* (Pseudococcidae: Hemiptera):** Mealybug species viz., *Phenacoccus solenopsis*, *Phenacoccus solani* and *Maconellicoccus hirsutus* have been recorded on cotton in India. **Recently, *P. solenopsis*** has emerged as a serious sucking pest of cotton due to decrease in the number of pesticidal applications in cotton agro-ecosystem. In India cotton mealy bug is known to be **introduced from Pakistan** since the three largest cotton players in Asia are China, India and Pakistan which share common borders that make them vulnerable to insect. The developmental period from immature crawler to adult stage was greater for males (18.7 days) compared to females (13.2 days), **probably due to the additional molt to the pupal stage in males.**
- **Cotton aphid, *Aphis gossypii* (Aphididae: Homoptera):** The Economic Threshold Level (ETL) is 10% affected plants. They are characterized by the presence of two tubes like structures called **cornicles** on the abdomen.
- **Cotton leaf hopper, *Amrasca biguttula biguttula* (Cicadellidae: Homoptera):** The leaves show symptoms of "**hopper burn**" such as yellowing, curling, bronzing, drying up and these symptoms are expressed differently depending on how the different varieties react to the toxic saliva of the insect. The vigour of the plants is impaired to a great extent. **The ETL is 2 Jassids or nymphs per leaf** or yellowing in the margins of the leaves.
- **Cotton whitefl , *Bemisia tabaci* (Aleyrodidae: Homoptera):** It is known to infest about 50 different species of plants but it becomes quite a serious pest of cotton in certain regions of the country. The ETL is **5-10 nymphs** or adults per leaf. A single female of this species lays about **70 stalked eggs** singly on the under surface of leaves, mostly on the top and middle leaves of plant.
- **Spotted bollworms, *Earias insulana* and *E. vittella* (Noctuidae: Lepidoptera):** The initial infestation takes place on **6 week old crop** in which the larva causes detopping (drooping and drying of the shoot) due to its feeding by boring into it. The moth of *E. vittella* has **green fore wings with a white streak** on each of them whereas that of *E. insulana* is **completely green.**
- **American bollworm, *Helicoverpa armigera* (Noctuidae: Lepidoptera):** American bollworm has a worldwide distribution in all the cotton growing

regions of the world. Moth is stout, medium sized with brownish/greyish **forewings with a dark cross band near outer margin** and dark spots near costal margins with a wing expanse of 3.7 cm. The spherical, yellowish **eggs are laid singly** on tender parts and buds of plants. The egg period lasts for 2-4 days. When full grown, they measure 3.7 to 5 cm in length. The full grown caterpillar **pupates in the soil in an earthen cell** and emerges in 16-21 days. Use of larval parasitoid such as ***Campoletis chloridae*** **(Ichneumonidae)**; *Eriborus angenteopilorus*; *Diadegma fenestalis*; *Bracon brevicornis*; *Peribaca orbata etc.*

- **Pink bollworm, *Pectinophora gossypiella* (Gelechiidae: Lepidoptera):** The larva enters into developing boll through the tip portion and the entrance **hole gets closed**. It feeds on the seeds and moves to adjacent locule by making a hole through the septum and makes **double seeds**. A typical **rosette-shaped bloom** shaped symptoms occur when examined will contain the larva. In **South India the insect is not known to hibernate in any stage of its development. Fumigation** of seeds with **methyl bromide** at 1.5 kg/100 cu. m. for 24 hours or with aluminum phosphide at 18 tablets/100 cu. m. or heat treatment for a few minutes at 60°C kills hibernating larvae in seeds.
- **Tobacco caterpillar, *Spodoptera litura* (Noctuidae: Lepidoptera):** In early stages, the caterpillars are gregarious and scrape the chlorophyll content of leaf lamina. Moth is medium sized and stout bodied with forewings pale grey to dark brown in colour having **wavy white crisscross markings**. Female lays about **300 eggs in clusters**. The eggs are covered over by brown hairs and they hatch in about 3-5 days. There are **eight generations** in a year.

Insect pests of pulses (Gram and Kharif pulses)

- **Plume moth, *Exelastis atomosa* (Pterophoridae: Lepidoptera):** The forewings are divided into two parts and **hind wings** into three parts and provided with a **fringe like border**. The larva has numerous rosettes of capitate spines and hair all over the body.
- **Redgram pod fl , *Malanagromyza obtuse* (Agromyzidae: Diptera):** Adult is a **black fly with strong legs and ovate abdomen**. Egg period is 3 days and eggs are laid singly or in **cluster inside the pod by piercing through the ovipositor**.
- **Stem fl , *Ophiomyia phaseoli* (Agromyzidae: Diptera):** Eggs are laid into tender part of the stem or petiole. **Maggot makes tunnels the stem** and feeds on the internal contents. **Drooping of the first two leaves and wilting** of the plants are observed.The maggot and pupae are present become swollen and start ribbing. Older plants show stunting but are not usually killed.
- **Spotted pod borer, *Maruca vitrata* (Pyralidae: Lepidoptera):** The larva feeds on seeds and the caterpillar with **short hairs on black warts** webs together the flowers and feeds on them. Moth is **dark brown forewings with white**

club shaped cross band along anterior margin and white hind wings with dark brown border.

- **Bean aphids, *Aphis craccivora* (Aphididae: Hemiptera):** It acts as a vector of **Rosette disease in groundnut and broad bean virus in pea.**
- **Blue butterfl , *Lampides boeticus* (Lycaenidae: Lepidoptera):** The adult moth is greyish blue with **prominent black spots in the hind wings and a long tail.**

Insect pests of pearmillet (*Pennisetum glaucum*)

- **Shoot fl , *Atherigona approximate* (Muscidae: Diptera):** It attacks the crop both in **seedlings and boot leaf** stage. It causes dead hearts in young plants and chaffy grains in the mature crop.
- **Stem borer, *Chilo partellus* (Crambidae: Lepidoptera):** It infests the crop a month after sowing and upto emergence of earhead. Central shoot withering leading to "**dead heart**" is the typical damage symptom.
- **Pink stem borer, *Sesamia inferens* (Noctuidae: Lepidoptera):** The adult moth is a **straw coloured moth with white wings.** The larva is pinkish brown with dark head. The larval period lasts for 25-54 days.
- **Blister beetles, *Mylabris pustulata* (Meloidae: Coleoptera):** The adults are medium to large, usually **black with large yellow spots and a red band across the abdomen** which sometimes changes into yellow spots. The antennae are orange or yellow. **The immature stages (larvae) do not feed on plants. They live in the soil and eat grasshopper eggs.**

Insect pests of sorghum (*Sorghum vulgare*) and maize (*Zea mays*)

- **Sorghum shoot fl , *Atherigona soccata* (Muscidae: Diptera):** The maggots bore into the shoot of young plants, a week after germination to about one month and as a result the central shoot dries up resulting in '**dead hearts**'. The adult is a small dark fly. It deposits **whitish eggs singly on the central surface of the leaves.** Use of higher seed rate of 12 kg/ha instead of normal rate of 10 kg/ha and removal of affected and extra plants at the time of thinning four weeks after sowing since shootfly affects only young plants of 4-5 weeks age.
- **Sorghum or maize stem borer, *Chilo partellus* (Crambidae: Lepidoptera):** Presence of 'dead heart' in the early stages is the main symptom. The bore holes may be visible in contrast to the dead-heart caused by the stem fly.
- **Sorghum midge, *Contarinia* (*Stenodiplosis*) *sorghicola* (Cecidomyiidae: Diptera):** The maggots feed on the developing grains and cause the developing grains to shrivel and severe infestation has a significant effect on the overall production of grains. **Larvae and pupae are parasitised by *Tetrastichus coimbatorensis.***

- **Sorghum earhead bug, *Calocoris angustatus* (Miridae: Hemiptera):** A single female lays 150-200 eggs @ 1-16 in each floret. **Bluish cigar-shaped eggs hatch in about 10 days.** A reduviid bug, *Reduviolus* sp., and lygaeid *bug, Geocoris tricolor* are predaceous on the sorghum earhead bug.
- **Red hairy caterpillar, *Amsacta albistriga*, *A. moorei* (Arctiidae: Lepidoptera):** *Amsacata albistriga* is predominant in South India while *A. moorie* dominates northern parts of **the country**. Moths emerge with the onset of monsoon from **diapausing pupae in June-July**. Eggs are cream or bright yellow and are laid in masses on young foliage or on the soil.
- **Corn worm or earworm, *Helicoverpa armigera* (Noctuidae: Lepidoptera):** Adult is brown coloured moth with a **'V' shaped speck** on fore wings and has a dull black border on the hind wing. The spherical, yellowish eggs are laid singly on tender parts and buds of plants.
- **Aphid, *Rhopalosiphum maidis* (Aphididae: Hemiptera)**: The aphid is dark bluish-green and somewhat ovate. It is 2 mm long, with black legs, cornicles, and antennae.

Insect pests of wheat (*Triticum aestivum*) and barley (*Hordeum vulgare*)

- **Brown wheat mite, *Petrobia latens* (Arachnida: Tetranychidae):** The brown wheat mite feeds on sap of leaves by inserting two needle-like stylets into the leaf thereby withdrawing nutrients from the plant.
- **Grain aphid, *Sitobion avenae* (Aphididae: Homoptera):** Sooty moulds develop on the honeydew which they secrete. This aphid is also a **vector of the barley yellow dwarf virus (BYDV).**
- **Shoot fl , *Atherigona naqvii* (Muscidae: Diptera):** The maggots bore into the shoot of young plants and as a result the central shoot dries up resulting in **'dead hearts'**. It occurs a week to one month after germination.
- **Ghujia weevil, *Tanymecus indicus* (Curculionidae: Coleoptera):** Adults are destructive and they cuts and feeds on the plumule of the young seedlings. In severe attack, it is necessary to resowing the crops. The damage results in symptoms like seedling stems severed and wilting plant lying on ground.
- **Armyworm, *Mythimna separata* (Noctuidae: Lepidoptera):** In the early stages, they feed on tender leaves in the central whorl and later feed on older leaves and skeletonize them totally. Pupation takes place in the soil at a depth of 0.5-5 cm but it may also occur under dry leaves of stubble or fresh tillers.

Insect pests of groundnut (*Arachis hypogaea*)

- **White grub or root grub, *Holotrichia consanguinea*, *H. serrata* (Scarabaeidae: Coleoptera):** Root grub is a **polyphagous pest** feeding on the roots of a wide range of plants like pulses, groundnut, sugarcane, vegetables *etc.* Adult is fully developed by the **end of November** and remains in the pupal cell. **Adult hibernates** till first rain. Adults emerge out of soil during first rain in evening, mate and feed on the leaves of the trees *viz.*, neem, drumstick, subabul *etc.*, and early in the morning get back and burrow into the soil and lay single egg per cell at a depth of 12 -15 cm. **Only one generation in a year.**

- **Red hairy caterpillars, *Amsacta albistriga* and *Amsacta moorei* (Arctiidae: Lepidoptera):** The larvae hatch out from the eggs feed gregariously by scarping the under surface of tender leaflets leaving the upper epidermal layer intact. Severely damaged crop presents the appearance as entire area has been grazed by cattle. After the receipt of heavy rains, the moths emerge from their earthen cells in the soil.
- **Leaf miner, *Aproaerema modicella* (Gelechiidae: Lepidoptera):** The moth is very small with dark brown wings and small distinct white spot on fore wings. It lays **shining, sculptured eggs, singly on tender leaves.**
- **Aphids, *Aphis craccivora* (Aphididae: Hemiptera):** The groundnut aphid also transmits **groundnut rosette virus and sometimes groundnut stunt virus diseases.**
- **Thrips, *Scirtothrips dorsalis* (Thripidae: Thysanoptera):** It consists of distortions of the young leaf lets and patchy areas of necrotic tissue that puncture and split as the leaflets grow. Nymphs and adults dark coloured with **fringed wings**. Female thrips lay 40-50 eggs inside the tissues of leaves and shoot.

Insect-pests of castor (*Ricinus communis*)

- **Capsule & Shoot borer, *Conogethes punctiferalis* (Pyraustidae: Lepidoptera):** The symptoms are **frassy matter at the bored shoots and webbed seed capsules covered with dark excreta.** Adult is medium sized with small **black dots on pale yellow wings**. It lays eggs on the developing capsules and egg period is 6 days.
- **Castor semi looper, *Achaea janata* (Noctuidae: Lepidoptera):** *A. janata* is a pale reddish brown moth, stoutly built with black hind wings having white band medially and three large white spots on the outer margins. The damage is caused by **both the caterpillar and adult moth**. The caterpillars feed voraciously on castor leaves. The **adult of this species are fruit sucking moths and cause serious damage to citrus crop.**
- **Castor butterfly or spiny caterpillar, *Ergolis merione* (Nymphalidae: Lepidoptera):** Brown butterfly with black wavy lines on the wings. Larva green coloured, spiny (spines branched at the tip) caterpillar with yellow stripe on the dorsal region.
- **Castor gallfl , *Asphondylia ricini* (Cecidomyidae: Diptera):** As a result of

feeding, the buds develop into **galls and produce no fruits and seeds**. Adults are **mosquito like small fl** . The female lays **eggs singly in the buds.**

- **Sesamum Leaf webber, *Antigastra catalaunalis* (Pyraustidae: Lepidoptera):** The larvae web together with top leaves or bore into tender shoots, capsules and feed on them.

Mustard (*Brassica campestris*)

- **Mustard aphid, *Lipaphis erysimi* (Aphididae: Homoptera):** Both the nymphs and adults suck the cell-sap from leaves, stems, inflorescence and the developing pods. They are louse like, pale-greenish insects abundant from December to March. **During summer,** it is believed to **migrate to the hills.**
- **Mustard sawfl , *Athalia lugens proxima* (Tenthredinidae: Hymenoptera):** The adult is black with yellowish femora and thorax. **The female possesses a saw-like ovipositor and inserts the eggs very near the leaf margin**. The larva is cylindrical and dark gray with three pairs of thoracic legs and **seven pairs of prolegs on abdominal segments 2-8.**
- **Painted Bug, *Bagrada cruciferarum* (Pentatomidae: Hemiptera):** Both nymphs and adults suck cell sap from the leaves and developing pods, which gradually wilt and dry up. The entire life cycle is completed in 19-54 days and it passes through **9 generations in a year**.

Insect pests of brinjal (*Solanum melongena*)

- Among the various pests brinjal shoot and fruit borer is highly monophagous and destructive which are necessary to grower go upto 30-40 rounds of sprays.
- **Shoot and fruit borer, *Leucinodes orbonalis* (Pyraustidae: Lepidoptera):** The larva bores into tender shoots in the early stage and causes **"dead hearts"**. The moth has brownish and red markings on the whitish fore wings. The female lays **250 eggs singly on tender shoots and developing fruits of brinjal.**
- **Epilachna beetle (or hadda Beetle) *Henosepilachna vigintioctopunctata* (Coccinellidae: Coleoptera):** They feed on the epidermal layers of leaves which get **skeletonized and gradually dry away**. The brownish hemispherical beetle has **12 - 28 black spots on the elytra**. The female lays **elongate, spindle-shaped yellowish eggs** in groups of 10-20 on the under surface of leaves. **Egg parasitoid, *Tetrastichus ovularum,*** larval parasitoid, *Uga menoni* control the population of pest during March-July.
- **Grey weevil, *Myllocerus subfasciatus* or *M. maculosus* (Curculionidae: Coleoptera):** The brownish weevil lays **500 eggs in the soil upto 8-10 cm deep.**
- **Aphid, *Aphis gossypii* and *Myzus persicae* (Aphididae: Homoptera):** The under surface of the leaves get crinkled and slightly curled backwards. Nymphs

of *A. gossypii* are greenish-brown or yellowish in colour while adults are yellowish-green to dark green in **colour,** little over one mm in length and have a pair of siphunculi in the posterior side of abdomen. **Adults of *M. persicae*** are usually of **green colour** but may be pale brown to pinkish, 1.5 - 2.5 mm long with long clavate siphunculi.

- **Brinjal stem borer, *Euzophera perticella* (Pyralidae: Lepidoptera):** It attacks on **stem of pencil thickness**. The entry of the caterpillar is near the ground level of the stem at leaf or branch axil and covers the hole with excreta and frass. **Creamy and scale-like, laid** singly in batches on young leaves, petioles and branches.
- **Leaf hopper, *Amrasca biguttula biguttula* (= *Empoasca devastans*) (Cicadellidae: Homoptera):** The plant become **stunted, leaves crinkle, turn yellowish and become cup shaped**. Brownish or reddish colour develops along the edges of the leaves. This is called the **'hopper burn'**.
- **Brijal lace wing bug, *Urentius hystricellus, U. sentis* (Tingidae: Hemiptera):** It is identified by their lace like network on veins of the forewings and lace like **embossed pattern on head and prothorax.**
- **Red spider mite, *Tetranychus telarius* (Tetranychidae: Acarina):** Red spider mites are found in large colonies on **underside of leaves protected by the silk webbing constructed by the female**. They are polyphagous infesting brinjal, bhendi and bean. Leaves present a characteristic blotches which become whitish then brown patches appear.

Insect pests of Okra (*Hibiscus esculentus*)

- **Okra shoot and fruit borer, *Earias vittella* (Noctuidae: Lepidoptera):** It is an **oligophagous pest** having okra and cotton as main hosts. It is also found feeding on a large number of malvaceous plants both wild as well as cultivated. The *E. vittella* moths measure about 2.5 cm across the wings and have a narrow light longitudinal **green band in the middle of forewing.** The female moth laid 200-400 eggs at night, singly on flower buds and tender leaves of okra. **Eggs are sky blue in colour.**
- **Red spider mite, *Tetranychus cinnabarinus* (Tetranychidae: Arachnida):** Colonies of mites comprising of eggs, nymphs and adults are found **feeding on ventral surface of leaves under protective cover of fine silken web** .
- This whitefly also acts as a **vector of transmitting the leaf curl virus.**
- **Management:** Spraying with 0.05% dimethoate 30EC or 0.01% imidacloprid 17.8SL or 0.1% acetamiprid.
- **Tomato fruit borer, *Helicoverpa armigera* (Noctuidae: Lepidoptera): Nature of damage:** Eggs are **laid singly**, generally on leaves and flowers but sometimes on fruits. They bore **circular holes** and thrust only a part of their body inside the fruit and eat the inner contents.

Cruciferous Vegetables (Cabbage, Cauliflower, Radish, Beetroot and Knolkhol)

- **Diamond back moth, *Plutella xylostella* (Plutellidae: Lepidoptera):** It is a well known pest of cruciferous crops throughout the world. The larvae **cause blisters on the leaves which dry away in course of time.** It attacks cabbage, cauliflower, radish and knol-khol. The moth is greyish brown with narrow wings having pale white markings anteriorly which are diamond-like white patches dorsally and wings are folded over back at rest.
- **Leaf webber, *Crocidolomia binotalis* (Pyraustidae: Lepidoptera):** The leaves are skeletonized by the larvae which remain on the under surface of **leaves in webs and feed on them**. The small moth with **light brownish fore wings lays the eggs in masses** and each mass containing 40-100 overlapping flat eggs.
- **Cabbage butterfl , *Pieris brassicae* (Pieriidae: Lepidoptera):** The pest passes winter in the plains and migrates to hilly regions during summer. As a result of their feeding the leaves are skeletonized, sometimes the caterpillars bore into the heads of cabbage and cauliflower. Eggs are **laid in clusters under surface of the leaf**. Eggs are **flask-sha ed about one mm long** and yellowish in colour. Adult butterflies have snow-white **forewings with black distal margins** more developed in females than in males; hind wings are also pure white with black apical spots.
- **Cabbage aphid, *Brevicoryne brassicae* (Aphididae: Homoptera):** Eggs are pale yellow with greenish tinge. Eggs are laid during November-December and they hatch in 20-22 weeks.
- **Mustard sawfl , *Athalia lugens proxima* (Tenthredinidae: Hymenoptera):** Mustard sawfly is one of the very few **hymenopterous insects reported** as crop pests and that too with chewing and biting habits. **Eggs are laid singly**, mostly during day time and inserted into leaf tissues near the periphery of leaves. When slight **disturbance they fall on the soil and feign death**. Females have a strong saw-like ovipositor hence it has been given the popular name sawfly.
- **Painted Bug *Bagrada cruciferarum* (Pentatomidae: Hemiptera):** The adults appear in field during October and their activity decreases with the onset of summer but is again appeared in autumn. Both nymphs and adults suck the cell sap from tender plant parts causing yellowing of leaves, gradually dry up and ultimately fall down exposing the plants to secondary invasion of bacteria and fungi. Adults are also **black and orange colour bugs similar** in colour pattern as nymphs that are why they have earned the common name of painted bugs. A single life-cycle is completed in 3-4 weeks and adults live for 16-18 days **with 6-8 generations in a year**.

Insect-pests of cucurbitaceous vegetables

- **Fruit flies, *Bactrocera cucurbitae* and *B. ciliatus* (Tephritidae: Diptera):** It is commonly called melon **fruit fl** . A puncture made by one female is often used by others also for **ovipositing the eggs**. On hatching the maggots feed inside the pulp of fruits and infested fruits identified by the presence of **brown resinous juice which oozes out from punctures.**
- **Pumpkin beetle, *Raphidopalpa foveicollis* (Galerucinae: Coleoptera):** *R. foveicollis* is found in all states of India and it is more abundant in northern states in association with *A. Lewisii* and in South India *A. cincta* is more common. The **beetles make irregular holes on leaves and also feed on flower** . They prefer young seedlings, tender leaves and damage at this stage may kill the seedlings. The **roots stem and fruits that come in contact with the soil are damaged by the grubs.** Adult beetles of R. *foevicollis* are 6-8 mm long having glistening **yellowish red to yellowish brown elytra** that are uniformly covered with fine punctures. Adults of *A. cincta* are similar in size and appearance except that the colour of elytra is greyish yellow to brownish grey with distinct palm margin all around. Adults of *A. lewisii* are slightly smaller (5-6 mm) with **blackish blue elytra.**
- **Serpentine leafminer, *Liriomyza trifolii* (Agromyzidae: Diptera):** The maggot punctures the epidermal layers and scraps chlorophyll content between the layers.
- **Pumpkin leaf caterpillar, *Diaphania indica* (Pyralidae: Lepidoptera):** Moth is medium with whitish wings, transparent with brown marginal patches.
- **Potato tuber moth, *Phthorimaea operculella* (Gelechiidae: Lepidoptera):** It is a pest of both in field as well as storage. Hibernation takes place in the larval or pupal stage in some cooler areas of the country. **Fumigate the tubers with CS_2 at 1kg/27 cu.m for 48 hours at 70°F or ethyl bromide at 1 kg/27 cu.m** for the least three hours before storage.

HORTICULTURAL CROPS PESTS

Insect pests of mango (*Mangifera indica*)

- **Mango hoppers (Cicadellidae: Homoptera):** In India, these hoppers are widely distributed in all mango growing regions. ***A. atkinsoni*** is more common in **North India. *I. clypealis*** is found all over India more predominant in **South Gujarat, Maharashtra and Karnataka. *I. niveosparsus*** is recorded from peninsular India. *I. niveosparsus* is slightly smaller with three spots on the scutellum and prominent white band across its light brown wings. *I. clypealis* is the **smallest with two spots on the scutellum and dark spots** on the vertex and is light brown in colour. *A. atkinsoni* is the largest and **light brown having two spots on the scutellum.**

- **Mango stem borer, *Batocera rufomaculata* (Cerambycidae: Coleoptera):** The **grub bores and tunnels through the bark of branches and stem feeding** on the inner contents. Adult is conspicuously long, brownish grey beetle measure ring and it has two pink dots and lateral spines on thorax, hard elytra and long horned.
- **Mango nut weevil or stone weevil, *Sternochetus mangiferae* (Curculionidae: Coleoptera):** This is the important pests in international export market. Eggs are laid singly on the epicarp of fruits or under the rind of fruits. There is only one generation in a year. Hot water treatment of fruit for specified period at **60°C kills** the weevil inside the stone.
- **Fruit fly *Bactrocera dorsalis* (Tephritidae: Diptera)**: It is widely distributed in the Orient region from Australia and Hawaii to Pakistan, hence it is also called **Oriental fruit fl** . The pest is active throughout the year in South India whereas in northern parts the pest **hibernates during winter (November to March) in pupal stage.**
- **Red tree ant, *Oecophylla smaragdina* (Formicidae: Hymenoptera)**: The ants web and stitch together a few leaves usually at the top of the branches and build their nests on citrus, jack-fruit, jamun, litchi, mango, sapota *etc.* Pupae are also pure white in colour.
- **Mango mealy bug, *Drosicha mangiferae* (Coccidae: Homoptera)**: There is a well established **sexual dimorphism** in the adult stage which is generally found **during April to June**. They lay eggs **about 5-15 cm deep** and in clusters of 300-400 eggs each. The **eggs laid in the soil take quite** a few months before they hatch. It is only **one generation during** the year. Application of a sticky band round the tree-trunk which check the nymphs from crawling up the trees.

Insect pests of citrus

- **Lemon butterfl , *Papilio demoleus* (Papilionidae: Lepidoptera):** It is widely distributed and **oligophagous pests**. The caterpillars **feed on the leaves and especially young seedlings** and trees are seriously affected. *P.demoleus* is a big beautiful butterfly with **yellow and black markings on all the four wings** having wing expanse of about 50-60 mm. ***P. polytes* males are black and females vary in form**. Freshly hatched caterpillars are dark brown and soon develop irregular white markings on their **body resembling bird's drop**. The full grown caterpillar is deep green and cylindrical in form and measures about 40-50 mm in length with a hump in front and has a **horn like structure on the dorsal side of the last body** segment. When the caterpillar is disturbed, it pushes out purple structure from prothorax called **osmeterium which emits a distinct smell**. Adult emerges in a week in summer and in 12-20 weeks in winter. **Hibernation is in pupal** stage and there are 2-4 overlapping generations per year.

- **Fruit sucking moths, *Othreis fullonica* and *O. materna* (Noctuidae: Lepidoptera):** This is the only group where the adult moths are harmful and damage the citrus fruits. Moth pierces the ripe fruit with its strong proboscis and sucks the sweet juice. *E. materna*: The moth is brownish black forewings with a white stripe and yellowish hind wings with a **circular black spot in the middle.** *E. fullonica*: The moth is brownish forewings and yellowish black hind wings with a **half moon or kidney shaped black spot**. *E. ancilla*: The forewings of moth are dark brown with a green band in the middle and hind wings are **yellowish with a kidney shaped black spot.**
- Eggs are lay singly on tender parts of wild foliage. Egg period is about 3-4 days.
- **Leaf miner, *Phyllocnistis citrella* (Gracillaridae: Lepidoptera):** The tiny red, apodous larva mines in between the epidermal layers of the leaf in a zigzag manner and feeds on chlorophyll which results in distortion of the leaf lamina. **The tunnel appears silvery white. Zigzag galleries below the epidermis of leaves**. Adult is a **silvery white small moth with brown striped forewings** having a prominent black spot near the tip and white hind wings. **Both pairs of wings are fringed with hairs**. The adult moth measures 6 mm in wing span.
- **Citrus psylla, *Diaphorina citri* (Psyllidae: Hemiptera):** This is widely distributed in India and is considered as a major pest of Punjab. The pest is active from February to November and **gives 16 generations per year**.
- **Bark eating caterpillars, *Indarbela tetraonis, I. quadrinotata* (Metarbelidae: Lepidoptera): Ribbon like or pipe like webbings on the stem near forks or angles of branches** and the tree trunk is the clear indication of its damage. Only **one caterpillar is seen in a gallery**. Cleaning the **tree trunks by removing all the webs** and placing cotton wool soaked in chloroform, kerosene, petrol and, carbon disulphide and sealing the same with mud or injecting into the hole with dimethoate 30EC 2ml/l or of dichlorvos 2 ml/l are effective.

Insect pests of pomegranate (*Punica granatum*)

- **Anar butterfly *Deudorix* (*Virachola*) *isocrates*) (Lycaenidae: Lepidoptera):** It is a **polyphagous pest having a very wide range** of host plants including aonla, apple, ber, citrus, guava, litchi, loquat, mulberry, peach, pear, plum, pomegranate, sapota and tamarind. The female lays **eggs singly on calyx of flowers or small fruits**. The bluish brown butterfly has an **orange spot on each of the fore wings and black spots on the hind wings**. The eggs are laid singly on **flowers and tender fruits**. Eggs are shiny white in colour and oval in shape. The fruits are covered with **polythene or paper bags** may escape infestation.
- **Woolly apple aphid, *Eriosoma lanigerum* (Aphidae: Hemiptera):** It is native of America and cosmopolitan in distribution. In India, **it was first recorded in 1889 at Conoor (Tamil Nadu)**. It attacks primarily **the underground roots**

but winged form also attacks trunk, branches, stems, twigs, leaf petioles and fruit stalks.

- **Codling moth, *Cydia pomonella* (Tortricidae: Lepidoptera):** The pest is distributed all over the **world, native of America and was introduced to India.**
- **San jose scale, *Quadraspidiotus perniciosus* (*Aspidiotus perniciosus*) (Diaspididae: Homoptera):** It was **introduced in India (Kashmir) from France in 1906** and it has been recorded on more than 150 host plants, including almond, apple, apricot, cherry, chestnut, citrus, crab apple, grapevine, gooseberry, mulberry, peach, pear, plum, quince, raspberry and strawberry. The **nymphs hibernate from December to March**, resume their activity around end of March and mature in about 4 weeks.
- ***Rhinoceros* beetle, *Oryctes rhinoceros* (Scarabaeidae: Coleoptera):** It is widely distributed in India and persistent in all coconut growing areas. The damage by the adult beetle which burrows by remaining in between leaf sheaths near the crown and thus cut across the leaf in its folded condition. The adult beetle is stout, black or reddish black, about 5 cm long and has a long horn projecting dorsally from the head in male; in the female the horn is short. The grub feeds on decaying matter and **larval stage lasts for 99-182 days.**
- **Red palm weevil, *Rhynchophorus ferrugineus* (Curculionidae: Coleoptera):** Few small holes with protruding chewed fibrous material and oozing out of a brown liquid from such holes, present in the tree trunk, indicate the early infestation by the pest. The reddish brown weevil has **six dark spots on thorax** and in male the conspicuous long snout has a tuft of hairs. The female lays eggs **in scooped out small cavities on palms of up to 7 years and on older trees deposited** in the wounds and other cut injuries found on the trunk crown.
- **Black-headed caterpillar, *Opisina arenosella* (Cryptophasidae: Lepidoptera):** The **larvae live on the under surface of leaflets within galleries of silk** and feed by scraping the green matter. Periodically releasing of parasitoids, *Goniozus nephantidis*, *Bracon brevicornis*, *Elasmus nephantidis* and *Trichospilus pupivora* are recommended to control the pest.
- Bark of fruit trees damaged from within and galleries formed; excreta webbed together by caterpillars visible externally are symptoms of damage caused by ***Indarbela* sp.**
- Polythene banding of tree-trunk in December-January and application of NSKE (5%) and *Beauveria bassiana* in January is effective against **Mango mealy bug.**

STORED GRAIN PESTS

- In India losses during post harvest handling and storage estimated at 15 % annually. **FAO estimate of total world losses in storage is 10% annually**.
- One group, internal feeders spends life feeding within the kernel of grain (*e.g.*

grain weevil, grain moth, lesser grain, borer, khapra beetle, pulse beetle and groundnut bruchid).

- The second group, external feeders spends its life feeding on the surface of grain or on the finished products of grain (*e.g.* **rice moth, red flou beetle, saw toothed beetle and cigarette beetle**).
- The insects that attack the healthy grains and cause damage to the stored grains are called primary pests (*e.g.* rice weevil, lesser grain borer, angoumois grain moth, pulse beetle, cigarette beetle, drug store beetle, sweet potato weevil, potato tuber moth, red flour beetle, Indian meal moth, rice moth and khapra beetle).
- The insect pests that cause damage to the grains or stored products that are already damaged by other insects are called secondary pests (*e.g.* saw toothed grain beetle, long headed flour beetle, flat grain beetle and grain lice).
- **Rice weevil, *Sitophilus oryzae* (Curculionidae: Coleoptera):** It infests the grain **both in storage and in fiel . Both adults and grubs** cause damage. At least **3-4 generations** are completed in a year.
- **Lesser grain borer or paddy borer beetle, *Rhyzopertha dominica* (Bostrichidae: Coleoptera):** Most characteristic feature is that the **head is deflexed downwards** to such an extent that is almost hidden in a dorsal view. It is important pest of whole cereals. **Both larvae and adults bore** into the grains and feed. There are **5-6 generations** in a year.
- **Khapra beetle, *Trogoderma granarium* (Dermestidae: Coleoptera):** It is originated from **India. Only larval** stage causes damage to storage grains. They usually found on **upper 50 cm layer of grains in a heap** or to the periphery in a sack of grains. The larvae are **positively thigmotactic,** they can be collected by merely placing gunny bags on a heap of grain. Life cycle is completed in 25 days under optimum conditions with about **12 generations** per year.
- **Red rust flou beetle, *Tribolium castaneum* (Tenebrionidae: Coleoptera):** *T. castaneum* is originated from India and it has spread throughout the tropical and subtropical world. ***T. castaneum* is common and *T. confusum* is found in imported food grains and in hilly regions**. The larvae and adults feed on a wide range of durable commodities and are important **secondary pests of cereals**.
- **Pulse beetle, *Callosobruchus chinensis* (Bruchidae: Coleoptera):** It is a pest of **pulses** both in the field at the time of harvest and in storage. Abdomen is conspicuously swollen. **Elytra do not cover the abdomen completely. Seven to eight overlapping generations** are completed per year.
- **Cigarette or tobacco beetle, *Lasioderma serricorne* (Anobiidae: Coleoptera):** Both grubs and adults bore into tobacco products *viz.*, cigarettes, cheroots and chewing tobacco. **Seven to eight overlapping generations** are completed in a year.

- **Sweet potato weevil, *Cylas formicarius* (Curculionidae: Coleoptera):** The grubs and adults bore into the tender vines, stems and tubers. Adults attack the leaves, vines and tubers. It is a pest that causes damage **both in the field and storage.**
- **Rice moth, *Corcyra cephalonica* (Galleriidae: Lepidoptera):** It is spot **free uniformly pale buff brown colour**. Larva is only responsible for damage and adult is free. It pollutes food grains with frass, moults and dense webbing. Grain converted to web mass and damaged grain or flour with bad odour.
- **Indian meal moth or meal worm moth, *Plodia interpunctella* (Phycitidae: Coleoptera)**: The adult forewing is basal half silver white or greyish with **outer 2/3 portion is reddish, copper bronze lusture with irregular bands. Only larva causes** damage and crawling caterpillars completely web over the surface of heap of grains with silken threads. **Primary pest** causes serious damage to ear and grain of maize contaminates the grain with excrement, cast skins, webbing, dead individuals and cocoons.
- **Drug store beetle, *Stegobium paniceum* (Anobiidae: Coleoptera):** It is similar to *Lasioderma serricorne* but can be distinguished by its antennae. In *Stegobium* the last three segments form a large loosely segmented club antennae.
- **Warehouse moth or almond moth, *Ephestia cautella* (Phycitidae: Lepidoptera):** It is also known as the dried currant moth and found in all over worldwide. Wings are **dirty white to greyish in colour** with indistinct black bands about **4 mm from head**. Only **larval stage is harmful**. It mainly feeds on germ portion leaving the rest of the kernel undamaged.
- **Preventive or prophylactic measures:** Threshing floor should be clean, free from insect infestation and away from the vicinity of villages or granaries.
- **Curative measures:** The infestation of stored grain insect pests can be controlled by the following methods like non chemical and chemical control measures.
- **Moisture content of grain:** Moisture is the critical factor in safe storage of food grains. Grains are stored at 10% moisture content escapes from the attack of insects (**except khapra beetle**).
- **Entoleter**: A direct method of controlling insect infestation is by removing insects from the infested commodity. Entoleters are used primarily in flour mills.
- **TNAU insect probe trap:** The use of trap is relatively a new method of detecting trapping insects in stored grains.
- Use of luminium phosphide for cover fumigation 3 tablets of 3 grams each per tonne of grain; for shed fumigation 21 tablets of 3 grams each for 28 cu. metres and **3g aluminium phosphide tablets releases 1g of phosphine gas**.
- Methyl bromide is very effective fumigant against all stages of insects used for export commodities and use for stock fumigation 28.32 g per m^3 and period of fumigation 24 hrs.

INTEGRATED PEST MANAGEMENT (IPM)

Introduction

- Michelbacher and Bacon (1952) coined the term '**integrated control**'.
- Integrated control was defined by Stern *et al.* (1959) as applied pest control which combines and integrates the biological and chemical control.
- The **idea of managing pest population was proposed by Geier and Clark 1961** who called their concept as protective management which later was shortened as pest management.
- In 1962, when "Silent Spring" was published by Rachel Carson, serious concerns about the disadvantages of pesticide use were widely raised. Carson and others suggested that pest control methods other than chemical pesticides should be used in order to protect wildlife, human health and the environment.
- Geier (1966) coined the term '**pest management**'.
- According to **FAO (1967), IPM** was defined as "a pest management system in the context of associated environment and population dynamics in pest species. It utilizes all suitable techniques and methods in as compatible manner as possible and maintains the pest population at levels below those cause economic injury".
- Later **Smith and Van Den Borsch in 1967** mentioned that the determination of insect numbers is broadly under the influence of total agro ecosystem and the role of principle element is essential for integrated pest management.
- **In 1972, the term IPM was accepted by CEQ** (Council of Environmental Quality) where IPM includes **I** - Integration that is harmonious use of multiple methods to control the impact of single pest as well as multiple pests; **P** – Pest or any organism that is detrimental to humans including vertebrates and invertebrate or weed or pathogens and **M** - Management refers to a set of decisions or rules based on ecological principles, economic and social consideration.
- The backbone of management of pest in an agricultural ecosystem is the concept of economic injury level (It is the level **of the pest up to which the damage can be tolerable**)
- In 1989, IPM Task Force was established and in 1990, IPM Working Group (IPMWG) was constituted to strengthen implementation of IPM at international level.
- In 1997, **Smith and Adkisson were awarded the World Food Prize** for pioneering work on implementation of IPM.
- **Economic Injury Level (EIL):** Lowest population at which pest will cause economic damage. EIL is usually expressed as the number of insects per unit area
- **Economic Threshold Level (ETL):** ETL is defined as the population density at which control measures should be applied to prevent increasing pest population

from reaching the economic injury level. Relationship between EIL and ETL can be expressed as **when no action is taken at ETL the population reaches or exceeds EIL.**

- **General equilibrium position (GEP):** It is the average population density of insect over a long period of time unaffected by temporary interventions of pest control.
- Development of resistance in insects against insecticides. *e.g.* **OP and synthetic pyrethroid resistance in *Helicoverpa armigera*.**
- Outbreak of secondary pests. *e.g.* **Whiteflies emerged as major pest when spraying insecticide against *H. armigera*.**
- Resurgence of target pests. *e.g.* **BPH of rice increased when some OP chemicals are applied.**
- **Subsistence phase:** Only natural control, no insecticide use.
- **Exploitation phase:** Applying more pesticides, growing high yielding varieties and get more yield and returns.
- **Crisis phase:** Due over use pesticides, problem of resurgence, resistance, secondary pest outbreak and increase in production cost.
- **Disaster phase:** Due to increased pesticide use, no profit, high residue in soil and collapse of control system.
- **Integrated Management Phase:** IPM integrates eco friendly methods to optimize control rather than maximize it.

Cultural control

- The manipulation of the cultivation practice is therefore known as **cultural control.**
- Since these control tactics usually modify the relationships between a pest population and its natural environment, they are also known as **ecological control methods.**
- **Tilling**, or turning of the soil at appropriate times can destroy soil-inhabiting insects at different stages either by crushing, burying them deep, occluding emergence tunnels, or exposing them to the full radiation of the sun. *e.g.* Pupae of moths, roots grub *etc.*
- **Crop sanitation:** Destruction or elimination of crop residues by burning, removal of diseased or insect infested fruit and plant parts is a good sanitation practice whereby the residual pest population is destroyed; and those surviving remain without shelter for hibernation.
- **Crop rotation**: Rotating crops in the field allowing an area to lie fallow may act as a deterrent to the increase of pest populations. **Groundnut with non leguminous crops is recommended for minimizing the leaf miner incidence.**

- **Polycropping or mixed cropping:** Crop mixtures are a characteristic of many traditional agricultural systems. Pest attack is theoretically reduced because appearance of the host plant is less and provided that the mixture is chosen to avoid crops attacked by the same pest. *E.g.* **Garden peas and sunhemp.**
- Early planting of paddy in *kharif* and late planting in rabi minimize the infestation of **rice stem borer**. Delaying the sowing of sunhemp till the onset of South-west monsoon avoids sunhemp hairy caterpillar, *Utethesia lotrix* attack. **Early sown sorghum in kharif reduces the infestation of shoot fl .**
- **Trap cropping** involves planting a suitable crop in a small area to attract the pest. The trap crop can be grown earlier or along with the main crop. Trap crop may also attract natural enemies thus enhancing natural control.

Trap crop	Main crop	Insect pest
Castor	Chillies	Tobacco caterpillar, *Spodoptera litura*
Tomato	Citrus	Fruit sucking moths, *Otheris spp*
Marigold	Cotton	American bollworm, *Helicoverpa armigera*
Mustard	Cabbage	DBM in cabbage

- **Trimming field buds:** Grasshopper eggs are laid in field bunds are destroyed by trimming field bunds.
- **Adaptation of high seed rate** in sorghum and later removal and destruction of shoot fly affected plants.

Physical control

- **Moisture manipulation:** Well dried grains (**moisture content <10%**), the survival of stored grain pests becomes impossible. *e.g.* Rice weevil, pulse beetle.
- **Vapour Heat Treatment (VHT):** Heated air is saturated with water (>RH 90%) for specified period of 6 to 8 hours for raising pulp temperature to 43-44.5°C in case of mango against fruit flies.
- **Oxygen stress and carbon dioxide concentration:** Air tight containers small volume of air is enclosed, the available oxygen is quickly utilized by insects and raise concentration of carbon dioxide. High concentration of carbon dioxide leads to death of stored products insects. Male insects can be made sterile by exposing them to gamma radiation or by using chemicals. By sterilizing the pupae of **screwworm (*Cochliomyia hominivorax*) with** radiations to make sterile males were obtained.

Mechanical control

- **Hand destruction:** as the name implies to destroy an insect by hand. Hand destruction also includes picking larval forms or egg masses from a plant and destroying them by some physical means. *e.g.* Hand-picking of lemon butterfly

(*Papilio demoleus*) larvae from citrus seedlings, red hairy caterpillar, American boll worm, egg masses of *Spodoptera.*

- **Provision of preventive barriers:** Digging of 30-60 cm wide and 60 cm deep trenches or erecting 30 cm height tin sheets barriers around the fields is useful against pests like **hairy caterpillars.**
- **Bagging or wrapping** of pomegranate and mango fruits in paper bags avoids the infestation of **pomegranate butterfl , *Virachola isocrates* and mango fruit fl , *Bactrocera dorsalis.***
- **Use of an alkathene band** around the tree trunks of mango to check the migration of first instar nymphs of **mealybugs and red ants.**
- **Sticky bands** around tree trunks against **red tree ant, *Oecophylla samaragdina.***
- **Mechanical traps** or collecting devices are usually employed for insect collecting or monitoring rather than control measures and depend on additional means to attract insects such as lights or chemicals. *e.g.* **Use of pheromone trapping** in isolated citrus orchards can be a good way of exterminating fruit fly populations.
- **Light traps** are arranged for attracting the insects, which are trapped by keeping water or oil in a container or a killing bottle below the light trap. *e.g.* **Most of the moths and beetles.**
- **Flame thrower** is a compressed air sprayer with kerosene oil for producing flames. There is a lance, which is fitted with a burner. Used for burning **locust populations, congregation of caterpillars, patches of weeds** *etc.*

Biological control

- **Biological control** defined as 'the manipulation of parasites, predators and pathogens to manage the density of insect populations'.
- Insect species often become **'pests' when this ecological balance is disrupted by natural events or human intervention.**
- First quarantine Act in USA came into operation in 1905. Government of India passed an act entitled **"Destructive Insects and Pest Act in 1914"** to prevent introduction into our country of any insect, fungus or other pests destructive to crops.
- **Foreign quarantine:** Prevents introduction of pests from foreign countries, takes care of pests of quarantine concern in India and restricts or prohibits import of plant materials.
- **Domestic quarantine:** It is the legislation to prevent of spread of already established pests from one part of the country to another. *e.g.* **Quarantine stations** at Mettupalayam, Shenbaganur and top hill monitor and prevent the spread of fluted scale, *Icerya purchase* in Tamil Nadu.

- **Enforcement of pest control measures:** It is the legislation to prevent damages and spread of already established pests. *e.g.* Cottony cushion scale, *Icerya purchase*; coffee berry borer, *Hypothenemus hampei*; sugarcane top borer, *Scirpophaga novella*; red hairy caterpillar, *Amsacta albistriga* and coffee stem borer, *Xylotrechus quadripes.*
- **Quality control:** It is the legislation to prevent the adulteration and misbranding of insecticides. The government of India enforced **'The Insecticide Act 1968'** which empowers to regulate the import, manufacture, sale, transport, distribution and use of insecticides.
- **Legislation to regulate activities of personnel engaged in pest control:** It ensures safety to workers engaged in plant protection operations and to public (Governed by Insecticide Act 1968).
- **Foreign pests introduced into India:** In past days, there were no restrictions on the transport of plants from one country to another and because the danger involved was not either realized or appreciated. *e.g.* **Cottony cushion scale, *Icerya purchase*; wooly aphid, *Erisoma lanigera*; potato tuber moth, *Pthorimaea operculella*, golden nematode of potato, *Globodera rostrochinensis*; serpentine leaf miner, *Liriomyza trifolii* and coffee berry borer, *Hypothenemus hampei.***
- **Foreign pests which have not entry into India: Grapevine phylloxera, cotton boll weevil, Mediterranean fruit fl , wheat stem sawfl , alfalfa aphid.**
- **Genetic control** refers to the control of pests by manipulating the genetic make-up of either the pest or the host. An example of manipulation of genetic material in the pest is such as the one where an insect is subjected to radiation, chemicals and other procedures. This process is often called **autocidal control**.

HOST PLANT REGISTANCE (HPR)

- Definition of HPR "Those characters that enable a plant to avoid, tolerate or recover from attacks of insects under conditions that would cause greater injury to other plants of the same species" **(Painter RH, 1951)**.
- Lesser damage than average damage is taken as resistance while more damage than average damage constitutes susceptibility.
- **Host evasion**: A host plant may pass through the most susceptible stage quickly or at time when insects are less in number. *e.g.* **Early planting of paddy in *kharif* minimize the infestation of stem borer, *Scirpophaga incertulas.***
- **Induced resistance** is a form of temporarily increased resistance as resulting from some conditions of plant or its environment such as changes in the amount of nutrients or water applied to the crop. *e.g.* **Application of potassium fertilizers.**

- **Host escape**: It refers to lack of infestation or injury to the host plant because of transitory circumstances like incomplete infestation thus finding of uninfested plant in a susceptible population does not necessarily mean that it is resistant.
- **Monogenic resistance** is controlled by a single gene is known as monogenic resistance.
- **Oligogenic resistance** is governed by a few genes is known as oligogenic resistance.
- **Polygenic resistance** is governed by many genes is known as polygenic resistance. This is also termed as **horizontal resistance**.
- **Major gene resistance** is controlled by one or few major genes. Major genes have a strong effect and these can be identified easily. This is also called **Vertical resistance**.
- **Minor gene resistance** is controlled by a number of minor genes and each contributing a small effect. It is called **minor gene resistance**. This is also referred to as **horizontal resistance**.
- **Vertical resistance:** A series of different cultivars of a crop show different reactions when infested with different insect biotypes is called vertical resistance.
- In other words, when infested with the same insect biotype and some cultivars show a resistant reaction while others show a susceptible reaction. It is also referred to **qualitative or biotype specific resistance**. Vertical resistance is generally but not always of a high level and is controlled by a major genes or oligogenes. It is **considered less stable**.
- **Horizontal resistance:** Horizontal resistance describes the situation where a series of different cultivars of a crop show no differential interaction when infested with different biotypes of an insect.
- All resistant cultivars show similar levels of resistance to all biotypes. This type of resistance is called **biotype-non-specific resistance, general resistance or quantitative resistance**.
- **Cross resistance:** Variety with resistance incorporated against a primary pest and confers resistance to another insect.
- **Multiple resistances:** Resistance incorporated in a variety against different environmental stresses like insects, diseases, nematodes, heat, drought and cold.
- **Sympatric resistance**: Acquired by coevolution of plant and insect (gene for gene) and governed by major genes.
- **Allopatric resistance:** Not by co-evolution of plant and insect and governed by many genes.
- **Antixenosis** is associated with insect behavior. 'Anti' means against while 'xenosis' means guest. Therefore, antixenosis means against a guest or to keep away a guest. In this case the guest is the unwelcome insect pest. The insect pests are kept away by breeding plants that possess characters which are

unattractive to insects. Examples of such characters are **colour, palatability, hairiness, waxiness, gummosis** *etc.* **Trichomes** in cotton resistant to **whitefly; wax bloom** on crucifer leaves deter feeding by **DBM;** plant shape and colour also play a role in non preference and open panicle of sorghum supports less *Helicoverpa armigera.*

- **Antibiosis** literally means against life. The host plant manufactures a chemical that interferes with the normal physiological functions of the pest. The chemical is ingested by feeding and effects may be on insect like death of immature, abnormal growth, failure in metamorphosis, reduced fertility of adults and abnormal reproductive behaviour. Classical example of **host plant resistance is DIMBOA (2,4 Di hydroxy-7-methoxy-1,4 benzaxin-3) content in maize** which imparts chemical defense against the **European corn borer, *Ostrinia nubilalis*.** Nutrionally related antibiotic effect in rice variety **Mudgo which is resistant to BPH**. When young females fed on variety Mudgo then ovaries of BPH are underdeveloped and contain few mature eggs due to less quantity of amino acid asparagine content in the resistant variety.
- **Tolerance** refers to the ability of a host plant to repair damage incurred by the pest or to increase growth and produce a crop in spite of pest attack. Soyabeans are one example where pest feeding produces a thinning that is compensated for by additional plant growth. Many traditional, long-term rice varieties have a high tillering capacity that enables them to compensate for stem borer damage.

CHEMICAL INSECTICIDES

Classification of insecticide

- Control of insects with chemicals is known is **chemical control**. The term pesticide is used to those chemicals which kill pests and these pests may **include insects, animals, mites, diseases or even weeds**.
- Chemicals which kill insects are called as **insecticides**. Insecticide may be defined as a substance or mixture of substances intended to kill, repel or otherwise prevent the insects.
- Chemicals used to control rats are known as rodenticides. *e.g.* **Zinc phosphide**.
- Chemicals used to control insects are known as insecticides. *e.g.* **Malathion**.
- Chemicals used to control mites are known as acaricides. *e.g.* **Dicofol**.
- Chemicals used to control birds are known as avicides. *e.g.* **Anthraquinone**.
- Chemicals used to control **snails and slugs** are known as **molluscides. *e.g.* Metaldehyde.**
- Chemicals used to control nematodes are known as nematicides. *e.g.* **EDB**.
- Chemicals used to control plant diseases are known as fungicides. *e.g.* **Copper oxy chloride**.

- Chemicals used to control weeds are known as herbicides. *e.g.* **2, 4- D.**
- When toxicant ingested acts on digestive system and brings about kill is known **stomach poison**. *e.g.* Malathion.
- When toxicant contacted to organism to bring death is known as **contact poison.** *e.g.* Fenvalerate.
- Toxicant enter into tracheal system as vapour and brings about death is known as **fumigant. *e.g.* Aluminium phosphide.**
- Chemicals penetrate into plant tissues and translocated through vascular system and cause death of insect feeding on plant is known as **systemic poison**. *e.g.* Dimethoate.
- The toxicant exerts physical effect is known as **physical poison. *e.g.* Activated clay.**
- The toxicants precipitate the protein is known as **protoplasmic poison. *e.g.* Arsenicals.**
- The toxicant inactivates the respiratory enzymes are known as **respiratory poison. *e.g.* Phosphine.**
- The toxicant inhibits the impulse conduction is known as **nerve poison. *e.g.* organophosphate and carbamate compounds.**
- **The toxicant inhibits the chitin synthesis is known as chitin inhibitor. *e.g.* Diflubernzu on.**
- **Inorganic compounds**: Compounds of mineral origin. *e.g.* Sulphur, zinc phosphide, fluorine compounds.
- **Animal origin:** *e.g.* **Neurotoxin,** isolated from marine annelids, *Lumbricorneris heteropoda* (Cartap).
- **Plant origin**: *e.g.* Nicotine, pyrethrum, rotenone, azadiractin.

Synthetic organic insecticides

- **Synthetic organic insecticides**: Organochlorines, organophosphate, carbamate insecticides *etc.*
- **First generation**: Inorganics and botanicals.
- **Second generation**: Synthetic organics.
- **Third generation**: Recent chemicals for reproductive control, IGRs and JH mimics.
- **Fourth generation**: Synthetic pyrethroids.
- **Fifth generation:** Synthetic pyrethroids like alfamethrin 10 EC, fenpropathrin 10 EC, ethofenpron 10 EC and neonecotinoids.
- **LD 50 (Lethal Dose):** The dose or amount of pesticide which kill 50% of the test populations/ animals. It is expressed in terms of milligrams of the substance of toxicant per kilogram body weight (mg/kg) of the test animal.

- **LC 50 (Median Lethal Concentration):** The concentration of a single pesticide which is lethal of 50% of the test animals.

Insecticide formulation

- **Emulsifiable concentrate (EC):** The active ingredient is dissolved in an organic solvent and is clear (but often with color). The spray mixture must be constantly agitated or the pesticide will separate from the water. *E.g.* Endosulfan 35 EC.
- **Flowable (F):** Finely divided solid particles in a liquid carrier forming a thick but pourable opaque product. Designed to be mixed with water at the time of application.
- **Liquid (L):** This is variously used by different companies for products with somewhat different properties. It is a suspension of microfine solid particles in an aqueous (water-based) carrier. Designed to be mixed with water at the time of application. *E.g.* Sevin 4 L.
- **Suspension concentrate (SC):** Finely divided solid particles in a liquid carrier. Similar to a flowable formulation. Designed to be mixed with water at the time of application. *E.g.* Spinosad 45 SC.
- **Granular (G):** This formulation appears as small pellets or granules similar to a WG or WDG. This formulation is designed to be used right from the container and without mixture with water. The granular formulation is ideal for products designed as soil insecticides as gravity helps pull the granules to the soil surface where they can be mechanically incorporated (with annual crops). *e.g.* Carbofuran 3 G, phorale 10G.
- **Soluble powder: (S or SP):** Regrettably, this abbreviation refers to two quite different formulation types. Sevin 80S is an 80% dry powder designed to be mixed with water to form a suspension and therefore is like a wettable powder. It is a totally soluble powder that dissolves when mixed with the spray water. *e.g.* Cartap hydrochloride 50 SP.
- **Wettable powder (WP):** This formulation comes as a dry, dust-like material. However, it is designed to be mixed with water at the time of application. The active ingredient does not dissolve in water but with thorough mixing becomes suspended. *e.g.* Diflubenzuron 25 WP, Carbaryl 50WP.
- **Water dispersible granules (WG or WDG):** This formulation appears as small pellets or granules. It is easier and safer to handle and mix than wettable powders. When the granules are mixed with spray water they break apart and with agitation, the active ingredient becomes distributed throughout the spray mixture. *e.g.* Thiamethoxam 25 WG, Clothianidin 50WG.
- **Fumigants:** Fumigants are pesticides in the form of poisonous gases that kill when absorbed or inhaled. *e.g.* Aluminium phosphide.

- **Smoke generators:** They are used in the form of coil like strips containing toxicant for the control of mosquitoes. When ignited, these coils release vapours. *e.g.* **Pyrethrum.**
- **Aerosols:** Aerosols contains a small amount of pesticide driven through a fine opening by a chemically inactive gas under pressure when nozzle is triggered or by burning toxicant or vaporising it with heat.
- **Commercial formulations of neem are available in 10000 ppm, 1500 ppm and 300 ppm the market.** Some of the neem formulations are Margosan, Achook, Bioneem, Neemazal, Neemax, Nimbicidine, Neemgold *etc.*

Plant origin insecticide

- **Nicotine:** Nicotine is found in the leaves of ***Nicotiana tabacum*** **and *N. rustica*from** 2% to 14%. Nicotine sulphate has been mainly used as a contact insecticide with marked fumigant action in the control of sucking insect's *viz.*, aphids, thrips, psyllids, leafminers and jassids.
- **Rotenone:** It is extracted from the roots of **Derris plant** which many contain 4 to 11% rotenone depending on the variety. Though rotenone is reported from 68 species of leguminous plants, principal commercial sources are *Derris elliptica, D. Malaccensis* from Malaysia and *Lonchocarupus utilis* and *D. uruca* from South America. Rotenone occurs in **Derris** roots (4-9%), Lonchocarpus (8-11%).
- **Plumbagin:** Plumbagin is naturally occurring napthoquinone of plant origin from **the roots of *Plumbago europea*** (Plumbaginaceae).
- **Pyrethrum:** It is extracted from dried **flowe heads of *Chrysanthemum cinerariaefolium*.**

Organochlorine insecticide

- **Diphenylmethane compound:** *e.g.* DDT, DDD, dicofol, methoxychlor *etc.*
- **Monocyclic compound:** *e.g.* BHC.
- **Bicyclic or chlorinated terpenes**: *e.g.* Toxaphene.
- **Polycyclic compound**: *e.g.* Chlordane, heptachlor, aldrin, dieldrin, endosulfan *etc.*
  - **DDT** (Dichloro diphenyl trichloro ethene) is efficient contact poison and phytotoxic to cucurbits.
  - **BHC** (Benzene hexa chloride/Hexa chloro cyclo hexane) is very good contact insecticide and available in 10 D or 50 WP.
  - **Endosulfan:** Safer to bees and available in 35 EC.

Organophosphate insecticide

- **Phosphorohalide and cyanides**: *e.g.* Sarin, tabun *etc.*
- **Phosphorothioates or phosphorodithioates:** *e.g.* Malathion, phorate, dimethoate, ronnel *etc.*
- **Phosphates**: *e.g.* Dichlovos, monocrotophos, phosphomidon *etc.*
- **Phosphorothiolate:** *e.g.* Oxydemeton-methyl, profenophos *etc.*
- **Phosphorothionate:** *e.g.* Chlorpyrifos, quinalphos, diazinon *etc.*
- **Phosphoroamidate:** *e.g.* Acephate, phosphoran *etc.*

Synthatic pyrithoids

- I generation synthetic pyrethroids: *e.g.* **Allethrin.**
- II generation synthetic pyrethroids: *e.g.* **Resmethrin, tetramethrin** *etc.*
- III generation synthetic pyrethroids: *e.g.* **Permethrin, cypermethrin, deltamethrin and fenvalerate.**
- IV generation synthetic pyrethroids: *e.g.* **Flucythrinate, cyfluthri** *etc.*

Neonicotinoids

- **Neonicotinoids:** They represent a novel and distinct chemical class of insecticides with remarkable chemical and biological properties. Neonicotinoids are similar to nicotines in activity partially to structure. These are selective and safe to non target organisms.
- They are used mainly against sucking pests like aphid, leaf hopper, whitefly, mealybugs *etc.,*
- **Imidacloprid (17.8 SL, 70 WS):** The trade names are confidor, gaucho and provado.
- **Acetamiprid (20 SP):** The trade names are pride, profil and supreme.
- **Thiamethoxam (25 WG, 70 WS):** The trade names are actara, cruiser and rinova.

Other insecticide group

- **Clothianidin (50 WG):** The trade names are dantos and celeso.
- **Thiacloprid (36 WG, 70 WG):** The trade names are calypso, bariard and alanto.
- **Phenyl pyrazoles (Fiproles):** The GABA receptor is the target site for fipronil. Blockage of GABA gated chloride channel reduces neuronal inhibition which leads to hyper exitation of the central nervous system, convulsions and death of a target pest.
- **Fipronil (0.3 G, 5 SC):** The trade names are reagent, tremidor, zoom and bilgran.

- **Spinosad (45 SC, 2.5 WSC):** The extract of the fermentation broth that contains spinosad is produced by the microorganism, *Saccharopolyspora spinosa.* Spinosad is a contact and stomach poison with some translaminar movement in leaf tissue. The trade names are tracer, spintor, precise, success, naturalyte and boomerang.
- **Avermectins:** Avermectins are produced by the soil microorganism *Streptomyces avermitilis*. These form a new class of compounds having nematicidal, miticidal and insecticidal activity. Avermectins are **action on GABA gated chloride channel**, causing an inhibitory effect results in the insect's death. This channel normally blocks reactions in some nerves, preventing excess stimulation of CNS. Emamectin benzoate and abamectin are the two major compounds in this group.
- **Emamectin benzoate (5 EC, 5 SG):** A trade name is **Proclaim.**
- **Abamectin (1.8 EC):** The trade names are Avid, Agrimec, Vertimec, Argi-mek and Avert.
- **Indoxacarb (14.5 SC, 30 WDG):** The active ingredient indoxacarb works by inhibiting sodium ion entry into nerve cells, resulting in paralysis and death of targeted pests. Indoxacarb is a stomach poison with slight contact action. The result is impaired nerve function, feeding cessation, paralysis and death. The trade names are avaunt, steward and torando.
- **Diafenthiuron (50 WP):** It is an inhibitor of oxidative phosphorylation by disruption of ATP formation (inhibitor of ATP synthase). The trade name is polo.
- **Pymetrozine (50 WDG):** It is a new insecticide highly active and specific against sucking insect pests. It has high degree of selectivity, low mammalian toxicity and safety to birds, fish and non-target arthropods. The trade names are full fill and chess.
- **Chlorfenapyr (3 SC):** It is mainly stomach poison and has contact action. The trade names are pirate and pylon.
- **Chlordimeform (50 SP, 4 EC):** It has marked translaminar and systemic activity. The trade names are galecron, fundal and spike.
- **Amitraz (50 SP, 20 EC):** It is a non systemic insecticide and acaricide with contact and respiratory action. The trade names are acarac, amitraze and baam.
- **Spiromesifen (2 SC, 4 F):** Spiromesifen is effective against whitefly, spider mites and psyllids. The trade names are Oberon and forbid.
- **Spirodiclofen (2 SC):** Spirodiclofen is a selective, non-systemic foliar insecticide and acaricide. The trade name is envidor.
- **Spirotetramat (14.5 SC, 22.4 SC):** Spirotetramat is effective against aphids, whiteflies, scales, mealybugs, psylla, phylloxera, thrips and mites. The trade names are movento and ultor.

- **Chlorantraniliprole (200 SC, 35 WG):** Chlorantraniliprole is a novel anthranilic diamide insecticide and efficacious for control of lepidopteran insect pests. The trade names are coragen and altacor.
- **Flubendiamide (240 WG, 480 SC):** It acts by ingestion and disrupts Ca^{2+} **balance in the nervous system** resulting in rapid cessation of feeding and extended residual control of important lepidopteran pests. The trade names are **fame and belt**.

Rodenticides

- The poisons used for the control of rodents are either acute poisons (single dose and quick acting) or chronic poison (multiple dose and slow acting).
- **Zinc phosphide:** It is a greyish black powder having garlic odour of phosphine. Its toxicity is due to the evolution of phosphine gas from the molecule.
- **Aluminum phosphide** is an inorganic phosphide used to control insects and rodents in a variety of settings. Aluminum phosphide is available in pellet and tablet form.
- **Anticoagulant rodenticides:** Warfarin was the **first compound** developed as anticoagulant rodenticide. Bromadiolone was first registered in India and is in use since 1988. Anticoagulants constitute more than 95 per cent of total rodenticide usage.
- **Multidose anticoagulants: Warfarin and fumarin** are used at 0.5 per cent concentration as fresh baits.
- **Single dose anticoagulants**: Bromadiolone used as fresh bait (0.25%) ready to use bait (0.005%).

Acaricides

- The substances exercising toxic effects on mites are specifically called as **miticides or acaricides.**
- **Dicofol (18 EC):** It is a hydroxylated product of DDT. It is non-systemic acaricide effective against all stages of mites and trade names are kelthane and hifol.
- **Tetradifon (Tedion 20 EC):** Persistent acaricide and kills all stages of mites except adults.
- **Propargite (57 EW):** It is **non-systemic acaricide**, used for control of mites on fruits, grain, vegetables, nut and crops. The trade names are comite and **omite**.

Insecticides Act 1968

- An act to regulate the import, manufacture, sale, transport, distribution and use of insecticides with a view to prevent risk to human beings on animals.

- Inter-departmental or ministerial or organizational co-ordination is achieved by a high level advisory board **"Central Insecticides Board" with 24 members (to be raised to 29 by an amendment)** drawn from various fields having expert knowledge of the subject.
- **Board members of CIB:** Director General Health Services, Chairman; Drugs Controller, India; Plant Protection Adviser to the Government of India; Director General, ICAR; Director General, ICMR.
- **In this board (CIB) having totally 24 members.**
- **The Insecticides Rules, 1971:** There are nine chapters in the insecticide rule, 1971 relating to the functions of CIB, RC, CIL, grant of licenses, packing, labelling, first aid, antidote, protective clothing *etc.*

Sprayer

- **Hand sprayer:** The tank capacity is **0.5 to 1.0 litres.** It also used for spraying small kitchen garden.
- **Knapsack sprayer:** It consists of a piston pump, a lever to operate, a built in the tank **(10 to 14 litre capacity)**, hose, lance and a nozzle. Suitable for spraying field crops.
- **Rocker sprayer:** It consist of pump assembly, platform, operating lever, pressure chamber, suction hose with strainer, delivery hose and an extension rod with spray nozzle. This is used for spraying fruit trees and tall crops.
- **Foot sprayer (pedal pump):** This sprayer is also suitable for tall crops. It requires **500-1000 litres water per hectare** depending upon the growth of a crop and four persons are needed to operate it. It consists of a plunger assembly a stand a suction hose, a delivery hose and an extension rod with a spray nozzle.
- **Ultra low volume sprayer (ULV)** (Hand carried, battery operated spinning dis-sprayer): Here the pesticides are applied **less than 5 litres** spray fluid produces fine droplets (80 μm).These are light weight sprayers (less than 3 kg) have a rotary atomiser (spinning disc) powered by an enclosed DC motor with a plastic spray head, a liquid reservoir, a handle and a power supply unit.
- **Electrodyn sprayer (EDS):** Electrodyn sprayer is completely a new system of spraying for the controlled droplet application of chemicals (CDA).It makes use of electrical energy for droplet formation and propulsion towards target crops. The EDS consist of a spray stick and a unique combination of bottle plus nozzle. The nozzle contains ready formulated chemical for immediate application to crops.
- **Rotary duster:** It consists of a blower with gearbox and a hopper with a capacity to hold **4-5kg dusts**. The duster is operated by rotating a crank and the motion is transmitted through the gear to the blower. The efficiency is 1-1.5 ha per day.

- **Knapsack duster:** It consists of dust container of **2-5 kg capacity** through which air current is blown by means of bellows which are worked by hand liner attached to one side of the container.
- **Power operated dusters:** They work by means of a petrol engine generally 1 to 3 HP. Larger dusters may have engines of 15 to 75 HP and are mounted to tractors. A power duster with a single outlet can cover 2 hectare per day.
- **In India first major accident in 1958 kill more than 100 people in Kerala** due to consumption of wheat flour and sugar contaminated with parathion leakage during shipment from Bombay to Cochin.
- More than 3000 people died due to inhaling of vapours of **methyl isocyanate, leaked from a carbaryl manufacturing plant in Bhopal in 1984.**

Antidotes

- **Universal antidote:** It is a mixture of **7 g activated charcoal, 3.5 g magnesium oxide and 3.5 g tannic acid** in half a glass of warm water may be used to absorb or neutralize poisons.
- Atropine is the usual antidote for **organophosphate and carbamate poisoning.** It can be given orally and in severe cases, injections are given. Repeated injections may be required.
- 2 PAM: It is injected intravenously as an antidote in **organophosphate poisoning.** It should not be used in case of carbamate poisoning.
- Calcium gluconate is recommended as an antidote for some **organochlorine insecticides.**
- Vitamin K is the preferred antidote for anticoagulant poisoning such as **warfarin.**
  - Dimercaprol (BAL) is recommended for **arsenic poison**.

INSECT HORMONES

- **Hormones** are chemical that alter normal growth and development by interfering with **the insect endocrine systems**. Synthetic compounds possessing activities of juvenile hormone and moulting hormone of insects often termed as **mimics or insect growth regulators.**
- **Juvenile Hormone (JH) Mimics:** The possibility that JH analogs may have potential as insect control was first recognized by **Williams (1965)**.
- **Brain hormone:** They are also called **activation hormone (AH).** It is secreted by **neuro secretory cells (NSC)** which are neurons of central nervous system **(CNS).** Its role is to activate the thoracic glands to produce moulting hormone (MH).
- **Juvenile hormone (JH):** It is also called **neotinin hormones.** It is secreted by **corpora allata** which are paired glands present behind insect brain. Their role is to keep the larva in juvenile condition.

- **Ecdysone:** It is also called **moulting hormone (MH).** Ecdysone is a steroid and is secreted by **prothoracic glands (PTG)** present near prothoracic spiracles. *e.g.* Juvabione (fernesol), methoprene (altosid), hydroprene (gentrol, altozar) and kinoprene (enstar)
- **Anti JH or JH Agonists or Precocenes:** Anti-juvenile hormones found in plants that induce reversible precocious metamorphosis and sterilization in insects by suppressing the function of the corpora allata gland. *e.g.* **Fenoxycarb** (Insegar 50 % WP) **and Pyriproxyfen** (Tiger 10 % EC).
- **Ecdysone or Moulting Hormone (MH) Agonists:** These compounds are synthetic analogues of natural ecdysone. When applied in insects, kill them by formation of defective cuticle. The development processes are accelerated bypassing several normal events resulting in integument lacking scales or wax layer.v *e.g.* **Tebufenozide** (Mimic, Confirm and Romdan), **Halofenozide** (Mach) and **Methoxyfenozide** (Prodigy)
- **Chitin synthesis inhibitors:** These are compounds which have ability of inhibiting chitin synthesis in insects by blocking the activity of the enzyme chitin synthetase. **Diflubenzu on** (Dimilin) is the most effective control of number of pests like cabbage looper, cotton leaf folder, gram pod borer and tobacco caterpillar. **Buprofezin** has been found promising against several sucking pests.

INSECT PHEROMONES

- **Pheromones** are chemicals secreted into the external environment by an animal which elicit a specific reaction in a receiving individual of the same species.
- Pheromones are **volatile in nature** and they aid in communication among insects.
- Pheromones are **exocrine in origin** (*i.e.* secreted outside the body). Hence they were earlier called as ectohormones.
- **In 1959, German chemists Karlson and Butenandt** isolated and identified the first pheromone, a sex attractant from silkworm moths. They **coined the term pheromone.**
- **Mass-trapping: (Male annihilation technique):** Large numbers of pheromones baited traps can be used in the fields to capture male moths of newly emerged and reduce the number of males for mating.
- **Example of sex pheromone:** Grandlure for cotton boll weevil, *Anthomonas grandis;* gossyplure and hexalure for pink bollworm, *Pectinophora gossypiella;* helilure for *Helicoverpa armigera*; gypsy moth, *Porthesia dispar* sex pheromone is gyplure and disparlure; cabbage looper, *Trichoplusia ni* sex pheromone is looplure and tobacco cutworm, *Spodoptera litura* sex pheromone is spodolure and litlure.

- **In *Trogoderma granariam*** mixture of fatty acid esters and methyl and oleate function as aggregation pheromones.
- In **ants, *Formica rupa* formic acid, while termites, *Zootermopsis nevadensis* hexanoic acid** functions as the trail marking pheromone.
- ***Dolichoderine* ants release a fruity odour by** the worker that results in an erratic behaviour of workers, when this is discharged into mandibles onto intruding insects that becomes marked as aggressor.

OTHER INSECT CONTROL METHODS

- Chemicals that induce avoiding (oriented) movements in insects away from their source are called **repellents**. They prevent insect damage to plants or animals by rendering them unattractive, unpalatable or offensive.
- Essentials oils of **Citronella** and cedarwood act as repellents. **'Odomos'** uses citronella oil extracted from lemongrass is commercial mosquito repellent. Pyrethrum extracted from **Chrysanthemum** is a good repellent and has been used against tsetse fly, *Glossina morsitans.*
- Synthetic repellents: **Dimethyl pthalate** for mosquito and ticks; **Deet** (diethyl-m-toulamide) for mosquito, blackflys and ticks; Benzyl benzoate for ticks and chiggers; **pymetrozine for Homoptera**.
- **Antifeedants** are chemicals that inhibit feeding in insects when applied on the foliage (food) without impairing their appetite and gustatory receptors or driving (repelling) them away from the food. They are also called gustatory repellents, feeding deterrents and reject ants.
- **Triazenes: AC 24055** has been the most widely used triazene which is an odourless, tasteless, non-toxic chemical which inhibit feeding in chewing insects like caterpillars, cockroaches and beetles.
- **Organotins: Triphenyl tin acetate** is an important antifeedants in this group effective against cotton leaf worm, colarado potato beetle, caterpillars and grass hoppers.
- Antifeedants from non-host plants of the pest can be used for their control. *E.g.* **Pyrethrum extracted from flowers of *Chrysanthemum cinerarifolium*, neem extracted from leaves and fruits of neem, *Azadirachta indica*** are an antifeedant against many chewing pests and desert locust in particular.
- Chemicals that cause insects to make oriented movements towards their source are called insect **attractants**. They influence both gustatory (taste) and olfactory (smell) receptors.
- **Sinigrin** which is present in crucifers which attracts cabbage butterfly, *Pieris brassicae.*
- **Trimedlure** attractant for Mediterranean fruit fly, *Ceratitis capitata.*

- **Methyl eugenol** attractant for oriental fruit fly, *Dacus dorsalis*.
- **Cue-lure** chemical against melon fruit fly, *Dacus cucurbitae.*
- Geraniol and eugenol for Japanese beetle, *Popillia japonica*.
- Any chemical which are interfere with the reproductive capacity of an insect that is called **chemosterillant**. *e.g.* **TEPA, METEPA, Chloro ethylamine, 5-Fluororacil and amithopterin**.
- **Triazines** ***e.g.*** **Tris-s-triazine** and **Triphenyl tin chloride.**
- **Sterile male release technique:** When a sterile male mates with normal female there will be no progeny. If adequate number of vigorous and competitive sterile males is introduced systematically into natural population the population will soon cease to exist.
- This theory of **male sterile technique** was conceived by **E F Knipling as early as 1937 and was published in 1955.** *e.g.* Screw worm (*Cochliomyia hominivorax*) a cattle pest was completely eradicated from Curacao Islands and south eastern parts of USA by male sterilization with gamma rays (CO^{60}).
- **Hybrid sterility:** In some insect cross-types or races which produce fertile females but sterile males among progeny. These sterile hybrids are excellent material for use in insect control. These sterile males are more vigorous and competitive than the sterile males produced after radiation or chemosterilization.
- **Population replacement:** The ability of disease transmission of vectors *i.e.* replacement of specific vector populations can as well be changed by genetic methods.
- **Auto-sterilization:** Sterilization of native insects in their natural environment by using chemosterilants along with the species specific attractants/lures/bait traps.

BIOLOGICAL CONTROL

Histry of biological control

- Suppression or prevention of a pest outbreak due to purposeful manipulation of natural enemies is known as **biological control**.
- In ancient times **Chinese used nests of Pharoach's ant, *Monomorium spp.* to** combat stored grain pests and *Oecophylla spp*. to control foliage feeding insects.
- In **1762, the Mynah bird was** imported from India to Mauritius for control of locust.
- The **first successful biological control attempt was taken in 1887-88**, when the developing citrus industry in California was seriously threatened by the scale *Icerya purchasi.* Dr. C.V. Rilly suggested that, the original home of the scale was Australia or New Zealand and that natural enemy of the scale should be introduced into USA from these countries.

- Then **Albert Koebele** was sent to Australia and he found a **small beetle** known as **Vedalia beetle,** ***Rodalia cardinalis*** attacking the scale in the Adelaide area in 1888 that beetle were brought and released into California.
- **First introduction of natural enemy into India was in 1898 when a coccinellid beetle,** ***Cryptolaemus montrouzieri*** was imported in India from Australia and released against coffee green scale, *Coccus viridis.*
- A parasitoid, ***Aphelinus mali*** **introduced from England into India in 1920 to** control woolly aphid on apple, *Eriosoma lanigerum.*
- The parasitoid, ***Prospatella perniciosus*** **imported from China to India** in 1958-60.
- **A parasitoid,** *Aphytis diaspidis* imported from USA to India in 1960 against used to control **apple sanjose scale** (*Quadraspidiotus perniciosus).*
- The **egg parasitoid,** ***Telenomus*** **sp. imported from New Guinea to India in 1964** to control castor semilooper, *Achaea janata.*
- In **1965, predator,** ***Platymeris laevicollis*** **introduced from Zanzibar to control coconut rhinoceros beetle,** ***Oryctes rhinoceros.***
- **Control of cottony cushion scale,** ***Icerya purchasi*** **on fruit trees by its predatory vedalia beetle,** ***Rodolia cardinalis*** **in Nilgiris**. The predator was imported from California in 1929 and Egypt in 1930 and multiplied in the laboratory and released. Within one year the pest was effectively checked.
- The **parasitoids,** ***Bracon brevicornis, Parasierola nephantidis, Trichospilus pupivora*** **were used at 10:10:20 adults** per tree against coconut black head caterpillar in 1960.
- The weevil, ***Cyrtobagous salviniae*** **was imported from Australia to India in 1982 for the biological suppression of water fern,** ***Salvinia molesta.***
- Water hyacinth (***Eichhornia crassipes***) was successfully controlled with a weevil, ***Neochetina eichorniae*** (Curculionidae: Coleoplera) **in Kerala and Karnataka**. The larvae tunnel and feed inside the petioles and crown of the plant. Adults are grey with brown mottles.

Parasitoids

- **Parasitoids:** An organism that mostly attacks another organism of the same taxonomic class, kills its host, requires are host for development and parasitic during immature stage and adult is free living. The parasitoids comes **90% under order Hymenopteran and 10% in Diptera**.
- **Egg parasitoids:** *Trichogramma chilonis* use for control of sugarcane internode borer @ 40,000 per hectare and releases six time from 4th month at 15 intervals. *T. japonicum* used for the control of rice stem borer @ 1 lakh per hectare per release and releases two times on 30 and 37 DAP.

- **Larval parasitoids:** *Bracon brevicornis* (Braconidae); *Eriborus trochantgeratus* (Ichnemoidae) *Parasierola nephantidis* (Bethylidae) ; *Platygaster oryzae* (Platygasteridae) ; *Sturmiopsis inferus* (Trachnidae) and *Spoggossia bassiana.*
- **Larval pupal paratiloid**: *Isotima javensis* use for control of top borer in sugarcane and *Eucelalotria bryani* (Tachinidae) use for control of early shoot borer in sugarcane.
- **Pupa parasitoids:** *Tetrastichus israeli* (Eulophidae) use for control of black headed caterpillar of coconut and *Brachymeria nephatidis* (Chalcidae).
- **Nymphal adult parasitoids:** *Aphelinus mali* use for control of apple woolly aphid and *Encarsia formosa* use for control of cotton whitefly.
- ***Epiricania melanoleuea* (Epircanidae: Lepidoptera) is single parasitoid of order Lepidoptera** that use for control of sugarcane leaf hopper.
- **Predators:** Are bigger in size, kill the prey, free living throughout their life, requires more than one prey to complete development, adults and larvae feed upon similar insects.
- **Fishes: *Gambusia alfinis* use for control of mosquito larvae.**
- Coccinellidae: ***Coccinella septumpunctata*** on aphids; ***Menochilus sexmaculata* on** mealy bugs and scales; ***Rodalia cardinalis* on cottony cusion scale and *Cryptolaemus montrozieri*** on grapevine mealy bug.
- Carabidae: ***Paederus fuscipes*** on rice brown plant hopper and leaf folder.
- Reduvidae: ***Rhiocoris fuscipes*** on *H. armigera* and ***Platymeris laevicollis*** on Rhinoceros beetle.
- Miridae: ***Cytothinus lividepennis*** on rice hoppers.
- Chrysopidae: ***Chrsoperala carnea* release 50,000** per hectare in cotton for control of aphids, scales, bollworms.

Parasitism

- **Simple parasitism** is irrespective number of eggs laid by the parasitoid attacks the host only once time. *e.g.* ***Apanteles taragamae* on the larvae of *Opisina arenosella* and *Goniozus nephantids***
- **Super parasitism** phenomenon is parasitization of an individual host by more larvae of single species. *e.g.* ***Apanteles glomeratus* on *Pieris brassica* and *Trichospilus pupivora* on *Opisina arenosella*.**
- **Multiple parasitisms** phenomenon is simultaneous parasitization of host individual by two or more different species of primary parasitoids at the same time. *e.g.* ***Trichogramma, Telenomous* and *Tetrastichus* attack on eggs of paddy stem borer, *Scirpophaga incertulas*.**
- **Hyper parasitism** is phenomenon when a parasite itself is parasitized by another parasite. *e.g.* ***Goniozus nephantidis* is parasitized by *Tetrastichus israeli*.**

- **Primary parasite:** A parasite attacking an insect which itself is not a parasite (beneficial to man).
- **Secondary parasite:** A hyperparasite attacking a primary parasite (harmful to man).
- **Tertiary parasite:** A hyperparasite attacking a secondary parasite (beneficial to man).
- **Quaternary parasite:** A hyperparasite attacking tertiary parasite (harmful to man).

MICROBIAL CONTROL

- Steinhaus (1949) coined the term **microbial control** when microbial organisms or other products (toxins) are employed by man for the control of pests on plants, animals or man.
- Over **400 species of fungi and more than 90 species of bacteria** which infect insects have been described including *Bacillus thuringiensis*, varieties of which are manufactured and sold throughout the world primarily for the control of caterpillar pests and more recently mosquitoes and black flies.
- Among fungal pesticides, five have been introduced since 1979 and three in 1981. So far, more than **40,000 species of *Bacillus thuringiensis* have been isolated and identified as belonging to 39 se otypes.**
- These organisms are active against **Lepidoptera or Diptera or Coleoptera. *Bacillus thuringiensis* products account for 90-95 percent of the total biopesticide market.**

Virus

- NPV (*Borrellina* virus): **About 300 isolates of nuclear polyhedral virus** have been isolated from the order Lepidoptera. Among these viruses baculoviruses (Baculoviridae) are successful in IPM. The NPV is observed to affect 200 species of insects like *Corcyra cephalonica, Pericallia ricini, Amsacta albistriga, Spodoptea litura* and *Heliothis armigera*, by ingestion.
- **Mode of action virus:** As an infection progresses, the **caterpillars internal organs are liquefied, its cuticle (body covering) discolors and eventually ruptures**.
- **NPV symptoms of infection:** Insects become dull in colour, feeding rate reduced and larvae **become pinkish white** especially in the ventral side due to accumulation of polyhedra. Diseased larvae hang upside down from the plants. This is called **tree top disease or wipfel krankeit.**
- **Nuclear Polyhedrosis virus (NPV):** It is **obligate pathogens**. The virus consists of a **proteinnaceous polyhedral occlusion body** inside which the virions or

virus rods are embedded. NPV specific to *H. armigera* (HANPV), *Spodoptera litura* (SI NPV) and *Amsacta albistriga* (Aa NPV).

Bacteria

- More than 100 pathogenic bacteria were recorded and in which *Bacillus thuringiensis* (*Bt*) is important and is isolated from flour moth, ***Ephestia kuhniella*** **by Berliner (1915).**
- **The spore forming bacteria are obligate (*Bacillus popillae* caused milky disease) and facultative (*B. thringiensis*).** The commercial formulations of Bt are Dipel, Thuricide, Bactoperin, Delfin, Spicthurin and Biobit.
- **Non spore forming bacteria** are *Pseudomons* sp. and *Serratia* sp.
- Green muscardine fungus, *Metarhizium anisopliae* on **rhinoceros beetle**
- White muscurdine fungus, *Beauveria bassiana* on **silk worm**
- White muscurdine fungus, *Cephalosporium lecanii* **on coffee green scale**

Protozoa

- Their mass production is difficult. They infect insect orders like Lepidoptera, Coleoptera, Orthoptera, Hemiptera and Diptera. *e.g.* ***Farinocystis tribolii*** **on red flou beetle and *Nosema bombycis* on silk moth**.
- **Rice moth, *Corcyra cephalonica*** is a potential host or prey insect for rearing number of parasitoids and predators.

INSECT ECOLOGY

- The word ecology (oekologie) (coined by German scientist **Ernst Haeckel, 1866)** was derived from the **Greek 'oikos' meaning household and logos meaning science**.
- Ecology is closely related to physiology, evolutionary biology, genetics and ethology.
- **Odum (1953) define** ecology is the study of the structure and functions of nature (or environmental biology).
- **Autecology** is the study of individual organisms or an individual species in relation to the environment.
- **Synecology** is the study of the group or groups of organisms associated in a community in the same environment *i.e.,* in relation to various other species living in the same environment.
- **Immigration** is one way inward movement from one place to another.
- **Emigration** is one way inward movement from one place to another.
- **Migration** is mass movement of entire population, where some insects return again to the moved area.

- **Biosphere** is the term used for all of the earth's ecosystems functioning together on the global scale.
- **Life table** is number of insects dying over a period of time are often represented in life tables a tabular form of accounting for deaths. Life table is a numerical aid used in the study of insect population to record in a systematic fashion. There are two types of life table.
- **Age specific (horizontal) life table** is based on observations made on single generation in different regions.
- **Time specific (vertical) life table** is based on observations on overlapping or multiple generations.
- Biotic factors also known as **'Density dependent factors'**.
- Abiotic factors known as **'Density independent factors'**.
- The process of repeated eating and being eaten is known as **food chain** and a network of inter related food chains forms a **food web**.
- **Reproductive potential** is the ability of the insects to multiply in a given time in the absence of all environmental resistance'.
- **Environmental resistance** is the sum total of physical and biological restraints in an environment that prevents a species from realizing its biotic potential.
- Thus, the **actual abundance** = biotic potential – environmental resistance.
- The insects are all **'poikilothermic'** ***i.e.*** **they have no precise mechanism for regulating the temperature of their bodies**. Insects require optimum range of temperature for development and survival.

Competition

- The type of interaction in which two individuals or species for a limited amount of food, water, nesting space, mates or other resources. If they do occur in the same habitat, they exploit different resources and remain active at different times and frequently displace one another. This is called **Gause's hypothesis/ principle** or **competitive exclusion principle** after the Russian biologist **G.F. Gause** who observed first such separation in 1932.
- More than one species consuming the same food materials is known as **interspecific competition**. *e.g.* ***Trichogramma*** **and** ***Chrosperla*** compete for *Helicoverpa* eggs in cotton.
- **Cannibalism:** When insects feed on the individuals of the same species. *e.g. Helicoverpa armigera, Tribolium confusem* (large larvae and adults feeds on egg and small larvae).

Accidental introduction of foreign pests

- Immature and adult stages of certain insects adhere closely to the plants such as scales and aleurodids and those which bore into the tissues of plant parts such as leaf miners, stem borers, gall insects *etc.*, and are more liable to be introduced into other countries. Such insects introduced into India from foreign countries are the diamond back moth, *Plutella xylostella* on cruciferous vegetables, Sanjose scale, *Quadraspidiotus perniciousus*on fruit trees on hills, green mealybug, *Coccus viridis* on coffee, potato tuber moth, *Phthorimoea operculella,* cotton cushiony scale, *Icerya purchasi*, serpentine leaf miner *Liriomyza trifolii,* spiralling whitefly, *Alerodicus disperses* and coconut mite, *Aceria guerreoronis*.

Survey

- Regular survey activity is necessary for successful surveillance programmes. An insect pest survey is a detailed collection of insect population information at a particular time in a given area. These surveys are both qualitative and quantitative.
- The **qualitative surveys** aim at the pest detection, employed with newly introduced pests and often precedes quantitative survey.
- The **quantitative surveys** attempt to define numerically the abundance of an insect population in time and space. It is useful to predict future population trends and to assess damage potentials.
- **Sampling techniques:** Surveillance requires suitable sampling techniques. A sampling technique is the method used to collect information for a single sample. The sampling techniques include direct counts (*in situ* counts), knock down, netting, trapping, (use of light trap, pheromone trap, sticky trap), and extraction from soil.
- **Fixed plot survey:** In an acre plot, five micro plots of m^2 area are marked one each in four corners (1 m away from bund) and fifth in the centre of the chosen field. Periodical assessment is done in these micro plots.
- **Roving survey:** This is conducted every week in randomly selected field plots along the prescribed route of the survey. Observations are recorded from the west corner by a diagonal walk to 100 m.

Level of pests

- **Economic damage** is the amount of injury which justify the cost of artificial control measures.
- **Economic injury level** is defined as the lowest population of insects that will cause economic damage.
- **Economic threshold level**: It is the level at which management action should be taken to prevent population reaching EIL.

- The word **'Pest'** derived from the Latin word **'Pestis'** meaning Plague. Pest is defined as any living organism that will cause the economic damage to human being.
- Pests that cause less than 5% loss in yield is said to be negligible. Insects which normally cause a loss ranging from 5 to 10% are said to be **minor pests** and those which cause a loss of 10% or more in general called as **major pests**.
- **Regular pest**: Occur most frequently (regularly) in a crop and have close association with that particular crop. *e.g* .Chilli thrips, *Scirtothrips dorsalis*, brinjal shoot and fruit borer, *Leucinodes orbonalis*.
- **Occasional pests**: Here a close association with a particular crop is absent and they occur infrequently. *e.g.* Rice case worm, *Nymphula depuctalis,* castor slug caterpillar, *Parasa lepida* and mango stem borer, *Batocera rufamaculata.*
- **Seasonal pests**: Occur mostly during a particular part of the year and usually the incidence is governed by climatic conditions. *e.g.* Red hairy caterpillar on groundnut and rice grasshoppers in June-July.
- **Persistent pests**: Occur on a crop almost throughout the year. *e.g.* Scales and mealybugs on many crops, thrips on chillies and paddy stem borer.
- **Sporadic pests**: Occur on a few isolated localities. *e.g.* Rice earhead bug, *Leptocorisa acuta* and castor slug caterpillar, *Latoia lepida.*
- **Epidemic pest**: Occur in a severe form in a region or locality at a particular season or time only. *e.g.* Rice hispa, *Dicladispa armigera* and rice leaf roller, *Cnaphalocrocis medinalis.*
- **Endemic pest**: Occur regularly and confined mostly to a particular area or locality. *E.g.* Red hairy caterpillar, *Amsacta albistriga* on groundnut in Kurnool, Ananthapur, Kadapa, Chittoor, Srikakulam and Vizag districts and stem borers of rice, paddy gall fly in Warangal districts.

MITES

- The study of mites and ticks is **called acarology**.
- Acari (or Acarina) are a taxon of arachnids that contains **mites and ticks.**
- The diversity of the Acari is extraordinary and its fossil history goes back to at least the early **Devonian period**.
- Mites are the most diverse representatives of an ancient linkage in phylum Arthropoda, **subphylum Chelicerata and subclass Acari**.
- Mites are arachnids and as such evolved from a segmented body with the segments organised into two tagmata **prosoma (cephalothorax) and an opisthosoma (abdomen).**
- This anterior body region is called the capitulum or **gnathosoma** and the remaining of the body is called the **idiosoma** and is unique to mites.

- The cephalothorax functions are **sensing, feeding and locomotion**.
- **Antennae, mandibles and maxillae are absents on mites.**
- Those members of the subclass Acari which feed on plants are known as **phytophagous mites**.
- In Tenuipalpidae also known as false spider mites or flat mites, **about 800 species have been described in over 25 genera**. Only *Brevipalpus*, *Tenuipalpus* and *Raoiella* and few other genera become pests of economic plants mainly on tropical fruit crops and ornamental plants.
- **Virus transmission** has only been documented in some **Eriophyidae or Tenuipalpidae species**.
- The two spotted spider mite is a very common pest of greenhouse crops as well as field crops and is found throughout the world. **Two spotted spider mite has often been referred to as a red spider mite**.
- The most common mites found in India infesting agriculture crop plants are rice leaf mite *Oligonychus oryzae* (Family: Tetranychidae), rice panicle mite *Steneotarsonemus* (Family: Tarsonemidae), jowar mite *Oligonychus indicus* (Family: Tetranychidae), cotton leaf mite *Tetranychus telarius* (Family: Tetranchidae), wooly mite of cotton *Aceria gossypii* (Family: Eriophyidae), red gram mite *Aceria cajani* (Family: Eriophyidae), chilli mites *Polyphagotarsonemus latus* (Family: Tarsonemidae), citrus rust mite *Phyllocoptruta oleivora* (Family: Eriophyidae), coconut eriophyid mite *Eriophyes gurreronis* (Family: Tetranychidae), vegetables mite *Tetranychus telarius* (Family: Tetranchydae) and tea mite *Oligonychus coffeae* (Family: Tetranychidae).
- **Wheat mite, *Petrobia latens* (Acarina: Tetranychidae): It is the only prominent wheat mite that webs in the leaves. Growing of tolerant wheat variety such as C-306.**
- **Coconut mite, *Aceria (Eriophyes) guerreronis*** (Acarina: Eriophyidae): Coconut mite has been serious problems to the production of coconut from 1997 onwards. When the infestation is severe, nut fall may occur. The nuts which continue to remain in the bunches are undersized and malformed, the degree of malformation varying with infestation.
- Mango malformation is caused by **Mite**
- What is the scientific name of wheat mite **Petrobia latens**
- In adult stage the four pairs of legs are present in **Mite**

RODENT PESTS

- The name **"rodent"** comes from the **Latin** word **'rodere'** which means **to gnaw**. They constitute more than **42% of the known mammalian species**.
- Of the rodents that live with man, commonly called **commensals**, the mole rat (*Bandicota bengalensis*), the house rat (*Rattus rattus*), the brown rat (*Rattus*

norvegicus), the house mouse (*Mus musculus*), the grass rat (*Millardia meltada* (Gray) and the bandicoot rat (*Bandicota indica*) are very important economically because of the damage to stored food supplies besides serving as reservoirs for some important **human diseases** as **plague, murine typhus, leptospirosis, food-poisoning** *etc*.

- According to Panse Committee's report **about 2.5% losses** are caused due to rodents in storage annually.
- In India, rodents have been estimated to cause 5 to 10% losses in **rice**.
- Most rodents have no more than **22 teeth though** one exception is the silvery mole-rat from central and east **Africa which has 28**.
- Burrow system of ***T. indica*** **is of a simple 'Y' shaped** type with one or two surface openings whereas *M. hurrianae* tend to congregate in complicated and extensive burrow systems with numerous surface openings.
- Pregnant period for rate is **19-21 days**; post-partum estrus can give birth every 3 weeks and young reach maturity around 6 weeks of age.
- One crop per year then one breeding season and two crops two breeding seasons.
- One female can potentially produce **36 young in one cropping season**.
- **Mole rat, *Bandicota bengalensis***: It is the most destructive pest of field crops and is distributed throughout India except the extreme arid zones.
- **Soft furred (Field rat or Grass rat), *Rattus (Millardia) meltada***: It is a medium sized rodent and ranks second in damage to different field crops. It is mainly a pest of irrigated cropping system.
- **Indian gerbil rat, *Tatera indica***: It is reddish gray in colour with white underside and it equals the common house rat in size with about 18 cm long head and body and a hairy tail little longer than the head and body.
- **Indian field mouse, *Mus booduga***: It is about 5 to 8 cm long with 5 cm long tail. It is brown in colour with a white belly.
- **House mouse, *Mus musculus***: The length of head and body is 75-100 mm and tail 65-90 mm respectively. Body weight about 15-20 g. The dorsum is dark brownish-grey and the venter light grey to creamy white.
- **Roof rat, *Rattus rattus***: Head and body 140-220 mm and tail 160-250 mm uncoloured. The body is slender, the muzzle sharp, the ears prominent and mostly bare of hair.
- **Biocide: Salmonella-based rodenticide** (*e.g.* Biorat); this is used in Cuba, Central America and some countries in SE Asia. This product is claimed to be specific to rodents.
- **Electromagnetic repellents:** This ultrasonic mouse repellent uses both ultrasound and electromagnetic action to disturb the habits of mice. Simply plug into any ordinary household socket. For large houses we would recommend one for upstairs and another for downstairs.

- **Single dose poison: Zinc phosphide and Barium carbonate** are some of the compounds which have been used as rat poisons. These are also called **acute poisons** as these are highly toxic in nature *i.e.*, they show immediate fatal results.
- **Preparation of bait:** Zinc phosphide-2 part, shredded food grains (as case)-96 parts and an edible oil- 2 parts.
- For control of rats in burrows, poisoned baits should be placed in paper cones and pushed into the holes without any prebaiting. This is called **"Torpedo or Hole Baiting"**.
- **Multiple dose poison:** The modern way to kill rodents in houses or godowns is by using **anticoagulants. The first anticoagulants (warfarin, pindone, diphacinone and clorophacinone)** are commonly known as the first-generation anticoagulants or multiple feed rodenticides.
- **Second-generation anticoagulants: Brodifacoum** is the most potent rodenticide currently available for commensal rodents.
- **Bait shyness:** Regardless of the formulation used to deliver the toxin, you should be aware that mice and rats can learn to identify and avoid toxic baits; this tendency is known as **"bait shyness."** To avoid bait shyness, it may be necessary to **prebait (offering rodents similar but nontoxic bait prior to applying the toxic bait)** an area prior to using toxic baits.
- Agricultural ornithology may be defined as the science of **birds in relation to agriculture.**
- It is interesting to note that only 05 species of about 1000 species of birds found in India (*i.e.,* only 2.1%) have been reported to inflict damage to crops and fruits.
- Community structure of birds at a grain store in Punjab that among 06 species recorded at the store, house sparrow, ***Passer domesticus*** **was the most abundant species having 77% dominance in the community.**
- **Snails and Slugs:** Only **one difference between slugs and snails is that snails carry coiled shells on their backs while slugs don't.**
- **Snails and slugs are hermaphrodites**, possessing both male and female genitalia. In some species, an individual may behave as male for a while, then as a female.
- **The radular mouths of these gastropods are directed downward** so that food can be taken from the surface over which they travel.

Important Facts of Entomology

Quiz

- True salivary glands of Hymenoptera are ***Labial glands***.
- Chetotaxy is essential for using immature stages for species identification in **Lepidoptera**.

- Wings are kept in upholding condition, when insects rest is **Butter fl** .
- The family of coleoptera that possess haplogastran abdomen is **Scarabaidae.**
- The mole cricket belongs to family **Grylloblatidae**
- Stridulatory organs are located on hind femur in which family **Acrididae**
- The males are winged and females are wingless in family **Pseudococcidae**.
- The number of stylets in female mosquitoes is **six.**
- In Dictyoptera and Orthoptera, the forewing which appears as sclerotized is known **Tegmina.**
- Example of an exopterigote insect order having pupal instar in its life stages in **Thysanoptera**.
- The hind wings are modified in to halters in which order **Diptera.**
- The wing modified in to elytra in which order **Coleoptera.**
- The mouth parts of thrips are **Asymmetrical type.**
- Hypognathous head is found in **Locust.**
- Johnston organ found in **Pedicel.**
- The mandibles in social bee are used for **Nest construction.**
- Number of stylets in mouthparts of head louse is **Three.**
- Type of wing coupling in Mecoptera is **Jugofrenate.**
- Newly emerged active immature forms of a scale insect are **Crawlers.**
- The out covering of an insect egg is **Chorion.**
- A spine like structure found at the tip of the abdomen of a chrysalis is **Cremaster.**
- Hook like structures or claws at the tips of prolegs of Lepidoptera larvae are **Crochets.**
- Hamuli wing coupling apparatus is present in which order **Hymenoptera.**
- A pad like structure present at the base of each tarsal claw is **Pulvillus.**
- A clasp like structure associated with the furcula of Collembola is **Tenaculum.**
- Mouth parts are mandibulate type in adults of Lepidoptera family **Micropterygidae.**
- Amplexiform wing coupling apparatus present in which **Papilionidae.**
- Father of insect physiology is **Sir V. B.Wigglesworth.**
- Hamuli wing coupling apparatus is present in which order **Hymenoptera.**
- The chitin is abundant in **Endocuticle.**
- Digestive system of insect from mouth to anus called as **Alimentary canal.**
- Fore gut and hind gut is originated from **Ectoderm.**
- The mid gut is also known as **Mesentron.**
- Salivary gland is also called as **labial glands.**

- Filter chamber is a modified from which part of digestive system **Mid gut.**
- Honey chamber is a part of alimentary canal in which worker bees store nectar in modified from **Crop.**
- Usually the reproductive opening in male is located in **9th segment.**
- Taenidea does not come out at the time of moulting from **Tracheoles.**
- The terrestrial insect is excreting the substances is **Uric acid.**
- The excretory product in red cotton bug is **Allantoin.**
- The resting potential of nerve axons in insect is **-70mv.**
- **Striated type** of muscles found in insects.
- Haemocoelous viviparity is prominent in the insect order **Strepsiptera.**
- Most terrestrial insects excrete 80-90% of their nitrogenous waste as **Uric acid.**
- Gastrulation is the process by which mesoderm and endoderm invaginated within the ectoderm.
- Haemocytes are derived from embryonic **Mesoderm.**
- Haemoglobin is found in the larvae of **Chironomous larvae.**
- The honey bee capable of flying at the speed of **25-30 km.**
- The scientific name of Indian honey bee is **Apis indica.**
- The scientific name of rock or wild honey bee is **Apis dorsata.**
- Honey bees rearing place is called **Apiary.**
- Communication in honey bee takes place by round and waggle dance is described **Kal Von Frisch.**
- Polarized light helps in finding food source **Honeybee.**
- In honey bee worker, the royal jelly is secreted by **Pharyngeal gland.**
- The silk moth belongs to family **Bombycidae.**
- The lac insect belongs to order **Hemiptera.**
- Which plant should be planted in the vicinity of apiary is **Castor.**
- Silk moth can be reared on which plant **Mulberry.**
- The pebrin disease is caused by **Protozoa.**
- Acarine disease is caused by **Mite.**
- Drooping of tender leaves and yellowing are the damage caused by **stem fl** , ***Ophiomyia phaseoli.***
- The alkaloid in cotton prevents the attack of pink bollworm is **Gossypol**.
- The scientific name of rice stem borer is ***Tryporyza incertulus***.
- Hairy leaf blade in cotton prevents the attack of ***Aphids***.
- Terminal rolling and drying of leaves from tip to base in both rice nursery and mainfield is **Thrips.**

- ***Nephotettix nigropictus*** is acts as a vector for rice tungro virus.
- ETL for GLH in tungro endemic area is **2 number/ hill.**
- Alternate host of *Stenochaetothrips biformis* is ***Echinochloa* sp.**
- The scientific name of white leaf hopper is ***Cofana spectra.***
- Paddy stem borer belongs to **Pyraustidae** family.
- White ear in rice is due to attack of **Stem borer.**
- Gall midge belongs to **Cecidomyiidae** family.
- Floating of tubular cases in rice field is due to ***Nymphula depunctalis*** pest.
- Scorching symptom of the entire rice field is due to ***Cnaphalocrocis medinalis.***
- Which of the following pest causes onion needle or silver shoot symptoms in rice is ***Orseolia oryzae.***
- Cutting of seedling tip before transplanting is a good management technique for controlling **of Stem *borer.***
- Passing of rope over the crop is done to dislodge the pest **is Caseworm.**
- Backward rolling of leaves is the typical symptom of ***Pelopidas mathias.***
- **Cecidogen** is a chemical responsible for gall formation in rice.
- ***Trichogramma chilonis*** egg parasitoid can be released against rice stem borer.
- Write the scientific name of rice mite ***Oligonychous oryzae.***
- Trimming of bunds is recommended for the control of **grasshopper.**
- The scientific name of blue rice leaf hopper is ***Empoascanara maculifrons.***
- Hopper burn' is the symptom of ***Nephotettix nigropictus, Sogatella furcifera, Nilaparvata lugens*** in rice.
- Use of synthetic pyrethroids may cause resurgence of sucking pest in rice is **True.**
- Tip drying and orange discoloration of rice leaves is typical symptom of ***Recilia dorsalis.***
- Which of the following pest causes damage to the grain at milky stage is **Earhead bug.**
- **Nilaparvata lugens** act as a vector ragged stunt and wilted stunt.
- **Nilaparvata lugens** act as a vector for transitory yellowing diseases.
- Zigzag striped leafhopper belongs to **Cicadellidae** family.
- Green mirid bug which feeds on BPH is **Cyrtorhinus lividipennis.**
- Alternate wetting and drying is a good management technique for controlling of **Brown plant hopper**.
- Early sowing of sorghum prevents the attack of **Sorghum shoot fl .**
- Phyllody disease of sesamum is transmitted by which insects ***Orosius albicinctus.***
- Mustard saw fly pupates in **Soil.**

- Lablab or cowpea is sown as an intercrop to minimise **stem borer** damage.
- Maize stem borer undergoes **4-6** generation per year.
- **Oats and barley** are immune to wheat gall nematode.
- Presence of shot holes and dead heart is the damage caused by ***Chilo partellus*** in maize.
- A bunchy root with galls in wheat is due to the attack by **Cyst nematode, *Heterodera avenae.***
- Winged forms of wheat aphid migrate to ***Cynodon dactylon*** for breeding.
- Rabi wheat suffers more from the attack of ghujia weevil ***Tanymecus inidcus***.
- Flowers and young sesame capsules with gall like swelling is the typical symptom of **Gall fl .**
- Gingelly phyllody is transmitted by ***Orosius albicinctus.***
- ***Asphondylia sesame*** is the scientific name of Gingelly gallfly.
- The insect where larvae feed on castor leaves voraciously and adult cause damage to citrus fruits is ***Achaea janata.***
- Water soaked spots on the castor leaves become yellow and dried are symptoms of **White fl .**
- Which one of the following is a serious but sporadic pest on castor is **Castor butterfl .**
- Presence of holes in the leaves of mustard is a symptom of ***Plutella xylostella.***
- **Sawfl** is a hymenoptran insect without crochets on its pseudolegs.
- Saw fly is having **8** pairs of abdominal pseudo legs.
- A hymenopteran insect having saw like ovipositor on its abdomen is **Sawfl .**
- **Painted Bug, *Bagrada hilaris*** excretes a sort of resinous material that spoils the mustard pods.
- Larva with branched spines in castor **spiny caterpillar, *Ergolis merione.***
- Which larva feeds on silk and developing maize grains? **Earworm.**
- Site of pupation for ash weevil is **Soil.**
- Severely affected groundnut filed looking as if grazed by cattle is the typical symptom of **Red hairy caterpillar.**
- Which life stage of red hairy caterpillar undergoes diapauses is **Pupa.**
- Groundnut shows burnt appearance when severely infested with **Leaf miner.**
- Fecal pellets seen on groundnut leaves and on the ground is an indicator of **Tobacco caterpillar** pest incidence.
- Cannibalism is common in ***Helicoverpa armigera*** larvae.
- Digging out a trench around the field is to prevent **Red hairy caterpillar** attack.
- Groundnut leaves with holes resulting in skeletonization and defoliation is due to **Semilooper.**

- Sphenoptera perotetti belongs to **Buprestidae** family.
- **Soybean** is an alternate host of groundnut leaf miner.
- ***Elasmolomus sordidus*** is scientific name of groundnut pod bug.
- Affected safflower buds with small bore holes; rotten with a foul smelling ooze coming out of the apices, giving the buds a soaked appearance is symptom of **Bud fly *Acanthiophilous helianth)***
- **Helilure** is a pheromone trap is used to monitor, attract and kill male moth of *Helicoverpa armigera.*
- Young groundnut pods showing bore holes with discoloured pulps is the symptom of **Earwig pod borer** in groundnut.
- Sunflower leaves with holes leading to skeletonization and defoliation is due to **Semilooper, *Trichoplusia ni***
- **Safflowe caterpillar, *Perigea capensis*** is a serious pest of safflower throughout India.
- The concentration of the gas per unit volume x time of the exposure is **Critical value.**
- The impact machine is used to kill insects in stored product is known as **Entoleters.**
- Pusa bin is used to the control of **Khapra beetle**.
- Zinc phosphide insecticide is **acute poison**.
- Webbing of rice grains is characteristic symptoms of **Rice moth**.
- The groundnut stored grain pests which attach both kernals and pods is ***Caryedon serratus***.
- The first stored grain pest that is resistance to insecticide is ***Tribolium castaneum***.
- The irregular holes in the grains due to the attack of **Lesser grain borer**.
- Which is a secondary pest ***Tribolium castaneum.***
- Which seed is least damaged by insects in the store **Paddy.**
- Presence of irregular holes of 1.5 mm diameter on grains of rice, sorghum, wheat, barley, maize in storage is due to attack by ***Sitophilus oryzae.***
- Dark brown beetle with head bent under the thorax; covered by a crenulated hood shaped pronotum and the posterior abdominal end blunt is ***Rhyzopertha dominica.***
- Brownish grey colored beetle with truncate elytra having elevated like spots near the middle of dorsal side is ***Callosobruchus maculatus* (*chinensis*).**
- Adult is slender ant like with a long snout and shiny black with reddish brown thorax and leg ***Cylas formicarius.***
- Name of two lepidopterous storage pest are ***Sitotroga cerealella* and *Plodia interpunctella.***

- Damage by potato tuber moth (*Gnorimoschema operculella*) occurs both in **field and storag** .
- Gaseous quinones released to the medium produce a readily identifiable acid odour in heavy infestations of ***Tribolium castaneum.***
- Name the family to which Angumois grain moth and potato tuber moth belongs to **Gelechiidae.**
- Name the pests that occur both in the field and storage **Rice weevil, pulse beetle, tamarind beetle, Angoumois grain moth, sweet potato weevil and potato tuber moth.**
- Merely placing the gunny bags on the heap of grains helps in the collection of ***Trogoderma granarium.***
- Adults of ***Rhizopertha dominica*** are powerful fliers and can move across godowns.
- Angoumois grain moth is an **internal feeder.**
- Rice moth *Corcyra cephalonica* is an **external feeder.**
- Early attack by the **sorghum shoot fl** on main shoot induced the production of a few synchronous tillers that grow rapidly and survive to produce harvestable ear heads.
- When population of one species is benefited and other is not affected is known as **Commensalism.**
- Number added to the population per unit time is **Natality rate.**
- Absolute population estimate can be made through **Quadrat method.**
- Tritrophic interaction involves in **Natural enemy.**
- **Kogan and Ortman** coined the term 'antixenosis'.
- Economic threshold level (ETL) is significantly low or zero in **Vector.**
- The main control measures in integrated pest management are **Non chemical control.**
- The control measures should be applied at **ETL** stage.
- The rotenone extract is obtained from which part **Root of Derris sp.**
- Insecticide which can be safely used in ware houses or godowns **Malathion.**
- The insecticide can be used with seed dressing **Imidacloprid.**
- Which insecticide is based on the toxin production by the marine annelids **Cartap hydrochloride.**
- Most widely used pheromone is **Sex pheromone.**
- In integrated pest management, pheromones are used for **Mating disruption.**
- Juvenile hormones are absent in which stage **Egg.**
- Moulting hormone (Ecdysone) is secreted from which gland **Prothoracic gland.**
- In Wiseman's rings, which organ is present **Neurone.**

- The Wiseman's ring is formed by organs **corpora cardiac; prothoracic gland and corpora allata**.
- Chemicals released by threatened aphids elicit defensive behavior among the ants that tend these aphids. Such chemicals would be best described as **Allomones.**
- Substances in the frass produced by bark beetles attract predators and parasites to trees that are infested by these beetles. Such substances would be best described as **Kairomones.**
- The chemical which does not stimulate the feeding is called **Deterrent.**
- The *Trichogramma* sp is a parasitoid of **Lepidopterans.**
- The lace wings, Chrysoperla carnea is feed on **Aphids.**
- The larva predators which suck the sap of lac insects are **Eublemma sp.**
- Grasping type of leg is present in which insect **Mantid.**
- The insects are ***Poikilothermic.***
- When organism live in narrow range of temperature are called as ***Stenothermic.***
- Favorable temperature for insect growth and development is ***13- 35 °C.***
- Under favorable condition of temperature, the egg laying capacity of insects is ***Increase.***
- Inter locking pattern of food chain is called ***Food web.***
- In trophic level, the plants are come in which category ***Producers.***

AGRICULTURAL ENTOMOLOGY OBJECTIVE QUESTIONS (MCQ)

1. The function of haemolymph in insect is

   A. Transpiration of nutrients B. Transpiration of oxygen
   C. Transpiration of excreta D. None of these
2. The eggs present in the ootheca of cockroach is

   A. 20 B. 16 C. 10 D. 24
3. Among the following which pest shows single generation in a year

   A. *Helicoverpa armigera* B. *Holotrichia consanguinea*
   C. *Trogoderma granarium* D. None of these
4. The property of DDT was first discovered by

   A. Paul Muller, 1930 B. Paul Muller, 1949
   C. Paul Muller, 1939 D. None of these
5. The malaria is first discovered in which state

   A. Bangalore B. Jaipur C. Hyderabad D. Nagpur
6. Among the following insecticide which is plant origin

   A. Endosulfan B. Imidacloprid C. Malathion D. Rotenone
7. The commercial formulation of *Bacillus thuringiensis* is

   A. Elcar B. Doom C. Dipel D. None of these

8. Insectivorous plant is
   A. Sunflower plant B. Pitcher plant
   C. Safflower plant D. None of these
9. According to the Classical Theory on head segmentation the insect head comprises
   A. Five segments B. Six segments C. Four segments D. Eight segments
10. The food channel in siphoning type of mouthpart is formed by
    A. Maxillary palpi B. Hypopharynx C. Modified galea D. Labium
11. The cement layer of the epicuticle is formed by secretion of
    A. Dermal gland B. Epidermal cells
    C. Trichogen cells D. None of these
12. Exarate type of pupa is common in
    A. Lepidoptera B. Coleoptera C. Diptera D. None of these
13. The female external genitalia grasshoppers are borne on:
    A. The 8th and 9th abdominal segments B. The 8th abdominal segment
    C. The 9th abdominal segment D. The 9th and 10th abdominal segments
14. The basic element of the nervous system is:
    A. Neuron B. Vannus C. Cuticle D. Nerve fiber
15. In India, the maximum honey is obtained from hives of which bee
    A. Indian bee B. Little bee C. Rock bee D. European bee
16. In India, which silk moth is monopoly for production
    A. Eri silk moth B. Munga silk moth
    C. Mulberry silk moth D. Wild silk moth
17. The main host of Erisilk worm is
    A. Mulberry B. Aak C. Caster D. None of these
18. Colour and taste of honey is varying due to
    A. Type of honey bee B. Day of collection
    C. Source of food D. None of these
19. The silvery shoot in onion is caused by
    A. Onion thrips B. Aphids C. White fly D. None of these
20. Light production in fire flies is due to the reaction of ....and ....
    A. Serotonin and water B. Luciferin and potassium
    C. Luciferin and magnesium D. None of these
21. The insects was evolved about
    A. 100 million years ago B. 300 million years ago
    C. 600 million years ago D. 1000 million years ago
22. The insects are physio-chemicaly
    A. Warm blooded B. Both of these C. Cold blooded D. None of these

23. The Binomial nomenclature was first given by
    A. Darwin B. Fabrious C. Linnaeus D. Snodgrass
24. The insect cuticle is secretion from
    A. Mesoderm B. Basement membrane
    C. Epidermis or Hypodermis D. None of these
25. Which of the following does not a pest of stored wheat grain
    A. *Sitophilus oryzae* B. *Callosobruchus chinensis*
    C. *Trogoderma granarium* D. *Tribolium castaneum*
26. The American multinational company, Monsanto has produced an insect resistant cotton variety. A toxic gene from which one of the following bacteria has been transferred to this transgenic cotton
    A. *Bacillus subtilis* B. *Bacillus thuringiensis*
    C. *Bacillus globlii* D. *Bacillus amyloiquifanciens*
27. The scientific name of sorghum midge is
    A. *Chilo partellus* B. *Contarinia sorghicola*
    C. *Spodoptera litura* D. None of these
28. The scientific name of rice mealy bug is
    A. *Phnococcus solenopsis* B. *Ripersia oryzae*
    C. *Maconellicoccus hirsutus* D. None of these
29. The newly introduced mealy bug in cotton is
    A. *Phnococcus solenopsis* B. *Ripersia oryzae*
    C. *Maconellicoccus hirsutus* D. None of these
30. The newly reported parasitoid of mealy bug, *Phenococcus solenopsis* in India is
    A. *Anagyrus* sp B. *Aenasius bambawalei*
    C. *Leptomastix dactylopii* D. None of these
31. Honey bee queen produces ....thousand eggs per day
    A. 0.5 – 1.5 B. 2 – 3 C. 4.0 – 4.5 D. 5.0 – 5.5
32. The wings of butterfly and birds are;
    A. Homologous organs B. analogous organs
    C. Vestigial organs D. Circulatory organs
33. The legless young of housefly is known as
    A. Caterpillar B. Maggot C. Nymphs D. Grubs
34. The mouth parts of plant bugs are
    A. Biting and chewing type B. Coiled and suctorial type
    C. Piercing and suctorial type D. Suctorial proboscis type
35. Which animal acts as natural tiller of soil
    A. Termite B. Earthworm C. Snake D. Millipede

36. For biological control of tomato fruit borer, the ideal trap crop is

A. Marigold B. Mustard C. Tobacco D. Garlic

37. What is the quantity of carboxyl 50% WP required to prepare 500 litres of spray fluid of 0.2 % concentration on active ingredient basis

A. 1.0 kg B. 1.5 kg C. 2.0 kg D. 3.0 kg

38. Which one of the following functions of mouth parts of insects makes them effective vectors of plant viruses

A. Biting and chewing B. Piercing and sucking

C. Rasping D. Punching

39. The term *Entomology* has been derived from the words, *entomon* means insect and *logos* means study

A. Latin B. Greek C. Roman D. Arabic

40. The insects are able to maintain the balance while walking due to

A. Small size B. Wings C. Exoskeleton D. Hexapod

41. The exchange of gases in respiration in insects takes place through

A. Spiracles B. Lungs C. Book lungs D. Cuticle

42. The color of grasshopper blood is

A. Red B. White C. Blue D. Colorless

43. In the reduction of a pest population, the balance of the natural enemies is maintained in

A. Pest management B. Pest control

C. Pest eradication D. None of these

44. The strain of the lac insect which survives on all hosts is known as

A. Kusumi B. Rangini C. Lakhini D. Namkumi

45. White grub beetles emerge from soil in between

A. 7.30- 8.00 PM B. 8.00- 9.00 PM C. 4.00- 5.00 AM D. 10.0- 11.0 AM

46. The pupation of fruit flies takes place in

A. Soil B. Fruit C. Stem D. Rind of fruit

47. Which of the following is a true fish

A. Silver fish B. Flying fish C. Jelly fish D. Cray fish

48. Which insect can survive at higher temperature and low humidity

A. Khapra beetle B. Rust red flour beetle

C. Rice weevil D. Potato tuber moth

49. The stability of ecosystem is maintained by

A. Homeostasis B. Diversity of species

C. Food web D. Trophic levels

50. Which is the larvicidal fish

A. *Gambusia* sp B. *Barbus* sp C. *Labeo* sp D. *Catla* sp

51. Bagging is recommended for the control of
    A. Anar butter fly B. Lemon butter fly
    C. Mango stone weevil D. Ber fruit fly
52. The highest degree of zenith of evolution found in
    A. Lemon butterfly B. White grub
    C. Honey bee D. Grasshopper
53. Super heating at 500° C is recommended for the control of
    A. *Pectinophora gossypiella* B. *Earias fabia*
    C. *Dysdercus cingulatus* D. *Gnorimoschema operculella*
54. Protein *hydrolysate* is used as an attractant for the control of
    A. Fruit fly B. White fly C. Butter fly D. Saw fly
55. The ideal conditions for survival and breeding of white grub are
    A. Light sandy well drained soils B. Light sandy water logged soils
    C. Heavy clay well drained soils D. Heavy clay water logged soils
56. The potato tuber moth was introduced in India in 1900 from
    A. Italy B. Sri Lanka C. Europe D. USA
57. The double boll/double seeds in cotton is casused by
    A. *Spotted* bollworm B. *American* bollworm
    C. *Pink* bollworm D. Aphid
58. Which one of the following is a vector for transmitting leaf curl viruses in vegetable crops?
    A. Aphids B. Thrips C. Honey bees D. White fly
59. Which of the following trap crop at flowering stage attracts 80-90% adults of *diamond back moth* is cabbage
    A. Mustard B. Fenugreek C. Castor D. Marigold
60. The ideal host for mass multiplication of *Trichogramma* is
    A. Almond moth B. Rice moth C. Wax moth D. Indian meal moth
61. The first record of sericulture dates back to
    A. 4700 BC in China B. 8000 BC in China
    C. 2500 BC in India D. 2500 BC in Sumeria
62. The first record of bio-control using predatory ants in citrus orchards of China was in
    A. 400 AD B. 200 BC C. 300 AD D. None of the above
63. The first major success with imported bio-control agents; *Cryptochaetum iceryae* and *Rodolia cardinalis* from Australia for the control of cottony cushion scale of citrus was achieved in the year
    A. 1880 B. 1888 C. 1890 D. 1901
64. Digestion and absorption is occurred in
    A. Fore gut B. Hind gut C. Mid gut D. Malpighian

65. The end of superfamily is
    A. Idae B. Ini C. Oidea D. None of these
66. The tomato leaf curl virus is transmitted by
    A. *Bemisia tabaci* B. Aphids
    C. *Trialeurodes vaporarium* D. *Trialeurodes abutilonia*
67. Insect eye can affectively perceive a spectrum of light ranging from
    A. 100-250 nm B. 700- 1000 nm C. 250- 700 nm D. None of these
68. The property of fumigant is
    A. High vapour pressure B. Increase temperature
    C. Low vapour pressure D. None of these
69. The insect order with panorpoid origin is
    A. Zoroptera B. Orthoptera C. Hemiptera D. Mecoptera
70. The following are the orthopteroid character
    A. Hind wing dominant in flight B. Malphigian tubules numerous
    C. Nymphs with ocelli D. All of these
71. The caste of termite having an median frontal rostrum and jaws are vestigial
    A. Nasute type B. Soil dwelling type
    C. Both of these D. None of these
72. The commercially used juvenoid is
    A. Methoprene B. Fenoxycarb C. Hydroprene D. None of these
73. The first insect order in which wing flexing mechanism developed
    A. Ephemeroptera B. Plecoptera C. Odonata D. Orthoptera
74. The chemical that mediate interaction between two individuals that belong in to different species is
    A. Hormone B. Allelochemicals
    C. Pheromone D. None of these
75. The toxic secretion of bombardion beetle is
    A. Canthradin B. Acidic acid C. Hydroquinone D. None of these
76. Amino acid present in silk is
    A. Glysine B. Glutamine C. Proline D. None of these
77. Posterior part of hind gut is closely associated with foregut in
    A. Gelechidae B. Pentatomidae C. Membracidae D. None of these
78. The 1 kg sorghum grains, the number of *Corcyra* eggs added
    A. 3000 B. 1500 C. 1000 D. 4000
79. The inverted Y-shaped suture on the insect head is known as
    A. Post occiput suture B. Epicranial suture
    C. Gena D. None of these

80. The meso and meta thorax are collectively known as
A. Prothorax B. Metathorax C. Mesothorax D. Pterothorax

81. Crochets are a part of the ....of caterpillars
A. Legs B. Prolegs C. Wings D. Antennae

82. The abdomen has a pair of terminal pincers of forceps in which order
A. Embioptera B. Thysanura C. Derpeptera D. Odonata

83. The finest branches of insect respiratory system are termed as
A. Tracheae B. Tanidia C. Air sac D. Tracheoles

84. Plumose antennae are found in which insect
A. Mole crickets B. Male mosquito C. Cockroaches D. Bees

85. Hemi-elytra is the fore wing of
A. 7th abdominal segment B. 8th abdominal segment
C. 9th abdominal segment D. 10th abdominal segment

86. The number of pairs of spiracles on the thorax of a grasshopper are
A. 3 B. 1 C. 2 D. None of these

87. Which of the following is not an aquatic insect
A. Dragon fly B. House fly C. May fly D. Dobson fly

88. The human louse belongs to the order
A. Hemiptera B. Diptera C. Siphonaptera D. Anoplura

89. The mustard saw flies belong to the family
A. Aleurodidae B. Syrphidae C. Tenthridinidae D. Apidae

90. The mudgo variety of rice is resistant to which pest
A. Brown plant hopper B. Gall midge
C. WBPH D. All the above

91. The first isolation and identification of an insect pheromone (silkworm moth) occurred in
A. 1969 by British scientists B. 1959 by German scientists
C. 1949 by Russian scientists D. None of the above

92. The mode of action of Bt is complex; the delta-endotoxin produce protein of 130 kDa in reductive environment of the insect mid gut that has a pH value of:
A. 6.0-7.0 B. 7.0-8.0 C. More than 9.5 D. Less than 9.5

93. Match the following hairy caterpillars of Arctiidae to their common names:

Scientific Name	Common Names
1. *Amsacta moorei*	A. Black hairy caterpillar
2. *Diacrisia oblique*	B. Sun hemp hairy caterpillar
3. *Utethesia pulchella*	C Red hairy caterpillar
4. *Pericallia ricini*	D. Bihar hairy caterpillar

Select and arrange the answers in proper sequence:

A. 1-C; 2-D; 3-B; 4-A B. 1-D; 2-C; 3-A; 4-B

C. 1-D; 2-A; 3-C; 4-B D. 1-B; 2-A; 3-C; 4-D

94. Match the given damage symptoms (Col. 1) with insect pests mentioned (Col.2):

Symptom of damage to the crop	Likely insect pest
1. Silvering or whitening of leaf surface due to removal of cell contents below the epidermis in onion	(A) Weevils
2. Leaves riddled with small cut holes in the leaves of cabbage, mustard and radish by feeding	(B) Brown Plant Hopper
3. Regular notches along the edges of leaves of a variety of crops	(C) Flea Beetles
4. Upward curling and necrotic brown lesions on paddy leaves	(D) Thrips

Select and arrange the answers in proper sequence

A. 1-A; 2-B; 3-C; 4-D B. 1-D; 2-C; 3-A; 4-B

C. 1-D; 2-A; 3-C; 4-B D. 1-B; 2-A; 3-C; 4-D

95. Match the given class of spray (Col. 1) with the droplet particle size (Col. 2):

Class of spray	Particle size (microns)
1. Coarse spray	(A) 100-400
2. Fine spray	(B) 50-99
3. Mist	(C) 0.1-49
4. Aerosol and Fog	(D) 400

Select and arrange the answers in right sequence:

A. 1-A; 2-B; 3-C; 4-D B. 1-D; 2-C; 3-A; 4-B

C. 1-D; 2-A; 3-B; 4-C D. 1-B; 2-A; 3-C; 4-D

96. Match the insect orders (Col. 1) to their significance (Col. 2):

Insect Orders	Significanc
1. Plecoptera	(A) Straight wings
2. Hymenoptera	(B) Purely wingless
3. Orthoptera	(C) Plaited wings
4. Zoraptera	(D) Membrane wings

Select and arrange the answers in right sequence:

A. 1-D; 2-A; 3-C; 4-B B. 1-D; 2-C; 3-B; 4-A

C. 1-C; 2-D; 3-A; 4-B D. 1-B; 2-C; 3-C; 4-D

97. Match the given insects (Col. 1) to their respective families (Col. 2):

Insect Names	Insect Families
1. Pigmy locust	(A) Cercopidae
2. Ambush bugs	(B) Cicindelidae
3. Tiger beetles	(D) Tetrigidae
4. Cuckoo spit bugs	(D) Phymatidae

Select and arrange the answers in proper sequence:

A. D; 2-A; 3-B; 4-C
B. 1-B; 2-C; 3-A; 4-D
C. 1-D; 2-C; 3-B; 4-A
D. 1-C; 2-D; 3-B; 4-A

98. Match the living arthropod classes (Col.1) with their examples (Col. 2):

Arthropod classes	Common Examples
1. Insecta	(A) King crabs or Horse shoe crabs
2. Crustacea	(B) Butterflies / Beetles / Bugs
3. Arachnida	(C) Millipedes and Centipedes
4. Myriapoda	(D) Shrimps / Prawns / Lobsters

Select and arrange the answers in right sequence:

A. 1-B; 2-D; 3-A; 4-C
B. 1-B; 2-C; 3-A; 4-D
C. 1-D; 2-C; 3-B; 4-A
D. 1-C; 2-D; 3-B; 4-A

99. Match the honeybee brood diseases (Col. 1) to their causal organisms (Col. 2):

Brood diseases of honey bees	Causal Organisms
1. European foul brood	(A) *Ascosphaera apis*
2. Chalk brood	(B) *Aspergillus flavus*
3. Stone brood	(C) *Streptococcus pluton*
4. American foul brood	(D) *Bacillus* sp.

Select and arrange the answers in right sequence:

A. 1-C; 2-A; 3-B; 4-D
B. 1-B; 2-C; 3-D; 4-A
C. 1-A; 2-B; 3-C; 4-D
D. 1-C; 2-D; 3-A; 4-B

100. Match the insect families (Col. 1) to the appropriate characters (Col. 2):

Families	Distinguishing characters
1). Coccinellidae	A. Phytophagous beetles with long antennae capable of being reflexed backwards over body; usually 11 segmented; tarsi pseudo-tetramerous.
2). Sphingidae	B. Moths with trapezoidal fore wing; antennae rarely with basal pectin; Rs + M1 in hind wing stalked.

3). Gelechiidae	C. Moths with antennae gradually thickened distally; apices pointed and always nearly hooked; exhibit crepuscular, nocturnal and rarely diurnal habit.
4). Cerambycidae	D. Oval to hemispherical beetles, strongly convex; elytra glabrous or with pubescence; short legs and antennae; apical segment of maxillary palpi securiform; basal sternite with post coxal plates.

Select and arrange the answers in right sequence:

A. 1-A; 2-B; 3-C; 4-D B. 1-D; 2-C; 3-B; 4-A

C. 1-A; 2-C; 3-B; 4-D D. None of the above

AMSWERS

1. B	2. B	3. B	4. C	5. C	6. D	7. C	8. B
9. B	10. C	11. B	12. B	13. A	14. A	15. C	16. B
17. C	18. C	19. A	20. C	21. B	22. C	23. C	24. C
25. B	26. B	27. B	28. B	29. A	30. B	31. B	32. B
33. B	34. C	35. B	36. A	37. C	38. B	39. B	40. D
41. A	42. D	43. A	44. B	45. A	46. A	47. B	48. A
49. B	50. A	51. A	52. C	53. A	54. A	55. A	56. A
57. C	58. D	59. A	60. B	61. A	62. C	63. B	64. C
65. C	66. A	67. C	68. A	69. D	70. D	71. A	72. A
73. B	74. B	75. C	76. A	77. C	78. B	79. B	80. D
81. B	82. C	83. D	84. B	85. C	86. C	87. B	88. D
89. C	90. A	91. B	92. C	93. A	94. B	95. C	96. C
97. D	98. A	99. A	100. B				

PLANT PATHOLOGY

TERMINOLOGY

Abiotic: Related with non-living factors or things without life. *e.g.* heat intensity, moisture imbalance, lack of nutrients that causes disorders (*e.g.* Khaira of rice).

Abiotic disease: A disease due to non-living agents or abiotic factors, also called ***disorder*** (*e.g.* Khaira of rice).

Acervulus (pl. acervuli): A boat-shaped (or cushion-like) asexual fruiting body with conidia and conidiophores. *e.g.* produced by *Colletotrichum.*

Acidophilous: Acid lovers *i.e.* growing on acidic medium, *e.g.* lickens on peaty soils.

Acid rain: Rain with pH of less than 5.6. H_2SO_4 is formed by the reaction of SO_2 and H_2O in atmosphere.

Acropetal: Chain of spores, in which the youngest at the tip and oldest at the base. *e.g. Alternaria.* (In fungicides, acropetal means upward movement through xylem)

Actinomycetes: (also called - ***Ray fungi***). These are filamentous bacteria & intermediate between fungi & bacteria.

Adaxial: Refers to the upper surface of a leaf, or directed towards the stem of a plant.

Adelphogamy: Pseudomictic copulation of mother & daughter cells *e.g.,* yeasts.

Aeciospores: A dikaryotic spore formed in aecium on alternate host in macrocyclic rusts (*e.g. Puccinia*), stage-I.

Aecium: The first fruiting body formed after plasmogamy having dikaryotic aeciospores.

Aerobic: Living only in the presence of oxygen; organism is requiring oxygen for respiration.

Aetiology (etiology): The study of causal factor and nature of pathogen (*e.g.* study of *Albugo candida*).

Aflatoxin It is a mycotoxin produced by fungus – *Aspergillus flavus.* It's a highly carcinogenic (cancer causing) toxin.

Agar agar: It is a solidifying agent and it forms gel below about 40 ^{0}C. It is a mixture of polysaccharide derived from red sea algae (*e.g. Gelidium, Gracilaria, Gigartina, Pterocladia*).

Alternate host: A host (other than main host) which is essentially (Obligatory) required to complete life cycle of any pathogen ex. heteroecious rust fungus requires alternate host to complete its life cycle. A host on which stage -0 and stage-I are produced, *e.g.*, Barberry for *Puccinia graminis tritici.*

Alternative host: A plant other than the main host attacked by the pathogen, not essential for completion of life cycle of the pathogen.

Anamorph (imperfect stage): Asexual stage of the fungus.

Anthracnose: Means charcoal like appearance, dark colored,,necrotic, sunken lesions produced by acervuli forming fungi *e.g. Colletotrichum.*

Antiseptic: A substance that prevents or kills microorganisms.

Akinete: A thick walled, non-motile spore of Cyanobacteria (BGA).

Archaebacteria: An ancient line of bacteria; found in extreme environment as methanogenic **(methane producers)**, halophilic (salt lovers) and thermoacidophilic (higher temp. & acid lovers). Differ from bacteria in their rRNA, lipids and wall composition (wall having ***pseudo-peptidoglycan***).

Asci (sing.- ascus)**:** Sac like cells in which meiosis occurs and generally contain 8 ascospores.

Atrophy: Under-development or reduction in size of a body.

Attenuation: Lessening of the capacity of a pathogen to cause disease, *e.g.*, reduction of its virulence.

Autoclaving: Sterilization by steam at 121.6 ^{0}C under pressure (15 psi -pounds per square inch) for 15 minutes. Moist heat has more penetration power than dry heat. Media and liquids are sterilized.

Autoecious: A rust fungus, capable of completing its life cycle only on single host (*e.g. Melampsora lini*: causing linseed rust).

Autolysis: Self-digestion resulting in cell breakdown and often liquefaction or death of cells by their own metabolites (enzymes).

Auxotroph: A mutant/strain of a microorganism that requires specific nutrient, which are not needed by its parent (prototroph).

Avirulent: Non-pathogenic, *i.e.* unable to cause disease.

Bakane: A disease of rice caused by *Gibberella fujikuroi* (*Fusarium moniliforme*). It is also known as foot rot or or foolish seedling (seedlings grow fastly due to gibberellic acid) or Fusarium blight. It was first reported in **Japan by Hori in 1898** and detail by **Kurosawa in 1926**.

Basipetal: A chain of conidia, in which youngest at base and oldest at tip, *e.g. Albugo.* (In case of fungicides, basipetal means down movement through pheolm).

BCG vaccine: *Bacillus Calmette Guerin.* Attenuated strain of *Mycobacterium tuberculosis,* used to immunize human being against TB.

Berberis (barberry bush)**:** Alternate host of *Puccinia graminis* pv. *tritici* (causing black rust of wheat). Basidiospores infect barberry and after infectin.spermatiospores (earlier called pycniospores) and aeciospores are produced on barberry. These aeciopspores infect wheat (in those areas where barberry is functional). Barberry is non-functionla in India that's why, infection in India on wheat, is caused by uredospores (so called repeating spores).

Bioluminescence: Emission of light by living organisms. It is a form of chemiluminescence. It occurs in some insects, fungi, bacteria *etc.,* Light emitting pigment Luciferin and ènzyme Luciferase (Luciferase is encoded by *Lux gene*) involved in this reaction (*e.g.* mushrooms: *Armillaria mellea,, Panellus, Mycena etc.,*; Bacetria: *Vibrio harveyi, V. fischeri*).

Black heart of potato: A non-parasitic disorder of potato. Generally, large sized tubers show black colored cavities. Black color is due to ***melanin***. It may be due to ***high temperature and low oxygen*** or ***poor ventilation*** or ***im-proper storage*** or ***sub-oxidation*** during storage and transit (transport) or in field.

Broth: A liquid medium without solidifying agent (*i.e.* medium without agar-agar or gelatin).

Blight: General and rapid death of plant foliage. It casues extensive spots, discoloration or destruction of leaves, flowers, stems or entire plants.

Brandspore: Uredospore or smut spore are called brandspore.

Bunt: A disease caused by the fungus *Tilletia* is called bunt. *e.g.* Karnal bunt of wheat, it gives rotten fish like smell in filed due to TMA (Tri-methyl-amine).

Colony Forming Unit (Cfu): One living cell forms one colony. So, we can estimate number of living cells in a sample.

Chemotherapy: Treatment with chemicals (*e.g.* by use of fungicides *etc.,*).

Chlamydospores: A thick walled asexual spore formed from hyphal cells that serve as overwintering or over-seasoning spore.

Chlorosis: Yellowing due to loss of chlorophill.

Cigar end: A disease in banana caused by *Verticillium theobromae* fungus.

Collateral host: A host (other than main host) of the same family on which a pathogen just survive during off season, not required for completing life cycle.

Cold sterilization: Method of disinfecting the culture media without heating them to avoid the risk of heat damage. It is done by using membrane filter and propylene oxide.

Conidium (pl. conidia): A thin walled, asexual spore that is borne exogenously on conidiophores.

Coprophilous: An organism growing on dung (***Dung loving***).

Cross protection: The phenomenon in which plant tissues infected with one mild strain of a virus are protected from infection by other more severe strains of the same virus. *e.g.,* Citrus Tristiza Virus

Cryotherapy: Cold treatment (*i.e.* treatment with low temperature).

Cyanobacteria (BGA): Prokaryotic, chlorophillous organism, often capable of fixing nitrogen, *e.g. Anabaena, Nostoc etc.,*

Cucurbitacin: A compound produced by cucurbits (mostly bitter cucurbits). It has anti-nematode action.

Damping off: Toppling down of sedlings on ground due to infection. Caused by *Pythium, Phytophthora, Fusarium* and *Rhizoctonia* sp.

Diazotrophs: Nitrogen fixing prokaryotes are called diazotrophs, *e.g. Rhizobium, Azotobacter, Anabaena etc.,*

Diazotrophy: Fixation of atmospheric nitrogen by prokaryotic organisms is known as diazotrophy.

Die-back: Progressive death of shoots, branches and roots starting from at the tip (*i.e.* death from top to down).

Dikaryotic (n+n): Mycelium or spores containing two sexually compatible nuclei per cell, common in the Basidiomycetes.

Disease: A malfunctioning process that is caused by continuous irritation by a pathogenic agent or environmental factor and leads to symptoms development.

Disinfectant: A physical or chemical agent that frees a plant from infection (*i.e.* this remove/kill pathogen, present inside plant tissue).

Disinfestant: An agent that kills or inactivates pathogens on the surface (or in the environment) of a plant or plant organ before infection takes palce (*i.e.* these remove/kill pathogen present on plant surface).

Disorder: A harmful **non-pathogenic** deviation from the normal growth and metabolism (*e.g.* Khaira of rice).

Dithane M-45 (now **Indofil M-45):** Trade name of **mancozeb** (maneb + zinc ion) fungicide (used against leaf spot diseases), generally, it has 75 % a.i (active ingredient).

Dithane Z-78: Trade name of **zineb** fungicide, mostly, it is used to control **fruit rots** (post harvest diseases), generally, it has 76 % a.i.

Downy mildew: A cottony tangled growth on lower surface of leaves, casued by oomycetes (order ***Peronosporales***).

ELISA: Enzyme Linked Immuno Sorbent Assay. A **serological test** which uses enzyme linked **antibodies.** Generally, it is used to detect viruses, bacteria *etc.*, (ELISA kits are also used to determine pesticides in water & soil samples).

Enation: A small out growth on plant surface, caused by certain viruses. *e.g.* Pea enation mosaic virus.

Epidemic: A widespread and severe outbreak of a disease. (wheat rusts, late blight *etc.,*).

Endemic: A diseseas present in a particular area. *e.g.* **wart disease** of potato in Darjeeling district of West Bengal and **golden nematode** in Nilgiri hills in Tamilnadu.

Heteroecious: A fungus requiring two different types of hosts to complete its life cycle. Rust fungi (*Puccinia*).

Hypertrophy: Over-growth due to increased ***cell size*** *e.g.* white rust (***stag head symptoms***).

Hyperpalsia: Over-growth due to increased ***cell number***.

Hypovirulence: Reduced virulence of a pathogen due to transmissible **dsRNA**.

Imperfect fungus/Deuteromycetic fungus/Mitosporic fungus: A fungus in which sexual reproduction is not known (*i.e.* sexual spores are not produced).

Infection: The establishment of a parasite within host plant.

Inoculum: The pathogen or its parts that can cause infection.

In vitro: In culture or outside the host. In glass (*e.g.* experiments in lab.).

In vivo: In the host (*e.g.* experiments in field).

Mollicutes: (Molli-soft, cute-skin), Prokaryotic cells without cell wall. *e.g.* mycoplasma, MLO/phytoplasma, spiroplasma.

Monocyclic: Having one cycle per season. *e.g.* Wilts, cereal cyst nematode, root rots *etc.*,

Mosaic: Inter-mingled patches of normal green and light green or yellowish color *e.g.*, viral diseases symptoms.

MLO (Mycoplasma Like Organism): Microorganisms found in phloem cells of diseased plants, similar to mycolpasma (except artificial culture on media). Now they are known as Phytoplasma & Spiropalsma.

Mycoplasma: Pleomorphic, prokaryotic microorganisms without cell wall (causes disease in animals).

Mycorrhiza: A symbiotic association between fungus and higher plant roots. Mycorrhiza's supply P to plants.

Mycotoxin: A toxin (poison) produced by fungus.

Necrotroph: Feeding only on dead organic tissues.

Obligate parasite: Grow and multiply only on living organisms.

Oomycete: A fungus-like chromista that produces oospores (sexual spore); a water mold.

Parasite: An organism living on/in another living organism (host) and draw nutrition from the host.

Pathogen: An entity that can cause disease *e.g.* bacteria, fungi and virus *etc.*,

Pathogenicity: Ability of a pathogen to cause disease.

Pathovar (pv.): In bacteria, a subspecies of strains that can infect only plants of certain genus or species.

Perfect stage: The sexual stage in the life cycle of fungus; it is also known as teleomorph.

Phage (Bacteriophage): A virus that attacks bacteria.

Phyllody: Conversion of floral parts into leaf-like structure *e.g.* Sesamum phyllody by Phytoplasma.

PR Proteins (Pathogenesis Related Proteins): Proteins produced in a cell within minute or hours after inoculation, but all being more or less toxic to pathogens *e.g.* β- 1,3 glucanase and chitinase.

Polycyclic: Completes many cycles in one year *e.g.* powdery mildew, downy mildew, rusts diseases.

Polyetic (Multi year): Requires many years to complete one life cycle or disease cycle *e.g.* Dutch elm, cedar apple rust, white pine blister rust, citrus tristeza.

Rust: Rusty appearance on the plant surface due to rust fungi (order - ***Uredinales***)

Sign: The pathogen or its parts seen on the infected plant, *e.g.* powdery mildew, downy mildew, rust *etc.,*

Smut: A black powdery mass of spores produced by smut fungi (belongs to order - *Ustilaginales*).

Sterilization: The elimination/killing of pathogens and other living organisms from soil, containers *etc.,* by means of chemical or physical agents (heat).

Virion: A complete or mature virus particle.

Virulent: Strongly pathogenic or capable of causing a severe disease.

Wilt: Loss of rigidity and drooping of plant parts, due to insufficient water (abiotic) in the plant and some times due to pathogens (biotic).

Important Famines

1. **Irish Famine (1845)** →In **Ireland** due to Late Blight of Potato (caused by fungus *Phytophthora infestans*).

   -1.25 Million people died due to starvation (1.5 million migrated)
2. **Bengal Famine (1942-43)**→In **Bengal** of India (Now some Part in Bangladesh).

   Due to Brown spot of rice (caused by fungus *Helminthosporium oryzae - Drechslera oryzae*), (Perfect stage.-*Cochliobolus miyabeanus*). About 2 million people died.

   In lieu of this disease, CRRI, Cuttack (Odisha) was established.
3. **Ceylon epidemic (1870)** → In **Ceylon** (Now in Srilnka). It was due to coffee rust (by fungus *Hemileia vastatrix*). Then coffee was replaced by Tea.

HISTORY OF PLANT PATHOLOGY

PA Micheli (Italian botanist)	• First time studied fungi, illustrated their reproductive structures & saw their spores on watermelon. • Proposed that fungi arise from their own spores. • He wrote a book: ***"Nova Plantarum Genera***". • He is known as **"Father of Mycology"**.
Anton de Bary (Germany)	• Gave **"*role of enzymes in plant disease production*"**. • Established **"*role of fungi as plant pathogen*"**. • **Demonstrated "Heteroecious Nature"** of *Puccinia graminis* **(causing black rust of wheat, 1865)**. • He discovered relationship between barberry & wheat. • He is known as **"Father of Plant Pathology"** (World wide).

Robert Koch (Germany Microbiologist)	• Established **"Koch's Postulates"** for proving pathogenicity of microorganism associated with diseased plant. • Gave use of "**Gelatin**" in microbiological culture media as solidifying agent. • Established causal agent of **anthrax disease** (a bacterial disease in animals). • Gave etiology of **T.B.** (Tuberculosis). • He is known as **"Father of Bacteriological Techniques"**. • Introduced the "Pour Plate Method"in 1882.
Anton Leeuwenhoek (1675)	• He discovered **bacteria** and called them ***"Animalcules".*** • **First** observed living cell of animals.
T. J. Burrill (Illinois, USA)	• First time reported **"Plant Bacterial Disease"** *i.e.* Fire blight of pear and apple, caused by *Erwinia amylovora.* Did not use of pure culture or Koch's postulates.
PMA Millardet (France botanist)	• He accidently noticed (in 1882) that vines treated with a "poison" ($CuSO_4$ + Lime + Water) by the growers (to prevent the grapes from being stolen) retained their leaves, whereas the untreated vines defoliated due to **downy mildew disease** caused by ***Plasmopara viticola*** fungus. • In **1885,** he published his work & gave final proof to control DM of grapes. • So, he is the discoverer of **"Bordeaux Mixture" (BM).** • ***Plasmopara viticola*** fungus is related with the discovery of BM. • He was a Professor of Botany (University of Bordeaux, France). • BM is known as "**first universal fungicid** ".
Adolf Mayer (Dutch, 1886)	• First described **"Plant Viral Disease"** *i.e.* Tobacco Mosaic disease. • Gave the name **"Mosaic"** (***Mosaikkrankheit***) for the disease. • Reported that disease could be **transmitted by injecting juice/sap** of diseased plant (Sap transmission of viruses). It caused by toxin from bacterium or new small **bacterium** that could pass through a filter.
E F Smith (USA)	• Worked on bacterial wilt of cucurbits & crown gall of apple. • Gave final proof that bacteria could be cause of plant diseases. • He is called as "**Father of Phytobacteriology".** • **He added 4th Rule of proof to Koch's Postulates (**1 to 3rd by Koch himself in 1876).

D Ivanowski (1892)	• Showed that causal agent of tobacco disease could pass through **Bacteria-Retaining Filter** (*e.g.* Chamberl and Filter). • Gave filterable nature of viruses.
MW Beijerinck (1935)	• He proved that Tobacco Mosaic was caused by ***"Contagium vivum fluidum"*** (*i.e.* living infectious fluid), not by bacterium or toxin. • He used the term "***Virus***" for contagious living fluid (*i.e.* poison). • He is knows as **"Father of Virology".**
WM Stanley (1935)	• He got **TMV in purified & crystallized form** as protein (*i.e.* crystallization of virus particles). • He got **Noble Prize** for Chemistry.
Swingle	• Used **"Hot Water Treatment"** for Loose Smut of Wheat.
Jensen	• First gave **"Hot Water Treatment"** for the control of Late Blight of Potato.
Knol & Ruska	• Invented **Electron microscope.**
H H Flor	• He proposed **"Gene-for-Gene Theory"** (GFG Theory) in flax/linseed rust caused by ***Melampsora lini*** fungus for disease resistance & susceptibility. • It is **Autoceious fungus** (*i.e.* complete life cycle on single host).
Doi *et al.* (1967)	• Discovered **"Mycoplasma like Organisms"** (MLOs) as causal agent of plant diseases in phloem cells of mulberry causing "yellows type symptoms" (now MLO known as **Phytoplasma**).
T O Diener (1971)	• Discovered **"Viroids"** as causal agent of **"potato spindle tuber disease"**. • Gave the term **"Viroid"**. (Viroid - naked infectious ssRNA particle).
Prusiner (1972)	• Discovered **"Prions"** (means Proteinaceous Infectious Particles -PIP.) • Got **Noble Prize** for Chemistry in 1997.
JC Walker	• Discovered disease resistance **phenolic compounds** (*i.e.* **Catechol & Protocatechuic acid**) in **Red scale onion** against **Smudge Disease of Onion** caused by *Colletotrichum circinans.* • He is known for giving **"Biochemical Basis of Resistance in plants"**.
JE Vanderplank (1963)	• Gave the concept of **"Horizontal & Vertical Resistance"** in plants. • He is known as **"Father of Epidemiology".**

Mason	• Introduced **"Burgundy Mixture (Soda borax)"**.
VB Singh	• Developed "**Chaubattia Paste**" in Almora district of UP (Now in UK) as wood dressing fungicide.
Van Schmeling & Marshal Kulka(1966)	• Developed "Systemic Fungicides" as Oxathiin group (*e.g.* carboxin-Vitavax).
Alexander Fleming (1929)	• He obtained first antibiotic '**Penicillin**' from fungus, ***Penicillium notatum***
Walther Hesse, (1882)	• Introduced the use of **"Agar Agar"** in microbiological methods as a solidifying agent. • Agar agar obtained from **Red Sea Algae** (*Gelidium, Gracilaria.*) & added @ 1.5-2%.
E.J.Butler (1918)	• He wrote a monograph Pythiaceous and allied fungi. • Book- **"Fungi and Diseases in Plants"** • He began the journal **"Review of Applied Mycology"** with S.G. Jones. • He wrote a book **"Plant Pathology"** (1949)
Oscar Brefeld	• He gave **"Pure Culture Technique"** for isolation & culturing of **fungi.**
Clark & Adams	• Developed **"ELISA" technique** for detection of **viruses.** • **ELISA**: Enzyme Linked Immuno Sorbent Assay.
Dye *et.al.*	• He introduced **"Pathovar Concept"** in bacterial taxonomy for sub-species (in short, it is written as **pv.**).
A.B. Frank (1885)	• Dicovered **Mycorrhiza** (symbiotic association between fungi & roots of higher plants).
Nocard & Roux	• Discovered **Mycoplasma** (PPLO- Pleuro Pneumonia Like Organism).
Twort & de'Herelle	• Discovered **Bacteriophage (Viruses that kill bacteria).** • The term **"Bacteriophage"** was given by **de'Herelle** in 1917.
Gierer & Schramm(1956)	• Reported that "**infective** part of virus is **nucleic acid"**.
F.C. Bawden & Pierie (1936)	• Reported that "**Virus**" (TMV) is made up of **"nucleic acid & protein"** (*i.e.* made up of **Nucleio-protein**). TMV- 95% proten+5% nucleic acid.
Kassanis (1966)	• Discovered **Satellite viruses.**
B B Mundkur	• Established **"Indian Phytopathologiacal Society"** in India (IPS). • Wrote the book "**Fungi & Plant Diseases**".
H. Stolp (1962)	• Discovered **"*Bdellovibrio*".** *Bdellovibrio* is a Gram –ve bacteria that parasite another Gram –ve bacteria & live in ***periplasmic space*** (*i.e.* space between cell wall & cell membrane).

Hewitt *et.al*	• First reported the **"transmission of plant virus by nematode".** • The disease was fan leaf grape vine, caused by *Xyphinema index* nematode **(Daggar nematode).**
Simpson *et al.*	• Fisrt **sequenced the full genome of a bacterial plant pathogen,** ***Xylella fastidiosa*** **clone 9a5c,** causing citrus variegated chlorosis, a serious disease of orange trees.
HC Gram 1884	• Developed **"Gram staining technique"**.
JR Petri 1887	• Invented **"Petri plate"** while working as an assistant of **Robert Koch.**

Father Associated with Plant Pahology & Others

S. No.	**Subject**	**Father**
1.	Father of Plant Pathology Father of Experimental Plant Pathology	Anton de Bary (At world level)
2.	Father of Plant Pathology in India or Father of Mycology in India First Director of International Mycological Institute	E J Butler (At India level)
3.	Father of Mycology	P A Micheli
4.	Father of Virology	M W Beijerinck
5.	Father of Phyto-bacteriology	E F Smith
6.	Father of Plant Bacteriology in India	M K Patel
7.	Father of Indian Rust	K C Mehta (Agra College)
8.	Father of Pure Culture Techniques	Oscar Brefeld
9.	Father of Microbiology	Antony van Leeuwenhoek
10.	Father of Green Revolution	Norman E. Borlaug
11.	Father of Nematology	N A Cobb
12.	Father of Biological Control	S D Garrett
13.	Father of Forest Pathology	R Hartig

14.	Father of Plant Epidemiology	J E Vanderplank
15.	Bible of Plant Pathology in India	Book **"Plant Diseases"** by R S Singh
16.	Father of Modern Bacteriology	Louis Pasteur
17.	Father of Tropical Plant Pathology	H M Ward
18.	Linneous of Mycology	Elias Fries
19.	Father of Modern Plant Pathology	J G Kuhn
20.	Father of Indian Phyto-bacteriology	M K Patel
21.	Father of Medical Bacteriology or Father of Bacteriological Technique	Robert Koch
22.	Father of Bt cotton in India	C D Mayee

Important books of Plant Patholohgy

S.No.	Book	Writer
1.	Enqiry into Plant	Theophrastus (370-286 BC)
2.	Vraksha Ayurveda	Surpal (1500)
3.	Synopsis Methodica Fungorum	C.H. Persoon (1801)
4.	Nova Plantarum Genera	P.A. Micheli (1729)
5.	Species Plantarum	C. Linnaeus (1757)
6.	Systema Mycologicum	E.M. Fries (1821)
7.	Pinax Theatri Botanici	Gaspard Buhin (1923)
8.	The Diseases of Cultivated Crops, Their Causes and Their Control	J.G. Kuhn (1858)
9.	Important Diseases of Forest Trees	Robert Hartig (1874)
10.	Fungi and Diseases in Plants	E.J.Butler(1918)
11.	Fungi and Plant Diseases	B.B. Mundkur (1949)
12.	Principles of Fungicidal Action Fungicides and Their Action Plant Pathology An Advanced Treatise	J.G.Harsfall (1956)
13.	Plant Disease Epidemics and Control	J.E. Vanderplank(1963)
14.	Diseases of Crop Plants in India	G. Rangaswami (1972)
15.	Bacterial Diseases in India	G. Rangaswami
16.	Biological Control of Plant Pathogens	K.F. Baker and R.J. Cook(1974)

17.	Principles of Plant Infection	J.E. Vanderplank(1975)
18.	Plant Diseases	J.G. Horsfall and E.B. Cowling (1977)
19.	Host Plant Pathogen Interaction in Plant Diseases	J.E. Vanderplank (1982)
20.	Disease Resistance in Plants	J.E. Vanderplank (1984)
21.	Fungicides in Plant Disease Control	Y.L.Nene & P.L.Thapliyal(1987)
22.	Plant Virology	REF Methews (1991)
23.	Introductory Mycology	CJ Alexopolous, CW Mims & M Blackwell (1996)

KOCH POSTULATES: (Postulate-Principle) was given by Robert Koch (1876).

Means proving of pathogenicity, *i.e.* principles to say that this particular microorganism is a causal organism of that disease.

- Organism must always be associated with disease in all diseased plants.
- Oraganism must be isolated and established in pure culture.
- Inoculation of a healthy plant of the same variety must reproduce exactly the same symptoms.
- Organism must be re-isolated from inoculated plant & its identity confirmed as the same as the original isolate (**This 4th postulate was given by EF Smith, 1905**).

In short, 4 Koch's postulates can be remembered as:-

I. Association	II. Isolation & purification	III. Inoculation	IV. Re-isolation
Association of M.O. with diseased host	Isolation & purification of M.O.	Inoculation of M.O. on same healthy host	Re-isolation of M.O.

Differences between prokaryotic and eukaryotic cells

S. No.	Characters	Eukaryotic cell	Prokaryotic cell
1.	**Meaning**	***Eu*** means true or good. So, eukaryotic means having a true nucleus.	***Pro*** means before or false. So, prokaryotic means before a nucleus (*i.e.* without nucleus)
2.	**Nucleus**	True nucleus present	Primitive nucleus present (known as ***nucleoid/ prokaryon/Incipient nucleus/Genophore etc.)***
3.	**Cell organelles**	Present	Absent (membrane bound cell organelles absent)

4.	**DNA**	Linear	Circular (**Except** mycoplasma, it has linear DNA)
5.	**Histone protein**	Present Histone + DNA (known as chromosome)	Absent Naked DNA (G-C rich) (*i.e.* naked genetic material) (known as nucleoid)
6.	**Chromosome No.**	One to many	Only one dsDNA molecule
7.	**Ribosomes** (Ribosomes found in all type of cells)	***80 S in cytoplasm.*** **But 70 S** in mitochondria & chloroplast (*i.e.* in organelles).	***Only 70 S***
8.	**Protein : RNA ratio in ribosomes**	2:1 (about 60%+40%)	1:2 (about 40%+60%)
9.	**Cell wall**	Present in plants (made up of **cellulose or chitin**)	Present (except in mollicutes) (made up of Peptidoglycan or pseudo-peptidoglycan)
10.	**Photosynthesis**	In chloroplasts	In carboxisomes or lamelles or chlorosomes in BGA (in Bacteria chromatophores membrane infoldings)
11.	**Protein synthesis**	Transcription in nucleus & Translation in cytoplasm on ribosomes	In cytoplasm
12.	**Nucleolus**	Present	Absent
13	**Respiratory enzymes**	In Mitochondria & cytoplasm	In Cell membrane in aerobic cell (In Bacteria-in mesosomes)
14	**DNA base ratio**	40	23.70
15	**Similar organelle**	Ribosome	Ribosome
16	**Cell division**	Mitosis and meiosis	Amitosis/ Binary fission
17	**Multiplication rate**	Slower	Faster
18	**Sexual reproduction**	Yes, involves meiosis	No, (transfer of DNA only)

19	**Example**	Animals, Plants, Insects, Nematodes, Algae, Protozoa, Fungi *etc.,*	***MBBS*** - **M**ycoplasma, **B**acteria, **B**lue green algae, **S**piroplasma (Bacteria- True bacteria, Archaea, Actinomycetes, Rickettsia) (Mollicutes-**M**ycoplasma, phytoplasma/MLO & **S**piroplasma)

Important Plant Pathogenic organisms/factors

Biotic & Abiotic Causes of Plant Disease or Disorders

S. No.	**Living/Biotic/Animate causes (Cellular)**		**Meso-biotic causes (acellular)**		**Non-Living/ Abiotic/ Inanimate causes**
	Means pathogens of living nature. **Each cell having both DNA & RNA as nucleic acid (NA)**		Means causal agents are neither living nor non-living in nature.		Means non-living factors that cause damage to plants.
	Eukaryotic	**Prokaryotic (MBBS)**	**With Nucleic acid** (either RNA or DNA never both)	**Without NA**	-
1	**ALGAE** (Red rust of tea, mango)	**MYCOPLASMA** (PPLO) (Disease in animals)	**VIRUS** (Tobacco mosaic disease)	**PRIONS (slow virus)** (Not in plants, only in animals)	**Nutrient deficienc** (*Khaira* of rice by Zn def.)
2	**FUNGI** (Rust of wheat)	**BACTERIA** (Fire blight of pear)	**VIROID** (potato spindle tuber disease)	-	**Light** (Etiolation by light def.)
3	**PROTOZOA** (phloem necrosis of coffee)	**BLUE GREEN ALGAE - BGA** (Cyanobacteria, N fixer)	**VIRUSOID** (*e.g.* Virusoid with Velvet tobacco mottle virus)	-	**Moisture stress** (Physiological wilt)

4	**NEMATODES** (root knot of vegetables)	**SPIROPLASMA** (corn stunt & citrus stubborn)	**SATELLITE VIRUSES** (*e.g.* Tobacco necrosis sat. virus)	-	**Temperature (High temp.** - sun scald of tomato, chilli, **water core of** apple & BHP) **(Low temp. -** Frost damage, freezing injury)
5	**PHANEROGAMIC PARASITES** (**CLOS-** *Cuscuta, Loranthus, Orobanche, Striga*)	**PHYTOPLASMA** (earlier called MLO) (sesame phyllody, little leaf of brinjal -LLB)	**Satellite RNA**	-	**Lack of O_2** (BHP-**Black heart** of potato by low O_2 & **Internal browning** of apple)
6	**INSECTS** (Litchi rust by mites & Wilt of pineapple by mealy bug)	-	-	-	**Air pollutants** **(Black tip of mango)**
Mollicutes - (1) Mycoplasma (2) Phytoplasma (3) Spiroplasma *(Trick for Prok.:* ***MBBS****-Mycoplasma, Bacteria, BGA, Spiroplasma)*					**Nutrient excessiveness** **Toxicity of pesticides** *(eg. Finger leaf disease of cotton due to 2,4-D toxicity)* **Improper cultural practices** **Soil pH (acidity, salinity, alkalinity)** **Akochi disease of rice due to H_2S toxicity.**

Classification of Plant Diseases: On the basis of regularity, severity and area covered.

(1) Sporadic diseases - A disease occurs at irregular intervals & locations and in few cases.

e.g. Angular leaf spot of cucumber

(2) Endemic diseases - A disease that permanently present in one locality (area) in mild to severe form.

e.g. **Wart disease of potato** in Darjeeling (WB) & **Golden nematode of potato** in Nilgiri Hills (TN).

(3) Epidemic diseases - A disease occurs in wider area & in severe form.

e.g. Wheat rusts, coffee rusts.

(4) **Pandemic diseases** - A disease occurs all over the world in severe form.

e.g. Late blight of potato.

Rich man's disease - *e.g.* Rusts and mildews (powdery mildew & downy mildew). Rusts and mildews are **rich man's diseases** flourishing at **high nitrogen** levels (as rich man applies more N).

Poor man's disease - *e.g.* Brown spot of rice (by *Drechslera oryzae -Helminthosporium oryzae*), because disease is closely associated with abnormal or poor soil conditions. **Glume blotch of wheat** (by *Septoria* sp.) is also a poor man's disease, because it occurs severely under-nutritional deficiencies.

Parasitism and saprophytism:

1. **Obligate parasites (Biotrophs):** Organisms that can draw nutrition only from living plant cells.
   e.g. Rusts, PMs, DMs, Viruses, phytoplasma, nematodes.
2. **Facultative parasites:** Basically these are saprophytes, but can live as parasites, if given the chance to attack weak or senescing plant tissues.
3. **Obligate saprophytes:** Organisms that can draw nutrition only from dead plant tissues. These are secondary invaders and can not cause disease.
4. **Facultative saprophytes:** Basically theses are parasites, but can survive as saprophytes, if necessary.

BIOTROPHS (Obligate parasites)	**NECROTROPHS (Facultative parasites/facultative saprophytes)**
Have narrow host range	Have wide host range.
Can not grow as saprophytes	Can grow as saprophytes.
Attack healthy host tisses at any stage.	Attack young, weak or senescing tissues.
Kill host cells slowly.	Kill host cells fastly by producing toxins or enzymes.
Penetrate directly or via natural openings.	Penetrate through wounds or natural openings.
Common diseases caused by biotrophs are: • Rusts • Powdery Mildews • Downy Mildews • Viral diseases • Phytoplasmal diseases • Nematode diseases	**Common diseases caused by necrotrophs are:** • Anthracnose • Cankers • Fruit rots & root rots • Leaf spots • Blights • Vascular wilts

IMPORTANT DISEASES OF RICE:

(A) Fungal diseases

S.No.	Name of disease	Causal organism	Nature/ survival	Specific control	Remarks
1	Brown leaf spot of rice	*Bipolaris oryzae* *(Cochliobolus miyabeanus)*	Seed & soil-borne		Responsible for **"Bengal Famine"** in 1942-43. (2 million died)
2	Blast of rice	*Pyricularia grisea* (perfect stage: *Magnaporthe grisea*)	Seed & soil-borne		-
3	Stem rot of rice	*Sclerotium oryzae* (perfect stage: *Leptosphaeria salvini*, now –*Magnaporthe salvini*)	Soil-borne (by sclerotia)	Destruction of debris	-
4	False smut of rice	*Ustilaginoidea virens* (P.s.- *Claviceps oryzae-sativae*)	Soil-borne (by sclerotia)	Use of sclerotia-free seed & crop rotation	-
5	Bunt of rice	*Tilletia barclayana*	Seed & soil-borne	Spray of propiconazole	-
6	Leaf smut of rice	*Entyloma oryzae*	Soil-borne	Clean cultivation	-
7	Sheath blight of rice	*Rhizoctonia solani* (In India, anastomosis group-1, having 3-16 nuclei per cell)	Seed & soil-borne (by sclerotia)	Application of silicon	-

(B) Bacterial diseases

S.No.	Name of disease	Causal organism	Nature/ survival	Management	Remarks
1	BLB of rice (Bacterial leaf blight)	*Oryzae* pv. *oryzae*	Seed & soil-borne	Antibiotic, hot water	**Kresek** phase appears
2	BLS of rice (Bacterial leaf streak)	*Xanthomonas oryzae* pv. *oryzicola*	Seed-borne	Antibiotic, hot water	

(C) Viral diseases

S.No.	Name of disease	Causal agent	Transmission	Management	Remarks
1	**Rice Tungro** (Cancer of rice) (Disease caused **by two viruses**)	By two viruses: Rice tungro spherical virus (RTSV-ssRNA virus-*Waikavirus*) & Rice tungro bacilliform virus (RTBV-dsDNA virus-*Badnavirus*)	Vector - *Nephotettix cincticeps* (-*N. virescens*), Green leaf hopper (in semi-persistent manner)	Roguing, Insecticides, resistant var.	Tungro means– Degenerated/ yellow-orange growth
2	**Hoja Blanca** of rice (white tip)	*Hoja blanca virus*	Vector- *Sogatodes oryzicola* (plant hopper)	Roguing, Insecticides, resistant var.	Circulative, propagative, trasnovarial transmission (*i.e.* persistent)
3	**Rice grassy stunt disease**	*Rice grassy stunt virus*	Vector- *Nilaparvat lugens* (plant hopper)	Roguing, Insecticides, resistant var.	
4	**Rice dwarf disease**	*Rice dwarf virus*	Vector - *Nephotettix cincticeps* (-*N. virescens*)	Roguing, Insecticides, resistant var.	

(D) Phytoplasmal/Spiroplasma (MLO) diseases

S.No.	Name of disease	Causal organism	Transmission	Management	Remarks
1	Rice yellow dwarf	*Ca. Phytoplasma*	*Nephotettix virescens*	Use of tetracycline & insecticides	

(E) Disorders (physiological disorders)

S.No.	Name of disorder	Causal factor	Remarks
1	**Khaira** of rice	Zn deficiency	First reported by YL **Nene** et al, 1965 from Tarai region of UP.
2	**Akiochi** of rice	H_2S toxicity	-
3	**Parrot beak** of rice (or straight head disease)	By **arsenic** contamination in soil as herbicide application	Heads remain upright at maturity, hence the name "Straight Head"
4	**Akagare** of rice	Due to toxicity of Iodine, organic acids & K deficiency, boggy/muck soils	Akagare means red ,withering

IMPORTANT DISEASES OF WHEAT:

(A) Fungal diseases

S.No.	Name of disease	Causal organism	Nature/ survival	Management	Remarks
1	Black rust	*Puccinia graminis tritici (Pgt)*, Having UG-99 race	Survive on hills, infection in India by uredospores	Spray of zineb, Sulphur, oxycarboxin	Alternate host - Barberry
2	Brown rust	*Puccinia recondite*			Alternate host - *Thalictrum*
3	Yellow rust	*Puccinia striiformis*			No alternate host known
4	Loose smut	*Ustilag segetum* var. *tritici*	Internally seed-borne (dormant mycelium in embryo)	Solar energy, Hot water, Systemic fungicides - carboxin	Hot water treatment by Swingle, Solar energy treatment by Luthra & Sattar
5	Karnal bunt (New bunt/ partial bunt)	*Tilletia indica (-Neovossia indica)*	Seed & soil-borne	-	Foul smell due to TMA (Tri-methyl-amine)
6	Flag smut (stem smut/ stripe smut)	*Urocystis agropyri (-U. tritici)*	Seed & soil-borne	Seed treatment with tebuconazole (Raxil)	Increasing in Rajasthan.
7	Powdery mildew of wheat	*Blumeria graminis* f.sp. *tritici (-Erysiphe graminis)*	Soil-borne by cleistothecia	Dinocap (Karathane)	-

(B) Bacterial diseases

S.No.	Name of disease	Causal organism	Nature/ survival	Management	Remarks
1	**Tundu** of wheat (yellow ear rot, gumming disease)	*Rathayibacter tritici (-Clavibacter tritici)*	Bacteria survive in galls in Soil or seeds. Vector- *Anguina tritici* nematode	Antibiotic, Brine solution (15-20%)	Disease is associated with ear cockle nematode (*Anguina tritici*)

(C) Viral diseases

S.No.	Name of disease	Causal agent	Transmission	Management	Remarks
1	Wheat mosaic disease	Wheat mosaic virus	*Polymyxa graminis* (soil-borne fungus)	Roguing, Insecticides, resistant var	Soil-borne disease

(D)Nematode disease (*i.e.* diseases caused by animals)

S.No.	Name of disorder	Causal agent	Remarks
1	Ear cockle of wheat (*Sehun rog*)	*Anguina tritici* (Seed gall nematode)	First nematode on plants. By Needham in England in 1743.
2	Molya disease of wheat & barley	*Heterodera avenae* (Cereal cyst nematode)	Reported from Neem Ka Thana, Sikar (Rajasthan) by Prasad *et.al* (1959).

IMPORTANT DISEASES OF BARLEY:

(A) Fungal diseases

S.No.	Name of disease	Causal organism	Nature/survival	Management	Remarks
1	Black rust	*Puccinia graminis tritici (Pgt)*,	Survive on hills, infection in India by uredospores	Spray of zineb, Sulphur, oxycarboxin	-
2	Brown rust	*Puccinia hordei*			-
3	Yellow rust	*Puccinia striiformis* f.sp. *hordei*			-
4	Loose smut	*Ustilag segetum* var. *hordei (-U. nuda)*	Internally seed-borne (dormant mycelium in embryo)	Solar energy, Hot water, Systemic fungicides - carboxin	-
5	Powdery mildew of baley	*Blumeria graminis* f.sp. *hordei (-Erysiphe graminis)*	Soil-borne by cleistothecia	Dinocap (Karathane)	-
6	Stripe disease of barley	*Drechslera graminea*	Seed & soil borne	Mancozeb	-

(B) Bacterial diseases

S.No.	Name of disease	Causal organism
1	Black chaff & bacterial streak	*Xanthomonas transluscens* pv. *transluscens*

(C) Viral diseases

S.No.	Name of disease	Causal agent	Management
1	Barley yellow dwarf	Barley yellow dwarf virus	Roguing, Insecticides, resistant var

(D) Nematode disease (*i.e.* diseases caused by animals)

S.No.	Name of disorder	Causal agent	Remarks
1	Molya disease of barley (Cereal cyst nematode)	*Heterodera avenae*	Reported from Neem Ka Thana, Sikar (Rajasthan) by Prasad *et.al* (1959).

Important Plant Diseases of National Importance

Field Crops Diseases	
Rice	Rice Blast (*Magnaporthe grisea*)
	Bacterial blight (*Xanthomonas oxyzae* pv. *oryzae*)
	Sheath blight (*Rhizoctonia solani*)
	Brown spot (*Helminthosporium oryzae*)
	Foot rot (*Fusarium moniliforme*)
	Bunt of Wheat (*Tilletia caries, Tilletia foetida*)
Maize	Northern leaf blight (*Exserohilum turcicum*), Southern leaf blight (*Bipolaris maydis*)
	Stalk rot (*Erwinia carotovora*)
	Banded leaf and Sheath blight (*Rhizoctonia solani*)
	Ear rot (*Fusarium vertcilloides*)
	Gray leaf spot (*Cercospora zeae-maydis*)
Finger millet	Blast (*Pyricularia grisea*)
	Cercospora leaf spot (*Cercospora eleusine*)
Potato	Late blight of potato (*Phytophthora infestans*)
	Early blight (*Alternaria solani*)
	Bacterial wilt (*Ralstonia solanacearum*)
	Rhizoctonia rot (*Rhizoctonia solani*)
	Wart (*Synchytrium endobioticum*)
	Viral diseases (Potato virus X, Y)
Tomato	Wilt (*Ralstonia solanacearum*)
	Late blight (*Phytophthora insfestans*)
	Early blight (*Alternaria solani*)

	Root-knot of Potato (*Meloidogyne* spp.)
	Damping off (*Pythium* spp., *Fusarium* spp.)
Cruciferae	Club root (*Plasmodiophora brassicae*)
	Alternaria leaf spot (*Alternaria brassicae*)
	Damping-off of seedlings (*Pythium* spp., *Rhizoctonia* spp., *Fusarium* spp. and others)
	Stalk rot (*Sclerotinia sclerotiorum*)
	Black rot (*Xanthomonas campestris* pv. *campestris*)
	White rust (*Albugo candida*)
Cucurbits	Powdery mildew (*Erysiphe cichoracearum*)
	Downey mildew (*Peronospora cubensis*)
Pea nut	Tikka disease (*Cercospora personata* and *Cercospora arachidicola*)
Lentil	Wilt (*Fusarium oxysporum*)
	Root rot (*Fusarium solani*)
	Collar rot (*Sclerotium rolfsii*)
	Gray mould (*Botrytis cinerea*)
Soybean	Rust (*Phakopsora pachyrhizi*)
	Anthracnose (*Colletotrichum glycines*)
	Bacterial pustule (*Xanthomonas campestris* pv. *glycines*)
Sugarcane	Red rot (*Colletotrichum falcatum*)
Fruits Plant Diseases	
Citrus	Huang lung bin/Greening (*Candidatus* Liberibacter asiaticus)
	Tristeza (CTV)
	Citrus decline (disease complex)
	Foot and root rots (*Phytophthora* spp.)
	Powdery mildew (*Oidium* spp.)
	Pink disease (*Pellicularia salmonicolor*)
	Anthracnose (*Colletotrichum gloesporoides*)
	Scab (*Elsinoe fawcetti*)
Apple and temperate fruit	Scab (*Ventunia inaequalis*)
	Pink disease (*Pellicularia salmonicolor*)
	Root not (Complex)
	Powdery mildew (*Oidium* spp.)
Mango	Mango malformation (Complex)
	Anthracnose (*Colletotrichum gloesporioides*)

Banana	Wilt (*Fusarium* spp.)
	Sigatoka leaf spot (*Mycosphaerella* spp.)
	Bunchy top (Viral)
Papaya	Leaf curl (viral)
	Ring spot (viral)
Tea	Blister blight (*Exobasidium vexans*)
	Blight (*Pestalosia theae*)
Coffee	Rust (*Hemileia vastatrix*)
Caradamom	Rhizome not (Complex)
	Chhirke and Phoorke (Viral)
Zinger	Rhizome rot (Complex)

Differentiate host – specific and non-host specific toxins

S.No.	Host specifi	Non-host specifi
1.	Selectively toxic only to susceptible host of the pathogen	No host specificity and can also affect the physiology of those plants that are normally not infected by the pathogen
2.	Primary determinants of disease	Secondary determinants of disease
3.	Produce all the essential symptoms of the disease	Produce few or none of the symptoms of the disease
4.	Ex: Victorin, T- toxin	Ex: Tentoxin, Tabtoxin

About Plant Pathogens:

	Fungi	Bacteria	Viruses	Nematodes
Survival	Crop residue Soil	Crop residue Soil Alt. hosts Insect vectors	- - Alt. hosts Insect vectors	Crop residue Soil - -
Dispersal	Wind Rain Insects	Wind Rain Insects	- - TMV Insects	Tillage Equipment Water run-off
Infection	Directly Wounds Insect feeding	- Wounds Insect feeding	- - Insect feeding	Directly - -

(1) Fungus:

- The study of mushrooms or fungi is called **Mycology** (Greek word *Mykes* - mushroom, *logos* - discourse).
- Father of Mycology - **P.A. Micheli**
- **Father of Indian Mycology- E. J. Butler.**
- The term "**Fungus**" is singular and "**Fungi**" is plural.
- The discovery of fungus as etiological agent (*i.e.* **fungus as a plant pathogen**) was established by **Anton de Bary** (1853), while studying with late blight of potato.

Definition: Fungi are eukaryotic, **achlorophyllus**, spore-bearing organisms that reproduce by means of sexually and/or asexually, and with usually filamentous, branched somatic structures are surrounded by cell wall made up of chitin or cellulose or both.

- Fungi are beneficial (*e.g.* mushrooms and yeast) and harmful (*e.g.* causes disease in plants).
- Fungi are cellular organism (*i.e.* made up of cells), mostly multi-cellular but yeast is unicellular.
- Each cell has both **DNA & RNA;** proteins, carbohydrates (80-90%), lipids *etc.*,
- Cells contain 80S ribosomes which are **free** in cytoplasm (while mitochondria have 70S ribosomes).
- Cells have **Mitochondria** for energy production.
- Some fungi also have **Plasmids** (*e.g.* yeast). *Plasmid means "Extra-chromosomal DNA"* in a cell.
- Fungal cells have **double membrane-bounded cell orgenelles** like nucleaus, mitochondria, tubular ER and the Golgi bodies (ribosomes).
- Zoospores of *Chytridiomycetes* contain **"Nuclear Cap"** (ribosomal cap) and several **"Gamma Particles"** which store protein. Gamma particles are single-membrane-bound, cap-shaped, dense bodies found only in zoospores. *e.g. Allomyces, Blastocladiella.*
- Fungal nucleus (with double -layered envelope - made up of lipoprotein) contains dsDNA with histone proteins.
- **dsDNA** in mitochondria is **without histones.**
- Ribosomes of fungus are structurally **distinct**. They are **S-shaped** and may be up to 80 (in number) in some fungi.
- Flagellar type is **9+2** (*i.e.* outer 9 microtubules + inner 2 microtubules).
- Diploid number of chromosome in *Neurospora crassa* is 14.
- *Neurospora* also known as **"Drosophilla of plant kingdom and red or pink bread mould".**

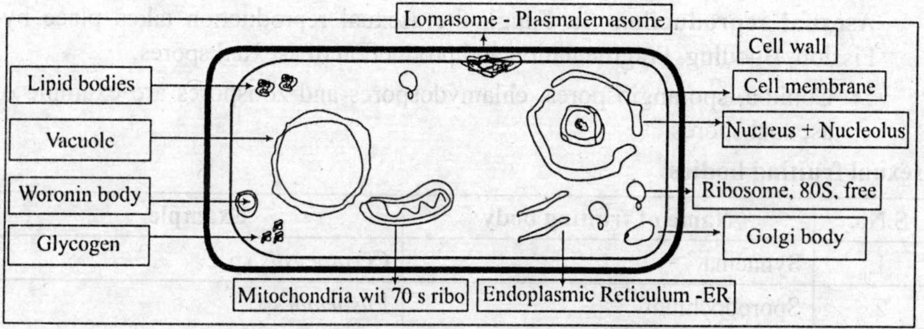

Fig. A typical diagram of fungal cell:

Fungal cells have double membrane-bound cell organelles like nucleus & mitochondria, tubular ER, the golgi bodies & ribosomes. In fungal cell, ribosomes are lying free in cytoplasm & not attached to the ER (like other eukaryotic cells).

Characteristics of fungi-

1. **Nutrition-** Heterotropic (due to absence of chlorophyll) and absorptive (Holotype - whole body of fungus absorb food). Glycogen and oil is storage food. Fungi are lived as Saprobes, Parasites and symbionts.
2. **Cell wall-** Cell wall of fungi is made up of glucan & **chitin** (polysaccharide carbohydrate) but glucan & cellulose in Oomycete fungi (*e.g.,* downy mildew fungi, *Pythium, Phytophthora, Albugo etc.,*).
3. **Thallus:-**
   - **Eucarpic thallus-** Thallus is differentiated into vegetative and reproductive structures (*i.e.* some part of thallus is conveted into reproductive part *e.g., Pythium*).
   - **Holocarpic thalls-** Thallus is not differentiated into vegetative and reproductive structures (*i.e.* whole thallus is conveted into reproductive part *e.g.,* yeast).
   - **Coenocytic hyphae (non-septate hyphae)** – Hyphae without septa and nuclei scattered in the cytoplasm. *e.g. Oomycota & Zygomycota.*
   - **Septate hyphae-** Septa present in hyphae. *e.g., Ascomycota, Basidiomycota and Deuteromycetes.*
   - **Homokaryotic mycelium-** Mycelium contains identical nuclei.
   - **Heterokarotic mycelium-** Mycelium contains nuclei of different genetic make up.
   - **Haustorium-** Food absorbing organ and mostly in obligates.
   - **Appressorium / infection peg-** Attachment organ with host for infection.
4. **Reproduction-** Fungi reproduce either asexually (without nuclear fusion) or sexually (with nuclear fusion) or both. Reproductive bodies of fungi are known as Spores.

Asexual reproduction- In Fungi, the asexual reproduction takes place by- Fission, Budding, Fragmentation and production of asexual spores.

- Conidia, sporangiospores, chlamydospores and zoospores are example of asexual spores.

Asexual fruiting bodies

S.No.	Name of fruiting body	Example
1.	Synnema	*Ceratocystis* sp.
2.	Sporodochium	*Fusarium* sp.
3.	Acervulli	*Colletotrichum*
4.	Pycnidia	*Macrophomina* sp.

Fungi in which sexual reproduction is not known (or fungi that reproduce by asexual spores) are known as ***fungi imperfecti*** or ***mitosporic fungi*** or ***asexual fungi*** or ***anamorphic fungi.***

- Imperfect fungi belong to *Deuteromycetes*.

Some imperfect fungi and their perfect stage

S.No.	Imperfect stage	Perfect stage (sexual stage)
1.	*Fusarium moniliforme*	*Gibbererlla moniliforme*
2.	*Pyricularia oryzae*	*Megnaporthe grisea*
3.	*Colletotrichum falcatum*	*Glomerella tucumanensis*
4.	*Helminthosporium oryzae*	*Cochliobolus miyabeanus*
5.	*Rhizoctonia solani*	*Thanatephorus cucumeris*
6.	*Cercospora arachidicola*	*Mycosphaerella arachidicola*
7.	*Cercosporidium personatum.*	*Mycosphaerella berkeleyii*
8	*Phomosis vexans*	*Diaporthe vexans*
9.	*Ustilaginoidea virens*	*Claviceps oryzae - sativa*
10.	*H.sigmoideum*	*Leptosphaeria salvinii*
11.	*H.teres*	*Pyrenophora teres*
12.	*H. gramineum*	*P. graminea*
13.	*Geotrichum candidum*	*Endomyces geotrichum*
14.	*Aspergillus nidulans*	*Emericella nidulans*
15.	*Histoplasma capsulatum*	*Emmonsiella capsulata*
16.	*Penicillium vermiculatum*	*Talaromyces vermiculatus*
17.	*Tubercularia vulgaris*	*Nectria cinnabarina*
18.	*Crytococcus neoformans*	*Filobasidiella neoformans*
19.	*Rhizoctonia bataticola*	*Macrophomina phaseolina*

20.	*Ascochyta pinodes*	*Mycosphaerella pinodes*
21.	*Colletotrichum musae*	*Glomerella musarum*
22.	*Phellinus noxius*	*Phomes noxius*
23.	*Helminthosporium sacchari*	*Drechslera sacchari*
24.	*Cryptosporiopsis malicorticis*	*Gloeosporium peprennas*
25.	*Monilinia fructicola*	*Sclerotinia fructicola*

Sexual reproduction

Fungi in which sexual reproduction is known (or fungi that reproduce by sexual spores) known as **perfect fungi** or **meiosporic fungi** or **sexual fungi** or **telomorph fungi**.

Sex organ of fungi is known is **gametangia.** Male sex organ is called **anthridium** and female sex organ is known as **oogonium (in Oomycota)** and **ascogonium (in Ascomycota).**

- Paragynous anthredium- *Pythium* and *Albugo*.
- Amphigynous anthredium- *Phytophthora.*

Features of fungi & fungal-like organisms

Phylum	Features			
	Hyphae	Cell walls	Spores	
			Sexual spores	Asexual spores
FUNGAL-LIKE				
Oomycota	Coenocytic (no septa)	Glucan, **cellulose** & hydroxyproline	Oospores	Zoospores, sporangiospores, conidia
TRUE FUNGI				
Zygomycota	Coenocytic (no septa)	Glucan & chitosan-**chitin**	Zygospores	Sporangiospores (aplanospores)
Ascomycota	Septate	Glucan & **chitin**	Ascospores	Conidia
Basidiomycota	Septate	Glucan & **chitin**	Basidiospores	Conidia (Rarely)
Deuteromycota	Septate	Glucan & **chitin**	None. Only asexual cycle is found	Conidia, chlamydospores

Sexual spores of fungi (trick – *ZOBA*; Z-zygospore, O-oospore, B-basidiospore, A-ascospore)

SNo	Sexual spores	Phylum	Remark
1.	Zygospore (2n)	*Zygomycota*	One zygospore produced after fusion of two gametangia. Zygospore is only dipolod phase of these fungi (meiosis occurs during germination of zygospore & rest life is haploid)

2.	Oospore (2n)	*Oomycota*	Majority of life cycle is diploid (only gametic phase is haploid)
3.	Basidiospore (n)	*Basidiomycota*	Usually 4 in number (externally on basidium)
4.	Ascospore (n)	*Ascomycota*	Usually 8 in number (internally in ascus)

Types of sexual fruiting bodies

S.No.	Name of fruiting body	Description	Example
1.	Cleistothecium	Completely closed ascocarp	*Aspergillus, Penicillium* & Powdery mildew fungi
2.	Perithecium	Flask shaped ascocarp	*Claviceps sp.*
3.	Apothecium	Cup shaped ascocarp	*Sclerotinia sclerotiorum.*

Parasexual cycle or parasexuality: It was discovered by **Pontecorva & Roper in 1952 in *Aspergillus nidulans.***

Fungi and their specific name

S.No	Fungi/order/family	Specific nam
1.	*Pythium*	Damping off fungi
2.	*Saprolengia*	Water mould
3.	*Saprolengia*	Primary sugar fungi
4.	*Albugo*	White rust fungi
5.	*Peronosporales*	Downy mildew fungi
6.	*Mucorales*	Pin mould & sugar fungi
7.	*Erysiphales*	Powdery mildew fungi
8.	*Uredinales* (now *Pucciniales*)	Rust fungi
9.	*Ustilaginales*	Smut fungi
10.	*Pilobolus*	Shot gun fungi & hat thrower fungi
11.	*Neurospora*	Red or pink mould & Drosophilla of plant kingdom
12.	*Sclerotinia sclerotiorum*	White mould
13.	*Aspergillus niger*	Weed of laboratory
14.	*Colletotrichum*	Anthracnose fungi
15.	*Basidiomycota*	Club fungi, mushroom, ray fungi & highly evoved fungi
16.	Agaricus bisporus	Button mushroom
17.	*Aspergillus flavus*	Aflatoxin producing fungi
18.	*Armillaria mellea*	Honey mushroom
19.	*Mucor & Rhizopus*	Bread mould

20.	*Botrytis* sp.	Grey mould
21.	*Sclerotinia sclerotiorum, Rhizoctonia solani & Claviceps* sp.	Sclerotia producing fungi
22.	*Fusarium & Claviceps* sp.	Produced two type conidia (Macro and micro conidia)
23.	*Coprophilous fungi*	Grow on dung
24.	*Aspergillus niger*	Black molds
25.	*Penicillium* spp.	Green molds

Some important toxins and their producer organism

S.No.	Toxin	Produced by
1.	Fusaric acid	*Fusarium* sp.
2.	Lycomarasmin	*F. oxysporum* f.sp. *lycopersici*
3.	Piricularin	*Pyricularia oryzae*
4.	Alternaric acid	*Alternaria solani*
5.	Ten toxin	*Alternaria tenuis*
6.	Amylovorin	*Erwinia amylovora*
7.	Fumaric acid	*Rhizopus* sp.
8.	**Wild fi e toxin or tab toxin**	***Pseudomonas syringae* pv. *tabaci***
9.	**Victorin**	***Helminthosporium victoriae***
10.	Cochliobolin	*H. oryzae*
11.	AM toxin	*Alternaria alternata*
12.	HC toxin	*Cochliobolus(Helminthosporium)carboanum*
13.	HS toxin	*Cochliobolus(Helminthosporium)sacchari*
14.	**HM/T toxin**	***Helminthosporium madis race-T***
15.	**Phaseo toxin**	***Pseudomonas syringae* pv. *phaseolicola***
16.	Colletotin	*Colletotrichum falcatum*
17.	Afla toxin	*Aspergillus flaves*
18.	Oxalic acid	*Sclerotium* and *Sclerotinia* spp.
19.	Ceratoulmin	*Ceratocyststis ulmi*
20.	PC toxin	*Periconia circinata*
21.	PM toxin	*Phyllostica maydis*
22.	CC toxin	*Corynespora cassiicola*
23.	ACL toxin	*Alternaria citri* Race- Lemon
24.	ACT toxin	*Alternaria citri* Race-Tangerine
25.	Destruxin -B	*Alternaria brassicae*

(2) Bacteria:

(MBBS) - Bacteria – Prokaryotes with cell wall

- Bacteria were discovered by **Leeuwenhoek** (1676). He called them "**animalcules.**
- **Leeuwenhoek** is known for the discovery of **"Living cell".**
- First non living cell was discovered by **Robert Hooke (1665).**
- The name **"Bacterium"** was given by **Ehrenberg** (1828).
- Bacteriology - Study of bacteria is called **bacteriology.**
- Father of **Phytobacteriology - EF Smith.** (*Phyto* means plant).
- The term **"Bacterium"** is singular and **"Bacteria"** is plural.
- First plant disease caused by bacteria was **"Fire Blight of Pear"**& reported by **TJ Burrill** in USA in 1878.
- Each bacterial cell is an independent bacterium

Definition:

"Bacteria are prokaryotic, single celled, microscopic organisms which surrounded by cell wall (made up of peptidoglycan/murein) and ***divided by binary fission".***

- Bacteria are beneficial and harmful.
- Bacteria are **cellular organism** (not chemical particle like virus).
- Not have well organized nucleus, that's why they are known as prokaryotic cell (*i.e.* a cell without nuclear membrane).
- Cell wall is made up of **Peptidoglycan (also called Murein/mucopeptide).**
- **Peptidoglycan - Protein + sugar.**
- **Not have Mitochondria** for energy production but,
- Has **mesosome** (infolding of cell membrane) as respiration sites.
- Each cell has **both DNA & RNA**, proteins, carbohydrates, lipids *etc.,* DNA is **free** in the cytoplasm.
- Number of chromosomes in bacteria may be **1 to 3**, but **always as circular** (linear in mycoplasma).
- Cell contains **70 S ribosomes** (sub units are 30 S & 50 S). Ribosomes contain **rRNA.**
- Prokaryotic ribosomes contain **Protein** (40 %) and **RNA** (60 %) in the ratio of 1:2 (while in eukaryotes, it is around 1:1).
- RNA of smaller subunit (30 S) of ribosome is 16 S (*i.e.* **16 S rRNA** - used in recent classification).
- Some bacteria have **Plasmids** [one to several *i.e.* up to thousands, *e.g.*
  (i) ***Agrobacterium vitis*** S4 has **5 circular plasmids & 2 circular chromosomes,** *i.e.* it has **7 DNA molecules per cell,**

(ii) ***Agrobacterium tumefaciens*** strain C58 has **2 plasmids**, as **plasmid AT & plasmid Ti;** and 1 circular & 1 linear chromosome, *i.e.* it has **4 DNA molecules per cell**)].

Plasmids are also known as **Mini-chromosome** or **cccDNA** (covalently closed circular DNA).

- Most plasmids are circular but some are linear.
- Linear plasmids are found in *Streptomyces, Mycobacterium, Rhodococcus, Arthrobacter etc.,*
- Plasmids can be part of **Mobilome**; since they are often associated with conjugation, a mechanism of horizontal gene transfer.
- **Plasmidome (*i.e.* total plasmid content)**-Refers to the **actual plasmids** that were found and isolated from a certain microorganism.
- **Segrosomes** - Segrosomes are **protein complexes** that ensure accurate segregation (partitioning) of plasmids or chromosomes during bacterial cell division.
- **Episome** - Integrated form of plasmid and chromosome (**Episome - plasmid + chromosome**).
- Bacteria are **organisms** made up of just one cell (unicellular).
- Filamentous bacterium is ***Streptomyces*** (also known as **"Store House of Antibiotics"**).
- **Most plant pathogenic bacteria** are **Gram - ve** (while *Streptomyces, Clavibacter etc.* are G +ve).
- Recently bacteria are classified on the basis of the sequencing of **16 S ribosomal RNA** (16 S rRNA).
- Bacteria **not have** intracellular organelles (*i.e.* membrane-bounded structures in cytoplasm).
- Most plant pathogenic bacteria are **necrotrophs** (*i.e.* kill first & then feed upon dead tissues).
- Most bacteria are **non-chlorophyllous** but some have photosynthetic apparatus with pigments like chlorobium-chlorophyll, bacterio-chlorophyll and bacterio-viridin which help in photosynthesis.
- But some bacteria are **autotrophic** (self feeder), *e.g.* purple sulphur bacteria and green sulphur bacteria.
- All plant pathogenic bacteria are **chemo-heterotrophs** (which use a chemical as energy source & organic compound as carbon source).
- Most plant pathogenic bacteria do not form **spores** (*i.e.* they are non-spore formers).
- Spore formers (**endospore** formers) are *Bacillus* and *Clostridium* (causes food poisoning /botulism).

- **Gram staining** and **IF test** (Immuno Fluorescence test) is done for bacterial identification.
- Majority of bacteria have **one circular chromosome** (one dsDNA).
- In bacteria, the entire hereditary material is packed into a single, irregularly folded compact mass called **nucleoid** or **genophore** or **bacterial chromosome**. The DNA is in the form of a **double helix** which forms a closed ring or circle with no free ends. The bacterial chromosome lacks a protein coat and it is in direct contact with the cytoplasm, since a nuclear membrane is absent. A small amount of protein, mainly in the form of an enzyme called **RNA polymerase**, may be found associated with the bacterial chromosome.
- **Genophore** term was given by **Ris**.
- Since a single copy of the chromosome occurs in a cell, the bacterial cells are **haploid**.
- **Common media** for bacterial growth are **NA** (Nutrient agar), **YGCA** (yeast glucose chalk agar) *etc.,*
- Flagellar type is **0+1**.
- **Diazotrophs:** Fixation of atmospheric nitrogen by prokaryotic organism is known as "***Diazotrophy***" & the organisms are referred as "***Diazotrophs***".
- **Archaebacteria:** Archaebacteria lack of peptidoglycan in cell wall but have **pseudomurein**. Live in extreme environments (*i.e.*extremophiles), such as hydrothermal vents and terrestrial hot springs (thermophilic). Also found in a diverse range of highly saline (halophiles), acidic (acidophiles) and anaerobic environments (methanogens).

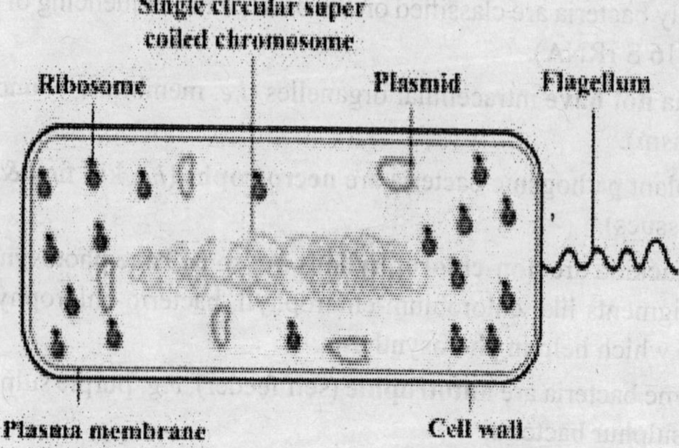

Fig. A typical diagram of bacterial cell:

Source: http://www.brainkart.com/article/Cell-Wall—Structure-and-Functions-of-Bacterial-Cell-Envelope_ 17824/

Plant diseases caused by bacteria are:

SNo	Disease	Caused by	History/Other
1	**RSD** of sugarcane	***Leifsonia xyli*** *(Clavibacter xyli)*	RSD- Ratoon stunting disease
2	Citrus Huanglongbing (HLB)	*Ca.* Liberibacter asiaticum	Earlier called **Citrus greening** **Vector-** ***Diaphorina citri***
3	Pierce's disease of grapevine (PDG)	*Xylella fastidiosa*	Discovered in 1892 by NB Pierce. **Vector-** Sharpshooter leafhoppers *(Graphocephala atropunctata*, blue-green sharpshooter, *Homalodisca vitripennis*, glassy-winged sharpshooter)
4	Yellow vine disease of cucurbits	*Serratia marcescens*	Vector - squash bugs, *Anasa tristis*

Morphological characters:

Shape of bacterial cells (cell morphology)

- Spherical or round shaped cells are called **Coccus / cocci** (eg. *Micrococcus*),
- Cylindrical or rod shaped cells are called **Bacilli** (*e.g., E. coli*),
- Spiral shaped cells are called **Spirillum** (*e.g., Spirillum*),
- Comma shaped or Curved-rods cells are called **Vibrio** (*e.g., Vibrio*),
- Club-shaped or match stick shaped (*Clavibacter*) and
- Filamentous shapes are called **Actinomycetes** (*e.g., Streptomyces*)

Motile (mobile) cells have long, whip-like flagella which may arise from one or both ends of the cell (polar) or from all over the cell (peritrichous).

On the basis of flagella, bacteria are classified into 6 groups (flagellum is 0+1 type)

(i) **Monotrichous** - Single flagellum at one end (*Xanthomonas, Vibrio*)

(ii) **Amphitrichous** - Single flagellum at both ends (*Nitrosomonas*, *Pseudomonas*)

(iii) **Cephalotrichous** - Several flagella at one end (*P. fluorescens*)

(iv) **Lophotrichous** - Several flagella at both ends (*Spirillum*)

(v) **Peritrichous** - Several flagella all over the cell (*Erwinia, Escherichia coli - E. coli, Bacillus, Clostridium*)

(vi) **Atrichous** - No flagellation (*Pasteurella*).

Sub-polar - Single sub-polar flagellum (*Agrobacterium*), (Like Monotrichous).

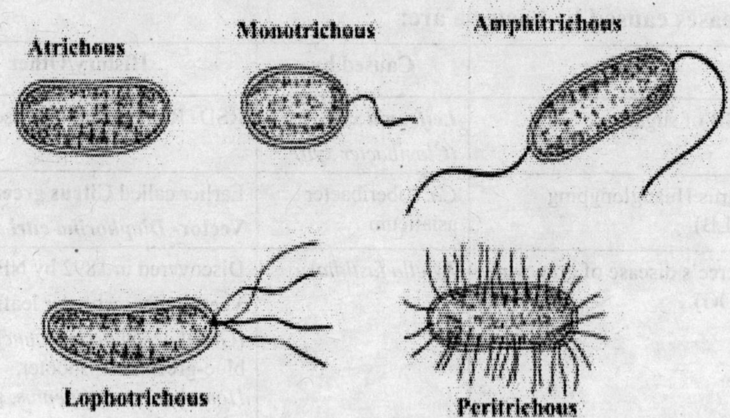

Flagellation types in bacteria

Difference between G +ve and G –ve bacteria

S. No.	Components	G +ve	G –ve
1	Cell wall thickness	Thicker : 20-80 nm	Thinner :10 nm
2	Peptidoglycan	High : 50-80 %	Low : 10-20 %
3	Di-amino-pimelic acid (DAPA)	In traces (or absent)	More (present)
4	Teichoic acid	Present : 20%	Almost absent
5	Lipid content (mycolic acid)	In traces (2-4 %)	More lipids (20-30 %)
6	Mesosomes (respiratory sites : work as mitochondria)	Normal	Very low
7	Pili	Absent	Present
8	No. of layers in cell wall only	1 (as peptidoglycan layer)	2 (as outer membrane & peptidoglycan)
9	Examples	*Clavibacter* (*Leifsonia*) *Streptomyces, Bacillus, Clostridium, Lactobacillus Rhodococcus*	*Agrobacterium, Xanthomonas, Pseudomonas, Erwinia, Serratia, Samsonia, Pantoea, Xylella, Ralstonia, Rhizobium*

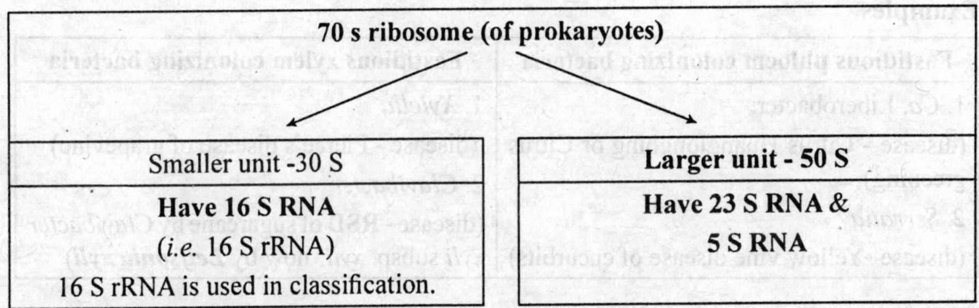

(3) Fastidious Vascular Bacteria (FVB)

FVB – Prokaryotes having special nutritional requirement

- Meaning of fastidious - fussy, choosy and particular.
- **FASTIDIOUS** means an organism having "***special nutritional requirement for growth***".
- **VASCULAR** means living in vascular tissues (xylem & phloem) only.
- **Bacteria** mean prokaryote with cell wall.
- Earlier known as RLOs (Rickettsia like organisms - these are un-cultivable).
- McCoy (1982) suggested the name FVB for RLOs.
- FVB limited to **phloem** were first observed by Windsor & Black (1972) in clover plant affected with "Clover Club Leaf Disease".
- FVB limited to **xylem** were also observed by **Windsor & Black (1973)** in grape plants affected with "Pierce's Disease of grapevine".

Definition "Bacteria that have specific nutritional requirement and live in vascular tissues of plants are called FVB". OR "FVB are fastidious in nature, limited to vascular tissues and depend on insect vector for infection & transmission".

- FVB are generally rod-shaped & without flagella.
- Mostly surrounded by a cell membrane and cell wall.
- Their outer layer of cell wall is undulating or rippled or wavy.
- FVB are sensitive to tetracycline & penicillin.

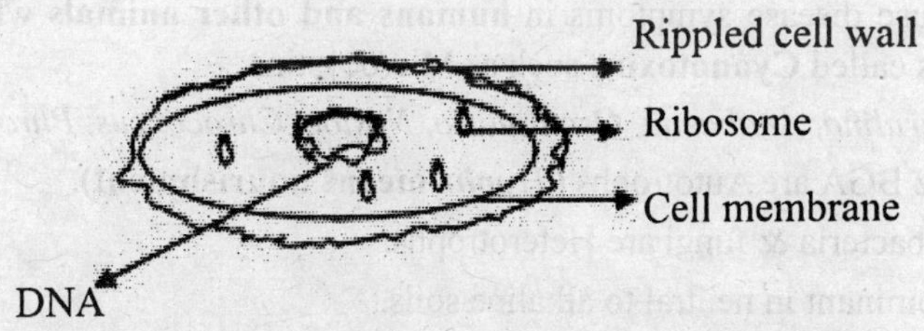

Fig. Fastidious Vascular Bacterial cell

Examples

Fastidious phloem colonizing bacteria	**Fastidious xylem colonizing bacteria**
1. *Ca.* Liberobacter (disease - Citrus Huanglongbing or Citrus greening) 2. *Serratia* (disease -Yellow vine disease of cucurbits)	1. *Xylella* (disease - Pierce's disease of grapevine) 2. *Clavibacter* (disease - **RSD** of sugarcane by *Clavibacter xyli* subsp. *xyli;* now by ***Leifsonia xyli***)

(4) Cyanobacteria (BGA):

- The term **"Cyanobacterium"** is singular and **"Cyanobacteria"** is plural.
- Also known as BG Algae, BG Bacteria, Cyanophyta.
- The name "Cyanobacteria" comes from the color of the bacteria (Greek: Kuavoc (Kyanos) means Blue).
- It is a **prokaryotic**, unicellular, aerobic & oxygenic **photosynthetic** bacterium.
- It forms **"Lichen"** with fungi. (Lichen = Fungi + Cyanobacteria). (Lichen - fungi + algae also).
- Having DNA, RNA, proteins *etc.,*
- DNA is free in the cytoplasm.
- **Not** have well organized nucleus, that's why they are known as **prokaryotic cell.**
- Cell wall is made up of peptidoglycan.
- These do not have flagella and do not reproduce sexually.

Definition BGA are unicellular, photoautotrophic prokaryotes containing "Phycocyanin" pigment in addition to chlorophyll.

- Nitrogen fixing Cyanobacteria is ***Anabaena*** (symbiotes - It is used as biofertilizer in rice fields mainly.
- **Heterocysts** are the sites of nitrogen fixation.
- Nitrogen fixation - ***Anabaena* + *Azolla*** - in rice field.
- Although, not usually considered as **pathogens**, Cyanobacteria have recently been implicated as agents of **intoxication,** *i.e.*, substances produced by them may cause disease symptoms in **humans and other animals** when ingested. **Toxin** is called **Cyanotoxins** such as Microcystins.
- *e.g., Spirulina, Anabaena, Oscillatoria, Nostoc, Chrococcus, Phromidium etc.,*
- Algae & BGA are Autotrophs **(*Trophe* means nourishment)**.
- Mostly bacteria & fungi are Heterotrophs.
- BGA dominant in neutral to alkaline soils.
- **BGA** (& most of the soil algae) act as **cementing agent** in binding soil particles and **check soil erosion** (as they secrete a substance known as **"Mucilage"**, which helps in increasing water retention capacity of soil).

(5) Virus:

Virus = Nucleic acid + Protein coat (Nucleo-protein) it is chemical particle not a cell

Definition:

"Viruses are submicroscopic entities which replicate only in the living cell and contain one type of nucleic acid (either RNA or DNA never both in one particle) and surrounded by protein coat (capsid)."

- Viruses are non-cellular/acellular (virus is **not** an organism).
- Chemically, ***virus is a particle***, made up of **nucleo-protein** (having nucleic acid + protein).
- Protein coat of a virus particle is called **"Capsid".**
- **Infectious part** of virus particle is **"Nucleic acid"** while **protein** acts as **protective coat.**
- **Majority of plant viruses, contain "+ssRNA"** (*e.g.,* TMV- ssRNA),
- *Fijivirus* has dsRNA.
- But some have DNA (eg. *Geminivirus* -ssDNA, *Caulimovirus*-dsDNA).
- Animal viruses mostly contain DNA.
- Viruses possess about 10 genes.
- Viruses use **"host's ribosome"** for **protein synthesis** (because viruses don't have own ribosomes).
- All plant viruses are transmitted from diseased plant to healthy plant by **insect vectors** (mostly) or by contact or by other biological agents (fungi, nematodes, phanerogams *etc.,*).
- Viruses are **obligate parasites**; because they multiply only in living host cells.
- Viruses can only be seen with an **electron microscope**.
- Although viruses **behave like microorganisms** and they **also behave like chemical molecules**.
- Mostly plant viral diseases are transmitted by insects (Mostly Aphids).
- **Antibiotics** have **no effect** on viruses (because virus is not a cell).
- Most drugs that destroy viruses also destroy the host cell.
- Most plant viruses have very small genomes *i.e.* 25,000 to 10,000 nucleotides.
- Most plant **viruses do not kill** their host's cells (as viruses are **obligate parasites**).
- Normal metabolism of the host is diverted towards the synthesis of virus nucleoproteins. Hence, virus infection normally leads to deficiency symptoms.

Some important facts about viruses:-

- **First viral disease of plants** (tobacco mosaic disease) was reported by **Adolf Mayer** (1886).

- The term **"Virus"** was coined by **Beijerinck** (1898). Also *'Contagium vivum fluidum"*.
- Virus means – **poison**.
- Father of Plant Virology - **Beijerinck.**
- Virus is an **"obligate parasite" (do not kill host cells). All viruses are parasitic** in cells.
- Viruses can be detected by **ELISA** (a serological technique).
- **"Crystallization of virus particles"** first obtained by **Stanley** (1935).
- **Stanley** got **Noble Prize (in** Chemistry in 1946) with other scientists.
- Plant viruses enter host cells only through **wounds** (made mechanically or by vectors or by infected pollen grain in to ovule).
- **Meristem tissues** are ***"virus free"*** part of plant.
- Viruses may cause local or systemic infection.
- Mostly, viral diseases are transferred by aphids followed by leaf hoppers.
- Viruses have 5-40 % nucleic acid & 60-65 % protein.
- TMV consist 5 % NA and 95 % Protein.
- Viruses don't produce enzymes **except Reo viruses and Rhabido viruses.**
- Mostly plant viruses have ss RNA (+ Ve).
- Gemini viruses have ss DNA.
- Caulimo viruses have ds DNA.
- Infectivity & protein coat synthesis are major functions of viral nucleic acid.
- **Amino acids are building blocks of protein.**
- Structural & chemical subunits of the protein coat are known as **Protein sub unit.**
- **Capsid** is composed of repeating sub units, also known as **capsomere.**

Functions of protein coat:-

- It acts as protective coat.
- In production of different types of symptoms.
- Increase infectivity of viruses.
- It plays a role in determining vector transmissibility of a virus.

Characters of Tobacco Mosaic Virus (TMV):

Size - **300×18** nm.	Protein sub units - 2130 Shape - Rod shaped.
Total RNA - **5** %.	Each protein subunit has **158 amino acids.**
Total protein - **95** %.	Each Nucleic acid has **6400** nucleiotide.

Plant diseases caused by viruses

S. No.	Disease	Caused by	History
1.	**Tobacco mosaic disease**	***Tobacco mosaic virus-TMV (Tobamo virus)***	**First viral disease of plants. Reported by Adolf Mayer (1886).**
2.	Leaf curl of cotton	*Cotton leaf curl virus (Geminivirus)*	Introduced in India from Pakistan.
3.	Leaf curl of tomato	*Tomato leaf curl virus (Geminivirus)*	-
4.	Yellow mosaic of legumes	Mungbean yellow mosaic virus (*Begomovirus)*	-
5.	Rice tungro disease (***Tungro*** -yellow-orange, dwarf)	*Rice tungro spherical virus (RTSV)* **and** *Rice tungro bacilliform virus(RTBV)*	Caused by two viruses: RTSV (*Waikavirus* - ssRNA) and RTBV(*Badnavirus* -dsDNA).
6.	Bunchy top of banana	*BBTV - banana bunchy top virus* (vector-Banana aphid – *Pentalonia nogronervosa*)	*Virus belongs to* ***Babuvirus*** *group (having ssDNA)*

Viruses do **not** carry out **respiration, digestion, or other metabolic** functions, therefore, they are **non-living organisms.** Viruses cannot grow or multiply outside the host cell that's why they are called **obligate parasites.** They can't multiply on their own, so they have to attack a 'host' cell and take over its machinery in order to be able to make more virus particles.

Why viruses are called as "Intermediate or Transition stage" between living & non-living organisms":

(A) Viruses are living because:

- They show growth and mutation (in nucleic acid).
- They have genetic material (either RNA or DNA never both).
- They have ability to infect (while deficiency disease not).
- They can be transmitted from diseased to healthy host (while deficiency disease not).
- They are capable to multiply/replicate (while deficiency disease not).
- They react to heat, chemical & radiation.
- The nucleoprotein of viruses has similarity with protein & NA of living organisms.

(B) Viruses are non-living because:

- They do not respire or excrete (while living organisms do like that).
- They can be crystallized (while living organisms not).
- They are inert outside the host.

- They depend on living organisms for replication/growth.
- They do not have energy producing enzyme system (while living organisms have; like Kreb cycle).
- Cell wall and cell membrane is absent.
- They do not show functional anatomy.
- Their sedimentation according to their molecular weight is like of non-living.

Unique characters found only in viruses (not in living organisms):

- Presence of only **one type of NA, (either RNA or DNA never both)**.
- They do not undergo binary fission.
- Capacity to reproduce from their sole Nucleic Acid (*i.e.* from total NA).
- Viruses make use of ribosomes of host cells.

Reasons for "VIRUS FREE MERISTEMS" of the infected plants:

(1) **Restricted movement:** Restricted virus movement due to lack of suitable vascular or plasmadesmatal connections.

(2) **Antiviral metabolites:** Presence of antiviral metabolites in dividing meristem cells.

(3) **Fast multiplication rate:** The higher multiplication rate of the meristem cells than spread of the virus.

Transmission of viruses:

Viral diseases may be transmitted by-

1. **By seeds-** Mostly seed transmission of viruses happens through the **Embryo.**
   e.g. - Cucumber mosaic virus & barley stripe mosaic virus.
2. **By Vegetative propogation-**
   By grafting- Citrus tristeza virus.
   By tubers- Potato virus X & potato virus Y.
   By suckers- Banana bunchy top virus.
3. **Mechanical or sap transmission-** TMV
4. **By fungi**

S.No.	Name of virus	Fungi
1.	Wheat mosaic virus	Polymyxa graminis
2.	Indian peanut clump virus	Polymyxa graminis
3.	Potato mop top virus	Spongospora subterranea
4.	Potato virus X	Synchytrium endobioticum
5	Lettuce big vein virus	*Olpidium brassicae*

5. **By nematode-**

- ***Longidorus, Paralongidorus*** **&** ***Xiphinema*** neamtodes transmit to polyhedral viruses, so they named as **NEPO viruses** (Nematode transmitted polyhedral viruses).
- ***Trichodorus*** **&** ***Paratrichodorus*** nematodes transmit to tubular viruses, so they named as **NETU viruses** (Nematode transmitted tubular viruses). They are also known as **TOBRA viruses.**

S. No.	Virus	Virus group	Nematode Vector
1.	**Grapevine fan leaf virus**	**NEPO viruses**	*Xiphinema index*
2.	Tomato ring spot virus	**NEPO viruses**	*X. americanum*
3.	Raspberry ring spot virus	**NEPO viruses**	*Longidorus elongatus*
4.	Pea early browning virus	**NETU viruses or TOBRA viruses**	*Paratrichodorus & Trichodorus*
5.	Tobacco rattle virus	**NETU viruses or TOBRA viruses**	*Paratrichodorus & Trichodorus*

6. **By insect vectors-**

- **Hashimoto-** First established relationship between insect vector and plant viruses, in case of *Rice dwarf virus* transmitted by leaf hoppers (*Racilia dorsalis & Nephottetix cencticeps).*
- **Watson & Roberts-** Coined the term non-persistent and persistent viruses.
- **Kinney *et.al*** - Coined the term **stylet-borne & circulating viruses** replacing **non-persistent** and **persistent,** respectively.
- **Harris-** Coined the term **non-circulating & circulating viruses**. He gave latest categorisation of viruses.
- **Chalfant & Chapman-** gave biomodal transmission. **Biomodal transmission** means a virus transmitted by both non-persistent & semi-persistent ways.
- *e.g.*- **Cauliflowe mosaic virus** is transmitted by *Myzus persicae* in non-persistent way & by *Brevicoryne brassicae* in semi persistent way.

Plant diseases and its vectors

S.No.	Name of disease	Vector
1.	Tomato yellow leaf curl disease	White fly (*Bemisia tabaci*)
2.	Bean golden mosaic disease	White fly
3.	Cotton leaf disease	White fly
4.	Bhindi Yellow vein mosaic disease	White fly
5.	Citrus tristeza disease	Aphid
6.	Potato leaf roll disease	Aphid

7.	Sugarcane mosaic disease	Aphid (*Aphis* sp.)
8.	Papaya ring spot disease	Aphid (*Aphis* sp.)
9.	Barley yellow dwarf disease	Aphid
10.	Cauliflower mosaic disease	Aphid
11.	Bunchy top banana disease	Banana aphid (*Pentalonia nigronervosa*)
12.	Sugarcane Fiji disease	Plant hopper
13.	Swollen shoot disease of cocoa	Mealy bug
15.	Rice tungro disease **(two viral disease)**	**Green leaf hopper (*Nephottetix virescens*)**
16	Pigeon pea sterility mosaic disease	**Eriophyid mite (*Aceria cajani*)**

Dipping in milk or spray of milk on agricultural tools and implements can check the spread of TMV in chilli, tomato, tobacco *etc.,* (other examples of viruses are Cucumber mosaic virus, Bean mosaic virus, Tobacco ring spot virus).

(6) Bacteriophage (or Phage):

- **Phage** means to eat. Therefore bacteriophage means to eat bacteria.
- Meaning –"***Eaters of bacteria***".
- Basically these are viruses.
- Virus that infects bacteria is called bacteriophage.
- First reported by **Twort (1915, England) and d'Herelle (1917, France),** independently.
- **The term "Bacteriophage"** was coined by **d'Herelle.**
- Nucleic acid in phages- DNA (in mostly, *e.g.,* T2 phages) and RNA (in rare cases).
- **Purity of the Ganga water** is due to bacteriophages (because phages kill water polluting bacteria).

BACULOVIRUSES (insect viruses): These are rod shaped with ds DNA comprising NPV (nuclear polyhedrosis virus) & GV (Granulosis virus). Among these, NPV has been exploited mostly for pest management (*i.e. Helicoverpa* sp.). Mostly used for control of Lepidopteron insects.

(7) Viroid:

Virois- naked ssRNA, it is a chemical particle (not a cell)

- Viroids discovered by **T.O. Diener** (1971) in potato (in Maryland, USA).
- **Viroid** means – **Virus-like.**
- Viroids are also known as "***Metaviruses" and "Miniviruses"***.
- First viroid disease was "Potato spindle tuber disease" **(means long pointed potato tubers).**

Definition: *"Viroids are small, low molecular weight, naked RNA molecules"* or *"Viroids are covalently closed, circular, ssRNA molecules of low molecular weight that multiply in plants and cause diseases".*

- Viroids are **SMALLEST** known plant pathogen.
- The size of RNA molecule is 250 to 370 nucleotides.
- Viroids contain **ONLY** ssRNA as Nucleic Acid.
- Viroids are self replicating, circular particles (*i.e.* viroid contains circular, ssRNA).
- Viroids not have protein coat (*i.e.* viroid are naked particles).
- Viroids are **NOT** found in animals.
- More than 40 viroids are known to infect plants.
- Viroids infect only plant cells, replicated in their nucleus, using the substances & enzymes of plant cells. Viroids are transmitted mainly by mechanical means (*i.e.* sap, vegetative propagation).
- Viroids do **not code for any protein** *i.e.* not have genes for encoding proteins (that's why they are naked *i.e.* without protein coat).
- Viroids can not be detected by immunoassays (like **ELISA**) because they not have protein (coat).
- Viroid can be detected by **R-PAGE** (**R**eturn **P**oly-**A**crylamide **G**el **E**lectrophoresis); **RT-PCR** (Reverse Transcription-Polymerase Chain Reaction) & Nucleic Acid Hybridization.

Example plant diseases caused by viroids:

SN	Disease	Caused by	History/remark
1	Potato spindle tuber disease	Potato Spindle Tuber Viroid **(PST Vd)**	First viroid disease of plants. Reported by **Diener**. RNA has **359 nucleotides**.
2	Coconut cadang cadang disease	Coconut cadang cadang viroid (CCC Vd)	First reported from Philippines in 1937. *Cadang* means **dying.** Disease developed very slowly (in 10-15 years).
3	Chrysanthemum stunt disease	Chrysanthemum stunt viroid	ssRNA **Fig. Viroid**
4	Citrus exocortis disease	Citrus exocortis viroid	
5	Dapple Apple disease	Apple scar skin viroid	
6	Tomato bunchy top dis.	Tomato apical stunt viroid	
7	Tomato planta macho disease	Tomato Planta Macho Viroid	TPMVd is the causal agent of a disease of tomato known as "Planta Macho" (male plant) disease.

***Why viroids can't be transmitted by vectors?**

- Viroids can't be transmitted as naked RNA, because of their susceptibility to **"Nuclease Enzyme"**. They are **protected** from this **enzyme-attack** by being **localized within the nuclei of infected cells** (Sanger, 1979).
- They can be transmitted by mechanical means with sap or vegetative means.

Why viroids lack protein coat?

- Viroids are the smallest infectious agents that multiply autonomously in plant cells; they consist only of small, circular RNA molecule that is too small to code for even one small protein & thereby lack a protein coat.
- Viroids cause plant disease following interactions with proteins *i.e.* interfering with host cell's gene regulation.
- *e.g.*, when they interact with an enzyme that disturbs protein synthesis, the growth of the host plant may be stunted.

(8) Prions:

Prion – Infectious protein; and it is chemical particle (not cell)

- Prions ***"The puzzling proteins"***.
- Prions are also known as **"Slow Viruses"** (because of their slow effect).
- Discovered by **Prusiner** (1982, in USA) in Scrapie disease of sheep & goats. **(FAQ).**
- The term ***"Prion"*** coined by **Prusiner** which comes from "**PIP** that lack nucleic acid".
- Prusiner got **Nobel Prize** in 1997 for Prions.
- Prions cause infectious diseases of nervous system. Prions are transmissible particles.
- So far, prions are **NOT** reported in **plants**, but they are known to infect **fungi**.
- Prions **NOT** have **Nucleic Acid**.
- This is the **only** pathogenic agent that **NOT** has any type of nucleic acid.

Definitio :

***"Proteinaceous Infectious Particles (PIP)"* are known as Prions.**

- Prions consist of a small (about 55,000 Da) protein, which is encoded by a chromosomal gene of the host. Prions are at first normal small protein molecules produced in nerve & other cells of the brain. Prions become pathogenic: because they cannot carry out their normal functions &, instead have adverse effects on the brain & cause disease.
- **ICTV (2005)** classified prions under "Subviral agents" along with viral satellites and viroids.

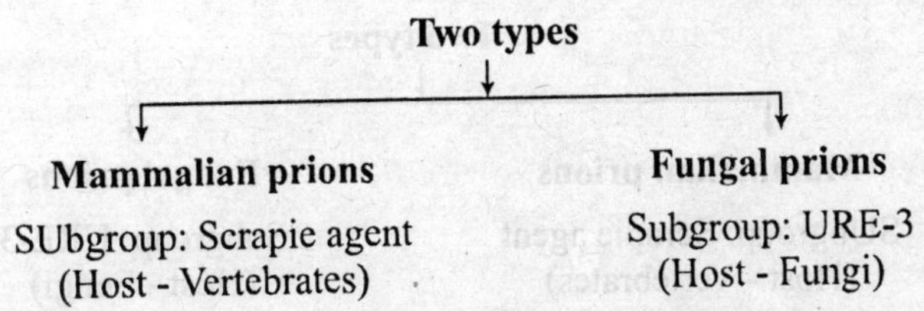

- **ICTV** stand for - **International Committee on Taxonomy of Viruses.**

Examples of diseases caused by prions

S. No.	Name of disease	
	In Human beings	**In Animals**
1	***Kuru***	Scrapie disease of sheep & goats
2	**CJD** (Creutzfeldt Jacob Diseases)	Mad cow disease (*i.e.* Bovine Spongiform Encephalopathy in cattle, BSE)
3	**GSS** (Gerstmann Strausslar Syndrome)	

* BSE of cattle might infect humans.

* Prions show high resistance to ionizing & UV radiations because radiations act mainly on **Nucleic Acids** (DNA/RNA).

* Prions can be destroyed by **Thermotherapy** (heat treatment). Heat causes coagulation/denaturation of nucleic acids & proteins.

(9) Virusoid

Virusoid – Extra small circular RNA in RNA viruses; it is **virus + Viroid** (*i.e.* viroid in virus).

- Discovered by **Randles *et al*** (1981) associated with velvet tobacco mottle virus in Australia. & coined the term "**Virusoid**".
- Virusoids are circular satellite RNAs, encapsidated in the helper virus. Thus, besides the own genome, the helpervirus also contain the ssRNA which are 220-338 nucloeotides long and are incapable of coding for any protein and possess ribozyme activity. Both RNAs, *i.e.* helper viral genome & Virusoid, are essential for infection. The virusoids depend on helper virus for replication.
- Virusoid is an infectious & found encapsidated in helper virus.
- Virusoid depend on helper virus for replication & encapsidation.
- Virusoids are similar to viroids except that the virusoids require helper viruses for the genome replication & encapsidation.
- The first Virusoid was associated with viruses causing velvet tobacco mottle virus disease.

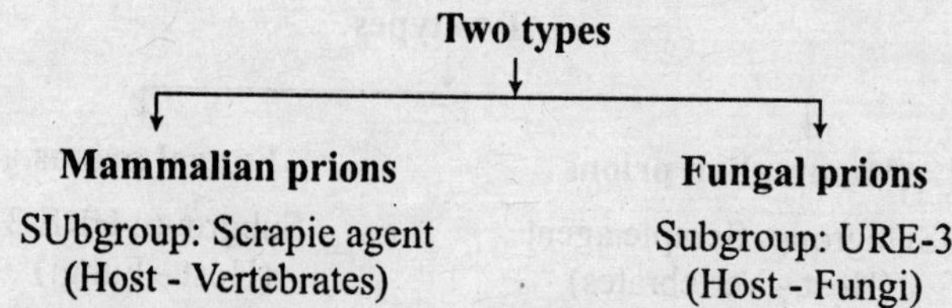

Fig. Virusoid

Similarities and dissimilarities between viroid and virusoid

	Viroid	Virusoid
(A) Similarities:	1. Structurally same. 2. Lack of coding capacity. 3. Replication through rolling circle amplication (RCA) mode.	1. Structurally same. 2. Lack of coding capacity (so, live in helper virus). 3. Replication through rolling circle mode
(B) Difference	1. Capable of causing infection alone. 2. No need of helper virus for replication. 3. Naked particle. naked ssRNA	1. Incapable of causing infection alone. 2. Need of helper virus for replication. 3. Encapsidated in virus particle. virusoid

(10) Phanerogamic PARASITES

Seed producing plants which feed on the other plants

- Also called **PHP (parasitic higher plants).**
- **Phanerogamic**-means seed producing plant & **Parasite** –means live on another living organism.
- Having DNA, RNA, carbohydrates, proteins *etc.,*
- Mistletoe is to be first plant parasite recognized by A. Magnus (1200 AD).

Definition:

"*Phanerogamic parasites are parasitic seed producing plant*".

OR

"The flowering plants that parasitize the cultivated plants are known as phanerogamic parasites".

- They parasitize their host by drawing nutrition and water.
- All such parasites produce seeds.

Phanerogamic parasites are categorized as follows:

(CLOS-HSHS-AMTM)

C- *Cuscuta*	**H-** Holo stem parasite	**A-**Alfalfa, Clover
L- *Loranthus*	**S-** Semi stem parasite	**M-**Mango
O- *Orobanche*	**H-** Holo root parasite	**T-**Tobacco, Tomato
S- *Striga*	**S-** Semi root parasite	**M-**Maize, Bajra

1. Total stem parasite or Holo stem parasite. *e.g.* – *Cuscuta* on clover & *Casssytha* on orange.
2. Partial stem parasite or semi stem parasite. *e.g.*- *Loranthus, Viscum* & (*Arceuthobium* on conifers).
3. Total root partial or holo root parasite. *e.g.* - *Orobanche* and *Rafflesia* (the largest flower).
4. Partial root parasite or semi root parasite. *e.g.*- *Striga and Santalum* (sandal/ *chandan*).

Attacked part	**Nature**	**Parasitic plant (other names)**	**Live On:**
Stem	**Total or holo or complete stem parasite** (because, not have chlorophyll)	***Cuscuta*** (Dodder, *Amarbel*, love vine, Strangle weed, Pull-down, Hell vine, Gold thread, Devil's hair, Hair weed, Ringlet, Devil)	Clover, alfalfa, flax, sugar beet, onion *(No. of seeds/plant up to 16,000)*
	Partial or semi stem parasite (because, it has chlorophyll)	***Loranthus (-Dendrophthae)*** (Banda, Giant mistletoe, Tropical mistletoe, Bird vine)	Mango, pear trees. **(Mistletoes also called "Water parasites" because, they use only water & nutrient from hosts)**
Root	**Total or holo or complete root parasite** (because, not have chlorophyll)	***Orobanche*** (Broomrape)	Mustard, chilli, tomato, tobacco, turnip, brinjal, crucifers
	Partial or semi root parasite (because, it has chlorophyll)	***Striga*** (Witch weed)	Bajra, sorghum, maize, sugarcane, rice *(No. of seeds/plant- 0.5 -5 lac)*

Rafflesia: It is a **total root parasite**. On *Tetrastiga* sp. (large vine of grapes family *i.e.* Vitaceae)

- It is the largest flower (10-11 kg weight with 1 m diameter).

***Santalum* (Sandal wood tree):** It is a **semi root parasite**. Up to one year, seedlings grow independently then make contact with roots of neighboring trees or shrubs and become semi root parasite.

(11) Algae:

- Algology/**Phycology** - The study of algae is called Algology/Phycology.
- The term "**Alga**" is singular and "**Algae**" is plural.
- Single cell has DNA, RNA, carbohydrates, proteins *etc.*,
- Caused **"Red Rust"** disease in plants.
- Algae are aerobic and photoautotrophic (*i.e.* carbon from CO_2 & energy from sun light).
- Dominant on the surface of soil or just below the surface layer (due to photosynthetic nature).
- The major types of algae, present in the soil are **green algae** and **diatoms.**
- "Algae are eukaryotic thallophytes with chlorophyll (b) as their primary photosynthetic pigments".
- Soil algae **increase aeration in submerged soils** by liberating large quantity of **oxygen** through photosynthesis in the soil environment.
- **Agar agar** is obtained from **red sea algae** (*e.g., Gelidium, Gracilaria*).
- Agar agar is good **hydrophilic agent** (water adsorber or water lover).
- **Agar agar is the best Imbibent** that adsorbs water, 99 times of its weight.
- **Imbibition-** The process of adsorption of water molecules on the surface of cells.
- Hydrophilic colloids have adsorption power in following sequence:
- Agar agar > Pectose > Protein > Starch > Cellulose.

Plant diseases caused by algae

S. No.	Disease	Caused by	Remark
1	**Red rust of tea**	*Cephaleuros parasitica* & *C. virescens*	*Cephaleuros* is green algae, living under leaf cuticle. They reproduce by means of zoospores in zoosporangium, which can be disseminated by wind, rain splash.
2	Red rust of mango	*Cephaleuros parasitica* & *C. virescens*	
3	Red rust of guava	*Cephaleuros parasitica* & *C. virescens*	
4	Red rust of citrus	*Cephaleuros parasitica* & *C. virescens*	**Why Red Rust? -** Fluffy, bright-orange **red spots** occur on leaves and stems that look like rust fungi. In fact, *Cephaleuros virescens* has the misleading common name **"Red Rust"**.
5	Red rust of coffee	*Cephaleuros coffee*	
6	Red rust of cacao	*Cephaleuros minimus*	

Other genera of green algae that infect plants are:- (i) *Rhodochytrium* and (ii) *Phyllosiphon* : on weeds & other cultivated minor important plants.

Control:

- Spray with BM and COC.

- Plant spacing & pruning to increase air circulation & light.
- Balanced use of fertilizers & irrigation to promote plant growth.

(12) Protozoa:

- **First** plant disease caused by protozoa is ***"Phloem necrosis of Coffee"***, discovered by **Stahel** in 1931.
- Flagellated protozoa were found to be associated with plant disease.
- **Plant Pathogenic protozoa** are placed in the genus – ***Phytomonas.***
- Protozoa such as the **malaria parasites** (***Plasmodium*** spp.), trypanosomes and leishmania, are also important as parasites and symbionts of multicellular animals.
- Each cell has DNA, RNA, proteins, carbohydrates *etc.,*

Singular	Plural
Protozoan	Protozoa
Protozoon	Protozoans

Definition

"Protozoa are eukaryotic, single celled & heterotrophic microorganism without cell wall, many of which are motile by flagella or by pseudopodia".

Plant diseases caused by Protozoa are:

S. No.	Disease	Caused by	Remark
1	Phloem necrosis of Coffee	***Phytomonas lepto-vaso-rum***	Discovered by **Stahel** in 1931.
2	Empty root of cassava	***Phytomonas*** **sp.**	-
3	Lethal yellowing of coconut (Hartrot of coconut palm)	***Phytomonas*** **sp.**	-
4	Sudden wilt (Marchitez) of oil palm	***Phytomonas*** **sp.**	-

Protozoa:

- Highly varied
- **Unicellular**
- **Heterotrophic**
- Mostly aerobic (there are some examples of facultatively anaerobic *protozoa*)
- Water living
- Typically **cell-wall-less**

- **Eucaryotes**
- It feed upon bacteria as well as particulate (as opposed, usually, to dissolved) nutrients
- Protozoa commonly range from 10 to 52 micrometers, but can grow as large as 1 mm, and are seen easily by microscope.

(13) Mollicutes:

(Molli-Soft; cute- Skin, *i.e.* covered only with cell membrane)

A group of prokaryotic organisms bounded by flexuous membrane and lacking cell wall.

I. Mycoplasma

II. Phytoplasma

III. Spiroplasma

- Animal Mollicute *e.g.,* Mycoplasma
- **Plant Mollicutes e.g. Phytoplasma** and **Spiroplasma**

(14) Mycoplasma:

mycoplsma is prokaryotes without cell wall

- Discovered by **Nocard and Roux** (1898).
- **Not** have well organized nucleus, that's why they are known as **prokaryotic cell** (Without nuclear membrane).
- Earlier called **PPLO** (Pleuro Pneumonia Like Organism).
- Each cell has both **DNA & RNA;** proteins *etc.,*
- DNA is free in the cytoplasm.
- Some mycoplasmas also have **Plasmids.**
- Cells contain 70S ribosome.
- Cells **not** have **Mitochondria.**
- **L.O. Kunke**l is known as **"Father of Plant and Insect Mycoplasmology."**
- Belongs to the class ***Mollicutes*** (**"molli"**- soft; **"cute"**- skin; soft skin means without cell wall).
- ***Mollicutes*** have the smallest known genome. *E.g.,* Bermuda grass Phytoplasma at 530 kb; in contrast, the genome of ***E. coli*** bacteria is about 4600 kb.
- Mycoplasma is also known as **"Jokers"** in the microbiological park.
- Mollicutes cause many diseases in animals, humans (like pneumonia) & >200 in plants.

Definition: "Prokaryotic, cell wall-less, pleomorphic microorganisms that can be cultured".

Disease causes in animals.

- Cells are bounded by cytoplasmic membrane (thus stain as Gram -ve), also known as "Triple layered unit membrane.
- On the culture media, colony gives "Fried egg appearance" (*i.e.* round with a halo).
- All spp. of mycoplasmas require Sterol / Cholesterol for growth.
- Mycoplasms are resistant to **enzyme "Lysozyme"** and **antibiotic "Penicillin"** (because of lack of cell wall).
- **Lysozyme & Penicillin** were discovered by **A. Flemming in 1922 & 1928,** respectively.
- Mycoplasms are **sensitive** to "**Tetracycline** & Chloramphenical antibiotics" which inhibit protein synthesis.
- Simple staining technique for mycoplasm is "Diene's Staining & DAPI (4,6-diamidino-2-phenylindole-2 HCl).

Wall-less prokaryotes

<table>
<tr><td colspan="2">Wall-less prokaryotes (means without cell wall, i.e. Mollicutes)
Mollicutes are surrounded by Triple layered lipid membrane (or Triple layered unit membrane).</td></tr>
<tr><td rowspan="3">Causing disease in animals & humans.
Can grow on media.
Earlier known as PPLO.

e.g., Mycoplasma.

Disease in animals:
Pneumonia</td><td>1. Causing disease in plants.
2. May or may not be grown on media.</td></tr>
<tr><td><table><tr><td>1. Round or elongated in shape.
2. Can not be grown on media.
3. Earlier known as MLO.
e.g., Phytoplasma</td><td>1. Spiral or helical in shape.
2. Can be grown on media.
3. Earlier known as MLO.
e.g., Spiroplasma.</td></tr></table></td></tr>
<tr><td>Ribosome (70S)
DNA
Membrane
Fig. Phytoplasma cell   Fig. Spiroplasma cell</td></tr>
<tr><td></td><td><table><tr><td>Plant diseases caused:
1. LLB
2. Sesamum phyllody
3. Sandal spike
4. Pear decline
5. GSD of sugarcane</td><td>Plant diseases caused:
1. Corn stunt
2. Citrus stub born</td></tr></table></td></tr>
</table>

(15) Phytoplasma:

It is plant pathogenic prokaryotes without cell wall

- **Doi** ***et al.*** (1967) discovered in **phloem tissues** of mulberry dwarf affected plants & named as **MLO.**
- Phytoplasma earlier called **MLO** (Mycoplasma Like Organism) and cells are 150-300 nm in diam.
- Phytoplasma means **phytopathogenic mollicutes,** *i.e.* belongs to **class Mollicutes.**
- Phytoplasma infects only plants & inhabit **phloem sieve** elements.
- Most plant MLOs are transmitted bt psyllids & leaf hoppers.
- Can not yet be grown on artificial nutrient media.
- Controlled by **Tetracycline** antibiotic.
- MLOs were renamed as **"Phytoplasm"** in 1994 at the 10th congress of the International Organization of Mycoplasmology by **Sears & Kirkpatrick.**
- In 2004 the genus name **"Phytoplasma"** was adopted and is currently at ***Candidatus*** status (*Ca*.)
- Father of Plant & Insect Mycoplasmology is **L.O. Kunkel.**
- **Definition** "Mollicutes that infect plants and can not be grown in culture media are called Phytoplasma". **OR** Those mycoplasmas that infect plants are called Phytoplasma.
- **Phytoplasma** are specialized bacteria that are obligate parasites of plant phloem tissue and transmitted by insects.

Diseases caused by Phytoplasma

S. No.	Disease	Caused by	Remark
1	**Mulberry yellows disease**	Phytoplasma	Discovered by **Doi *et al.*** (1967)
2	**Little Leaf of Brinjal (LLB)**	Phytoplasma	Transmitted by Leaf hopper
3	**Sesamum phylody**	Phytoplasma	Transmitted by Leaf hopper
4	Sandal spike	Phytoplasma	Transmitted by Leaf hopper
5	Pear decline	Phytoplasma	-
6	Coconut lethal yellowing	Phytoplasma	-
7	Big bud (stolbur) of tomato	Phytoplasma	-
8	GSD of sugarcane	Phytoplasma	-
9	**Rice yellow dwarf**	Phytoplasma	Transmitted by Plant hopper.
10	**Phyllody** of bottle gourd, potato, cotton, soybean, & cucumber		

- **DAPI test** is more useful in diagnosis of Phytoplasma. DAPI binds to DNA & causes it to fluorescence under UV.
- The G+C content of Phytoplasma is low (*i.e.* 23-29 mol %).
- **Yellows** are important group of diseases caused by Phytoplasma & Spiroplasma.

(16) Spiroplasma:

spiral shaped prokaryotes without cell wall

- **Davis *et al.*** (1972) discovered Spiroplasma in corn stunt disease.
- **Davis & Worley (1973)** named the causal agent as **"Spiroplasma".**
- Spiroplasma infects only plants & found in **phloem sieve tubes.**
- Can be grown in culture media.
- Controlled by **Tetracycline** antibiotic (*i.e.* metabolism attacking antibiotic).

Definition "Spiral shaped, wall less prokaryotes that can be grown in culture media are called Spiroplasma".

- Spiroplasmas are helical Mycoplasma.
- Spiroplasma cells are bounded by single, triple layered membrane.
- Spiroplasma can infect & multiply within the insects.

Diseases caused by Spiroplasma

S. No.	Disease	Caused by	Remark
1	**Corn stunt disease** Corn stunt was first named by Kunkel (1946).	*Spiroplasma kunkelii* (can't be cultured)	Transmitted by Corn Leaf hopper (*Dalbulus elimatus & D. maydis*)
2	**Citrus stubborn &** Horseradish brittle root	*Spiroplasma citri* (can be cultured)	Transmitted by Beet leaf hopper (*Circulifer tenellus*)

Difference between Phytoplasma and Spiroplasma

S. No.	Phytoplasma	Spiroplasma
1	**Round** in shape.	**Spiral** in shape.
2	Can **not** be grown in culture media.	Can be grown in culture media.
3	It is cell wall less prokaryote.	It is also cell wall less prokaryote.
4	Causes **LLB** disease.	Causes **Corn stunt** disease
5	Discovered by Doi *et al.* (1967)	Discovered by Davis *et al.* (1972)

Comparison between mycoplasma, phytoplasma and spiroplasma

S.No.	Mycoplamsa	Phytoplasma	Spiroplasma
1.	Causes disease in animals	Causes disease in plants	Causes disease in plants
2.	Round or elongated in shape	Round or elongated in shape	Helical or spiral in shape
3.	Can grow on media	Can't grow on media	Can grow on media

4.	Formerly known as PPLO	Formerly known as MLO	-
5.	Animal diseases (*e.g.*,Pneumonia)	**Sesame phyllody,** little leaf of brinjal	Corn stunt, citrus stubborn

(17) Nematode:

- Nematodes (*e.g.,* eelworms) are non-segmented **INVERTEBRATE** animals.
- Most plant-parasitic nematodes cause root knots and lesions and can severely retard growth.
- Some nematodes feed on plants with their stylets (spears).
- Nematodes produce eggs and larvae which undergo several molts before becoming plant-pathogenic adults.
- Nematodes are also troublesome because they can act as very efficient vectors of 2 groups of plant viruses.
- **Father of Nematology** is **NA Cobb.**

Diseases caused by nematodes (by animals)

S. No.	Disease	Caused by	Remark
1	Ear cockle of wheat/ Sehun disease	*Anguina tritici*	Transmit *tundu/yellow ear rot* of wheat
2	Root knot of vegetables	*Meloidogyne* sp.	-
3	*Molya* disease of barley	*Heterodera sp.* (Cereal cyst nematode)	First reported by Vasudeva from Neem Ka Thana, Sikar, Rajasthan

Nematodes belong to the **Animal kingdom and phylum** Nematoda. Most of the important parasitic genera belong to the **order Tylenchida** and few under Dorylaimida.

Important plant parasite species are

- *Meloidogyne incognita* (root knot nematodes),
- *Heterodera* and *Globodera* (cyst nematodes),
- *Tylenchulus* semipenetrans (citrus nematode),
- *Pratylenchus* (lesion neamtode),
- *Ditylenchus* (stem and bulb nematode),
- *Anguina* (seed-gall nematode),
- *Aphelenchoides* (foliar nematode) *etc.,*

(18) Soil Microflora:

- Cultivated soils have more population of microorganisms than the fallow soils.
- Cultivation and tillage operations increase microbial activity.

- The earthworms constitute the major portion of the total weight of organisms in soil.

Bacteria:

- Bacteria are the most dominant group of micro organism in the soil & equal to half of the microbial biomass present in soil.
- Their population ranges from 100,000 to several hundred millions per gram of soil.
- Bacterial classification of taxonomic basis is given in "Bergey's Manual of Systematic Bacteriology".
- The most common soil bacteria are *Pseudomonas, Arthrobacter, Clostidium, Bacillus etc.,*

Actinomycetes (also called *Ray fungi*):

- These are regarded as an **intermediate** between bacteria and fungi.
- Actinomycetes are numerous in soil and they are second only to bacteria.
- The percentage of actinomycetes in the total microbial population increases with soil depth.
- The most common genera (with abundance) in soil are Streptomyces (upto 70%), Nocardia, Micromonospora, Actinomyces *etc.,*
- Favorable pH for actinomycetes ranges from 6.5 to 8.0.
- Their poplation ranges from 10000 to 100000000 per gram of soil.

Fungi:

- More in surface layers of well aerated, cultivated soils.
- The most common fungi are heterotrophic and dominant in acid soils.
- Most common fungi in soil are *Aspergillus, Mucor, Penicillium, Trichoderma, Alternaria, Rhizopus, Fusarium etc.,*
- Fungi range from few thousand to a few millions per gram of soil.

Protozoa:

- Population ranges from 10,000 to 100,000 per gram of soil.
- Mostly feed on soil bacteria (predating soil bacteria).
- Protozoa are abundant in the upper soil layer.

Algae:

- Their number ranges from 100 to 10,000 per gram of soil.
- Live on soil surface visible with naked eyes have photosynthetic pigment.
- Algae are **photo-autotrophic**. Green algae are dominating in arid soils.

Population of micro-organism in soil (per gram of soil)

S No.	Microorganism	Number (per gram soil)	Rank (as per population in soil)
(A) Prokaryotes			
1	Bacteria	100,000 to several hundred millions	I
2	Actinomycetes	10,000 to 100,00,0000 (10^4 - 10^8)	II
(B) Eukaryotes			
1	Fungi	Thousand to few millions	III
2	Protozoa	1000 to 100,000	IV
3	Algae	100 to 10,000	V

(Source: Agril. Microbiology by Rangaswami & Bagyaraj, 2010)

World wise crop loss due to major pests (total loss)

S. No.	Loss due to	Per cent loss	Rank
1	Plant diseases	14.1	I
2	Weeds	12.2	II
3	Insect-pests	10.2	III
	Total loss	36.5	-

(Source: Plant Pathology by GN Agrios, 2005)

Deficency symptoms observed of nutrients on old leaves

1. **Nitrogen-**
   - Available at 6-8 pH.
   - Older leaves show **V-shaped chlorosis** or yellowing at tip of older leaves.
   - Either show uniform chlorosis.
   - Essential for protein.
   - Play an important role in auxin synthesis.
2. **Phosphorus-**
   - Pre mature leaf fall.
   - Bronzing on older leaves.

- Later develop red purple discoloration.
- Required for grain maturity.
- Essential constituent for nucleic acid and phytin.
- Most essential function is energy storage and transfer of energy (ATP & ADP).
- Known as key of life.
- **Increase root growth and cell formation.**

3. **Potassium-**
    - Most important function is **stomata regulation.**
    - Increase disease resistance.
    - **Translocation of sugars.**
    - The term **luxury consumption** is related to K.
    - Spot of dead tissues appear at tips.
    - Scorching or burning on margin of bottom leaves.

List of disorders/diseases due to nutrient deficienc

Nutrient	Disorder
Boron (B)	**Black tip**/ Koeli disease/necrosis of mango
	Browning of cauliflower
	Heart rot of sugar beet
	Internal cork of apple
	Brown heart of cabbage
	Hollow stem of brassica
Zinc (Zn)	Khaira disease of rice
	Little leaf of apple
	White bud of corn
	Mottle leaf of citrus
Calcium (Ca)	Blossom end rot of fruits
	Black heart of celery
	Bitter pit of mango
Manganese (Mn)	**Phala blight** of sugarcane
	Grey speck of oat
	Marsh spot of pea
	Top yellowing of gram
Nitrogen (N)	Red leaf of cotton
	Buttoning and hollow stem of califlower
	Pan such and tip burn of rice

Potassium (K)	Tip drying of onion Marginal necrosis of palm spongles of bidi tobacco
Phosphorus (P)	Tip burn and marginal necrosis of mango
Copper (Cu)	Die back and gummosis of citrus Wither tip of apple Reclamation of cereals Exanthema in pear
Iron (Fe)	Ivory white of paddy White eye of rice Leaf blanching in sugarcane Green netting in citrus
Magnesium (Mg)	Sand drown of tobacco
Oxygen (O)	Black heart of potato
Water (H_2O)	Jonathan spot of apple

Disorder due to nutrient toxicity

Nutrient	Disorder
Manganese (Mn)	Crinckle leaf of cotton
Potassium (K)	Leaf scorching of mango

Symptomology and Diseases of crop plants

Rust- A disease giving a rusty appearance to a plant and caused by one of the *Uredinales* fungi (rust fungi).

Name of disease	Causal organism
Rust of bean	*Uromyces phaseolitypia*
Pea rust	*U. fabae*
Gram rust	*U. cicerisarietini*
Rose rust	*Phragmidium tuberculata*
Soyabean rust	*Phakopsora pachyrrhizi*
Linseed/ flax rust	*Melampsora lini*
Bajra rust	*Puccinia penniseti*
Sorghum rust	*P. purpurea*
Groundnut rust	*P. arachidis*
Wheat stem/black rust	*P. graminis tritici*
Wheat brown/orange/leaf rust	*P. recondite*

Wheat stripe/yellow rust	*P. striiformis*
Rust of sunflower	*P. helianthi*
Coffee rust	*Hemiliea vastatrix*

Wilt- Loss of rigidity and dropping of plant parts, generally caused by insufficient water in the plant.

Name of disease	**Causal organism**
Cotton wilt	*Fusarium oxysporum* f. sp. *Vasinfectum*
Pigeon pea wilt/Fusarium wilt of arhar	*F. udum*
Linseed wilt	*F. o.* f. sp. *Lini*
Panama wilt of banana	*F. o.* f. sp. *Cubensis*
Wilt of gram	*F. o.* f.sp.*ciceris*
Wilt of castor	*F. o.* f.sp. *ricini*
Verticillium wilt of cotton	Verticillium dahlia
Sugarcane wilt	*Cephalosporium sacchari*

Downy mildew- A plant in which the sporangiophores and sporangia of a fungus appear as a downy growth on the lower surface of leaves and stems, fruits *etc*.,, cused by fungi in the family *Peronosporaceae.*

Name of disease	**Causal organism**
DM of pea	*Peronospora pisi*
DM of crucifers	*Peornospora parasitica*
DM of cucurbits	*Psuedoperonospora cubensis*
DM of grapes	*Plasmopara viticola*
DM/green ear of bajra	*Sclerospora graminicola*
Brown stripe DM of maize	*Sclerophthora rayssiae* var. *zeae*
DM of sugarcane	*Sclerospora sachhari*
Philippine DM of maize	*Sclerospora philippinensis*
DM/leaf shredding of sorghum	*Sclerospora sorghi / Peronosclerospora sorghi*
DM of isabgol	*Pernopspora plantaginis*
DM of sunflower	*Plasmopara helianthi*

Powdery mildew (PM)- Chlorotic or necrotic areas are covered with superficially growing mycelia and fructifications of the pathogen giving a dusty or powdery appearance.

Name of disease	**Causal organism**
PM of grapes	*Uncinula nector*

PM of rose	*Sphaerotheca pannosa*
PM of cucurbits	*S. fuliginea / Erysiphae cichoracearum*
PM of sunflower	Erysiphe cichoracearum
PM of cereals	*Erysiphe graminis*
PM of leucerne	*E. polygoni*
PM of pea	*Leveillula taurica*
PM of mango	*Oidium mangiferae*
PM of wheat	*Blumeria graminis tritici*
PM of apple	*Podosphaera leucotricha*

Smut and bunt diseases

Smut- A disease caused by smut fungi (*Ustilaginales*) characterized by masses of dark, powdery and sometimes odorous spores.

Bunt- A disease of wheat caused by the fungus *Tilletia* in which contents of the wheat grains are replaced by odorous smut spores.

Name of disease	Causal organism
Head smut of maize	*Sphacelotheca reiliana*
Common smut of maize	*Ustilago maydis*
False smut of rice	*Claviceps oryzae-sativae*
Loose smut of rice	*Ustilago tritici*
Flag smut of wheat	*Urocystis tritici*
Whip smut of sugarcane	Ustilago scitaminea
False smut of paddy	*Ustilaginoidea virens*
Black smut/rice smut/ kernel smut/ rice bunt	*Neovossia horrida*
Leaf smut of rice	*Entyloma oryzae*
Smut of oat	*Ustilago avenae*
Covered smut of barley	*U. hordei*
Head smut of sorghum	*Sphacelotheca reiliana*
Grain/covered/kernel/short smut of sorghu	*S. sorghi*
Loose smut	*S. cruntea*
Long smut	*Tolyposporium eherenbergii*
Karnal bunt of wheat	*Nevossia indica / telletia indica*
Hill bunt / stinking bunt	*Telletia caries / T. foetide / T. controversa*

Anthracnose- A disease that appears as black sunken leaf, stem or fruit lesions and caused by fungal pathogens that produce their asexual spores in an acervulus.

Die back- Death of shoot tissues that begins at the tip and progress backward toward main stem.

Disease	**Causal organism**
Anthracnose of cotton	*Colletotrichum indicum*
Red rot	*C. falcatum*
Anthracnose/ ripe fruit rot/die back of chilli	*C. capsici*
Anthracnose of mango	*C. gloeosporoides*
Anthracnose of guava	*C. psidii*
die back of citrus	*C. gloeosporoides*
Anthracnose of bean	*C. lindemuthianum*
Anthracnose of papaya	*C. papaya*
Anthracnose of banana	*C. musae*
Anthracnose of sorghum	*C. graminicola*
Anthracnose of soyabean	*C. truncatum*

Diseases of crops

Crop	**Disease**	**Causal organism**
Maize	Common smut	*Ustilago maydis*
	Ear or kernel rot and seed rot or seedling blight	*Diplodia maydis, Fusarium* spp., *Gibberella zeae, Aspergillus* spp., *Penicillium* spp.
	Goss wilt or leaf freckles	*Clavibacter michiganensis* subsp. *nebraskensis*
	Head smut	*Sphacelotheca reiliana*
	Holcus spot	*Pseudomonas syringae* pv. *syringae*
	Rust	*Puccinia sorghi*
	Northern leaf blight	*Setosphaeria turcica*
	Brown stripe downy mildew	*Sclerophtora rayssiare* var. *zeae*
	Stalk rot of maize	*Erwinia cartovora* pv. zeae
	Brown spot of maize	*Physoderma zeae maydis*
	Sugarcane downy mildew	*Pernospora sacchari*

Cotton	Fusarium wilt	*F. oxy.* f.sp. *vasinfectum*
	Verticillium wilt	*Verticillium dahliae*
	Root rot	*Rhizoctonia bataticola* or *Macrophomina phaseolina*
	Black root rot	*Thielaviopsis basicola*
	Angular leaf spot	*Xanthomonas campestris pv.malvacearum*
	Dwarfing and yellowing	Deficiency of sulphur
Sugarcane	Downy mildew	*Sclerospora sacchari*
	Ring spot	*Leptoshaeria sacchari*
	Whip smut	*Ustilago scitaminea*
	Red rot	*Colletotrichum falcatum*
	Wilt	*Cephalosporium sacchari*
	Red stripe	*Pseudomonas rubrilineans*
	Ratton stunting disease (RSD)	*Clavibacter xyli* (*Leifsonia xyli*)
	Grassy shoot	*Phytoplasma*
	Pinaple disease	*Ceratocystis paradoxa*
Gram	Ascochyta blight	*Ascochyta rabiei* (Perfect Stage-*Didemyella rabiei*)
	Wilt	*F.oxy.* f.sp.ciceris
	Rust	*Uromyces ciceris-arientini*
Tomato	Blossom end rot	*Calcium deficiency*
	Bacterial spot	*Xanthomonas campestris* pv. *vasicatoria*
	Bacterial speck	*Psuedomonas pv. tomato*
	Bacterial canker	*Clavibacter michiganensis* sub sp. *michiganensis*
	Pith necrosis	*Psuedomonas corrugata*
	Bacterial wilt	*Ralstonia solanacearum*
	Alternaria blight	*Alternaria solani*
	Collar rot	*Sclerotium rolfsii*
	Leaf mould	*Cladosporium fulvum*
	Damping off	*Pythium sp.*
	Fusarium wilt	*Fusarium oxysporum* f.sp. *lycopersicii*
Groundnut	Collar rot/dry root rot	*Aspergillus niger*
	Early leaf spot	*Cercospora arachidicola*
	Late leaf spot	*C. personatum*
	Stem rot	*Sclerotium rolfsii*

Potato	Late blight	*Phytophthora infestans*
	Early blight	*Alternaria solani*
	Wart	*Synchytrium endobioticum*
	Pink rot	*Phytophthora erythroseptica*
	Stem canker/black scurf	*Rhizoctonia solani*
	Black leg/soft rot	*Erwinia cartovora* pv. *cartovora*
	Brown rot/ bacterial wilt	*Psuedomonas solanacearum*
	Common scab	*Streptomyces sacbies*
	Purple top rot	*Phytoplasma*
	Witches broom	*Phytoplasma*
	Leaf roll	*Potato leaf roll virus*
	Potato spindle tuber	*Potato spindle tuber viroid*
	Black heart	O_2 deficiency

Standard endings of TAXA:

- Phylum ends with ***mycota***
- Sub- phylum ends with **mycotina**
- Class with *mycetes*
- Sub- class with *mycetidae*
- Order with *-ales*
- Family with ***-aceae***
- No special ending for genus and species.

Plant disease epidemiology

Major plant disease epidemics

Year	Disease	**Causal organism**	**Locality**
1845	Late blight of potato	*Phytophthora infestans*	Ireland
1870	Coffee rust	*Hemiliea vastatrix*	Sri Lanka
1878	Downy mildew of garpe	*Plasmopara viticola*	France
1916-17	Wheat rust	*Puccinia sp.*	USA
1921	Bunchy top of banana	BBTV	Australia
1936	Red rot of sugarcane	*Colletotrichum falcatum*	U.P., India
1943	Brown leaf spot of rice	*Helminthosporium oryzae*	W.B., India
1951	BLB of rice	*Xanthomonas campestris pv. oryzae*	Khapoli area of Maharastra

1956	Ergot of pearl millet	*Claviceps fusiformis*	MH, India
1984	Apple scab	*Venturia inequalis*	J & K, India
1999	Rice tungro	RTBV & RTBV	Punjab, India

Computer stimulation programmes for disease forecasting

- First computer stimulation programme called **EPIDEM**, for **early blight of potato and tomato in 1969.**

Programme	Disease
MYCOS	Blight of chrysanthamum
EPICORN	Southern blight of corn
EPIVEN	Apple sacb
EPIDEMIC	Stripe rust of wheat
SIMCAST	Late blight of potato
BLIGHTCAST	Late blight of potato
TOMCAST	Early blight of tomato
CUFAST	Early blight of tomato
FAST	Early blight of tomato
PLASMO	Downy mildew of grapes
PLAM	Tikka disease of ground disease
EPIVET	Potato diseases of viral
TOMEX	Diseases of tomato

Expert system used in plant pathology

Expert system	Detail
PLANT/ds	First expert system (1983) used for 17 soyabean diseases in USA
POMME	Apple diseases. First expert system incorporate decision making process
PASCO	Apple diseases
POTATOES	Late blight of potato
EPIPRE	Wheat rust

Difference between monocyclic and polycyclic diseases

S.No.	Monocyclic disease	Polycyclic disaese
1.	Only one cycle occurring per season	Completes many cycles in a year
1.	Saturation type disease curve	Sigmoid type curve
2.	It also known as simple interest disease	Also known as compound interest disease
3.	Example- fusarium wit and smut	Example- powdery mildew and wheat rust

Integrated disease management

- Management suggests need for continuous adjustment in the cropping system, while control suggests final solution of disease.
- Exclusion means- keeping out.
- Legal restriction of a disease or pathogen in an area is known as- Quarantine.
- IPCC stands for- International plant protection convention (1951).
- First quarantine act was enforced in France, 1660.
- In USA, Quarantine act was passed in 1912.
- In India, **Destructive insect pest act (DIPA)** passed in 1914.
- DPPQS stands for- Directorate of plant protection, quarantine and storage, it situated at Faridabad (Harayana).
- DPPQS handles bulky material quarantine clearance.
- Small material(Research material) handled by NBPGR, New Delhi.
- National Plant Quarantine Station (**NPQS**) - Rangpuri (New Delhi).
- National bureau of plant genetic resources (**NBPGR**), New Delhi. It handles germplasms and transgenic planting material.

Disease	Causal organism	Native Place	Year of introduction
Coffee rust	*Hemileia vastatrix*	Srilanka (Ceylon)	1879
Late blight of potato	*Phytophthora infestans*	England	1883
Flag smut of wheat	*Urocystis tritici*	Australia	1906
Downey mildew of grapes	*Plasmopara viticola*	Europe	1910
Downey mildew of maize	*Sclerospora philippinensis*	Java	1912
Rice blast	*Pyricularia oryzae*	1918	South East Asia
Black rot of crucifers	*Xanthomonas campestris* pv. *Campestris*	Java	1929
Powdery mildew of rubber	*Oidium heveae*	Malaya	1938
Fire blight of pear		England	1940
Crown gall of Apple/pear	*Agrobacterium tumefaciens*	England	1940
Bunchy top of banana	*BBTV*	Srilanka	1940

Wart of potato	*Synchytrium endobioticum*	Netherlands	1953
Bacterial blight of paddy	*Xanthomonas oryzae pv. oryzae*	Philippine	1959
Golden Nematode of potato	*Globodera rostochinensis*	Europe	1961

Types of quarantine measures-

1. **Domestic quarantine-** Restricts the movement of a disease from one place to another with in a country example as given below–

Disease	Exist in
Wart disease of potato	Darjeeling (West Bengal)
Mosaic of Banana	Maharashtra and Gujarat
Bunchy top of banana	Kerla, Assam, West Bengal and Orrisa
Golden nematode of potato	Nilgiri hills (Tamilnadu)
Apple scab	J&K, Himachal Pradesh and Uttrakhand

2. **International Quarantine-** Restricts the movement of a disease from one country to another country.
3. **Total embargoes-** Total restriction on import and export of agricultural commodities for Fiji disease of sugarcane, *e.g.,* embargoes on import of sugarcane cutting from Australia, Fiji and Phillipines. Total ban on import of cocoa from Africa and Sri Lanka for Swollen shoot of cocoa.
   - Crop rotation reduces the disease by breaking the life cycle of pathogen.
   - Crop rotation means repeated sowing of the same crop in the cycle at periodical intervals.
   - Crop rotation useful in managing- Fusarium wilt disease, Molya disease of wheat and barley and red rot of sugarcane. In short we can say crop rotation is useful in management of soil borne plant diseases.
   - Sanitation means keep clean the field from diseased plant parts or plant debris.
   - Sanitation useful in managing- Apple scab, Red rot of sugarcane, Bacterial wilt of cotton, Powdery mildew of pea and cereals and early & late blight of tomato & potato.
   - Rouging means physical removal of unwanted or diseased plants from the field.
   - Rouging useful in managing- Smut of wheat & barley, ergot of pearl millet and viral diseases.
   - Barbery is an alternate host of *Puccinia graminis tritici.*

- Leersia and sugarcane are collateral host of *Pyricularia oryzae.*
- Root rot of cotton and be controlled by mixed cropping with moongbeen & mothbean.
- Downy mildew diseases are favoured by wet & cool weather.
- Powdery mildew diseases are favoured by dry & warm weather.
- Time of sowing check positive matching between host and pathogen.
- Early sowing of pea useful in managing powdery mildew disease.
- Early sowing of mustard managing white rust disease.
- Late sowing of gram managing root rot disease.
- Damping off disease and stem rot mustard severe due to high plant population and more irrigation.
- Diseases caused by *Xanthomonas, Phytophtora, Pythium* & *Sclerotinia* favoured by frequent irrigation.
- Deep summer ploughing is beneficial in controlling soil borne plant pathogens by causing coagulation of protein & nucleic acid in microorganisms.
- **Panama wilt of banana** can be reduced by **floodin** of field.
- Protective nets of more than 40 mesh are used in cage houses to prevent entry of insects.
- Hot Water Treatment given by Jenson (1889) for control of late blight of potato.
- **Swingle (1892)** gave hot water treatment (HWT) for loose smut of wheat.
- Soil solarisation technology first time develoed in Isarael. First descriped by Katan *et al.* in 1976.
- Katan also known as father of soil solarisation.
- In soil solarisation clear & transparent (not black or coloured) polythene sheets should be used.
- Solar Heat Treatment (SHT) was given by **J. C. Luthra & Sattar (1934)** for control of loose smut of wheat.
- Post-harvest diseases are controlled by refrigeration and radiation.
- Father of biological control is **S.D. Garrett.**
- **Weindling (1932)** first time suggested use of *Trichoderma* as a biocontrol agent.
- **New and Kerr-** First time control crown gall disease by *Agrobacterium radiobacter* strain K-84.
- Mycofumigation was described by **Koitavashy (2005).**
- **Suppressive soil-** Soils that suppress certain diseases due to presence in soil of microorganism antagonistic to the pathogen.
- **Condusive soil-** Soils that favour certain soil borne diseases.
- BCAs act as reduce diseases by three mechsnisms- Antagonism (Antibiosis, competition, exploitation), Hypovirulence and Induced host resistance.

Some of antibiotics produced by BCAs

S.no.	Antibiotic	Source
1.	2, 4-diacetylphloroglucinol	*Pseudomonas fluorescens* F113
2.	Agrocin 84	*Agrobacterium radiobacter*
3.	Bacillomycin D	*Bacillus subtilis* AU195
4.	Bacillomycin and fengycin	*Bacillus amyloliquefaciens* FZB42
5.	Gliotoxin	*Trichoderma virens*

- Hypovirulence is the phenomenon of reduced virulence of a pathogen strain than normal ones developed as a result of its infection by double standard RNA (dsRNA).
- **Kloepper** *et al.* (1980) - first demonstrated importance of siderophore in biocontrol.
- The function of siderophores is to supply iron to the cells of produce organism.
- ***Psuedomonas putida*** is also known as **oil scavenger and super bug**
- Parasites have some etiological relationship with its host and exploited host slowely. While Predators not have etiological relationships with its host by direct feeding.
- *Glicocladium, Coniothyrium, Trichoderma* & *Darluca* are common examples of mycoparasites.
- PGPR stands for- Plant Growth Promoting Rhizobacteria.
- Commercially available BCAs, their trade name (s) and target pathogen

S.No.	BCA	Trade name	Target pathogen
1.	*Agrobacterium radiobacter* strain K 84	Galltrol	*Agrobacterium tumeifaciens*
2.	*Agrobacterium radiobacter* strain K 1026	Nogall	*Agrobacterium tumeifaciens*
3.	*Bacillus subtilis*	Kodiak	*Rhizoctonia* and *Fusarium*
4.	*Bacillus subtilis*	Quantum 4000	*Alternaria* and *Fusarium*
5.	*Burkholderia cepaci*	Deny	*Pythium* and *Bacillus subtilis*
6.	*Psuedomonas fluoroscens*	Blight Ban	Frost damge and *Erwinia amylovora*
7.	*Psuedomonas fluoroscens*	Doggar G, Bioshield	*Rhizoctonia* and *Pythium*
8.	*Streptomyces griseoviridis*	Mycortop	*Pythium*
9.	*Peniophora gigantean*	PG Suspension	*Heterobasidion*

10.	*Gliocladium virens*	Glio gard	*Pythium* and *Rhizoctonia*
11.	*Ampelomyces quisqalis* m 10	AQ10 Biofungicide	Powdery mildew fungi
12.	*Trichoderma viride*	Antagon TV, Biocon, Biogaurd	Soil borne fungi
13.	*Trichoderma harzianum*	F- stolp, Binap- T, root shield and trichodex	Soil borne fungi
14.	*Fusarium oxysporum* (Non-pathogenic)	Fusaclean	*Fusarium oxysporum*

- Thiram, mancozeb, zineb *etc.,* are organic sulphur fungicides. These are thiocarbamates. Thiocarbamates are derivatives of dithiocarbamic acid.
- Inorganic suphur marketed as **kolodust or kolofog**.
- **Quinone fungicides** are naturally occurs in plants.
- Club root of cabbage can be reduced by application of lime.
- Early sowing is beneficial for stem rust of wheat.
- Thiram was first discovered by **Tisdale (1934)**.
- Carboxin (Vitavax) especially used for smut diseases.
- Oxycarboxin (Plantvax) especially used for rust diseases.
- Sulphur, dinocap (Karathane) mainly used for control of powdery mildew diseases.
- Metalaxyl (Apron) mainly used for control of oomycetes fungi.
- Powdery mildew diseases can be controlled by sprinkler irrigation.
- Chestnut compound only used for soil drenching never used for foliar spray.
- Burgundy mixture is a modified form of Bordeaux mixture.
- Composition of Bordeaux mixture - lime + $CuSo_4$ + water (8:8:10).
- Chemicals used for manage the blast disease known as blasticides. *e.g.,* Blasticidin antibiotic.
- Mostly antibiotic are produced by *Streptomyces.*
- **Carbendazim** is **xylem mobile** fungicide.
- Fosetyl AL is ambi-mobile (xylem + phloem) fungicide.
- Tricyclazole (beam) is an antipenetrate fungicide and effective against rice blast.
- Benzimidazole inhibits cell division of fungus.

Q.1 Write 'T' for true and 'F' for false.

1. Late blight of tomato is caused by *Alternaria solani.*
   False *(By Phytophthora infestans)*
2. Phytoplasmal diseases of plants can be controlled by **tetracycline**.
   True (act on cell membrane)

3. Black rot of cabbage is caused by fungus. **False** (by bacterium - *Xanthomonas*)
4. *Xanthomonas* bacterium is responsible for blight of pomegranate. **True**
5. Powdery mildew diseases can be controlled by sulphur fungicides. **True**
6. **Black heart of potato** is an infectious disease. **False** (non-infectious)
7. Purple bacterium is responsible for causing **purple blotch of onion.** **False** (*Alternaria porri* fungus)
8. **Copper fungicides** can control diseases caused by both fungi and bacteria. **True**
9. **Coorg, Satsuma, pineapple & blood red** cultivars are resistant to citrus canker. **True**
10. *"Irish famine"* was due to late blight of tomato. **False** (LB of potato)
11. Late blight of potato was introduced in India from **Europe** in 1883. **True**
12. Black heart of potato is caused by **low O_2 and high temperature.** **True**
13. Panama Disease is also called wilt disease. **True**
14. Excessive application of nitrogen to crop plants makes them susceptible to diseases. **True**
15. Panama disease is associated with Banana. **True**
16. Bacterial diseases of plants can be controlled by streptocycline. **True**
17. Black tip of mango is severe in orchards far from brick kilns. **False** (near to brick kilns)
18. Cu deficiency is also associated with citrus die-back. **True**
19. Citrus canker bacterium is survived on twigs. **True**
20. "Irish famine" was due to late blight of potato. **True**
21. Black champa & Regina cultivars of grape are resistant to downy mildew disease. **True**
22. Powdery mildew of pea can be managed by adjusting date of sowing. **True**
23. Wilt disease causing pathogens are mainly soil-borne. **True**
24. The study of the cause of a disease is called Etiology. **True**
25. Carbendazim (Bavistin) is a systemic fungicide. **True**
26. In imperfect fungi, sexual reproduction is fully known. **False** (sexual reproduction is not known)
27. Fungicide **"Thiram"** is mainly used as seed treatment. **True**
28. The pathogen of **Phomopsis blight of brinjal** is a seed - & soil – borne. **True**
29. ***"Chimney disease"*** is associated with Mango. **True**
30. In India, **MMD** was first reported by Maries in 1891 from Darbhanga, Bihar. (**MMD** - Mango Malformation disease) **True**
31. **MMD** can be managed by De-Blossoming, spray of NAA (200 ppm) & Anti-malformins. **True**
32. **Viroids and virions** are same. **False**
33. **Phytoplasmal** cells have a well defined cell wall covering on it. **False**

Q.3 Fill in the blanks.

1. Powdery mildew of pea is caused by ....fungus. ***Erysiphe polygoni***
2. *Uncinula necator* is the pathogen of ....disease of grapes. **Powdery mildew**
3. Fungus causing black leaf spot of banana is ...***Mycosphaerella/Pseudocercospora***
4. Leaf curl of tomato is transmitted by.........................insect vector. ***Bemisia tabaci* (White fly**
5. In papaya, *Pythium* spp. is responsible for causing.........................**Foot rot**
6. Fungitoxic compound of Bordeaux mixture is.........................**Cu**
7. In India, Zn deficiency in guava is reported from........in Rajasthan. **Pushkar valley**
8. Powdery mildew of ber is caused by..................***Oidium erysiphoides f.sp. ziziphi***
9. *Pythium and Phytophthora belong* to..........................class of fungi. **Oomycetes**
10. Panama epidemic was due to.........................**Panama wilt of banana**
11. Bordeaux mixture is a.......................fungicide. **Copper Based**
12. Father of Plant Pathology is ....**Anton de Bary**
13. Insect vector of Leaf curl disease of tomato is .... ***Bemisia tabaci***
14. Citrus die back disease is caused by.............. ***Colletotrichum gloeosporioides***
15. *Hishimonas phycitis* is a vector of.........................of brinjal. **LLB**
16. First plant bacterial disease was discovered by....................**T J Burril**
17. Powdery mildews are favoured by...........................**Dry & warm weather**
18. Powdery mildew of rose is incited by........................... ***Sphaerotheca pannosa* f.sp. *rosae***
19. IDM stands for.....................**Integrated disease management**
20. Purple blotch of onion is incited by which fungus.............. ***Alternaria porri***
21. *Alternaria solani* belongs to.......................class of fungi. **Mitosporic fungi Deuteromycetes**
22. Little leaf of brinjal (LLB) is caused by.................**Phytoplasma**
23. "Father of Plant Pathology in India" is ....**E J Butler**
24. A fungal bio-control agent most commonly used in the fields is......***Trichoderma***
25. The insect transmitter of ring spot of papaya is......***Myzus persicae & Aphis gossypii* (Aphids)**
26. Downy mildew of grapes is caused by.............. ***Plasmopara viticola***
27. Fungus causing *Phomopsis* blight of brinjal is .... ***Phomopsis vexans***
28. The common name of Dithane M-45 (Indofil M-45) is.............**Mancozeb**
29. MOLs now known as................. ***Phytoplasma***

30. A well known text book on ***"Plant Diseases"*** is written by..........**RS Singh**
31. ***Hishimonaus phycitis*** is leaf hopper that transmit.......**Little leaf of brinjal**
33. Antibiotic used to control citrus canker is.................**Streptomycin**
34. Sexual spore of Oomycetes_is .... **Oospore**
35. The spray of NAA @ 200 ppm is effective against ....disease of mango. **Malforantion**
36. Black tip of mango is a ....deficiency disorder. **Boron**
37. Karathane (dinocap) fungicide is effective against...........disease. **Powdery mildews.**
38. Indofil M-45 (Dithane M-45) is..........fungicide. **Organic sulphur**
39. Downy mildew of cucurbits was introduced in India from........**Srilanka (1940).**
40. Cells of Phytoplasma (MLOs) are sensitive to..........**Tetracycline antibiotic (it acts on cell membrane).**
41. Yellow vein mosaic of bhendi is transmitted by......***Bamisia tabaci*** **(white fly)**
42. Guava wilt first reported from.......in the year......**UP, 1935.**

PLANT PATHOLOGY OBJECTIVE QUESTIONS (MCQ)

1. Conidia can be produced by

   A. All fungi and some bacteria B. Some fungi and some bacteria

   C. Algae, fungi and phytoplasma D. Cyanobacteria and lichen

2. How many species of bacteria are known?

   A. 400 B. 800 C. 1200 D. 1600

3. How many species of bacteria are known to cause diseases in plants?

   A. 40 B. 60 C. 80 D. 100

4. Most of the plant pathogenic bacteria are:

   A. Obligate parasites B. Obligate saprophytes

   C. Facultative parasites D. Facultative saprophytes

5. Which one is correct about the shape of bacterial cells?

   A. Rod shape, spherical, spiral, filamentous, bacilli

   B. Rod shape, spherical, spiral, octagonal, bacilli

   C. Rod shape, spherical, spiral, hexagonal, bacilli

   D. Rod shape, spherical, spiral, rectangular, bacilli

6. How many copies of plasmid is/are present in a single bacterial cell?

   A. Single B. Multiple

   C. Single or multiple D. None of the above

7. Which one is CORRECT about *Streptomyces*?

   A. Produces conidia, branched hyphae without cross walls, Gram positive

B. Not produces conidia, branched hyphae with cross walls, Gram positive
C. Produces conidia, branched hyphae with cross walls, Gram negative
D. Not produces conidia, branched hyphae without cross walls, Gram negative

8. In some basidiomycetes, primary mycelium is Monokaryotic, secondary mycelium is dikaryotic then tertiary mycelium is
A. Monokaryotic B. Dikaryotic C. Diploid D. Homokaryotic

9. Presence of clamp connection indicates the.............nature of the that hyphae
A. Monokaryotic B. Dikaryotic C. Diploid D. Homokaryotic

10. The word "Toadstool" is a distortion of the German word "Toadstul", which means
A. Toad on stool B. Chair C. Death chair D. None

11. A typical basidium is..............., which is formed on dikaryotic hyphae
A. Unicellular B. Bicellular C. Multicellular D. Acellular

12. Which one of the following is an example of a "non-self-limiting tumor"?
A. Wart disease of potato B. Crown gall of apple
C. Root knot of vegetables D. None of the above

13. RNA and DNA are
A. Nucleosides B. Polynucleotides C. Nucleoprotein D. Nucleosomes

14. Ri plasmids can be found in
A. Rhizopus B. Rhizomorph C. Agrobacterium D. Rhizobium

15. "*Pi-ta gene*" is associated with
A. Wheat B. Rice C. Potato D. Pineapple

16. "*Earthy odour*" at first rains is due
A. Panspermia B. *Serratia* C. *Streptococcus* D. *Streptomyces*

17. *Serratia marsescens* causes
A. Citrus HLB B. Yellow vine disease of cucurbits
C. Bacterial wilt of corn D. RSD of sugarcane

18. The G+C content of DNA in *Xanthomonas* is
A. 51-53 % B. 63-71 % C. 58- 70 % D. 59.41 %

19. In bacteria, secondary metabolites are formed at/during
A. Begining of Lag phase
B. Initiatin of Log phase
C. The end of the growth phase or stationary phase
D. The end of decline phase

20. Pierce's disease of grapevine is incited by
A. Phytoplasma B. Phloem limited bacteria
C. Xylem limited bacteria D. MLO

21. **"DAPI stain"** and **Diene's stain** developed and primarily used to stain

A. Cyanobacteria B. Phytoplasma C. Algae D. Virus

22. The pathogen of bacterial wilt of cucurbits mostly survives in the intestine of

A. Aphids B. Psyllids C. Leaf hoppers D. Beetles

23. Read the following statements regarding **"Antibiotics"**

1. Produced by microbes
2. One of the most significant discoveries of nineteenth century
3. "Anti" is a Greek word that means 'Against' & "Bio" means "life"
4. "Against life" with reference to human beings
5. Effectively kill or retard the growth of viruses
6. Discovered by A. Fleming while working on *Staphylococcus* bacteria
7. Chemotherapeutic use of penicillin was demonstrated by H. Florey & E. Chain

The correct statements are:-

A. 1, 2, 3, 6, 7 B. 3, 6, 7 C. 1, 2, 4, 5, 7 D. 1, 2, 3, 6, 7

24. Which are the sources of autonomous dispersal of plant pathogenic bacteria?

A. Seeds & soil B. Soil & Insects C. Air & seeds ' D. Water & wind

25. How many number of plasmid(s) may be found in a bacterial cell"?

A. Always one B. One to five only
C. One to several D. Always more than one

26. "Citrus *Huanglongbing*" is caused by

A. Nutritional deficiency B. Bacteria
C. Nutritional toxicity D. Phytoplasma

27. "L – form bacteria" can be generated in laboratory by treating the cells with

A. Lysozyme enzyme B. Ribozymes
C. Electric shock D. Liquid nitrogen

28. "Coconut (root) wilt" is caused by

A. Nutritional deficiency B. Bacteria
C. Nutritional toxicity D. Phytoplasma

29. Freeze-drying method is also known as

A. Tyndalization B. Lyophilization C. Coliphage D. NOTA

30. *Bdellovibrios* live in

A. Nematode B. Protozoa
C. Periplasmic space of bacteria D. Fungus

31. Which one is **incorrect** about symplastic translocation of fungicides?

A. Upward movement B. Down ward movement
C. Basipetal movement D. All the above

32. Quinoxyfen is a
    A. Nematicide B. Bactericide C. Viricide D. Fungicide
33. Salicylic acid, jasmonic acid, BABA, AABA, GABA is
    A. Red chemicals B. Green chemicals
    C. Viricide D. Fungicide
34. Which one statement is incorrect about bacterial leaf blight of rice?
    A. Kresek phase is appeared during flowering
    B. Kresek phase is appeared in nursery
    C. Blight phase is appeared in field
    D. Most damaging phase is kresek
35. The correct sequence of Gram staining chemicals during staining process is the application of
    A. Crystal violet, iodine, alcohol, safranin
    B. Iodine, crystal violet, safranin, alcohol
    C. Safranin, iodine, alcohol, crystal violet
    D. Alcohol, iodine, crystal violet, safranin
36. CGA 245704 is a?
    A. Plant defence activator B. Antibiotic
    C. Elicitor D. Fungicide
37. A book silent spring written by Rachel Carson's published in the year?
    A. 1952 B. 1962 C. 1972 D. 1982
38. First antibiotic used against MLO's ?
    A. Tetracycline B. Streptocycline C. Penicilline D. Aureofungin
39. Black tip of mango occurs only in:
    A. India B. Pakistan C. China D. Japan
40. In fruits, downy mildew diseases are controlled by
    A. Vitavax B. Dinocap
    C. Metalaxyl (Ridomil) D. Streptomycin
41. Which one microorganism is used as bio-control agent against plant diseases?
    A. *Pseudomonas solanacearum* B. *Ralstonia*
    C. *Trichoderma viride* D. *Puccinia graminis*
42. How much quantity of fungicide has to be dissolved in one litre of water for preparing spray solution of 1 % conc.?
    A. 1 mg B. 10 gm C. 100 ppm D. 100 gm
43. Usually wilt causing *Fusarium* spp. perennates or survive in soil as
    A. Conidia B. Oospores C. Cleistothecia D.Chlamydospores
44. In North Indian plains, fungi causing late blight of potato survives through
    A. Infected tubers lying in soil B. Infected tubers stored in cold storage
    C. On alternate hosts D. All the above

45. Apple scab is caused by
A. *Plasmopara viticola* B. *Venturia inaeuqalis*
C. *Spiroplasma sp.* D. *Taphrina deformans*

46. Downy mildew of cucurbits is caused by
A. *Pseudoperonospora* B. *Uncinula*
C. *Albugo* D. *Erysiphe*

47. In the standard formulation of 6:6:50 of Bordeaux mixture, the 6 is measured by the unit of
A. litre B. Pound C. PPM D. Gallon

48. Which one, serves as infective propagules in powdery mildew diseases
A. Zoospore B. Conidium C. Sporangium D. Oospore

49. Fungus producing micro- & macro-conidia and chlamydospores is
A. *Alternaria* B. *Fusarium* C. *Puccinia* D. *Pythium*

50. Which one is odd out?
A. Zoospores B. Ascospores C. Zygospores D. Oospores

51. Which is a suitable chemical to manage a viral disease in the fields?
A. Fungicide B. Insecticide (for vector control)
C. Bactericide D. None of these

52. *Fusarium oxysporum* var. *subglutinans* is associated as a major factor to cause a
A. Mango malformation B. Citrus canker
C. Wilt of guava D. Foot rot of papaya

53. Which of these fungicides is not effective in controlling powdery mildews?
A. Calixin B. Dinocap C. Sulphur dust D. Bordeaux paste

54. Which one of the following organisms is not used as bio-control agent against plant diseases?
A. *Pseudomonas solanacearum* B. *Pseudomonas fluorescens*
C. *Trichoderma viride* D. *Trichoderma harzianum*

55. Leaf curl of tomato is transmitted by
A. *Bemisia tabaci* B. *Hishimonas phycitis*
C. Aphid D. Nematodes

56. Little leaf of brinjal is caused by
A. Fungus B. Algae
C. Virus D. Phytoplasma (MLO)

57. Rhizome rot of ginger is caused by
A. *Uncinula necator* B. *Sclerospora* sp
C. *Pythium aphanidermatum* D. Nutrient deficiency

58. Leaf spot of banana is also known by the popular name of
A. Moko disease B. Panama disease
C. Wilt disease D. Sigatoka disease

59. Among the following, the cheapest method of plant disease control is
 A. Seed treatment B. Soil treatment C. Foliar spray D. Drenching
60. How much quantity of a product to be added to get 1 ppm solution is?
 A. 1 g/litre B. 10 mg/litre C. 100 mg/litre D. 1 mg/litre
61. The discovery of which one fungicide is associated with *Plasmapara viticola*
 A. Carboxin B. Carbendazim
 C. Bordeaux mixture D. Mancozeb
62. The statement: "*It happens only in India*", can be most fitted for which disease
 A. Leaf curl of papaya B. Black tip of mango
 C. Damping off of seedlings D. Bunchy top of banana
63. Downy mildew diseases can be controlled by
 A. Tetracycline & COC B. Dinocap & Thiram
 C. Mancozeb & metalaxyl D. Streptomycin
64. In India, domestic quarantine act applies to
 A. Mango malformation B. citrus canker
 C. Foot rot of papaya D. Bunchy top of banana
65. How much quantity of fungicide has to be dissolved in one litre of water for preparing spray solution of 0.25 % concentration?
 A. 2.5 mg B. 2.5 gm C. 2.5 ppm D. 25 gm
66. "Anthracnose of grape" is caused by
 A. *Colletotrichum sp* B. *Elsinoe ampelina*
 C. *Xanthomonas sp* D. *Plasmopara sp*
67. Guava wilt is caused by
 A. *Fusarium oxysporum* f. sp. *psidii* B. *Puccinia* sp.
 C. Phytoplasma D. *Albugo candida*
68. The pathogen causing "black rot of cabbage" is a
 A. Internally seed-borne B. Soil - borne
 C. Both 1 & 2 D. None of these
69. "*Irish famine*" is related to
 A. India B. France C. Srilanka D. USA
70. In the standard formulation of 5: 5: 50 of Bordeaux mixture, the 50 is measured by the unit of
 A. Litre B. Millilitre C. PPM D. Gallon
71. "Sigatoka" is a
 A. Fungal disease of Banana
 B. Leaf spot of Banana
 C. A town in Sigatoka Valley of Fiji Island
 D. All the above

72. Diseases caused by obligate pathogens are
A. Rusts B. Downey mildews
C. Powdery mildews D. All the above

73. Powdery mildew diseases are dangerous (favored by) in
A. Wet & cool weather B. Dry & warm weather
C. During hot summers D. None of these

74. Ridomil (metalaxyl +mancozeb) is an effective fungicide against
A. Rusts B. Downey mildews
C. Powdery mildews D. All the above

75. Black tip of mango was first time reported from
A. Bihar B. Punjab C. Tamilnadu D. MP

76. Food absorbing organ of many fungi is
A. Haustorium B. Appressorium C. Oospore D. Cleistothecia

77. Bordeaux mixture is a
A. Sulphur fungicide B. Copper fungicide
C. Mercuric fungicide D. None

78. The pathogen of "Anthracnose or die-back & fruit rot of chilli" is a
A. Seed-borne B. Soil - borne C. Both 1 & 2 D. None of these

79. "Hen & Chicken" disorder is found in
A. Pear B. Carrot
C. Grapes (B deficiency) D. None of these

80. '*Diplodia* stem end rot' is associated with
A. Plum B. Date palm C. Mango D. Bael

81. Wilt or Pananma disease resistant cultivar (s) of banana is
A. Poovan B. Cavendish C. Rajabale D. All the above

82. The typical *"target board effect" or bull-eye*" appearance is noticed in
A. Downy mildew of cucurbits B. Early Blight of tomato
C. Foot rot of papaya D. Late blight of potato

83. *"Honey Comb Appearance"* is seen in
A. Foot rot of papaya B. Leaf curl of papaya
C. Ring spot of papaya D. Downy mildew of cucurbits

84. This disease is also known as ***"Bird's Eye Spot"*** disease of grapes
A. Anthracnose B. PM C. DM D. None

85. Guava Wilt is more severe during
A. Summer B. Winter C. Zayad D. Rainy season

86. The most widely used biocontrol fungus Trichoderma releases the enzyme
A. Chitinases only B. Glucanases only
C. Cellulases only D. All the three enzymes

87. Total chromosome number in bacteria is
A. 4 B. 3 C. 2 D. 1
88. The red pigment producing bacteria is
A. Serratia B. Staphylococcus C. Micrococcus D. Monococcus
89. Detection and identification of a physiological race is done by
A. Host range B. Nested multiplex PCR
C. Selective culture medium D. ELISA
90. The 'Father of Seed Pathology' is
A. Trirumalachar B. S.B. Mathur C. Paul Neergard D. Borduex
91. Solidifying agent agar-agar is produced from
A. Blue green algae B. Green algae
C. Red algae D. Brown algae
92. Black leg disease is related to a vegetable crop of
A. Cruciferae B. Solanaceae C. Cucurbitaceae D. Leguminosae
93. The genome of TMV codes for coat protein subunits consisting of
A. 158 amino acids B. 160 amino acids
C. 162 amino acids D. 164 amino acids
94. Glume blotch is very commonly found in
A. Rice B. Wheat C. Maize D. Barley
95. The antibiotic produced from bacteria is
A. Tetracycline B. Polymixin C. Erythromycine D. Neomycin
96. Staining of Rickettsia and some protozoa is done with
A. Giemsa stain B. Acid fast stain C. Gram stain D. Cotton blue
97. Solar heat treatment to manage loose smut of wheat was prescribed by
A. J.C. Luthra B. V.V. Chenulu C. E.J. Butler D. B.B. Mundkur
98. A1 and A2 mating types are found in the fungus
A. Helminthosporium maydis B. Phytophthora infestans
C. Sclerotium rolfsii D. Fusarium oxysporum
99. Tospovirus is
A. Enveloped DNA virus B. Enveloped RNA virus
C. Non-enveloped DNA virus D. Non-enveloped RNA virus
100. The best indicator plant for potato spindle tuber viroid detection is
A. Tobacco B. Brinjal C. Tomato D. Chilly
101. Serological methods are not used for the detection of
A. Viroid B. Virus C. Fungi D. Bacteria
102. Genome of Banana bunchy top virus consists of
A. dsDNA B. ssDNA
C. ssRNA D. Both RNA and DNA

103. Gram Theory was introduced by
    A. Louis Pasteur B. E.F. Smith C. Robert Koch D. Robert Smith
104. Pleospora is the teliomorph of
    A. Stemphylium B. Cercospora C. Septoria D. Paecilomyces
105. Viroids are naturally transmitted by
    A. Mechanical means B. Seeds
    C. Aphids D. White flies
106. The most sensitive technique for the plant virus diagnostic is
    A. ELISA B. Western blotting
    C. Immunodiffusion D. Real time PCR
107. Candidates liberibacter asiaticus is a
    A. Phloem limited bacterium B. Phloem limited phytoplasma
    C. Xylem limited bacteria D. Xylem limited mycoplasma
108. Double stranded DNA is found in a) Mycovirus
    A. Luteovirus B. Mycovirus C. Caulimovirus D. Tospovirus
109. The virus causing sunflower necrosis disease is a
    A. Tospovirus B. Tobamovirus C. Ilarvirus D. Potyvirus
110. Who demonstrated first that the pathogen associated with mosaic disease of tobacco was filterable?
    A. Beijerinck B. Stanley C. Ivonavsky D. Paul Frosch
111. Which of the following is a cup-shaped structure?
    A. Apothecium B. Cleistothecium C. Zoosporangium D. Sporangium
112. Clark and Adams are known for development of
    A. PCR B. ELISA
    C. Real time PCR D. Nucleic acid hybridization
113. Which of the following viruses are bacilliform in shape?
    A. Banana streak Mysore virus B. Cucumber mosaic virus
    C. Tobacco streak virus D. Potato virus X
114. The fourth rule of Koch's postulates was prescribed by a) Robert Koch himself
    A. E.F. Smith B. Brefeld C. Leeuwenhoek D. Bordeaux
115. Resolving power of transmission electron microscope is
    A. 1 mμ B. 0.2 nm C. 1 nm D. 1 fm
116. Southern blotting is
    A. Transfer of protein from gel to nitrocellulose membrane
    B. Transfer of RNA from gel to nitrocellulose membrane
    C. Transfer of DNA from gel to nitrocellulose membrane
    D. Transfer of dyes from gel to nitrocellulose membrane

117. The first systemic account on Indian fungi was published by
A. Mundukar B. Butler and Bisby
C. Vasudeva D. Ramkrishnan & Subramanian
118. The "Herbarium Cryptogamiae Indian Orientalis" was first created by
A. E.F. Smith B. J.G. Kuhn C. E.J. Butler D. K.C. Mehta
119. A chemical compound that prevents multiplication without killing the fungus is
A. Mycocidal B. Antibiotic C. Fungicide D. Fungistatic
120. Who advanced the gene for gene concept of disease resistance and susceptibility?
A. Van der Plank B. Flor C. Ganmann D. Muller
121. The first plant virus whose genome was completely sequenced
A. CMV B. CaMV C. TMV D. PVX
122. Which of the following chemical fungicides acts by interfering with nuclear division?
A. Benzimidazoles B. Oxathiins C. Strobulirins D. Triadimefon
123. Mycoparasitism is associated with
A. Trichoderma B. Cercospora C. Septoria D. Alternaria
124. Phytoplasma are mainly transmitted by
A. Aphids B. Mites C. Leaf hoppers D. Thrips
125. Which one of the followings methods gives accurate estimate of the radius of the virus?
A. Electron microscopy B. X-ray crystallography
C. Mass spectrometry D. Serology
126. Which of the following is an internally borne pathogen?
A. Synchytrium endobioticum B. Ustilago nuda
C. Ustilago hordei D. Ustilago avenae
127. Ustilago nuda causes
A. Smut of oats B. Loose smut of wheat
C. Bunt of wheat D. Loose smut of barley
128. Cotton leaf curl virus and Groundnut bud necrosis virus is transmitted by which vectors?
A. Thrips and thrips B. Thrips and whitefly
C. Whitefly and thrips D. Thrips and aphids
129. "Murein B" contains what type of crosslinkage between the peptide side chains?
A. 2, 4 linkage B. 1,4 linkage C. 1,3 linkage D. 3,4 linkage
130. Dipicolinic acid is an important constituent of
A. Bacterial cell wall B. Bacterial protoplasm
C. Endospore D. Ascospore

131. F-factor remains as integrated within bacterial chromosomes of which type of bacterial cell?
A. F+ B. F' C. Hfr D. Pfr

132. Which one of the following bacterium contains peritrichous flagella?
A. Erwinia B. Xanthomonas C. Agrobacterium D. Bacillus

133. Serologically which class of immunoglobulins is important?
A. IgA B. IgD C. IgG D. IgM

134. The antibodies have a length of approximately
A. 100 nm B. 50 nm C. 10 nm D. 1 nm

135. Cucumber mosaic virus is transmitted
A. Only by mechanical means
B. Only by aphids
C. By both aphids and mechanical means
D. By seeds, aphids and mechanical means

136. Which of the following is an anti-oomycetes fungicide?
A. Carbendazim B. Benomyl
C. Thiophenate methyl D. Metalaxyl

137. Mycorrhiza" is the association of
A. Fungi and Higher plants B. Fungi and Algae
C. Fungi and Bacteria D. Fungi and Insects

138. "Lichens" shows which type of association
A. Symbiosis B. Commensalism C. Synergism D. Antibiosis

139. Filters used in Laminar air flow are known
A. HEPA -filters B. Asbestos-filters
C. Porcelin-filters D. Sintered glass filters

140. "Grassy shoot" disease of sugarcane is caused by
A. Fungi B. Bacteria C. Phytoplasma D. Virus

141. L-form bacteria do not possess
A. Cell membrane B. Cell wall C. Mesosome D. Ribosome

142. Which bacterium is aflagellate?
A. Pseudomonas B. Xanthomonas C. Agrobacterium D. Xylella

143. Typical "stem-pitting symptoms are produced in the disease
A. Citrus exocortis B. Citrus tristeza C. Citrus greening D. Citrus canker

144. The smallest pathogenic agent is:-
A. Virus B. Viroid C. Virusoid D. Prion

145. Teichoic acid found in
A. Fungi B. Gram (+) bacteria
C. Gram (-) bacteria D. Protozoa

146. Pronujcelium in rust fungi bears
    A. Basidiospores B. Aeciospores C. Uredia D. Telia
147. Rust of linseed and flax (autoecious rust) is caused by
    A. *Puccinia recondita* B. *P. striformis'*
    C. *P. graminis tritici* D. *Melantpsora lini*
148. The first Indian scientist who collected and identified fungi in India
    A. EJ Butler B. KR Kirtikar C. JF Dastur D. KC Mehta
149. In 2005 which pathological scientist got Borlaug award
    A. Rattan Lal B. VL Chopra C. CD Mayee D. S Nagarajan
150. All plant viruses (except cauliflower mosaic virus) contain
    A. DNA B. RNA
    C. Both RNA and DNA D. Either RNA or DNA

ANSWERS

1. B	2. D	3. D	4. D	5. A	6. C	7. A	8. B
9. B	10. C	11. A	12. B	13. B	14. C	15. B	16. D
17. B	18. B	19. C	20. C	21. B	22. D	23. B	24. A
25. C	26. B	27. A	28. D	29. B	30. C	31. A	32. D
33. B	34. A	35. A	36. A	37. B	38. A	39. A	40. C
41. C	42. B	43. D	44. B	45. B	46. A	47. B	48. B
49. B	50. A	51. B	52. A	53. D	54. A	55. A	56. D
57. C	58. D	59. A	60. D	61. C	62. B	63. C	64. D
65. B	66. B	67. A	68. C	69. B	70 D	71. D	72. D
73. B	74. B	75. A	76. A	77. B	78. C	79. C	80. C
81. D	82. B	83. A	84. A	85. D	86. D	87. D	88. A
89. A	90. C	91. C	92. A	93. A	94. B	95. A	96. A
97. A	98. B	99. B	100. C	101. A	102. B	103. C	104. A
105. A	106. D	107. A	108. C	109. C	110. C	111. A	112. B
113. A	114. A	115. B	116. C	117. B	118. C	119. D	120. B
121. B	122. D	123. A	124. C	125. A	126. C	127. D	128. C
129. A	130. C	131. C	132. A	133. C	134. C	135. D	136. D
137. A	138. A	139. A	140. B	141. B	142. D	143. B	144. D
145. B	146. A	147. D	148. B	149. D	150. B		

Chapter 7

AGRICULTURAL ENVIRONMENT AND ECOLOGY

- The term "ecology" was derived from two Greek words *i.e.*, **Oikos (house or dwelling place), and logos (the study of)**, to denote the relationship between organisms and their environment.
- The word ecology was given by **Ernst Haeckel**
- **Species**: A uniform interbreeding population spread over space and time
- **Vegetation**: Collective and continuous growth of plants
- **Flora**: Species content of region irrespective of numerical strength of each species
- **Population**: Group of individuals organisms of same species in a given area
- **Community**: Group of populations of different population in a given area
- **Habitat**: Particular set of environmental conditions suitable for successful growth of organism
- **Ecad**: Population of individuals which although belonging to same genetic stock but differ markedly in vegetative characters (environmentally induced).
- **Ecotype**: Population of individuals of species which are genetically different.
- **Ecotone**: Zone of transition between two ecosystems, rich in biodiversity **(edge effect)**.
- Ecosystem term proposed by **AG Tansley**.
- **Ecosystem**: Set of organisms interacting with each other and their physical and chemical factors extant in given space.
- **Succession**: Natural process by which different groups of communities colonize same area over a period of time
- **Primary succession**: Starts on primitive substratum by **pioneers** or primary colonizers. *e.g.* crustose lichen
- **Secondary succession**: Starts on previously built substrata. *e.g.* foliose lichen, herbs.
- **Autogenic succession**: A community modifies its own environment, causing its displacement.
- **Allogenic succession**: Replacement of existing community by **external causes.**

- **Climax community**: Final community that maintains itself in equilibrium with prevailing environment in an area
- **Ecological pyramids**: Graphic representation of trophic structure and function at successive trophic levels of ecosystem, this concept was given by **Charles Elton**. Pyramid of energy always remains upright.
- **Primary productivity:** Rate at which radiant energy is stored by producers (photosynthesis and chemosynthesis) as organic materials.
  - Gross Primary productivity
  - Net Primary productivity
- **Secondary productivity:** Energy stored at consumer levels
- **Effect of light on plants:** Chlorophyll production, heating, transpiration, stomatal movement, distribution *etc.,*
  - **Heliophytes**: Grow best in **full sunlight**
  - **Sciophytes**: Grow best in **shade**
- **Photoperiodism**: Total duration of light period to which plants are exposed
  - **Short day**: Require photoperiod less than critical maximum (12-14 hrs) for flowering. *e.g.* Cannabis, datura, green gram, rice, soybean, sugarcane *etc.,*
  - **Long day**: Require more than critical maximum *e.g.* wheat, pea, barley, turnip, rye, lettuce, onion, carrot, radish *etc.,*
  - **Day neutral**: Flowring not dependent on photoperiod. *e.g.* cucumber, potato, cotton, sunflower, maize *etc.,*
- **Phototropism**: Movement as regulated by light – towards (positive phototropism) or away from light (negative/skototropism).
- **Chilling injury**: Killing or injury of plants of hot areas due to exposure to low temperatures
- **Freezing injury**: Injury to cells due to freezing of water in intercellular spaces in plants
- **Classification of vegetation based on temperatu e conditions**:
  - **Megatherms**: high temperature year-round, tropical rainforests – dominant
  - **Mesotherms**: high and low temperature alternates, tropical deciduous forests-dominant
  - **Microtherms**: low temperature, mixed coniferous forests-dominant
  - **Hekistotherms**: very low temperatures (0-10ºC), alpine vegetation-dominant
- **Vegetation based on precipitation**:
  - **Evergreen forests:** Heavy rainfall throughout year
  - **Sclerophyllous forests:** Heavy rainfall in winter, low in summer
  - **Grasslands:** Mainly in areas with heavy rains in summer and low in winter
  - **Xerophytic vegetation:** Scanty rainfall throughout year

- **Biodiversity 3 levels:**
  - **Genetic diversity:** Genetic variation among species
  - **Species diversity:** Total range of species on earth
  - **Ecosystem diversity:** Variations in biological communities, ecosystems and the interactions among different levels.
- **At geographical scales biodiversity may be grouped as**
  - **Alpha diversity**: It refers to the diversity within a particular area or ecosystem, and is usually expressed by the number of species (*i.e.,* **species richness**) in that ecosystem
  - **Beta diversity**: It is a comparison of diversity between ecosystems, usually measured as the **change in amount of species** between the ecosystems.
  - **Gamma diversity**: It is a measure of the overall diversity for the different ecosystems within a region.
- **Endemism** is an ecological condition of a species being unique to a distinct geographic location, such as an island, nation, country or other defined zone, or habitat type; organisms that are native to a place are not endemic to it if they are also found elsewhere. A particular type of animal or plant may be endemic to a zone, a state or a country. The extreme opposite of endemism is cosmopolitan distribution.
- **Bioprospecting**: nations endowed with rich biodiversity explore molecular, genetic and species-level diversity to derive products of economic importance.
- **Keystone species**: Species which help in maintenance of an ecosystem. Any change in its population leads to major changes in presence of other species in that system. *e.g.* Tiger, Lion, Crocodile, Elephant (top predators)
- **Flagship species**: Species chosen for their vulnerability, attractiveness or distinctiveness to represent an environmental cause, such as an ecosystem in need of conservation. *e.g.,* Indian tiger, African elephant, leatherback sea turtle
- India is recognized as one of the **mega-diverse countries**.
- **IUCN** (International union of Conservation of Nature and Natural Resources) or **WCU** (World Conservation Union): Headquarters in **Morges, Switzerland**. Maintains **red data book or red list (started in 1963)** – a catalogue of species at risk of extinction. 8 categories of species are in this list:
  - Extinct
  - Extinct in wild
  - Endangered *e.g.* lion tailed macaque, Nilgiri langur, Gangetic dolphin
  - Critically Endangered
  - Vulnerable
  - Lower risk
  - Data deficient
  - Not evaluated

- **Biodiversity hotspot concept** was given by **Norman Myers**.
- Now, **34 hotspots** in world covering 1.4% of global land, supporting 60% of species.
- India has **4 hotspots**: Indo-Burma, Himalayas, Western Ghats and Sri Lanka, Sundalands (includes Nicobar islands).
- **Interactions between organisms :**
  1. **Positive interactions**
     - **Mutualism**: obligatory, often permanent association and beneficial to both populations. *e.g.* lichens, symbiotic nitrogen fixers, mycorrhiza
     - **Commensalism**: Only one species is benefitted, neither harmed. *e.g.* lianas, epiphytes
     - **Protocooperatiom**: Both benefitted, but not obligatory association. *e.g.* sea anemone attached to shells of hermit crab
  2. **Negetive interactions**
     - **Exploitation** *e.g.* cuckoo laying eggs in others birds' nest
     - **Parasitism** *e.g.* cuscuta (total stem parasite), orobranche (total root parasite), *Loranthus sp.* (partial stem parasite), sandal (partial root parasite)
     - Predation
     - **Competiton:** Compete for limited resources (light, nutrients, water, space). It may be interspecific or intraspecific
     - **Antibiosis:** both are not benefitted. Harmful chemicals produced by one species, inhibit other species growth. *e.g.* many actinomycetes and lichens produce substances inhibiting mold and bacterial growth.
- **Hardin** (1960) – **Competitive exclusion theory**: states that 'complete competitors cannot co-exist'.
- **Ecological** groups of plants
  1. **Hydrophytes: growing in abundance of water**
     - **Free floatin** *e.g.* Wolffia, Azolla, Eicchornia
     - **Rooted with floating leaves** *e.g.* Lotus, Trapa, Nymphaea
     - **Submerged floatin** *e.g.* Utricularia, Ceratophyllum
     - **Rooted submerged** *e.g.* Hydrilla, Vallisneria
     - **Rooted emergent** *e.g.* Sagittaria, Ranunculus
  2. **Terrestrial plants** - Plants grow on earth crust
  3. **Helophytes:** Marshy plants *e.g. Rhizophora mucronata, Sonneratia*
  4. **Oxylophytes :** Grow on acid soils
  5. **Psychrophytes :** Grow in cold soils
  6. **Halophytes**: Grow on saline soils
  7. **Lithophytes** : Grow on rocks

8. **Psammophytes** : Grow on sand
9. **Chersophytes:** Grow on wastelands
10. **Eremophytes:** Grow in deserts and steppes
11. **Psilophytes:** Grow on savannah

- **Ecological niche**: The physical space occupied and functional role of an organism in the community.
- Ecological equivalents: Organisms with similar niches in different geographical regions. *e.g.,* kangaroos of Australia and Bison of N. America
- Energy capture efficiency of autotrophs is only **0.1%.**
- **Ecological Footprint**: The quantity of biologically productive land and water required to sustain any entity, *viz.* a person, a community *etc.* indefinitely. It accounts for the demand on nature in terms of energy, food, water and materials consumed and wastes generated by that entity.
- Most productive ecosystems in terms of net primary production are **open oceans, tropical rain forests, savannahs and tropical seasonal forests.**
- **Protected area:** An area of land or sea dedicated to protection and maintenance of biological diversity natural and associated cultural resources, and managed through legal or other methods. *e.g.* reserves, national parks, monuments, protected landscape, habitat/species management area *etc.,*
- Conservation may be - *ex-situ* or *in-situ*
- ***Ex-situ* conservation**: Conservation outside the habitats. *e.g.* zoo, seed bank, botanical gardens
- ***In-situ* conservation**: Conservation in original habitat or man-made similar ecosystems. *e.g.* biosphere reserves, national parks and sanctuary

Comparison of different protected areas

	Sanctuary	**National park**	**Biosphere reserve**
Established	By IUCN category II protected areas	By IUCN category II protected areas	By IUCN category V protected areas
Purpose	Species oriented protection	Protect habitat of particular wild animal	Protection of all life forms of ecosystem
Interference	Limited	Only in buffer zone	Only in buffer zone
Tourism	Permitted	Permitted	Usually not permitted
Boundary	Not circumscribed	Circumscribed by legislation	Circumscribed by legislation

Research, scientific management	Lacking	Lacking	Managed
Number	551 (as of May, 2019)	104 (as of May, 2019)	18 (as of 2016)
Examples	Sessa Orchid Sanctuary, Salim Ali Bird Sanctuary, Indian Wild Ass Sanctuary, Gir Forest National Park	Blackbuck National Park, Clouded Leopard National Park, Desert National Park, Gulf of Mannar Marine National Park	Nilgiri Biosphere Reserve, Gulf of Mannar Biosphere Reserve, Agasthyamala Biosphere Reserve

- Biosphere reserve program was launched by UNESCO under the Man and Biosphere (**MAB**) program in 1971.
- **Nilgiri biosphere reserve** was the **first Indian reserve** established in 1986, spread in states of Kerala, Karnataka and Tamil Nadu.
- **Biofertilizer** is a substance it contains living microorganism, when it's applied to seed, plant, or soil, colonizes the rhizosphere or the interior of the plant and promotes growth by increasing the supply or availability of primary nutrients to the host plant. Biofertilizers are not fertilizers. Biofertilizers add nutrients through the natural processes of atmospheric nitrogen fixation, solubilizing phosphorus and stimulating plant growth through the synthesis of growth promoting substances.

N_2 fixing Biofertilizers

- **Free-living**: Azotobacter, Clostridium, Anabaena, Nostoc
- **Symbiotic**: Rhizobium, Frankia, *Anabaena azollae*
- **Associative symbiotic**: Azospirillum

P Solubilizing Biofertilizers

- **Bacteria**: *Bacillus megaterium var. phosphaticum, Bacillus circulans, Pseudomonas striata*
- **Fungi**: *Penicillium sp, Aspergillus awamori*

P Mobilizing Biofertilizers

- **Arbuscular mycorrhiza**: *Glomus sp., Gigaspora sp., Acaulospora sp., Scutellospora sp., Sclerocystis sp.*
- **Ectomycorrhiza**: *Laccaria sp., Pisolithus sp., Boletus sp., Amanita sp.*
- **Orchid mycorrhiza** : *Rhizoctonia solani*

Biofertilizers for Micro nutrients

- Silicate and Zinc solubilizers *Bacillus sp.*

Plant Growth Promoting: Rhizobacteria

- **Pollution:** Undesirable changes in physical, chemical or biological properties of air, water or soil that makes it harmful or create health hazards for any living organism. Pollutants *e.g.* deposited matter like soot, tar, dust, gases, acid, metals, chemicals,
- **Structure of atmosphere** : 5 layers-
  - **Troposphere:** Lowest layer, extending from earth's surface **to 8 km** in poles or 18 km at equator. Temperature decreases with height @1°C for 165 m (normal **lapse rate**). All major weather events occur in this sphere.
  - **Stratosphere:** Up to **50km** height. Temperature increases with height due to ozone layer which absorbs UV radiation.
  - **Mesosphere:** Third layer extending to 80 km, temperature decreases with height.
  - **Ionosphere:** Extends from 80 km to 400km, contains ions which help reflect radio waves back to earth and temperature increases with height.
  - **Exosphere:** Outermost layer extending beyond ionosphere.
- **Central Pollution Control Board:** Air quality standards for different sectors- National Ambient Air Quality Standards (NAAQS) for SO_2, NOx, SPM, Lead, CO, RSPM *etc.*,
- **Air pollutants and sources**:
  - **Industrial gases**: Compounds emitted depend on industry type. Refineries-SOx, NOx; stone crushers-dust, RSPM; mettalurgy-heavy metals and acids
  - **Automobiles**: Release hydrocarbons, particulate matter, CO, CO_2, Nitrogen oxides *etc.* Smoke intensity from diesel vehicles is measured on **Hartridge scale** (upper safety limit 65 units).
  - **Power plants**: Thermal power plants release fly ash, **SO_2**, hydrocarbons, particulate matter *etc.*
- **Acid rains**: due to oxides of sulphur and nitrogen producing **nitric acid** (30-40%) and **sulphuric acid** (60-70%) with rainwater. The following effects of acid rains in the nature:
  - Increase soil acidity, affect flora and fauna, increase availability of heavy metals, affects microbial population
  - Corrodes buildings, monuments and other structures
  - Water bodies acidification - death of fish
- **Photochemical pollutants**: like ozone, PAN, hydrogen peroxide is formed by light dependent reactions between hydrocarbons and nitrogen oxides. **Ozone near earth's surface affects crop growth negatively**.
  - **Photochemical smog/oxidising smog**: Formed in warm, sunny areas. Mixture of oxidising pollutants. *e.g.* Los Angeles smog. Adversely affects plant growth, humans and animals; leads to emphysema in man.

  - **Reducing smog:** Formed at humid, cool conditions. *e.g.* London smog
- **Ozone depletion:** In stratosphere, ozone plays a protective role by absorbing harmful UV radiations from the sun. Its depletion by reaction with CFCs, nitrogen oxides and hydrocarbons creates hole in ozone layer. 1% reduction in O_3 will increase UV radiation by 2% on earth.
  - **Vienna Convention** for the Protection of Ozone Layer, 1985
  - Montreal protocol on Substances that deplete the ozone layer, 1987: dealt with 8 Ozone depleting substances (ODS). Industrialized countries were to freeze CFC production at 1986 levels, followed by 20% reduction by 1995 and 50% reduction by 2000.
  - **Ozone day**: 16 September
- **Heavy metals:** Released into air and soil by various industries and other processes.
  - **Mercury:** Present in fungicides, paints, paper pulp *etc.*
  - **Lead:** Comes from leaded fuels (added to reduce knocking in vehicles)
  - **Zinc:** Emitted by refineries and smelters.
  - **Cadmium:** Produced by industries producing pesticides, battery, paints *etc.*
- **Aerosols:** also called suspended matter, they are liquid or solid particles floating in air. Consist of sulphates, soot, organic carbon, mineral dust, sea salt *etc.* They are reduce agricultural yield, alter temperature and monsoon pattern.
- **Global cooling/ global dimming:** Reduction in radiation reaching earth due to air pollution (particulates, aerosols) and cloud. The Indian Ocean hazels lead to slight cooling over Indian subcontinent.
- **Air quality can be measured via**
  - Emission assessment by stack sampling and meteorological measurement
  - Particle measurement through samplers, dustfall
  - Measurement of gases - grab/stack sampling
- **Control of air pollution**
  - Source correction- change raw material/ equipment and method
  - Collection and treatment of pollutants- Use setting chambers, cyclone filters, bag filters, wet collectors electrostatic precipitators, gas scrubbers, adsorption, incineration *etc.,*
- **Noise pollution**: The loudest noise a person can stand without much discomfort is 80dB while 45dB is the safe limit fixed by WHO for a city.

Zone	Permissible noise level standards in the daytime (dB)	Permissible noise level standards at night (dB)
Industrial Zone	75	70
Commercial Zone	65	55
Residential Zone	55	45
Silent Zone	50	40

- **Sources of water pollution**: Sewage, industrial effluents, agricultural discharges and run-off, discharges from power plants
- **Water quality measurement**
  - Biochemical Oxygen Demand (BOD): Amount of oxygen required for oxidation of organic wastes in a unit volume of water. Test done at **20°C for 5 days or 27°C for 3 days**. Measure of degree of water pollution.
  - Chemical Oxygen Demand (COD): Oxidizing agents usually Potassium dichromate is used to oxidise organics. **COD value is always more than BOD.**
  - Dissolved oxygen (DO): by oxygen probe or **Winkler method**.
  - Turbidity is measured by Turbidimeter.
- **Eutrophication**: Increase in productivity of water bodies due to addition of nutrients like nitrates and phosphates, leading to algal blooms. Excessive growth of these lead to reduction in oxygen levels in the water and may also release toxic chemicals, causing death of aquatic species.
- **Biomagnification/biological amplificatio** : Increase in concentration of chemicals/ metals/ compounds with successive trophic levels in a food chain.
  - **Rachel Carson**: Book **'Silent Spring'** (1962) - documented the detrimental effects on the environment of the indiscriminate use of pesticides on species along food chain.
- **Marine pollution**: By oil spills, sewage, garbage, agricultural discharges and plastic waste. Oil spills are the most serious causes of marine pollution. Spilled oil can harm living things because its chemical constituents are poisonous. This can affect organisms both from internal and external exposure to oil through ingestion or inhalation and through skin and eye irritation respectively. Oil can also smother some small species of fish or invertebrates and coat feathers and fur, reducing birds' and mammals' ability to maintain their body temperatures.
  - **Gulf of Mexico Oil spill (2010)**: 53,000 barrels were flowing into the gulf every day for three months.
  - **Exxon Valdez Oil Spill, (1989)**: 10.9 million gallons oil was spilled into Alaskan Coast line.

- **Gulf war oil spill, (1991)**: 330 million gallons of oil discharged into the Persian Gulf. Lot of marine life along with wildlife suffered due to oil spill that was about 5 inches thick.

- **Metal pollution:**
  - **Mercury- Minamata epidemic** in **Japan**. Methyl mercury is the toxic form. Sources- chloralkali plants, paper and pulp industries, batteries, thermometers, tube light
  - **Fluoride- Knock-knee syndrome**, dental fluorosis; affinity with calcium hence gets accumulated in bones, teeth. Natural deposits in some soils.
  - **Cadmium- Itai-itai disease**, Japan - Pain in spine and joints, softening of the bones and kidney failure.
- **Solid waste management**: 3Rs- Reduce, Reuse, Recycle; management through collection, disposal and recovery.
  - **Volume reduction**: Before disposal, volume may be reduced by incineration, shredding or pyrolysis
  - **Sanitary landfill** : Anaerobic decomposition of deposited wastes.
  - **Recovery/recycling**: Can be done for paper, metals, glass, organics
- Radiation may be Electromagnetic *e.g.* UV rays, gamma rays or Particulate *e.g.* beta particles, alpha particles
- Natural nuclear fuel: **Uranium-235**
- **Half-life**: Time taken for radioactivity to decline 50% from original value. *e.g.* Half-life of C^{14} (5568 years) and Strontium-90 (28yrs)
- **Nuclear disasters**
  - Three Mile Island, USA (1979)
  - Chernobyl, USSR (1986)
  - Fukushima Daiichi nuclear disaster, Japan (2011)
- **Bioremediation**: Use of living organisms for removing pollutants from biosphere. It may be natural/intrinsic bioremediation or enhanced bioremediation. Mainly uses microbes and plants.
- **Microbial bioremediation:** *e.g. Pseudomonas cepacia* - degrades chlorinated hydrocarbons; *Acetobacter liquefaciens* S-1 - consumes azo dyes; P*seudomonas putida, Citrobacter spp.* - heavy metals
- **Phytoremediation:** Categories- rhizodegradation, phytodegradation, phytoextraction, rhizofiltration, phytovolatalisation, phytostabilization. *e.g.* Thlaspi caerulescens accumulates 30-40% Zn in biomass; Indian mustard accumulate Cd, Cr; Typha latifolia removes Cd, Fe, Pb, Zn
- **Bioindicators**: Living organisms such as plants, planktons, animals, and microbes, which are utilized to screen the health of the natural ecosystem in the environment. They are used for assessing environmental health and biogeographic changes taking place in the environment.

- **Environmental impact assessment (EIA)** is the assessment of the environmental consequences (positive and negative) of a plan, policy, program, or actual projects prior to the decision to move forward with the proposed action.
- On 27 January 1994, the Ministry of Environment and Forests, under the **Environmental (Protection) Act 1986**, promulgated an EIA notification making Environmental Clearance (EC) mandatory for expansion or modernization of any activity or for setting up new projects listed in Schedule 1 of the notification. Since then there have been 12 amendments made in the EIA notification of 1994.
- 'A' category projects require environment clearance from the Centre and the 'B' category projects from the concerned State Level Environment Impact Assessment Authority.
- India has an estimated renewable energy potential of about 900 GW from commercially exploitable sources viz. Wind – 102 GW (at 80 meter hub height); Small Hydro – 20 GW; Bio-energy – 25 GW; and 750 GW solar powers, assuming 3% wasteland.

Sector	Total installed capacity in MW (2017)
Wind power	32279.77
Solar power	12288.83
Small hydro power	4379.85
Biomass power	8182
Waste to power	114.08

- **Climate change**: Variation in the mean state of climate or in its variables persisting for an extended period typically decades or longer.
- **Global warming** is the increase in earth's average temperature due to transmission of incoming short wave radiation from the sun and absorption of outgoing long wave radiation from earth by greenhouse gases. The process is known as greenhouse effect.
- **Causes of climate change**
  - **Natural:** solar variation, orbital variation, plate tectonics, volcano, glaciation
  - **Anthropogenic:** industrial processes, agricultural activities, deforestation, waste disposal, fossil fuel burning, use of CFCs
- **Major Greenhouse gases**
  - **Carbon dioxide**: Emitted through aerobic and anaerobic decomposition, burning of fossil fuels, residues, *etc.*, respiration.
  - **Methane**: Produced by anaerobic decomposition or fermentation. Sources: paddy fields, wetlands, enteric fermentation *etc.,*
  - **Nitrous oxide**: Produced in soil during aerobic (nitrification process) and anaerobic (denitrification process) conditions.

- **Impacts of global warming on crops**
  - Accelerates physiological and biochemical processes
  - Reduces photosynthesis
  - Enhances respiration
  - Enhances evaporation
  - Reduced crop growth duration and biomass
  - Increases CO_2: Benefit C3 plants more by reducing photorespiration, improved resource use efficiency, nutritional quality may reduce
  - Increase in temperature: Negative impacts by reducing yields and reducing crop duration, affect flowering and fruiting in several plants which require lower temperatures, loss of biodiversity (C3 crops may be replaced by C4 crops which have better thermal tolerance), shifts in crop growing areas to cooler regions.
- **Facilities to study impact of climate change include**
  - Free Air CO_2 Enrichment (FACE)
  - Free Air Temperature Enhancement (FATE)
  - Temperature Gradient Tunnel (TGT)
  - Open Top Chamber (OTC)
  - Phytotron
- United Nations conferences on Environment and outcomes

1. 1972: UN Conference on Human Environment, Stockholm, Sweden
   - United Nations Environment program (UNEP) was set up
2. 1982: Earth Summit, Nairobi, Kenya
   - 1983: World Commission on Environment and Development/Brundland Commission - Report **'Our Common Future'** (1987) - concept of **'Sustainable development'**
   - 1992: UN Conference on Environment and Development (UNCED) or **Earth Summit, Rio de Janeiro**: Outcomes - (1) Rio Declaration on Environment and Development/ Earth Charter (2) Agenda 21 (3) Convention on Forestry (4) Convention on Climate change and biodiversity.
   - United Nations Framework Convention on Climate Change (UNFCCC) an international environmental treaty that come into force on 21 March 1994.
   - Convention on Biological Diversity
3. 2002: World Summit on Sustainable Development (WSSD)/Rio+10, Johannesburg, South Africa. The height light of Rio+10 are:
   - Poverty eradication
   - Water and sanitation
   - Health

- Agriculture
- Energy
- Biodiversity
- Climate change

4. 2012: United Nations Conference on Sustainable Development /Rio+20, Rio de Janeiro
    - Non-binding document-'The Future We Want': poverty eradication, energy, water and sanitation, health, and human settlement.

- **UNFCCC**: Objective is to stabilize greenhouse gas concentrations in the atmosphere at a level that would prevent further chances of climate change. Presently, **195 countries are member of UNFCC**. They meet annually at the **Conference of Parties (CoP)** to assess progress in dealing with climate change since 1995.
    - **COP 8 - at New Delhi** - Delhi Declaration on Climate Change
- **Kyoto Protocol**, 1997: Signed at **CoP-3** and entered into force on 16 February 2005. There are currently 192 parties. Industrialized countries to reduce their greenhouse gas emissions **5.2% below 1990** levels in the period from 2008 to 2012. The targets apply to carbon dioxide, methane, nitrous oxide, sulphur hexafluoride, hydro fluorocarbon and per fluorocarbon. The Protocol is based on the principle of common but differentiated responsibilities. The US has not ratified the Kyoto Protocol, while Canada denounced it in 2012.
    - Doha Amendment, 2012: Only 66 countries have agreed, hence not entered into force. It was to extend the protocol period from 2013-2020.
- **Paris Agreement, 2015**: The Paris Agreement central aim is to strengthen the global response to the threat of climate change by keeping a global temperature rise this century well **below 2°C** above pre-industrial levels and to pursue efforts to limit the temperature increase even further to 1.5°C.
- 3 flexibility mechanisms can be used by Annex I Parties (developed countries and economies-in-transition) to meet their emission reducing commitment - International Emissions Trading (IET), the Clean Development Mechanism (CDM), and Joint Implementation (JI).
    - **Emissions Trading**: Allows Annex I Parties to trade their emissions.
    - **CDM**: Developed countries may invest in GHG reducing projects in developing countries. The emission reductions produced are called **Certified Emission Reductions (CERs)**. Most of these reductions are through renewable energy, energy efficiency, and fuel switching.
    - **JI**: Industrialized countries can invest in projects that reduce GHG emissions in other developed countries. Reductions produced are called **Emission Reduction Units (ERUs)**, which represents an emission reduction equivalent to one tonne of CO_2 equivalent.

- **Inter-Governmental Panel on Climate Change (IPCC)**: Established in 1988 by UNEP and WMO. The IPCC produces reports that support the UNFCCC. Reports the scientific, technical and socio-economic information relevant to understanding the scientific basis of risk of human-induced climate change, its potential impacts and options for adaptation and mitigation. It has published 5 assessment reports so far in 1990, 1996, 2001, 2007 and 2014. it consists of four reports:
  - Working Group I: The Physical Science Basis
  - Working Group II: Impacts, Adaptation and Vulnerability
  - Working Group III: Mitigation
  - Synthesis Report
- **IPCC's Fifth Assessment Report (AR5)**: Projections are based on "Representative Concentration Pathways" (RCPs). Summary:
  - Human influence on the climate system is understandable. It is extremely likely that human influence was the dominant cause of global warming between 1951-2010.
  - Without new policies to mitigate climate change, projections suggest an increase in global mean temperature in 2100 of 3.7 to 4.8 °C, relative to pre-industrial levels.
  - The average increase in the temperature of the Earth's surface has been 0.85°C in the period 1880-2012.
  - On a global scale, the ocean warming is largest near the surface, and the upper 75 m warmed by 0.11 [0.09 to 0.13]°C per decade over the period 1971 to 2010.
  - Over the period 1901 to 2010, global mean sea level rose by 0.19 [0.17 to 0.21] m.
  - Anthropogenic GHG emissions in 2010 have reached 49 ± 4.5 $GtCO_2$-eq/yr. Emissions of CO_2 from fossil fuel combustion and industrial processes contributed about 78% of the total GHG emissions increase from 1970 to 2010.

Major GHGs contributing to climate change:

Species	CO_2	CH_4	N_2O
Atmospheric lifetime		9.1	131
Pre-Industrial concentration (1750)	278 ppm	722ppb	270ppb
Global annual mean concentration (2019)	407 ppm	1803ppb	324ppb

  - From 1980 to 2011, the average annual increase in globally averaged CO_2 was 1.7 ppm yr^{-1}. The main contributors to increasing atmospheric CO_2 abundance are fossil fuel combustion and land use change.
  - The observed decrease in CH_4 growth rate from the early 1980s through

2006 indicates an approach to steady state where total global emissions have been approximately constant at ~550 Tg yr^{-1}.

- N_2O has increased at an average rate of ~0.75 ppb yr^{-1} since 1970s.
- Compared to the previous report, the lower bounds for the sensitivity of the climate system to emissions were slightly lowered, though the projections for global mean temperature rise (compared to pre-industrial levels) by 2100 exceeded 1.5 °C in all scenarios.

- **Impacts of climate change**
  - Changing pattern of precipitation or melting snow and ice, affecting water resources in terms of quantity and quality.
  - Many terrestrial, freshwater and marine species have shifted their geographic ranges, seasonal activities, migration patterns, abundances and species interactions in response to ongoing climate change.
  - Assessment of many studies covering a wide range of regions and crops shows that negative impacts of climate change on crop yields have been more common than positive impacts.
  - Some impacts of ocean acidification on marine organisms have been attributed to human influence.
  - Surface temperature is projected to rise over the 21st century under all assessed emission scenarios. It is very likely that heat waves will occur more often and last longer, and that extreme precipitation events will become more intense and frequent in many regions. The ocean will continue to warm and acidify, and global mean sea level to rise.
- Some laws and treaties made or signed by India for pollution control:
  - The Motor Vehicles Act, 1988
  - The Prevention of Food Adulteration Act, 1954
  - The River Boards Act, 1956
  - The Atomic Energy Act (Radiation Protection Rules), 1962
  - **The Water (Prevention and Control of Pollution) Act, 1974**
  - The Water (Prevention and Control of Pollution) Cess Act, 1977
  - **The Air (Prevention and Control of Pollution) Act, 1981**
  - **The Environment Protection Act, 1986**
  - **The Biomedical Waste (Management and Handling) Rules, 1998**
  - Plastic Waste Management Rules, 2016: minimum thickness of plastic carry bags should **be 50 microns**
- **Biodiversity conservation:**
  - Convention on International Trade in Endangered Species, 1973: Establishes lists of species whose international trade is to be controlled
  - Ramsar Convention on Wetlands of International Importance, 1971
  - UNESCO Biosphere Reserve Programme, 1971

- **The Wildlife Protection Act, 1972**
    - National Bureau of Plant Genetic Resources (NBPGR), New Delhi
    - National Bureau of Animal Genetic Resources (NBAGR), Karnal
    - National Bureau of Fish Genetic Resources (NBFGR), Lucknow
- **Bonn Convention, 1979-** Convention on conservation of Migratory species of wild animals
- International Convention for the Prevention of Pollution of the sea by Oil, 1954
- Basel Convention on the control of Trans-boundary movements of Hazardous Wastes and their Disposal, 1992: planned to reduce the movements of hazardous waste between nations and especially prevent transfer from developed to less developed countries.

MULTIPLE CHOICE QUESTIONS (MCQ)

1. Disease caused by inhalation of cotton dust
   A. Pneumoconiosis B. Silicosis
   C. Byssinosis D. Asbestosis
2. Content of methane in biogas
   A. 50-60% B. 30-40% C. 5-10% D.4-6%
3. Headquarters of WHO
   A. Geneva B. Nairobi C. Washington D. Rome
4. Ecological niche term was coined by
   A. Elton B. Tansley C. Odum D. Grinnel
5. Who played a vital role in UN Conference on Environment at Stockholm?
   A. Sunderlal Bahuguna B.Narasimha Rao
   C. Indira Gandhi D. MedhaPatkar
6. Waste from sugarcane mills used to generate electricity is
   A. Press mud B. Bagasse C. Molasses D. Plant waste
7. Sea water contains ____% salt
   A. 1.5 B. 2.5 C. 3.5 D. 4
8. Chasmophytes are ecological group of plants which grow on
   A. Acid soils B. Sandy soils C. Rock surface D. Rock crevices
9. Pollutants known as environmental estrogen
   A. Vinyl chloride B. Mercaptans C. Ethreal D. Dioxins
10. Recommended forest cover in hilly areas
    A. 30% B. 33% C. 45% D. 60%
11. World Habitat Day is celebrated on
    A. First Monday of October B. First Sunday of October
    C. Second Friday of October D. First Friday of October

12. DO level required for fishes to live in water
    A. 2ppm B. 4ppm C. 6ppm D. 8ppm
13. Forest types in India
    A. 16 B. 8 C. 12 D. 9
14. Which is an off-site conservation method
    A. Sanctuary B. National park C. Sacred groves D. Gene bank
15. Wilsons disease is caused by accumulation of which element
    A. As B. Hg C. Cd D. Cu
16. Aitken particle size is
    A. 2µ B. <0.2µ C. 0.2µ D. <2µ
17. Most abundant noble gas in Earth's atmosphere
    A. Neon B. Krypton C. Helium D. Argon
18. 'Operation Faith' was conducted to stabilize which compound that was responsible for Bhopal gas tragedy
    A. Methyl isocyanide B. Phosgene
    C. Ozone D. Nitrous oxide
19. Plastic recycling releases
    A. Dioxins B. HCN C. HCl D. All the above
20. Maximum permissible limit of chloroform in drinking water
    A. 0.001ppmor 1ppb B. 0.1ppm
    C. 0.5ppm D. 1ppm
21. Which is not a correct match?
    A. Nesslers reagent- Ammonia B. Devardas alloy – Cu
    C. Curcumin method – Boron D. Aluminon method - Cadmium
22. Which is not a criteria pollutant?
    A. Lead B. Carbon dioxide
    C. Ozone D. Particulate matter
23. According to the Human Development Report 2016 of UNDP, India's HDI rank is
    A. 119 B. 130 C. 131 D. 134
24. Which industry consumes maximum quantity of water in India?
    A. Paper B. Thermal C. Textiles D. Distillery
25. Which of the following is a submerged floating hydrophytes
    A. Wolfia B. Eichhornia C. Nymphea D. Ceratophyllum
26. Which act came into force as an aftermath of the Bhopal gas tragedy
    A. Environment Protection Act B. The Water Act
    C. The Air Act D. The Hazardous Waste Rules

27. Value of solar constant
    A. 2600 W/m^2 B. 81.6 W/m^2 C. 1370 W/m^2 D.1.94 W/m^2
28. Sulphur content of diesel
    A. 1-2% B. 0.1-0.2% C. 0.01-0.05% D. None
29. Silvering or bronzing of lower leaf surface particularly young leaves is an injury symptom of
    A. SO_2 B. O_3 C. PAN D. HF
30. In March 2017, which two Indian rivers were established as "living entities" by the high court ofUttarakhand
    A. Ganga and Indus B. Yamuna and Sutlej
    C. Alaknanda and Ganga D. Ganga and Yamuna
31. Term biodiversity was coined by
    A. WG Rosen B. Grinnel C. E.P. Odum D. R.D. Mishra
32. Most commonly used element for the fabrication of solar semiconductor cells is
    A. Carbon B. Germanium C. Silicon D. Platinum
33. UN Millennium Ecosystem Assessment Report was published in
    A. 1999 B. 2000 C. 2005 D. 2010
34. Plant hormone responsible for phototrophic movement
    A. Auxin B. Cytokinin C. Ethylene D. Gibberellins
35. Which of the following chemicals acts as chemical scavenger in plants exposed to SO2 pollution?
    A. Ascorbic acid B. Pyruvic acid
    C. Oxaloacetic acid D. Methionine
36. Most orchids that grow on other trees for support show which interaction
    A. Mutualism B. Commensalism
    C. Protocooperation D. Competition
37. Sunken stomata, reduced shoot/root ratio, ridged stem are adaptations seen in which type plants
    A. Hydrophytes B. Xerophytes C. Halophytes D. Oxalophytes
38. Dispersion pattern exhibited by most population
    A. Clumped B. Regular C. Random D. None
39. According to Raunkiaer's life form classification, plants with buds completely hidden in soil as bulbs and rhizomes are
    A. Phanerophytes B. Chamaephytes C. Cryptophytes D. Therophytes
40. The pyramid of numbers is always inverted in which food chain
    A. Grassland B. Pond C. Forest D.Parasitic
41. Brakes in a vehicle are source of which pollutant specifically
    A. Hydrocarbons B. CO C. SOx D. Asbestos

42. What percent of sunlight reaching atmosphere is used in photosynthesis?
    A. 0.01 B. 0.02 C. 0.5 D. 1
43. Plants absorb radiation in which range
    A. Green and red B. Green and red C. Yellow and blue D. Blue and red
44. Atmospheric fixation accounts for what percent of global nitrogen assimilation?
    A. 2 B. 5 C. 10 D. 20
45. National Wasteland Development Board was constituted in
    A. 1982 B. 1985 C. 1988 D. 1992
46. Which organization was created by the recommendation of the 1976 report of National Commission of Agriculture for comprehensive forest resource surveys?
    A. FRI B. FSI C. IIFM D. None
47. Monazite sands on Kerala shores are source of
    A. Uranium B. Barium C. Cesium D. Thorium
48. Geothermal energy production site in India
    A. Bombay high B. Gujarat C. Lakshadweep D. Ladakh
49. Calorific value of typical biogas sample (KCal/m3)
    A. 6123 B. 7200 C. 5871 D. 4937
50. Which cycle is disturbed when a volcano erupts?
    A. Carbon B. Water C. Nitrogen D. Phosphorus
51. Major forest types in India
    A. Tropical dry deciduous and tropical moist deciduous
    B. Tropical deciduous and temperate mixed
    C. Tropical evergreen and tropical dry deciduous
    D. Tropical moist deciduous and temperate mixed
52. Which unit of radioactivity gives an indication of biological damage?
    A. Curie B. Roentgen C. Rad D. Rem
53. Kaziranga wildlife sanctuary is famous for
    A. Elephant B. Rhino
    C. Liontailed macaque D. Deer
54. Single most important treaty to protect species at international level made in 1973
    A. CITES B. BDA C. WWF D. None
55. Blackbuck sanctuary is established as
    A. Kazhiranga, Assam B. Vansda, Gujarat
    C. Velvadhar, Gujarat D. Kanger Ghati, Chhattisgarh
56. Gir lion project was started in
    A. 1972 B. 1973 C. 1992 D. 2006

57. Round revolution was for
    A. Egg B. Potato C. Onion D. Oilseeds
58. Without naturally occurring greenhouse gases, Earth's average temperature would be near ……… instead of the much warmer 15°C
    A. 0°C B. -15°C C. -18°C D. -21°C
59. 99% of the total mass of atmosphere is within a height of ……km from earths surface
    A. 30 B. 20 C. 18 D. 45
60. Phosphatic fertilisers are source of
    A. As B. Cd C. Pb D. Hg
61. Which of these pairs are incorrectly matched?
    A. Minamata convention: mercury
    B. Stockholm convention: persistent organic pollutants
    C. Basel convention : lead
    D. None of above
62. The permitted maximum threshold concentration of sulphur dioxide in any industrial activity is
    A. 2ppm B. 3ppm C. 4ppm D. 5ppm
63. The boiler flue gas is source of
    A. Hal B. NO C. HF D. VOC
64. Most commonly used organism to indicate the water pollution is
    A. *Entamoebahistolytica* B. *Escherichia coli*
    C. *Eichhorniacrassipes* D. *Lemnapaucicostat*
65. The development of "anaerobic hyphoxia" conditions in aquatic water bodies is mainly due to presence of
    A. Bacteria B. Algae C. Fungi D. All of above
66. Development of Blue-baby Syndrome in children is due to drinking water polluted with
    A. Phosphorous fertilizer B. Nitrate fertilizer
    C. Calcium based fertilizer D. Fluoride
67. The number of categories used to divide air quality in India is
    A. 4 B. 5 C. 6 D. 7
68. Which of the following is not a marine pollutant?
    A. Oil B. Plastics
    C. Dissolved oxygen D. All of the above
69. Best indicator of Sulphur pollution in environment is
    A. Algae B. Lichen C. Protozoa D. Fungi

70. The ratio of BOD/COD will always be
    A. Equal to 1 B. Less than 1 C. More than 1 D. None of them
71. The National Air Quality Monitoring Programme (NAMP), prepared by Central Pollution Control Board, doesn't monitor which of the following pollutants?
    A. Carbondioxide B. Sulphur dioxide
    C. Oxides of nitrogen D. Suspended particulate matter
72. Which one of the following is a wrong statement?
    A. Greenhouse effect is a natural phenomenon
    B. Eutrophication is a natural phenomenon in freshwater bodies
    C. Most of the forests have been lost in tropical areas
    D. Ozone in upper part of atmosphere is harmful to animal
73. The Pollution Standard Index (PSI) scale has span from
    A. 0-200 B. 0-300 C. 0-400 D. 0-500
74. Pollutant associated with damaging ozone layer is
    A. Pollutant releasing chlorine B. Pollutant releasing bromine
    C. Both of above D. None of above
75. Molecule associated with biological amplification is
    A. Cadmium B. Lead C. Silicon D. Fluoride
76. Increase in concentration of a pollutant from the environment to the first organism in a food chain is
    A. Bioaccumulation B. Biomagnification
    C. Deposition D. None of these
77. Recently, the Supreme Court ordered that only vehicles with engines compliant with which emission standards can be sold from April 1, 2017?
    A. BS III B. BS IV C. BS V D. BS VI
78. YAMUNA ACTION PLAN was launched in
    A. 1985 B. 1986 C. 1993 D. 1995
79. Which of the following are not electromagnetic radiations?
    A. X-rays B. Cosmic rays C. Gamma rays D. Radio waves
80. Which organism produces a super-bioabsorbent composed of glucose and glucuronic acid that can absorb and hold water?
    A. *Pseudomonas cepacia* B. *Acetobacter liquefaciens*
    C. *Alcaligenes latus* D. None
81. Anti-malarial medicine 'quinine' is produced from
    A. Eucalyptus B. Aconite plant C. Cinchona D. Money plant
82. Which statement is not correct about coral bleaching?
    A. Occurs due to water pollution
    B. Paling of coral due to high water temperature

C. Decrease in zooxanthellae present in coral due to climate change
D. None of these

83. Kamuthi Solar Power Project was established in which state, making it the state with the highest installed solar-power capacity
A. Gujarat B. Rajasthan C. Maharashtra D. Tamil Nadu

84. Growing agricultural crops between rows of trees is known as
A. Jhum cultivation B. Social forestry
C. Taungya system D. None of these

85. India's position in global wind power installed capacity
A. 2nd B. 3rd C. 4th D. 5th

86. Nuclear energy contributes what percent to the total installed energy capacity
A. 0.27% B. 1.55% C. 1.83% D. 2.08%

87. EIA is now mandatory for how many categories of developmental activities which involve investment of more than Rs 50 crores
A. 25 B. 26 C. 29 D. 31

88. Which government made an eye-catching plea for climate change action by holding the world's first underwater cabinet meeting?
A. Maldives B. Mauritius C. Bali D. Japan

89. In context of environment, the term "dirty dozen" refers to
A. 12 most harmful greenhouse gases B. 12 ozone depleting substances
C. 12 persistent organic pollutants D. None of the above

90. The Wildlife Protection Act, 1972 provides for various categories of protected areas. Which is not included?
A. National parks B. Wildlife sanctuaries
C. Biosphere reserve D. Tiger reserves

91. World's first country to ban deforestation
A. Bhutan B. Norway C. Canada D. Mexico

92. WCU stands for
A. Wild conservation unit B. Wild conservation union
C. World communication union D. World conservation union

93. Ecosystem has two components
A. Plants and animals B. Weeds and trees
C. Biotic and Abiotic D. Frogs and men

94. Which one of the following terms does not reflect the idea of ecosystem?
A. Holocoene B. Cherposem C. Biosystem D. Geobiocoenosis

95. Environment planning organization is
A. CSIR B. NEERI C. CPHERI D. ICAR

96. CO_2 in soil and CO_2 in atmosphere differ
    A. By a factor of 2 B. By a factor of 1
    C. By a factor of 3.5 D. By a factor of 4
97. Trapping plume occurs when?
    A. Temperature increases with height
    B. Temperature decreases with height
    C. Temperature remains steady with height
    D. Plume lies in a lapse region and sandwiched between inversion layers
98. In all fresh water in India, the dominant ion is
    A. HCO_3^- B. Cl^- C. Ca^{+2} D. $SO4^{-2}$
99. Ozone toxicity to plants is generally due to the involvement of which unsaturated hydrocarbons
    A. Unsaturated fatty acid B. Methylene
    C. Ethylene D. None of the above
100. In vitro pollen germination is inhibited by
    A. SO2 B. O3
    C. Both of the above D. None of the above
101. Which one of the following is the most productive ecosystem?
    A. Oligotrophic lake B. High Sea
    C. Estuary D. Savanna
102. Energy flow in an ecosystem is
    A. Cyclic B. Multidirectional
    C. Sequential D. Unidirectional
103. The most abundant alkaline component of the atmosphere is
    A. Ammonia B. Nitrogen
    C. Sulphur dioxide D. Carbon Dioxide
104. The maximum permissible level of BOD (mg/l) in wastewater that is let into river is
    A. 20 B. 30 C. 40 D. 10
105. Which of the following is a recalcitrant?
    A. Sugarcane waste B. DDT
    C. Lignin D. Cellulose
106. Which waves carry most energy
    A. UV light B. Infra red light C. Microwaves D. Milimeter waves
107. Any particulate matter, gas or combination thereof other than water vapour is called
    A. Air curtain B. Air emission C. Air monitoring D. Air contaminant

108. Geostationary satellites orbit above the earth at about
   A. 10 km above the earth B. 500 km above the earth
   C. 1500 km above the earth D. 35786 km above the earth
109. BOD of the effluents discharged on land for irrigation should not exceed
   A. 30 mg/l B. 100 mg/l C. 300 mg/l D. 60 mg/l
110. Atmospheric radioactive window permits thermal radiation of which wavelength to leave the earth?
   A. 4.3 to 9.3 μm B. 9.5 to 10.6 μm C. 7 to 12 μm D. 7.3 to 10.3 μm
111. An objective of environmental audit is
   A. Raw-material & Waste minimization
   B. Energy conservation & monitoring
   C. To improve technical competency
   D. All of the above
112. Biome is a natural community of
   A. Plants in a geographical area
   B. Animals in a geographical area
   C. Plants and animals in a geographical area
   D. Plants in Arctic region
113. Who proposed that succession is not orderly and directional but is heterogeneous?
   A. Clements B. Egler C. Tansley D. Reiter
114. Methanization is carried out by the process of
   A. Incineration B. Alcoholic fermentation
   C. Thermo-chemical transformation D. Anaerobic fermentation
115. Development of 'Green Belts' around industries are
   A. To control ground leachates B. To mitigate the gaseous pollutant
   C. To increase bioaesthetic D. All of the above
116. For maximum biogas production following conditions are recommended
   A. Strict anaerobic B. Strict aerobic
   C. Only cellulose as a carbon source D. Semi aerobic
117. Molar concentration of pure water is
   A. 1 M B. 18 M C. 55.5 M D. 100 M
118. Zeitgeber means
   A. Time giving B. Photoperiod information
   C. Temp. information D. Rainfall information
119. Environmental Impact Assessment (EIA) seeks to achieve goals pertaining to
   A. Sustainable development B. Prevention of pollution only
   C. Conservation of resources only D. Social benefit

120. Tissue protein can be estimated by
   A. Gravimetric method B. Biurate method
   C. Caloriemeter D. Manometer
121. Which is the most abundant element in the hot water springs?
   A. Nitrogen B. Sulphur C. Carbon D. Phosphorous
122. Eutrophication of water bodies is triggered by
   A. Excessive growth of phtoplanktons B. Excessive growth of fishes
   C. Excessive inflow of nutrients D. Bright sunlight
123. Climatic stress is caused by insufficient and/ or excessive regime of
   A. Temperature B. Humidity C. Solar radiation D. All
124. Loss of biodiversity is primarily due to
   A. Green revolution B. White revolution
   C. Blue revolution D. Mining activity
125. Intake of lead may primarily cause the damage of
   A. Brain B. Lung C. Liver D. Kidney
126. Which one of the following is not a permitting cryoprotectant ?
   A. Glycerol B. DMSO C. Sucrose D. 1,2–propanediol
127. The aggregation of all eco systems on the earth is referred to as
   A. Atmosphere B. Ecosphere C. Stratosphere D. Ionosphere
128. Return of an ecosystem to a condition prior to disturbance refers to as
   A. Rehabilitation B. Restoration C. Rejuvenation D. Reclaimation
129. Energy flow in ecosystem is governed by
   A. First law of thermodynamics B. Second law of thermodynamics
   C. Planck's law D. Kirchoff's law
130. In Battele environment evaluation system the total parameter importance units is lowest for
   A. Ecology B. Environmental Pollution
   C. Aesthetics D. Human interest
131. Ecolabels are indicators of
   i. Acceptable level of environmental impact of a product.
   ii. Environmental performance of a product.
   iii. Claims of environmental friendliness of a product.
   Choose the correct answer from the codes given below:
   A. i, ii, iii B. i, ii only C. i only D. ii only
132. Ecosystem diversity can be best studied using the
   A. Topographical maps B. Geoinformatics
   C. Geodesy D. Geology

133. Bench marking in environmental management refers to
   A. Potential risk assessment
   B. Reporting of environmental performance
   C. Assessment of organization's business processes against the best-in-class operations to improve the performance
   D. Setting of environmental standards to be followed by environmental managers
134. Basel convention is related to
   A. Control of ozone depletion
   B. Control of water pollution
   C. Transboundary movement of hazardous wastes and their disposal
   D. Environmental auditing
135. Wind rose is a
   A. Graphical representation of wind velocity vector over a period of time in a polar diagram
   B. Graphical representation of wind velocity vector in a spherical coordinate system over a period of time
   C. Graphical representation of horizontal and vertical wind speeds over a period of time in polar diagram
   D. Graphical representation of instantaneous wind velocity at a particular time
136. Which one of the following protozoan is related to water borne disease?
   A. Spumella sp.   B. Entamoebahistolytica
   C. Paramoecium   D. Plasmodium vivax
137. Disaster Management Act in India came into existence in the year
   A. 2003   B. 2005   C. 1998   D. 2006
138. Who has been named Person of the Year by PETA?
   A. Naturo   B. Naruto   C. Tonaro   D. Ratono
139. IUCN released an updated ____ list for classifying different species on 4th Dec, 2017
   A. Green   B. Red   C. Blue   D. Yellow
140. India was awarded certificate of commendation for its effort to combat illegal trade in which of the following?
   A. Wild plants   B. Wild life
   C. Rare fossil fuels   D. Rare mineral
141. Boost in oxygen level between 445 and 485 million years ago caused which event?
   A. Great Ordovician Biodiversification Event
   B. Ice Age
   C. Cambrian Explosion
   D. Global Warming

142. Which emissions are turned to fuel using a membrane made of calcium, iron oxide and lanthium?
   A. Carbon B. Sulphur
   C. Oxygen D. Carbon and Sulphure
143. India and Greece have signed Air Services Agreement (ASA) and MoU on Cooperation (MoC) in which field?
   A. Oil and Gas B. Solar Power
   C. New and Renewable Energy D. Old and Non-renewable Energy
144. Which plant has been removed from the purview of the Indian Forest Act?
   A. Acacia B. Jacaranda C. Neem D. None of the above
145. Which car manufacturer has promised manufacture of BS-VI compliant cars by 2018?
   A. Mercedes-Benz B. Toyota Kirloskar
   C. Maruti Suzuki D. Tata Motors
146. Scurrying for solutions to fight the toxic air pollution, the government has said it plans to transport coal differently. How?
   A. via aircrafts
   B. via covered rail wagons and trucks
   C. via covered rail wagaons and open jeeps
   D. via bullock-carts
147. IUCN downlisted which species from its list of endangered animals on 14th Sept 2017?
   A. Pipistrelle B. Gazelle C. Mouse Deer D. Snow Leopard
148. Which foundation has launched the nationwide Rally For Rivers campaign?
   A. Isha Foundation B. Being Human Foundation
   C. ISCKON Foundation D. None of the above
149. Which environment quiz will be launched at national level on Teacher's Day by MoEF?
   A. Prakriti Shiskha B. Prakriti Sankalp
   C. Prakriti Sandarbh D. Prakriti Khoj
150. An organism which can monitor air pollution is
   A. Fungi B. Algae C. Bacteria D. Lichen

ANSWERS

1. C	2. A	3. A	4. D	5. C	6. B	7. C	8. D
9. D	10. D	11. A	12. B	13. A	14. D	15. D	16. B
17. D	18. A	19. D	20. A	21. D	22. B	23. C	24. B
25. D	26. A	27. C	28. C	29. C	30. D	31. A	32. C
33. C	34. A	35. A	36. B	37. B	38. A	39. C	40. D
41. D	42. B	43. D	44. A	45. B	46. B	47. B	49. B
49. *	50. D	51. A	52. D	53. B	54. A	55. C	56. A
57. B	58. C	59. A	60. B	61. C	62. D	63. B	64. B
65. B	66. B	67. C	68. C	69. B	70. B	71. A	72. D
73. D	74. C	75. A	76. A	77. B	78. C	79. B	80. C
81. C	82. A	83. D	84. C	85. C	86. C	87. C	88. A
89. C	90. C	91. B	92. D	93. C	94. B	95. B	96. D
97. D	98. A	9. B	100. B	101. D	102. D	103. B	104. B
105. B	106. A	107. D	108. D	109. A	110. B	111. D	112. C
113. B	114. B	115. D	116. A	117. C	118. B	119. A	120. C
121. B	122. C	123. D	124. D	125. A	126. B	127. B	128. B
129. A	130. C	131. B	132. B	133. C	134. C	135. A	136. B
137. B	138. B	139. B	140. B	141. A	142. D	143. C	144. D
145. A	146. B	147. D	148. A	149. D	150. D		

AGROFORESTRY

INTRODUCTION

- In India, the forest is cover **712,249 sq km** (71.23 million ha), which is **21.67 %** of its total geographical area (FSI, 2019).
- The tree cover of the country is **95,027 sq km** (9.50 million ha) *i.e.* **2.89 %** of the geographical area.
- The total forest and tree cover of India is **807,276 sq km** (80.73 mha) which is **24.56 %** of its total geographical area, which is far below the national goal *i.e.* **33%** as per National Forest Policy, 1988 (FSI, 2019).
- As compare to previous assessment (2017), there is an increased of **5,188 sq km** in the total forest cover of the country.
- Area wise, the state of **Madhya Pradesh** has the **largest forest cover** (77,462 sq km) in the country followed by Arunachal Pradesh (67248 sq km), Chhattisgarh (55586 sq km).
- In terms of percentage forest cover of India, **Mizoram with 85.41%** has highest followed by Arunachal Pradesh (79.63%), Meghalaya (76.33%), Manipur (75.46%) and Nagaland (75.31%).

HISTORY

- The origin of agro-forestry practices, *i.e.* growing trees with cereal crops and grasses, is believed to have been during Vedic era (1000 BC); whereas agro-forestry as a science is developed recently.
- As natural vegetation is cut for agriculture and other developmental activity, the benefits that trees provide are best sustained by integrating trees into agricultural system, that practice known as **agro-forestry**. This is also one of the options to reduce the pressure on natural forest.
- The systematic researches in agro-forestry were started after the establishment of "International Council for Research in Agro-forestry" **(ICRAF) in 1977;** it was renamed in 2001-02, as **World Agro-forestry Centre**.
- In India, research in agro-forestry was started **in 1983** by the establishment of AICRP on Agro-forestry (ICAR) at 20 centres and later establishment

of the National Research Centre for Agro-forestry (NRCAF) at **Jhansi (1988)**; recently it's renamed as **Central Agro-Forestry Research Institute (CAFRI).** At present 39 centres of agro-forestry are working in the country.

DEFINITIONS

- ***Agro-forestry*** is a collective name for land-use systems involving trees combined with crops and/or animals on the same unit of land. **It is the interaction of agriculture and trees, including the agricultural use of trees.**
- In agro-forestry systems interactions between different components *i.e.*, **tree, agriculture crop, animal** *etc.*
- Agro-forestry generally involves **two or more species** of plants (or plants/ animals), at least **one** of them is a woody perennial
- An agro-forestry system always has **two or more outputs**
- The cycle of an agro-forestry system is always more than one year; and agro-forestry system is **more complex** than mono-cropping system.

TRADITIONAL AGRO-FORESTRY SYSTEMS IN INDIA

- Agro-forestry is widespread in all ecological and geographical regions of India. It's ranged from apparently simple forms of shifting cultivation to complex home-gardens.
- **Shifting cultivation (slash and burn system):** It is one of the farming system in **north-eastern** high rainfall areas in which forests land is cleared by slash and burn method cropped with common arable crops for a few years and then left for regeneration of natural vegetation. Traditionally, it was uncultivated for **10-20 years** but in recent times *i.e.* **2 to 5 years** in many areas.
- **Taungya system:** The taungya (taung = **hill**, ya = **cultivation**) is a Burmese word coined in **1850**s. **Brandis** introduced the 'taungya system' into India **(1890)** and **the North Bengal and north-eastern hill region (1896)** is the place, where first taungya plantations were raised. In south India, the system is known as **'kumari'**. It is mainly practiced in high rainfall areas *i.e.*, above **1200-1500 mm**. It involves cultivation of crops in forests or forest trees in crop-fields.
- **Home-gardens/homesteads Cultivation:** In homestead a number of crops raised together with tree, livestock, poultry and/or fish production, mainly for basic needs of the farmer. It is an old-age practice in coastal states *i.e.*, **Kerala, Tripura, Assam**, West Bengal and North-eastern states. The home-gardens emerge to be a **multi-storey canopy**. The suitable species are coconut, areca nut, guava, mango, citrus, tamarind, jack-fruit, papaya, banana and other multipurpose trees.

- **Plantation based Agro-forestry systems:** Modern commercial plantation crops like rubber, coffee, poplar, eucalyptus and oilpalm represent a well-managed and beneficial land-use activity in tropics.
- **Scattered trees on farm lands:** The growing of agricultural crops under scattered trees on farm lands is old practice adopted in India from centuries. For example, *Prosopis cineraria* and *Zizuphus, D. sissoo* in **arid areas;** *Acacia nilotica* in **Indo- Gangetic** plains; *Grevia optiva* in hills of **Uttarakhand and HP**; *Eucalyptus globulus* in southern hills of **Tamil Nadu** and *Brossus flabellifer* in **peninsular coastal regions.**
- **Trees on farm-boundaries:** Trees grown on farm boundaries are also often used old practice. In northern parts of India, ***Eucalyptus* and *Populus*** are commonly grown along the field boundaries especially in Punjab and Haryana.
- **Wood lots:** In this system, farmers grow trees in separate blocks as wood-lots along with agricultural fields *e.g.* wood-lots of bamboo, Casuarina, poplar, *Eucalyptus*, *Leucaena leucocephala*, *Pterocarpus santalinus* and *Dalbergia sissoo* have become popular in many parts of the country.

Systems for soil conservation or amelioration:

- Approx. **178 Mha** waste land area in India, in which **63.85Mha** are degraded. The deep and narrow gullies may be controlled by putting them to **permanent vegetation** after closure to grazing. Suitable species are *Acacia nilotica, Azadirachta indica, Butea monosperma. Bambusa spp., Dendrocalamus, Dalbergia sissoo, Prosopis juliflora, Tectona grandis* and **grasses** are *Dichanthium annulatum, Bothriochloa pertusa, Cynodon dactylon* and *Sehima nervosum.*
- **Shelter belt:** In arid regions wind velocity is very high throughout the year. Farmers build shelter-belt (kana bundi) by local vegetation to **check wind velocity** within safer limits. *Crotolaria burhia, Leptadenia pyrotechnica* and *Aerva pseudotomentosa* bushes are planted **across the wind direction**. This also helps increase crop yields along the lines.
- **Trees on rangelands:** Raising the **protein rich species** such as *Salvadora oleoides, Capparis decidua, Acacia nilotica, A. leucophloea. P. cineraria, Prosopis juliflora etc.* on common community grazing lands.

Benefits from Agro-forestry

Environment Benefits The general ecological benefits include:

(i) Reduce pressure from natural forest

(ii) Protection of ecological systems.

(iii) Improvement and create microclimate

(iv) Efficient recycling of nutrients due to deep root system
(v) Improvement in soil structure
(vi) Reduction of surface run-off, nutrient leaching and soil erosion
(vii) Increment in nutrient status through leaf litter

Economic Benefits These benefits include:

(i) Increments in outputs of food, fuel-wood, fodder and timber;
(ii) Reduction in incidence of total crop failure, common to single-cropping or mono-cropping
(iii) Increase in levels of farm incomes

Social Benefits These benefits include:

(i) Improvement in rural living standards from sustained employment a incomes;
(ii) Improvement in nutrition and health due to increased quality and diversity of food output.
(iii) Stabilization and improvement of upland communities.

Classification of Agro-forestry Systcm

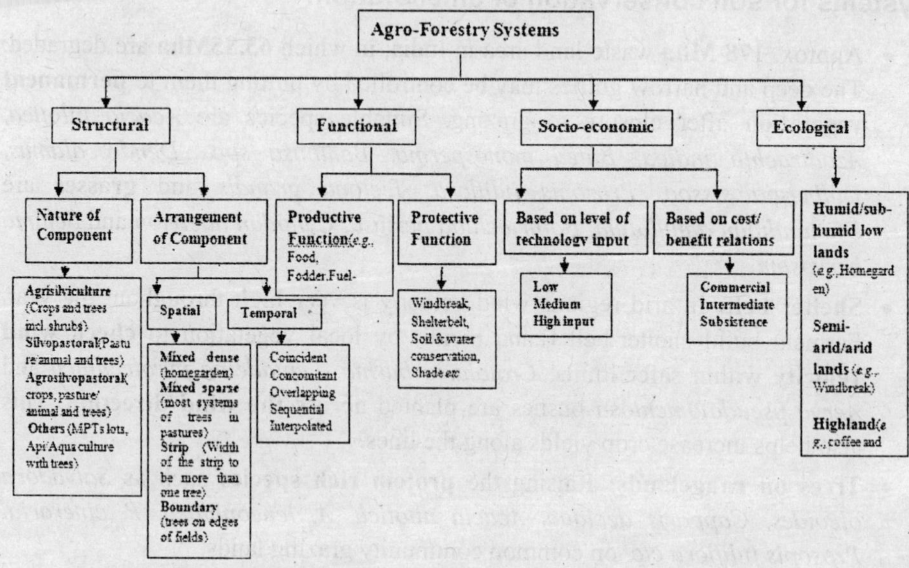

1. Agrisilvicultural System: (crops and trees including shrubs/vines and trees)

In this system, agricultural crops raised with tree crops and forest crops. Based on the nature of the components this system can be classified into various forms.

(i) Improved fallow species in shifting cultivation: This system is practiced extensively in the **north-eastern** hill region. It is called **'jhum'** in the north-eastern hill region and **'podu'** in AP and Orissa. The main feature of the improved fallow system is that the trees and shrubs are not grown with crops on the same plot at the same time. The main function of the fallow land is to improve the soil fertility and reduce soil erosion; some plants can be raised for their commercial value.

(ii) Taungya: This is a modified form of shifting cultivation in which the labour is permitted to raise crop in an area but only side by side with the forest species planted by it. The practice consists of land preparation, tree planting, growing agricultural crops for **1-3 years** and then moving on to repeat the cycle in a different area.

(iii) Multipurpose tree species garden: In this system, various kinds of tree species are grown mixed together. The main function of this AF system is production of food, fodder and wood products for home consumption and surplus sale for cash.

(iv) Alley Cropping (Hedgerow Intercropping): Alley cropping, which is typically regarded as the inter-cropping of forest trees and crops simultaneously, on same piece of land. Generally, free are planted in an East – West direction to decrease the shade effect. The spacing of 4 – 8m between rows and 25 cm to 2 m between tree within rows in followed. The closer spacing is generally used in humid areas.

(v) Multi-purpose Trees and Shrubs on Farmlands: In this system, various multipurpose tree species are raised in some systematic patterns or scattered on plot/field. The main components are multipurpose trees, horticultural trees and agricultural crops. Suitable tree species are *Acacia albida, A. senegal, Azadirachta indica, Cassia siamea, Casuarina equisetifolia, Cocos nucifera, Leucaena leucocephala, etc.*

(vi) Crop combinations with Plantation Crops: Perennial trees and shrub crops, such as coffee, tea, coconut and cocoa are combined into intercropping systems.

(vii) Agroforestry fuel wood production: In this system, planting of various fuel wood species on or around agricultural lands such as *Acacia nilotica, Albizia lebbek, Cassia siamea, casuarina equisetifolia, Dalbergia sissoo, Eucalyptus tereticornis, Prosopis juliflora etc. Melia dubia, M. Composita etc.,*

(viii) Shelter-belt: Shelter belts are belts consisting of several rows of trees established at **right angles** to the prevailing wind. The purposes are:

- To reduce the velocity and deflect air currents of prevailing winds,
- To provide general protection to leeward side against the effect of wind erosion
- To protect the leeward areas from the desiccating effect of hot air
- To provide food, fodder, timber *etc.*

The main characteristics of shelter-belts are:

- **Shape and composition:** Shelter-belts have a typical **triangular** shape. This can usually be raising tall trees in the centre
- **Density and width:** A certain degree of penetration by winds is planned as to raise a solid wall, the protection decreases very fast on the leeward side. Shelter-belts up to **50m width** are considered ideal under Indian conditions.

- **Height and spacing:** The ratio of height and width should be roughly **1:10**. However, this figure may vary with local conditions.
- **Orientation:** Orientation of shelter-belts depends on the direction and velocity of the prevailing winds. Shelter-belts may be raised in **quadrangles** if the wind direction tends to change very often.
- **Length:** The minimum length of a shelter-belt should be about **25 times** its height.
- **Choice of species: Grasses:** *Saccharum spontaneum, S. munja, Panicum antidotale, Cenchrus sps.* **Shrubs:** *Calotropis procera, Clerodendron phlomoides, Cassia auri-cufata, Dodonaia viscosa, Capparis aphylla, Lonnea coromandefica, Tamarix articulate etc.*

(ix) Wind-break: Windbreaks have been a **traditional agro-forestry** practice. Wind break or shelterbelts are characterized as linear plantings of trees and/or shrubs established primarily for environmental purposes such as reducing wind speed, snow accumulation, retaining soil moisture and buffering extreme temperatures. The height of a windbreak is directly related to the amount of leeward area protected by wind damage. These areas usually extend up to **10 to 15** times the tree height or up to 150 m with 10 m high trees. Some windbreaks system has shown to increase wheat yields by 15% and improve net economic benefits by 17%. In addition, wind-breaks can provide a wide range of useful products such as timber, fodder, fiber and mulch *etc.*

a. **Permeability:** A wind-break helps in filtering and breaking the force of the wind. The wind velocity in the protected area may be reduced to **25 and 75 per cent** of the open wind speed. The desired permeability can be obtained by selection of suitable tree and shrub species such as ***Eucalyptus*** **and *Casuarina*.**

b. **Orientation:** For best results, raise the wind-breaks, at **right angles** to the winds from which protection needed. Wind-breaks planted **north-south** are a good compromise as they provide can protection from winds coming from the western quarter.

c. **Height:** The wind-break height determines the size of the protection area. The taller the wind-break, the greater the area it protects.

(x) Soil Conservation Hedges: Trees can be planted on physical soil conservation works (grass strips, bunds and terraces), wherein they play two roles: to stabilize the structure and to make productive use of the land. The important tree species for soil conservation are *Acacia catechu, Acacia modesta, Alnus nepalensis, Grevillea robusta, Leucaena leucocephala, Pinus roxburghii, Prosopis juliflora etc.*

(xi) Riparian buffer: In present-day agriculture, facing some **adverse environmental impacts,** so riparian buffer systems require for the establishment and management of vegetation (*e.g.* trees, shrubs and grasses) along denuded waterways to buffer system losses and **reduce agrochemical pollution**. Suitable species for buffer systems are **willows (*Salix spp.*), poplars (*Populus spp.*), *Morus alba* and *Acer* spp.**

2. Silvo-pastoral System (trees + pasture and/or animals): Silvo-pastoral systems are characterized by integrating trees with forage and livestock production. This system is further classified into three categories: (i) Protein bank (ii) Living fence of fodder trees and hedges (iii) Trees and shrubs on pasture.

(i) Protein bank: In this silvopastoral system the various multipurpose trees (protein-rich trees) are planted on or around farmlands and rangelands for cut-and-carry fodder production to meet the feed requirements of livestock during **the fodder-deficit perio** in winter.

For humid and sub-humid areas: *Artocarpus spp., Anogelssus lattifolia, Bombax malabaricum, Cordia dichtotoma, Dalbergia jambolana, Samanea spp., Zizyphus etc.*

For dry regions: *Acacia nilotica, Ailanthus excelsa, Opuntia ficus, Prosopis spp., Rhus etc.*

(ii) Living Fence of Fodder Trees and Hedges: In this system various fodder trees and hedge are planted as live fences to protect the property from stray animals or other biotic influence. Suitable species are: *Acacia spp, Sesbania grandiflora, Gliricidia sepium, Erythrina abyssinica, Euphorbia spp., etc.*

(iii) Trees and Shrubs on Pastures: In this system various tree and shrub species are scattered irregularly or arranged according to some systematic pattern, especially to supplement forage production. **For humid and sub-humid regions**: *Derris indica, Emblica officinalis, Psidium guajava, Tamarindus indica.* **For dry regions**: *Acacia spp., Prosopis spp.* and *Tamarindus indica.*

3. Agrosilvopastoral System, (trees + crops + pasture/animals): This system has been grouped into two subgroups:

(i) Home garden: In every homestead has around **0.2-0.5ha land** for personal production, on which are grown trees for construction timber, fruits, vegetables, small plots of sugar-cane *etc.*

(ii) Woody hedgerows: In this system various woody hedges especially fast-growing and coppicing fodder shrubs and trees, are planted. The main aim of this system is production of food/fodder/fuel wood and soil conservation.

4. Other Systems: it includes-

(i) Apiculture with Trees: The main objective of this system is the production of honey.

(ii) Aquaforestry: In this system various trees and shrubs preferred by fish are planted on the boundary and around fish-ponds. The main aim of this system is fish production and bund stabilization around fish-ponds.

(iii) Multipurpose Wood Lots: In this system, Multi-purpose tree species are grown mixed or separately at special location for various uses such as wood, fodder, soil protection, soil reclamation *etc.*

Limitations of Agro-forestry

An integrated food-tree farming system is advantageous, but it has certain negative aspects.

Environment Aspects:

(i) Competition of trees with agricultural crops for space, light, moisture and nutrients which may **reduce** agricultural yield;

(ii) Damage to agricultural crop during **tree harvest** operation;

(iii) Potential of trees to serve as **hosts to** insect pests that are harmful to food crops;

(iv) Rapid regeneration by prolific trees, which may **displace** agricultural crops and take over entire fields.

Socioeconomic Aspects:

(i) Requirement for **more labour** inputs, which may causes shortage at times in other farm activities;

(ii) Competition between agricultural and forest tree crops, which could cause total **yields to be lower than** those of a single crop;

(iii) **Longer rotation** period of forestry crop and acquire an economic value;

(iv) Resistance by farmers **to shift agriculture** crops with trees, especially where land is limited;

(v) The fact agro-forestry is **more complex** and more difficult to apply, compared to single-crop farm.

Through skilful management practices, any or all of **these aspects can be controlled**. For example, once it is easy to adopt some or all of the following strategies:

- Select **short rotation, fast growing species** that have small or sparse crowns so that sufficient sunlight will reach the agricultural crop for photosynthesis;
- **Select legume** trees so that they will help in Nitrogen fixation and supplement the nutrients to non-legume agricultural crops;
- Select tree species that have **deep-rooted** so that they will absorb moisture and nutrients from the deeper layer of the soil;
- Select the tree species which have **no allelopathic effect** on agricultural crops;
- Select **multi-purpose tree species** (MPT's) so that they will provide various needs of farmer such as timber, fuel wood, fruits, fodder *etc.,*;
- Adopt wide spacing of the trees to reduce their competitive effects on the food crops. Further, the following factors should be keept in mind when selecting agro forestry tree species:

– Environmental adoptation
– Needs of farmers or local market
– Easy to maintain
– Availability of genetic material

Agro-forestry Diagnosis & Design (D & D)

Agro-forestry D & D is a family of procedures for the diagnosis of land management problems and potentials and the design of agro-forestry solutions. The **ICRAF** has developed an approach to support agro-forestry researchers and development fieldworkers to plan and implement effective research and development projects.

Key features of D & D:

- **Flexibility**: Flexible discovery of procedure which can be fit to available resources.
- **Speed**: Rapid appraisal, in depth follows up of the project.
- **Repetition**: D & D is an open ended learning process.

Criteria of a good agro-forestry design: A good agroforestry design should fulfill the following criteria

- Productivity
- Sustainability
- Adaptability

Procedures of AF Diagnosis and Design: Two levels:

- Macro D & D
- Micro D & D

Macro D & D: Objectives:

- To identify **broad issues and problems** constraining all land use systems in a given ecozone.
- To identify and **prioritize areas** for potential agroforestry interventions.
- To identify **research priorities** and formulate research programme.
- To identify **needs, opportunities and mechanisms** for inter-institutional collaboration for technology development.

Steps in Macro D & D exercise

- Identification of study ecozone
- Delineation of land use systems within ecozone.
- Description of land use system
- Analysis of land use system constraints and potential.
- Analysis of potential agro-forestry technologies.
- Definition of agro-forestry research needs.
- Inter institutional coordination.

Micro D & D

The objectives of micro D & D are similar but the micro is based on the potential in a specific ecozone.

Analysis of land use system and constraints:

- Prioritize the needs of the households.
- Identify the production constraints.
- Assessing potential for system development.

Design & evaluation of micro D & D

- **Technological specificatio** : It is a package of technology. It covers feasibility analysis, socioeconomic analysis and assessment by farmers. Successful technological design requires adequate research experience.
- **Ex-ante evaluation of technology**: It gives an idea of probability of impacts and implications. The analysis looks at benefits and conflicts or problem likely to arise. There are **4 types of analysis** is done.
- Economic viability
- Sustainability
- Farmer's acceptability
- Adoption potential

Type of research under AF: Four types of research:

- Basic research: Design to generate new technology.
- Strategic research to solve specific research problems.
- Applied research to create new technology.
- Adaptive research

Research Design Criteria: Trials are conducted to achieve the performance of technology.

Methodological considerations in D & D

- **Research Team**: Team of experts.
- **Research Domain and recommendation domain**: Modification of original research domain
- **Data collection methods**: Collection of minimum of raw data, maximum of useful information.
- **Analytical methods and the role of farmers**: Interaction with farmers, Policy makers and development agent to arrive at logical conclusion. The farmers should be treated as equal participants.
- **Logistical and operational aspect**: For macro D & D exercise, total staff time required from planning to conclusion is estimated to be **three months**. The survey with farmers is an important part of the operational aspect.

The factors responsible for the bad performance of agroforestry:

- Lack of institutional support mechanisms
- Lack of extension mechanisms
- Lack of quality planting materials and inadequate research on agroforestry models
- Lack of post-harvest processing technologies
- Insufficient marketing infrastructure and price discovery mechanisms
- Restrictive and non-conducive legal provisions for harvesting and transportation of trees grown on farmlands (***Arun Kumar Bansal Committee*** (2011))
- Mandate of agroforestry falls under the ambit of various ministries, departments, agencies, state governments, *etc.*

NATIONAL AGROFORESTRY POLICY (2014)

India became the first nation in the world to adopt an **agroforestry policy (2014).** Agroforestry was earlier covered as a part of other policies such as National Forest Policy (**1988**), National Agriculture Policy (**2000**), Planning Commission Task Force on Greening India (**2001**), National Bamboo Mission (**2002**), National Policy on Farmers (**2007**) and Green India Mission (**2010**).

The **Policy aims** to improve coordination, convergence and synergy between various elements of agroforestry, scattered across various existing missions, programme and schemes under different ministries *i.e.*, **agriculture, rural development and environment**.

Objectives of the National Agroforestry Policy 2014

1. To promote AF to **increase farm income and livelihoods** of rural households, especially the small and marginal farmers.
2. **To protect and stabilize ecosystems**, and promote resilient cropping and farming systems to minimize the risk during extreme climatic events.
3. Sustainable supply of **raw material to wood based industries.** That creates new avenues for rural employment, and reduces pressure on the forests.
4. To develop **capacity and strengthen research in agroforestry** and create a massive people's movement for achieving these objectives.

Salient features/strategy of Agroforestry policy 2014

- **Establishment of national level institutions to promote Agroforestry:** MOA, DARE, MOEF, Agricultural University, NABARD *etc.*
- **Research and capacity building**: carried out by ICAR and ICFRE
- **Simplification of regulatory mechanism**: The policy objectives takes into consideration Forest Rights Act (**FRA, 2006**), Panchayats (Extension to

Scheduled Areas) Act **1996** (PESA), JFM **(1990)**. Initially **20 MPTs** will be exempted from all restrictions pertaining to harvesting, transportation and marketing grown under agroforestry systems.

- **Institutional credit and insurance coverage**: To create Farmers Producers Organization (FPO) to promote the AF produce at higher level.
- **Improving farmer's access to quality planting material:** The policy encourages the participation of private sector to produce quality planting material.
- **Strengthening market access**: strengthen by help of private sector participation

AGROFORESTRY OBJECTIVE QUESTIONS (MCQ)

1. The practice of forestry with the object of producing wood and other forest products as a business enterprises is known as

   A. Protection forestry B. Social forestry

   C. Production forestry D. Agro forestry

2. Which of the following rights and concessions are availed by tribes in forest?

   A. Free grazing in open forest

   B. Right to live in forest

   C. Collection of grass for cattle and making huts

   D. All

3. Write the full form of ICRAF-.

   A. International centre for Research in Agroforestry

   B. International council for Research in Agroforestry

   C. Indian centre for Research in Agroforestry

   D. Indian council for Research in Agroforestry

4. Components of Agro-forestry are

   A. Animal, crop B. Crop, tree and animal

   C. tree, animal D. All

5. Total geographical of India is

   A. 329 mha B. 327 mha C. 326 mha D. 325 mha

6. The total Forest area of India (2015) is

   A. 24.16% B. 24.01% C. 23.15% D. 20.45%

7. First forest policy after independence come into force in

   A. 1952 B. 1947 C. 1950 D. 1988

8. President of India launched the much-needed National Agro-forestry Policy (NAP) in

   A. 2014 B. 2013 C. 2012 D. 2011

9. World forestry Day celebrated on

   A. 22 March B. 21 March C. 6 June D. 15 July

10. World Wildlife Day celebrated on
    A. 22 March B. 3 March C. 6 June D. 15 July
11. The Green India Mission (GIM) is one of the eight mission under the National Action Plan on climate change (NAPCC) started in year
    A. 2011 B. 2012 C. 2010 D. 2009
12. National Green Tribunal (NGT) Act comes into force in
    A. 2011 B. 2010 C. 2016 D. 2014
13. The year ..... celebrate as International Year of pulses
    A. 2010 B. 2016 C. 2014 D. 2011
14. Forests regenerated from seed are called
    A. Coppice B. High forests
    C. Seed forest D. None of the above
15. ....is also known as poor man timber
    A. Bamboo B. Ber C. Teak D. Ficus
16. .... is collective name for land use system involving trees combined with Crops/or animals on the same unit of land.
    A. Protection forestry B. Social forestry
    C. Production forestry D. Agro forestry
17. Agroforestry normally involves two or more species of plants woody at least one of which is always?
    A. Biennial B. Annual C. Perennials D. None
18. There are .... agro climatic regions of India
    A. 15 B. 10 C. 20 D. 12
19. According to Nair (1987) Agroforesty system can be classified on the basis of
    A. Structural, Functional, Socio economic and Ecological
    B. Socio economic and Ecological
    C. Functional and Socio economic
    D. Structural and Functional
20. The objective of improved fallow species in shifting cultivation is to recover depleted
    A. soil nutrients B. soil erosion C. soil moisture D. All
21. Shifting cultivation is practiced extensively in the .... Region
    A. North eastern B. Northern C. western D. N-W
22. The Taungya is a Burmese word coined in Burma in the year
    A. 1865 B. 1850 C. 1906 D. 1911
23. The Taungya system was introduction into India by Brandis in year
    A. 1890 B. 1906 C. 1911 D. 1850

24. The first Taungya plantation were raised in 1896 in
    A. North eastern B. North Bengal
    C. western Rajasthan D. N-W
25. Growing of tree with crops called ……..system
    A. Agrisilviculture B. Silvipasture C. Silvoagri D. None
26. Shifting cultivation is also called ……in north eastern hill region.
    A. Jhum B. podu C. kumari D. Hill
27. Shifting cultivation is also called ……. in Andhra Pradesh.
    A. Jhum B. Podu C. Kumari D. Hill
28. In southern India, taungya system of cultivation is called
    A. Jhum B. Hill C. Kumara D. Podu
29. Taungya system is practiced in area with an assured annual rainfall of over
    A. 1200-1500 mm B. Below 750 C. 3000 above D. 450-750mm
30. In taungya system '*taung*' means ……and '*ya* 'mean cultivation
    A. Hill B. Podu C. Kumari D. Jhum
31. In village taungya, each family has about 0.8 to 1.7 ha of land to raised trees and cultivate crops for
    A. 3 to 4 year B. One year C. 2 years D. 10 years
32. Taungya system are 3 types viz.
    A. Departmental B. Leased C. Village taungya D. All
33. Alley cropping is also known as
    A. Hedge row inter cropping B. Relay cropping
    C. Mixed cropping D. None
34. The primary purpose of alley cropping is to ………….. crop yield
    A. Maintaining B. Increase C. Both D. Decrease
35. Ideally hedge row intercropping should be positioned in …… direction
    A. East West B. N-E C. N-S D. N-W
36. In alley cropping, the spacing used in field is usually ……….. between tree rows
    A. 4 to 8m B. 1-2m C. 8-10 D. 10-12
37. In alley cropping the spacing used in field is usually …….. between trees within rows
    A. 4 to 8m B. 25cm-2m C. 1-2m D. 10-12m
38. On sloping land hedge rows should always be placed on the
    A. Contour B. Middle
    C. At distance 8-10 D. Any where
39. The system in which various multipurpose fuel wood/firewood species are inter planted on or arranged agricultural land called
    A. Farm forestry B. Alley cropping
    C. Energy plantation D. Agro-forestry fuel wood production

40. In shelter belt, there are several rows of trees established at ...... to the prevailing wind

A. Right angles B. Parallel C. Triangle D. All

41. The purpose of shelter belt is to ...... the velocity of prevailing wind

A. Reduce B. Increase C. No effect D. None

42. Shelter belt have a typical ......Shape

A. Right angles B. Parallel C. Triangle D. All

43. Ideal tree density in shelter belt's approximately

A. 50-60% B. 10-20% C. 15-25% D. 30-40%

44. Shelter belt upto ...... width are considered ideal under Indian condition

A. 50m B. 20m C. 30m D. 40m

45. In shelter belt, the ratio of height and width should b roughly ..... Ratio

A. 1:10 B. 2:5 C. 1:2 D. 2:15

46. The minimum length of a shelter belt should be about ..... times its height

A. 25 B. 10 C. 20 D. 50

47. The .... are strips of trees and/or shrubs planted to protect fields, homes, canals from wind.

A. Shelterbelt B. Alley cropping C. wind breaks D. None

48. A wind break 20 m tall will give some protection from .... on the upward side to ...... on the downwind side

A. 100, 500m B. 200, 400m C. 100, 200m D. 150, 300m

49. Wind breaks of ..... rows are more effective for most farm situation

A. 3-5 B. 1-2 C. 3-4 D. 5-6

50. IGFRI located at..........

A. Jhansi B. Dehradun C. Coimbatore D. Delhi

51. .... is defined as the elimination of branches in order to obtain trees with clean bole

A. Pruning B. Thinning C. training D. None

52. The elimination of branches by physical and biotic agencies of the environment is

A. Artificial pruning B. Natural pruning

C. Bud D. None

53. Removal of branches from the selected portions of the tree by mechanical mean's is

A. Artificial pruning B. Natural

C. Bud D. None

54. ....are the key feature of D&D

A. Flexibility B. Speed C. Repetition D. All

55. An Agroforestry research programme normally begins with ……..D&D exercise covering an entire ecological zone within a country

A. Macro B. Micro C. Mini D. None

56. The D&D team comprises ……… specialist

A. 5-10 B. 6-8 C. 2-4 D. 10-15

57. A macro D&D is usually completed in about

A. 3 Month B. 10 month C. 5 month D. 6 month

58. Tree species which given fuel, fodder, fibre, food and timber are called

A. Multipurpose trees B. Single use tree
C. Forestry tree D. Specific tree

59. In jhum cultivation, average number of member in a family varies from

A. 4-7 B. 10-15 C. 5-10 D. 6-8

60. In conventional shifting cultivation cycle involving 2-4 years of cropping with how much years of fallow cycle

A. 15 years B. 10 years C. 5 years D. 6 years

61. Trees for fuel wood should have

A. High calorific value and specific gravity
B. Low calorific value and specific gravity
C. High moisture
D. Hardest

62. How much percent of the total living biomass of the trees is in roots?

A. 20-25% B. 10-15% C. 5-10% D.60-70%

63. The term social forestry was used for the first time by

A. Westoby, 1968 B. Walker, 1967 C. Brandis,1865 D. Wilson,1906

64. Van Mahotsav was started in the year …. by K.M. Munshi

A. 1950 B. 1952 C.1947 D.1974

65. Rural fuel wood plantation scheme was introduced in the year/plan

A. 6th 5 year plan B. 8th 5 year plan C. 7th 5 year plan D. 4th 5 year plan

66. NRCAF located at……

A. Jhansi B. Dehradun C. Coimbatore D. Delhi

67. The term used for shifting cultivation in Brazil is

A. Tavi B. Melpa C. Roca D. Masole

68. The Javanese word used for home garden is

A. Pekarangan B. Karen C. Parakar D. Jumar

69. Which of the classification of agroforesty system is not based on arrangement of components

A. Hedge row B. Sparse
C. boundary plantation D. dense

70. Nutrient cycling in Agricultural system is
    A. leaky B. open C. closed D. All
71. Taungya system was first started in
    A. Australia B. India C. Berma D. Germany
72. Botanical name of babul is
    A. *Acacia nilotica* B. *Acacia samia*
    C. *Acacia auriculiformis* D. *Acacia catechu*
73. Poineer worker among the following of alley cropping is
    A. PKR Nair B. Budelman C. B.T. Kang D. Young
74. A large regional unit characterized by major vegetation type associated fauna of a specific climatic zone is called
    A. Ecosystem B. Biomass C. Biome D. landscape
75. A unit of land with a natural boundary having a mosaich of patches is called
    A. Landscape B. Biomass C. Population D. Vegetation
76. Bell shaped pyramid of age structure of a population indicate-
    A. Stable B. Expanded C. Decline D. None of the above
77. Pyramid of energy is always
    A. Inverted B. Always Inverted
    C. Stable D. Upright
78. Average energy content of 1 gram coal is equivalent to
    A. 11.5 Kcal B. 10 Kcal C. 7 Kcal D. 250 KCal
79. Miami model of productivity was proposed by
    A. Clements B. Lieth C. Glaeson D. Wittakar
80. Tolerance model of succession was proposed by
    A. Conell and Slatyer B. Elements
    C. Gleason D. Engler
81. Which species of Bamboos flowers every year without the death of plants?
    A. *Arundinacea* B. Bambusa C. Dendrocalamus D. All
82. Which species have became extinct due to lopping in many parts of india-
    A. *Anogeissus pendula* B. Sal
    C. Teak D. Butea
83. Gum trangacanth is obtained from
    A. Acacia B. Cassia
    C. *Astagalus gummifer* D. Butea
84. Phylogenetic system of plant classification has been given by
    A. Echinson B. Darwin C. Hutchinson D. C.Lenious
85. First Indian Forest Act was in acted in
    A. 1865 B. 1867 C. 1952 D. 1894

86. Which one is biological measure to conserve soil moisture?
A. contour farming B. Gully Plucking
C. Ley Farming D. Agro forestry
87. Sampling intensity for tropical wet evergreen forest should be
A. 20 % B. 25% C. 10% D. 4%
88. Which of the following tree fixes nitrogen through Frankia?
A. *Casuarina equisetifolia* B. Acacia
C. Teak D. sandalwood
89. Lignotubers are found in
A. Eucalyptus B. Sal C. Casuarina D. Rhizophora
90. Agrisilviculture System means growing of
A. Agriculture crop + forest tree
B. Fodder crop + forest tree
C. Agriculture crop + Fodder crop + forest tree
D. All
91. Indian Council of Forestry Research & Education (ICFRE) located at
A. Jhansi B. Dehradun C. Coimbatore D. Delhi
92. The Forest-PLUS programme started in
A. July 2012 B. July 2010 C. July 2011 D. July 2014
93. National Environment Policy come into force
A. 2006 B. 2010 C. 2011 D. 1986
94. On the basis of Agro-ecological zone India is divided in to how many zones
A. 15 B. 8 C. 36 D. 20
95. The felling made in an immature stand for the purpose of improving the growth and form of tree is termed as
A. Heading back B. Pruning C. Thinning D. Coppicing
96. CAZRI – Central Arid Zone Research Institute located at
A. Jhansi B. Jodhpur C. Bikaner D. Delhi
97. The production of woody plants combined with pasture is referred to as a system
A. Agrisilviculture B. Silvo-pastural C. Agripasture D. None
98. System in which honey bees are planted on the boundary, mixed with agriculture crop is called
A. Agrisilviculture B. Apiculture with trees
C. Aquaculture D. None
99. In .... system various trees and shrubs preferred by fish are planted on the boundary and around fish ponds.
A. Agrisilviculture B. Apiculture C. Aqua Forestry D. None

100. ....Agroforestry system is those where in the use of land is directed towards satisfying basic needs and is managed mostly by the owner/occupant and his family.
A. Intensive B. Subsistence C. Extensive D. Surplus
101. The word forest has its origin from...........word
A. Latin B. Greek C. Urdu D. Hindi
102. Taungya system was introduced to india by
A. Brandis, D. B. Tejwani, K.G
C. Chatterjee, B.N. D. None of these
103. Famous award given in field of Agroforestry is
A. Indra Priyadarshini Vrikshamitra Award (IPVA)
B. Rafi Ahmed Kidwai Award
C. Hari om Ashram Trust Award
D. Vasant Rao NaikASward
104. Which agroforestry tree is evergreen in nature?
A. Poplar B. Deodar C. Sisham D. Amaltas
105. Which tree is known as flame of forest?
A. *Butea monosperma* (Dak) B. *Prosopis jiuliflora*
C. *Prosopis cineraria* D. None of these
106. *Leucaena leucocephala* on an average fix about..........kg N/ha
A. 100 B. 125 C.150 D.200
107. Taungya system is practiced mostly in
A. Rajasthan B. Kerala C. Gujarat D. All
108. Mimosine an important toxicant is found in which of the following feed stuffs
A. Seena B. Subabul C. Neem D. None of above
109. Percernt mimosine content in the seed of subabul is
A. 10 B. 12 C.14 D.16
110. The important toxicant present in bamboo is
A. Tanin B. HCN C. Mimosine D. Saponin
111. Acacia catechu is mainly used for
A. Ingredient in preparing paan B. Gums
C. Resins D. None of these
112. The queen of timber
A. Shisham B. Vagai C. Teak D. Eucalyptus
113. Which tree species is called green gold?
A. Sal B. Sandal C. Bamboo D. Teak
114. Chipko movement was launched for protection of
A. Forests B. Grasslands C. Wetlands D. Live stocks

115. "Rangeland and Agroforestry" published from
A. New Delhi B. Jhansi C. Jodhpur D. Dehradun

116. The ratio of height and width of shelterbelt is
A. 1:10 B. 1:3 C. 1:5 D. 1:3

117. Area under forest in India according to National Remote Sensing Agency (NRSA) is
A. 75 m ha B. 100 m ha C. 50 m ha D.150 m ha

118. Major objective of the agro forestry is
A. Productivity B. Sustainability C. Adaptability D. All

119. Which is the faulty land use system?
A. Mixed cropping B. Multiple cropping
C. Shifting cultivation D. All

120. Which is fire resistant bamboo species is?
A. *Dandrocalamusha miltonii* B. *Melocanna bamboosoides*
C. Both D. None of these

121. Hard wood present in
A. Angiosperm B. Gymnosperm C. Both D. All

122. Soft wood present in
A. Angiosperm B. Gymnosperm C. Both D. All

123. Fodder from tree is usually considered as
A. Rainy fodder B. Best fodder C. Summer fodder D. Scarcity fodder

124. Which tree used for dye preparation?
A. Madhuca B. Casuarina C. Teak D. Eucalyptus

125. Scientific basis of the practice of the silviculture is
A. Forest community B.Forest biodiversity
C. Forest management D. Forest ecology

126. Summer wood also known as
A. Late wood B. Early wood C. Heart wood D. Alburnum

127. Spring wood also known as
A. Late wood B. Early wood C. Heart wood D. Alburnum

128. India has monopoly in which silk
A. Munga B. Mulberry C. Tasar D. Sri

129. Bamboo multiplied through
A. Corm B. Rhizomes C. Tubers D. Suckers

130. Which of the following is the lowest strata layer?
A. Herb layer B. Canopy layer
C. Understory layer D. Immature layer

131. Which refers to the development of an ecosystem in an area that has never had a community living within it?
A. Pioneering plants B. Primary succession
C. Coriolis effect D. Phytochemicals

132. Which of the following sheds its leaves during the dry season?
A. Deciduous B. Doldrums C.Megatherms D. Epiphytes

133. The main benefits of agroforestry include which of the following
A. Proper use of trees B. Biodiversity
C. Afforestation D. Wildlife conservation

134. Agroforestry shares principles with which of the following
A. Intercropping B. Ecology C. Mixed cropping D. Cover cropping

135. Strip cropping is similar to
A. Parkland B. Alley cropping
C. Fauna based system D. Crop-over-tree system

136. The best windbreaks approach is 100% crown density in the low-profile shurb, rows, and .... density in the tall rows
A. 50% B. 60% C.70% D. 80%

137. A windbreak with 50% crown density is more effective than one of either 100% or of _______ density
A. 15% B. 20% C. 25% D. 30%

138. *Acacia modesta* period of ripening of seed is
A. May - July B. September – November
C. February – April D. November

139. Teak is naturally found in
A. Tropical wet everygreen forest B. Subtropical forest
C. Tropical decidous forest D. Alpine

140. Fodder availability in India is
A. 450 mt B. 500 mt C. 600 mt D. 650 mt

141. Brick Planting mainly done in
A. Desert B. Hill C. Plain D. Any where

142. Height class of trees in forest inventory are
A. 1, 2 and 3 mt B. 1, 3 and 5 mt C. 2,4 and 6 mt D. 0,5 and 10 mt

143. Tetrazolium (TZ) test is used for
A. Germination B. Viability C. Growth D. Storage

144. Maximum used Volume calculation formula in India is...
A. Smelian B. Huber C. Quater girth D. Newton

145. Phulwara Butter is obtained from...............
A. *Madhuca indica* B. *Madhuca butyracea*
C. *Garcinia indica* D. *Melia azedarach*

146. Japan wax is the brand name of wax obtained from......

A. *Rhus succedanea* B. *Meliaazedarch*

C. *Rhus succedanea* + *Meliaazedarch* D. *Sapium sebiferum*

147. Sandal is

A. Partial stem parasite B. Complete root parasite

C. Complete stem parasite D. Partial root parasite

148. The Study of life history and general characters of trees with reference to environment-

A. *Silvics* B. *Synecology* C. *Ecology* D. *Autecology*

149. Oleoresin known as *ralorlal- dhuma* is commercially obtained from

A. Sal B. Chir pine C. Deodar D. Hopea

150. Restocking of balanks in an artificially regenerated area with fresh sawing or planting is

A. Enrich planting B. Pollarding C. Beating up D. Forking

ANSWERS

1. C	2. D	3. A	4. B	5. A	6. A	7. A	8. A
9. B	10. B	11. A	12. B	13. B	14. B	15. A	16. D
17. C	18. A	19. A	20. A	21. A	22. B	23. A	24. B
25. A	26. A	27. B	28. C	29. A	30. A	31. A	32. D
33. A	34. C	35. A	36. A	37. B	38. A	39. D	40. A
41. A	42. C	43. A	44. A	45. A	46. A	47. C	48. A
49. A	50. A	51. A	52. B	53. A	54. D	55. A	56. A
57. A	58. A	59. A	60. A	61. A	62. A	63. A	64. A
65. A	66. A	67. C	68. A	69. A	70. C	71. C	72. A
73. C	74. C	75. A	76. A	77. D	78. C	79. B	80. C
81. A	82. A	83. C	84. C	85. A	86. C	87. C	88. A
89. C	90. A	91. B	92. A	93. A	94. B	95. C	96. B
97. B	98. B	99. C	100. B	101. A	102. A	103. A	104. B
105. A	106. D	107. B	108. B	109. B	110. B	111. A	112. C
113. C	114. A	115. B	116. A	117. A	118. D	119. C	120. C
121. A	122. B	123. C	124. D	125. C	126. A	127. B	128. A
129. B	130. A	131. B	132. A	133. B	134. A	135. B	136. A
137. C	138. D	139. C	140. A	141. A	142. B	143. B	144. C
145. B	146. C	147. D	148. A	149. A	150. C		

Chapter

9

AGRICULTURAL STATISTICS

- **R.A. Fisher** is the father of statistics
- **Statistics (Defination)** Statistics is the science which deals with the
  - (i) Collection of data
  - (ii) Organization of data or Classification of data
  - (iii) Presentation of data
  - (iv) Analysis of data
  - (v) Interpretation of data
- **Primary data:** The data which is collected by actual observation or measurement or count is called primary data. *e.g.* household survey done by going into field for socio-ecomic survey
- **Secondary data:** The data which are compiled from the records of others is called secondary data. Data collected about area, yield and production from "**statistics at a glance**" released every year by government.
- An arrangement of data into rows and columns is known as **Tabulation.**

Exclusive classes	Inclusive classes
1-10	1-9
10-20	10-19
20-30	20-29

- A frequency distribution can be both discrete and continuous
- Frequency of a variable is always a **integer**
- Discrete series: 5,7,6,8,9,9,3,2
- Continuous series:

Marks	No. of students
1-10	5
10-20	6
20-30	2

Characteristics of Frequency Distributions: A frequency distribution is characterized by its location, scale and shape. Thus, for comparing different sets of data or frequency distributions we need to compare the followings characteristics:

(i) Location (Central or average values)

(ii) Scatter (Dispersion or variation or spread)

(iii) Shape (Skewness and Kurtosis)

- **One dimensional**: Bar and line diagram
- **Two dimensional**: Rectangular, Circular, Pie
- **Three dimensional**: Cubes cylinder and spheres
- **Non dimensional**: Pictogram
- **Ogive curve** corresponds to **cumulative frequency**

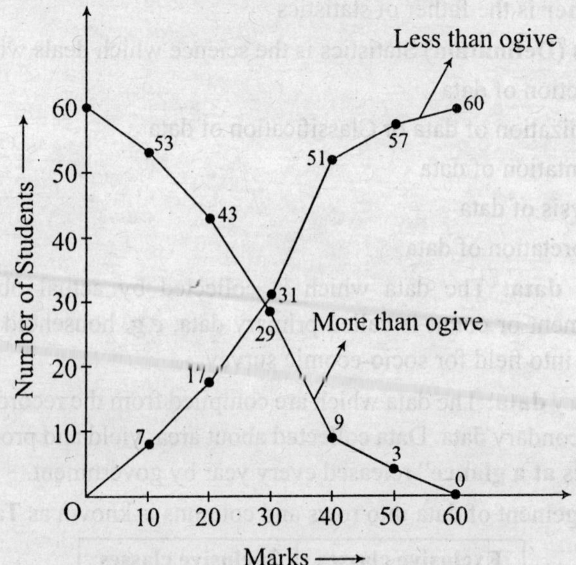

- Ogives for more than type and less than type distribution intersect at

(a) Mean	(b) median	(c) Mode	(d) origin
Ans: median			

Cumulative frequency is also known as Ogive which is useful for finding out quartiles, deciles and percentiles.

- Histograms can be drawn only for continuous frequency distributions and they can be used only when all class intervals are equal.
- Histogram can be drawn for equal and unequal classes.
- When the mid points of the tops of adjacent bars are joined then frequency polygon is developed.
- Frequency polygons can be developed with the help of histogram

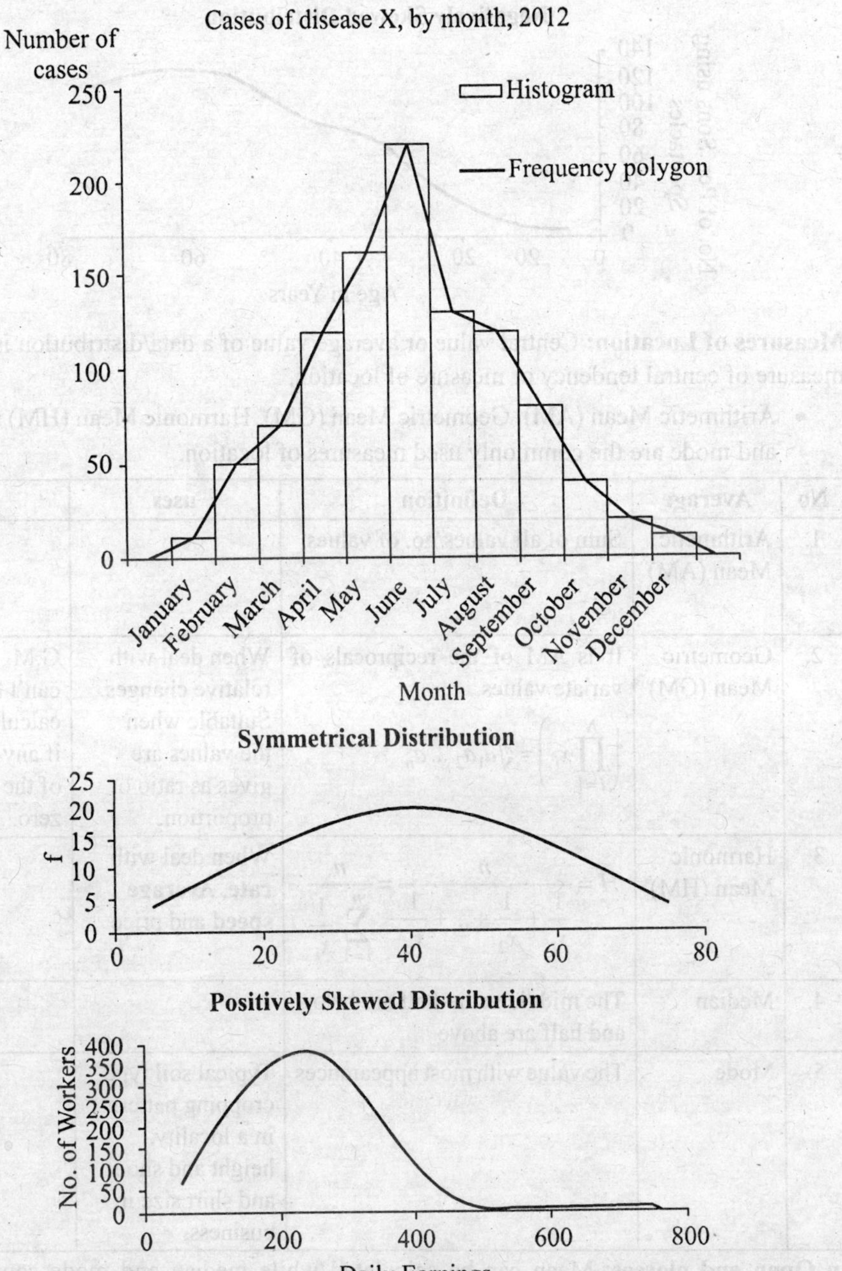
Cases of disease X, by month, 2012
Number of cases
250
200
150
100
50
0
Histogram
Frequency polygon
January
February
March
April
May
June
July
August
September
October
November
December
Month
Symmetrical Distribution
25
20
15
10
5
0
f
0
20
40
60
80
Positively Skewed Distribution
400
350
300
250
200
150
100
50
0
No. of Workers
0
200
400
600
800
Daily Earnings

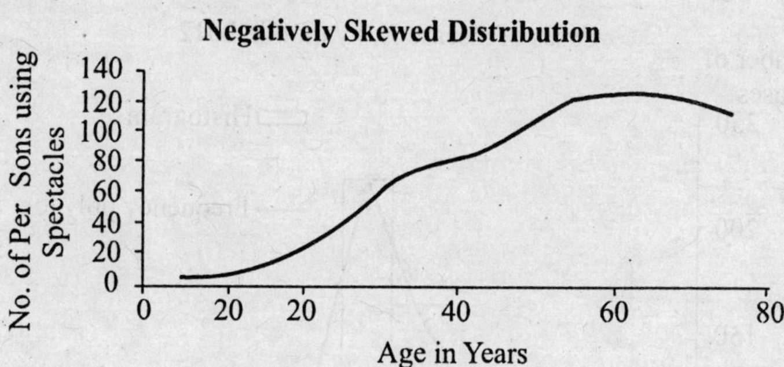

Measures of Location: Central value or average value of a data/distribution is called measure of central tendency or measure of location.

- Arithmetic Mean (AM), Geometric Mean (GM), Harmonic Mean (HM) median and mode are the commonly used measures of location.

No	Average	Definition	uses	
1.	Arithmetic Mean (AM)	Sum of all values/no. of values		
2.	Geometric Mean (GM)	It is AM of the reciprocals of variate values. $\left(\prod_{i=1}^{N} x_i\right) = \sqrt[n]{a_1 a_2 \ldots a_n}$	When deal with relative changes. Suitable when the values are gives as ratio or proportion.	G.M. can't be calculated if any value of the set is zero.
3.	Harmonic Mean (HM)	$H = \dfrac{n}{\dfrac{1}{x_1} + \dfrac{1}{x_2} + \ldots + \dfrac{1}{x_n}} = \dfrac{n}{\sum_{i=1}^{n} \dfrac{1}{x_i}}$	When deal with **rate, Average speed** and price	
4.	Median	The middle value-half are below and half are above		
5.	Mode	The value with most appearances	Typical soil type, cropping pattern in a locality, height and shoe and shirt size in business,	

In Open end classes: Mean can be calculated while median and mode cannot be calculated.

- Mean is based upon all the observations while median and mode are not.
- Mean can be put for further algebraic treatment while median and mode cannot be.

- Mean is less affected by fluctuations of sampling compared to median and mode.
- Mean is affected very much by extreme values while median and Mode is not at all affected by extreme values.
- The distribution can have multi-modes but not multi-mean and multi-median which means mode is not while mean and median are unique
- Second quartile is same as median.
- Geometric mean is antilog of the arithmetic mean of the logarithmic values of the variable.
- Harmonic mean is the reciprocal of the arithmetic mean of the reciprocal of the observations of the data set.
- Q2(second quartile)=50th percentile=5th decile
- Geometric mean is better than other means when, the data are in ratios and percentages.
- In a discrete series having 2k+1 observations, median is k+1 th term
- The first quartile divides a frequency distribution in the ratio of 1:3, second into 1:1 and third into 3:1
- Mode is the value around which the items tend to be most heavily concentrated.
- Sum of the absolute deviations about median is minimum
- Sum of the deviations about mean and sum of the square of the deviations about mean is zero.
- Histogram can be useful to find the mode graphically.
- Median is the suitable measure of central tendency for qualitative data.
- The eightieth percentile divides a frequency distribution in the ratio of 8:2 *i.e* 4.
- Harmonic mean gives less weightage to large values and vice versa.
- Weighted mean ≥ A.M. in terms of efficiency
- For a highly variable series the most suitable mean is Harmonic mean

MEAN

- Mean is affected by origin and scale. If five is subtracted from each observation, then the mean observation is reduced by five and if each value is divided by five, then the mean of observations is $1/5^{th}$ of the mean of original observations
- The algebraic sum of the deviations from arithmetic mean is zero *i.e.* **S(x-A.M.) = 0**
- Arithmetic mean cannot be obtained if a single observation is missing or lost
- Sum of squares of the deviations from mean is always zero.

MEDIAN

- Median can be calculated for grouped data with open end classes.
- The sum of the absolute deviations taken from median is minimum.
- In case of even number of observations median cannot be determined exactly. We merely estimate it by taking the mean of two middle terms.

QUARTILES

- The quartiles divide the distribution in four parts. There are three quartiles.
- The first (lower).quartile (Q1) marks off the first one-fourth, the third (upper) quartile (Q3) marks off the three-fourth. It may be noted that the second quartile is the value of the median and 50th percentile.
- Mean – Mode = 3 (Mean – Median)
- A.M>G.M.>H.M except when all the observations are same then A.M.=G.M. =H.M.
- G.M.= sqrt(A.M. X H.M.)
- Mode is suitable for qualitative data.
- Ans: True
- Decile divides the group in to ten equal parts.
- Ans: True
- Geometric mean can be calculated for negative values. Ans: False

DISPERSION

Dispersion means scattering of the observations among themselves or from a central value (Mean/ Median/ Mode) of data.

The various measures of dispersion are:

- Range
- Mean deviation
- Variance
- Standard deviation
- Coefficient of variation
- Quartile deviation
- Positive square root of variance is called **Standard Deviation**
- Standard deviation is absolute measure of **dispersion**
- Coefficient of variation is relative measure of dispersion.

Range: The difference between the value of the smallest item and the value of the largest item of the series is called range.

- Range=Largest item (L)– Smallest item(S) Co-efficient of Range = L – S/L + S

Inter-Quartile Range: Inter-quartile range represents the difference between the third-quartile and the first quartile. It is also known as the range of middle 50% values.

Inter-quartile range = Q3 – Q1

Percentile Range: It is the difference between the values of the 90th and 10 th percentile. It is based on the middle 80% items of the series.

Percentile Range = P90 – P10

Quartile Deviation / Semi–Interquartile Range: Quartile Deviation gives the average amount by which the two quartiles differ from the median. Quartile deviation is an absolute measure of dispersion. It is calculated by using the formula –

- Q.D. = Q3 – Q1/2
- Coefficient of Q.D. = Q3 – Q1/Q3 + Q1

Mean Deviation: It is the mean of the absolute deviations of variate values from a central value. Larger is the mean deviation higher is the dispersion in the data.

Variance: It is the mean of the squares of the deviations of variate values from their mean.

- Positive square root of variance is called Standard Deviation.
- **Coefficient of variation** It is the ratio of standard deviation to the mean and is expressed in %.
- C.V. = S.D. X 100/Mean
- Larger is the value of C.V. higher is the dispersion in the data. CV is free from the units of measurements and can be used to compare the dispersion in the set of data with same or different units of measurements where as all other measurers of dispersion discussed earlier are not free from units of measurements and can be used to compare dispersion in the sets of data having the same units only.
- Which measure is affected most by the presence of extreme values?

  (a) Range (b) Standard Deviation (c) Quartile Deviation (d) Mean deviation

  Ans: Standard Deviation
- If the CV of variety I is 40% and variety II is 20% then Variety II is more consistent.
- For the set of data 6, 6, 6,6,6,6 the Standard deviation value is zero.
- The absolute measures of dispersion will have the original units.
- The mean deviation value for a set of data cannot take negative value.
- Measure of dispersion suitable for comparing any two series is **Coefficient of variation**.
- Standard error of sample mean is standard deviation/sqrt(n)

Skewness:

- The departure from symmetry of a curve is called skewness
- For a symmetrical distribution mean, mode and median all are equal.
- Like coefficient of variation coefficient of skewness is also a pure number and have no units of measurements and thus can be used to compare the skewness in sets of data with same or different units of measurements.
- When the values of mean, median and mode are equal, there is no Skewness.
- When mean > median > mode, Skewness will be positive.
- When mean < median < mode, Skewness will be negative.

Kurtosis:

- **Leptokurtic,** β2 > 3 or Kurtosis > 0
- **Mesokurtic,** β2 =3 or the normal curve
- **Platykurtic,** β2 < 3 or When Kurtosis < 0,
- kurtosis (β2) is also pure number having no units of measurements
- It measures the relative peakedness or flatness of a distribution compared to the normal distribution

Basis	Dispersion	Skewness
Nature	Shows the spread of values from the central value	Shows whether series is symmetrical orasymmetrical.
Inference	All the measures of dispersion are positive.	Coefficient of skewness can be positive or negative.

Dependency on origin and scale:

Term	Origin	Scale
Mean	Dependent	Dependent
Variance/standard deviation	Independent	Dependent
Coefficient of correlation	Independent	Independent
Coefficient of Variation (C.V.)	Dependent	Independent
Regression coefficient	Independent	Dependent

Range

Term	Range
Correlation coefficient	-1 to 1
Partial correlation coefficient	-1 to 1
Multiple correlation coefficient	0 to 1
Coefficient of determination (R^2)	0 to 1
Regression coefficient	$-\infty$ to ∞

Karl pearson's coefficient of skewness	-3 to +3
Normal distribution	$-\infty$ to ∞
probability	0 to 1
Coefficient of contingency	0 to 1
T test	$-\infty$ to ∞
F test	0 to ∞
Z test	$-\infty$ to ∞
Chi-square	0 to ∞
Level of significance	0 to 1

TEST of Significanc

Condition	Size	Variance of distribution	Test used
Test for single/ Comparison for two means/proportions	For both large and small sample	known	(Always)
Z test			
Test for single/ Comparison for two means/proportions	n>30, large sample	Unknown	F test
Test for single/ Comparison for two means/proportions	n<30, small sample	Unknown	T test
Comparison for more than two means	For both large and small sample	known	(Always) Z test
Comparison for more than two means	For both large and small sample	Unknown	F test
Test the hypothetical value of the population variance	Small sample	Unknown	Chi-square
To test the goodness of fit	Small sample	Unknown	Chi-square
To test the independence of attributes	Small sample	Unknown	Chi-square
To test the significance of Sample correlation coefficient	Small sample	Unknown	T test
To test the significance of Sample regression coefficient	Small sample	Unknown	T test
To test the significance of partial correlation coefficient	Small sample	Unknown	T test
To test the equality of two population variance	Small sample	Unknown	F test
To test the significance of multiple correlation coefficient	Small sample	Unknown	F test
To test the equality of several mean	Small sample	Unknown	F test(ANOVA)

Hypothesis:

- **Null Hypothesis** ($H_{o)}$**:** The hypothesis which is tested for possible rejection under the assumption that it is true.
- **Alternative hypothesis**($H_{1)}$ **:** The hypothesis which complementary to the null hypothesis

Error in Sampling

- **Type I Error:** Rejected H_o when it is true.
- **Type II Error:** Accepted H_o when it is wrong. *i.e.,* Accepted H_o when H_1 is true.
- **P(Type I Error)=** α
- **P(Type II Error)=** β

REGRESSION AND CORRELATION

Basis	Correlation	Regression
Relationship	Correlation tells the degree of average relationship between two or more variables.	Using relationship between known variables and unknown variables, the unknown variable is estimated
Cause & Effect	Unable to tell which series is the cause and which is the effect, despite high degree of correlation.	The given independent variable is the 'cause' and dependent variable is the effect.
Change in scale	Unaffected by the change of origin and scale.	Regression coefficient is affected by change in scale.
Application	Limited Application.	Wider Application.

Regression Equation X on Y	Regression Equation Y on X
Original Form:	**Original Form:**
X = a + by	Y = a + bx
Formula :	**Formula :**
X - X = bxy (Y - Y)	Y - Y = byx (X - X)
Here -	Here -
X → Mean of x series.	X → Mean of x series.
Y → Mean of y series.	Y → Mean of y series.
bxy → Regression coefficient X on Y	byx → Regression coefficient Y on X

(i) If both the coefficients are positive correlation coefficient will be positive, and if both the coefficients are negative, the coefficient of correlation will also be negative.

(ii) Both the regression coefficients will have the same sign.

(iii) The product of both the regression coefficients cannot be more than 1.

Calculation of Correlation Coefficient f om Regression Coefficient

r = (bxy) X (byx)

bxy → Regression coefficient x on y

byx → Regression coefficient y on x

DISCRETE PROBABILITY DISTRIBUTIONS

1. Binomial Distribution: A random variable 'X is said to follow binomial distribution with parameter *n* and *p* if it assumes only nonnegative values and its probability mass function is given by

$$P(X = x) = p(x) = \begin{cases} {}^{n}c_x p^x q^{n-x}; & x = 0, 1, ..., n; q = 1 - p \\ 0 & , \text{ otherwise} \end{cases}$$

Where: *p* and *q* are probability of success and failure respectively.

Mean = *np*

Variance = *npq*

Mean > variance

2. **Poisson Distribution:** A random variable X is said to follow a Poisson distribution with parameter λ if it assumes only non-negative values and its probability mass function is given by

$$P(X = x) = p(x) = \begin{cases} \dfrac{e^{\lambda}\lambda^x}{x!}; & x = 0, 1, ..., ; \lambda > 0 \\ 0 & , \text{otherwise} \end{cases}$$

Mean = λ

Variance = λ

Mean = variance

- Poisson distribution is widely used as a possible model in biological sciences.

3. Continues probability distributions: The normal distribution is one of most useful theoretical distribution for interpretation of research data. Gauss discovered this distribution in 1809.

A continuous random variable 'X' is said to have a normal distribution with parameter μ(mean) and σ^2 (variance) if its probability density function is given by

$$f(x) = \frac{1}{\sigma\sqrt{2\pi}} \exp\left\{-\frac{1}{2}\left(\frac{x-\mu}{\sigma}\right)^2\right\}$$

Standard Normal Variate

A continuous random variable 'X' is distribution with parameter μ(mean) and σ^2 (variance) then the variable $z = \frac{x-\mu}{\sigma}$ is distributed as normal with mean 0 and variance 1 and *z* is called standard normal variate.

- The normal distribution curve is bell-shaped, symmetrical about the line $x=\mu$ and mean, median and mode of the distribution coincide.
- Areas covered by the various ranges of the normal random variable are given by:

  $p(\mu - \sigma < x < \mu + \sigma) = 0.6826$

  $p(\mu - 2\sigma < x < \mu + 2\sigma) = 0.9544$

  $p(\mu - 3\sigma < x < \mu + 3\sigma) = 0.9973$
- ***Q.D:M.D.S.D.***: 10:12:15
- Most of the distributions occurring in practice *i.e.*, Binomial, Poisson, *etc.*,, can be approximated by normal distribution.
- Many of the sampling distribution *i.e.,* Student *t*, *F*, Chi-square *etc.*,, tend to normality for large sample.

DEIGN OF EXPERIMENTS

Results of experiments are affected by

- Known cause (due to different treatment)
- Unknown cause (due to experimental material)
- The results of experiments are affected not only by the action of treatments, but also by extraneous variations which tend to mask the effect of the treatments.
- Experimental error includes all type of extraneous variations.
- Source of Experimental Error:

  (i) Inherent variability in experimental material

  (ii) Lack of informality in physical conduct in experiment.
- Purpose of design of experiment is to minimize the experimental error so that we can compare the known cause and unknown cause.
- If σ^2 is the error variance per unit and there are r replications, the error variance of the difference between the mean of two treatments is $2\sigma^2/r$.
- Three basic techniques fundamental to design of experiments are replication
- **Local control (Blocking):** blocking is the arranging of experimental units in groups (blocks) that are similar to one another for reducing experimental error
- Randomization
- Replication and blocking help to increase precision in the experiment, randomization is used to decrease bias.

Type of Design	application	Error D.F	remarks
Complete Randomized Design(CRD)	Homogeneous experimental unit (no variability in experimental unit)	N-n	Zero way elimination, Mainly used for lab conditions, Number of replications for different treatments need not be equal
Randomized Block Design (RBD)	Experimental material is not homogeneous, variability in experimental unit in one direction	(n-1) (r-1)	One way elimination, Mainly used for Field conditions
Latin Square Design (LSD)	Variability in experimental unit in Two direction	(n-1) (r-1)	Two way elimination, Mainly used for animal experiments and when the soil fertility gradient is in two directions
Split Plot Design	To study the effect of two or more factors. Certain factors may require larger plots than for others.	Main plot (A) error: (r-1)(a-1) Sub-plot (B) error: a (r-1) (b-1) Interaction: AB (a-1) (b-1)	Greater degree of precision is obtained for sub lot than the main plot
Strip plot design	Two factors each require large experimental units	Main plot(A) error: (r-1)(a-1) Main plot(B) error: (r-1) (b-1) Interaction: AB (r-1)(a-1) (b-1)	Allow for greater precision in the measurement of the interaction effect while sacrificing the degree of precision on the main effect.
Factorial Experiment	To study the effect of two or more factors and their interaction	2^n factorial experimental (r-1) (2^n-1)	Design provides information on the average effect of the individual factors as well as the interaction between these factors

Augmented Designs	Control treatments (checks) are replicated *r* times (adequate no of experimental unit is available) and other new treatments (test treatment) are replicated less than *r* times(limited no of experimental unit is available)		

Where, N= Total number of observations, n=number of treatments and r = number of replications

STATISTICS OBJECTIVE QUESTIONS (MCQ)

1. The third quartile is also called
   A. Lower quartile B. Median C. Mode D. Upper quartile
2. Which of the following is not a two dimensional diagram
   A. Square diagram B. Multiple bar diagram
   C. Rectangular diagram D. Pie chart
3. Non dimensional diagrams are also known as
   A. Histograms B. Pie-charts C. Pictograms D. Bar-charts
4. With the help of Ogive curves one can determine
   A. Variance B. Skewness C. Median D. Correlation
5. Ogives for more than type and less than type intersect at
   A. Mean B. Median C. Mode D. Origin
6. Cumulative frequencies are computed in order to draw a/an
   A. Histogram B. Frequency polygon
   C. Ogive curve D. Pie-chart
7. The components of a pie chart are computed for a factor by
   A. Percentages B. Angles C. Sectors D. Circles
8. In the following, which is a discrete variable
   A. Area B. Height
   C. Number of households D. Time
9. Geometric mean of 4 and 9 is
   A. 4 B. 5 C. 6 D. 7
10. The modal class is the class that includes
    A. Largest number of items B. Smallest number of items
    C. Large value of all items D. Smallest value of all items

11. The collection of all possible outcomes of an experiment is called
    A. A simple event B. A sample space C. A sample D. A population
12. Probability can never be greater than
    A. 0.2 B. 0 C. 1 D. 0.5
13. The distribution for which mean and variance are identical is
    A. Binomial B. Poisson C. Normal D. Geometric
14. If correlation coefficient of the variables X and Y is zero, then X and Y are
    A. Linearly related B. Not necessarily independent
    C. Not linearly related D. Dependent
15. If correlation coefficient is equal to 1, then the relationship between the two variables X and Y is
    A. Y is proportional to X B. Y is inversely proportional to X
    C. Y is equal to X D. None of these
16. In the regression line of Y on X, the variable X is known as
    A. Dependent variable B. Explanatory variable
    C. Explained variable D. Random variable
17. If an investigator selects states from a nation, districts from the state and panchayat samities from a district. Then such a sampling procedure is known as
    A. Two stage sampling B. Three-stage sampling
    C. Cluster sampling D. Stratified sampling
18. If a distribution has three local maximums, its frequency curve will be
    A. Bell shaped moderately skewed B. J-shaped
    C. Multimodal D. Unimodal
19. If one of the regression coefficients of X on Y is greater than unity, then the other must be
    A. Equal to 1 B. Greater than 1 C. Less than 1 D. Zero
20. The discrepancy between estimates and population parameters is known as
    A. Human error B. Enumeration error
    C. Sampling error D. All of thes
21. If the sample values are 1,3,5,7,9 then the standard error of sample mean is
    A. Sqrt(2) B. 1/sqrt(2) C. 2 D. 1/2
22. Which of the following statements does not hold good in case of stratified sampling?
    A. Stratified sampling is convenient
    B. Stratified sampling is always good
    C. It enables to gather information about different strata separately
    D. It reduces error for fixed cost

23. F-distribution is applied for
    A. Testing the equality of two population variances
    B. Testing the equality of two or more population means
    C. Testing the equality of several regression coefficients
    D. All of these
24. If n=sample size and N=population size, the finite population correction is
    A. (N-1)/N B. (n-1)/N C. (N-n)/N D. (N-n)/n
25. Which sampling design is most appropriate for cluster sampling
    A. Simple random sampling without replacement
    B. Simple random sampling with replacement
    C. Startified random sampling
    D. Quota sampling
26. Chi square distribution curve in respect of symmetry is
    A. Negatively skewed B. Symmetrical
    C. Positively skewed D. Any of these
27. T distribution is used to test
    A. The validity of a postulated value of population mean
    B. The significance of sample correlation coefficient
    C. The equality of two population means
    D. All of these
28. Chi-square distribution is used for the test of
    A. Goodness of fit
    B. Hypothetical value of population variance
    C. Both 1 and 2
    D. Neither 1 nor 2
29. Which of the following does not receive the application of chi- square distribution
    A. Testing the equality of variances of two normal distributions
    B. Testing the independence of attributes
    C. Testing the hypothetical value of the variance of a normal population
    D. Testing goodness of fit
30. To test the population variance in small samples, the test used is
    A. Normal test B. F test C. t-test D. Chi-square test
31. Degrees of freedom for Chi-square in case of contingency table of order 4X3 are
    A. 12 B. 9 C. 8 D. 6
32. Equality of several normal population means can be tested by
    A. Bartlett's test B. F test C. Chi-square test D. T-test

33. The term sampling error refers to
    A. Mistakes
    B. Bias
    C. Difference between statistic and parameter
    D. Both mistakes and bias
34. The expression n/N is called as
    A. Sampling frame B. Sampling fraction
    C. Sampling error D. Strata
35. Sampling frame represents
    A. List of random number
    B. List of voters
    C. List of sampling units of a population
    D. List of sampling methods
36. In which of the following situations cluster sampling is appropriate?
    A. When the units are situated far apart
    B. When sampling frame is not available
    C. When all the elementary units are not easily identifiable
    D. All of these
37. Sampling error can be reduced by
    A. Choosing a proper probability sampling
    B. Selecting a sample of adequate size
    C. Using a suitable formula for estimation
    D. All of these
38. A systematic sample does not yield good results if
    A. Variation in units is periodic
    B. Units at regular intervals are correlated
    C. Both (A) and (B)
    D. None of these
39. As a normal practice, sampling fraction is considered to be negligible if it is
    A. Less than 10 percent B. Less than or equal to 5 percent
    C. More than 10 percent D. More than 5 percent
40. The errors in survey other than sampling errors are called
    A. Formula errors B. Planning errors
    C. Non sampling errors D. None of these
41. The number of possible samples each of size n can be drawn out of N population units with SRSWOR is
    A. N^n B. n^2 C. NCn D. n!

42. An unordered sample of size n can occur in

    A. n ways B. n! Ways C. one way D. n2 ways

43. Replication in an experiment means

    A. The Number of blocks

    B. The number of plots

    C. The number of times a treatment occurs in an experiment

    D. Total number of experiments

44. The experiments in which treatments are allocated through random process to experimental units are categorized as

    A. CRD B. RBD C. LSD D. Factorial design

45. Analysis of experimental data means

    A. Estimation of treatment effects

    B. Dividing the total variance into component variances

    C. Testing of hypothesis about the parameters involved in the experimental mode

    D. All of these

46. The analysis of variance of an experimental data is based on the assumptions that

    A. The response variable is distributed normally

    B. The errors are independent

    C. The errors are homoscedastic

    D. All of these

47. Measures of dispersion include

    A. Mean, variance quartile deviation

    B. Mean deviation, standard deviation, range

    C. Range, geometric mean, median

    D. Variance, coefficient of variance, mode

48. A cycle in a time series is represented by the difference between

    A. Two successive peaks

    B. The end points of a converse portion

    C. The mid points of a trough and crest

    D. None of these

49. Increase in population in India represents which of the following components?

    A. Trend B. Seasonal Variation

    C. Cyclic variation D. Irregular variation

50. A time series consists of

    A. Two components B. Three

    C. Four D. Five

51. Secular trend is indicative of long term variation towards
    A. Increase only B. Decrease only
    C. Either increase or decrease D. None of these
52. Method of least squares to fit in the trend is applicable only if the trend is
    A. Linear B. Parabolic C. Both 1 & 2 D. Neither 1 & 2
53. The equation y=aBx represents
    A. Gompertz curve B. Exponential curve
    C. Logarithmic curve D. None of these
54. Linear trend of a time series indicates towards
    A. Constant rate of change B. Constant rate of growth
    C. Change in geometric progression D. All of these
55. Analysis of time series deals with
    A. Economic data
    B. Forecasting future values
    C. Knowing observed variables
    D. Knowing observed variables and forecasting future values
56. The value of b in the trend line y=a +bx
    A. Is always positive B. Is always negative
    C. Is both positive or negative D. Always take the value one
57. The mean of first' n' natural numbers is
    A. (n+ l)/2 B. (n+2)/2 C. (2n+l)/2 D. 2((n+l)/2)
58. The median of the following values : 7, 8, 5, 3, 2, 9, 4, 6 , 1
    A. 4 B. 5 C. 6 D. 7
59. 10 is the mean of a set of 7 observations and 5 is the mean of a set of 3 observations. The mean of the combined set is given by
    A. 15 B. 10 C. 8.5 D. 7.5
60. The variance of a set of numbers is 36. Its standard deviation is
    A. 5 B. 6 C. 7 D. 8
61. When coefficient of skewness is zero, the distribution is
    A. J-shaped B. U-shaped C. Symmetrical D. L-shaped
62. The coefficient of mean deviation about mean is
    A. M.D./Mean B. Mean/M.D.
    C. M.D./ Mean/Median/Mode D. M.D./ Mode
63. The geometric mean of two numbers 8 and 18 is
    A. 12 B. 13 C.14 D. 15
64. The coefficient of Quartile Deviation is calculated using the formula
    A. (Q3 – Q1)/(Q3 + Q1) B. (Q3 + Q1)/(Q3 – Q1)
    C. (Q3 – Q1)/2 D. (Q3 + Q1)/2

65. Data classified on the basis of 'time' is referred to as
A. Quantitative classification B. Qualitative classification
C. Chronological classification D. Geographical classification
66. Quartile deviation for normal distribution equal to
A. (2/3)S.D. B. (4/5)S.D. C. (3/2)S.D. D. (6/7)S.D.
67. Median is better suited for
A. Nominal level data B. Ordinal level data
C. Interval level data D. Ratio level data
68. The coefficient of variation of Poisson distribution with mean 4 is
A. ¼ B. 2/4 C. 4 D. 2
69. "Lack of Memory" property is associated with
A. Geometric Distribution
B. Binomial Distribution
C. Negative Binomial Distribution
D. Poisson Distribution
70. Select the most appropriate design when all factors are not of equal importance in experimentation
A. Augmented B. Strip plot design
C. LSD D. Split plot design
71. CCE crop cutting experiment recommended plot size is
A. 20m X 10m B. 10 m X 10m C. 10m X 5m D. 5 m X 5 m
72. The chi-square test is
A. Parametric test B. Non parametric test
C. Both 1 & 2 D. Not classified
73. When both the variables are not normally distributed, in such a case we may use
A. Partial correlation B. Multiple correlation
C. Rank correlation D. Both 1 &2
74. The relationship between independent and dependent variable is evaluated by
A. Regression coefficient B. Correlation coefficient
C. Both 1 & 2 D. None of these
75. The optimum plot size in the experiment is determined by
A. Maximum curvature method B. Fair field smith variance law
C. Both 1 & 2 D. Uniformity trial
76. The critical difference (CD) is a form of t test and its formula is given by
A. CD = t X SE(d) B. CD = t + SE(d)
C. CD = t/SE(d) D. CD = F X SE(d)
77. If two factors are involved in experimentation and both the factors requires long plot sizes, the design suited for this will be
A. Split plot B. Strip plot C. CRD D.RBD

78. Coefficient of Skewness and Kurtosis of Normal distribution are

A. -3, 3 B. 3,3 C. 3,0 D. 0,3

79. Assertion: Normal distribution is symmetrical

Reason: Mean, median and mode coincide in normal distribution

A. Assertion is correct but Reason is wrong

B. Both Assertion and Reason are wrong

C. Assertion is wrong but Reason is correct

D. Both Assertion and Reason are correct

80. Which of the following is true in respect of Normal Distribution?

A. β1 = o, β2 = □ B. β1 = 3, β2 = O C. β1 = 0, β2 = 3 D. None of these

81. Which of the following is true in case of Normal distribution

A. Sum of two independent normal variates is normal

B. Difference of two independent normal variates is normal

C. Both 1 & 2

D. None of these

82. The normal distribution is a limiting case of binomial distribution if

A. n tends to □, p tends to 0 B. n tends to 0, p tends to q

C. n tends to □, p tends to n D. n tends to □ and p or q is small

83. If A and B are independent events, then which of the following is correct?

A. and are independent B. □ and B are independent

C. and are independent D. All of the above

84. When the correlation coefficient is r ± 1, then the two regression lines

A. Are perpendicular to each other B. Coincide

C. Are parallel to each other D. Do not exist

85. The coefficient of correlation will have positive sign when

A. X is increasing, Y is decreasing B. both X and Y are increasing

C. X is decreasing, Y is increasing D. there is no change in X and Y

86. If one regression coefficient is greater than unity, then the other must be

A. Greater than the first one B. Equal to one

C. Less than one D. Equal to zero

87. Two independent variables are

A. Correlated B. Uncorrelated

C. partly correlated D. partly uncorrelated

88. Correlation coefficient between X and Y is

A. The geometric mean between the regression coefficients

B. the arithmetic mean between the regression coefficients

C. the harmonic mean of the regression coefficients

D. the average of the regression coefficients

89. Farther the two regression lines cut each other, the ---- will be the degree of correlation
   A. Greater B. Lesser
   C. neither greater nor lesser D. Zero
90. In the equation of a straight line Y =a +bX, if 'b' is negative, then the slope is
   A. Upward B. Downward
   C. Parallel to X-axis D. Parallel to Y-axis
91. The standard deviation of the sampling distribution of a statistic is known as
   A. Sampling unit B. Sample survey
   C. Sampling distribution D. Standard error
92. Test for single proportion, for large n, one can apply
   A. Normal test B. t-test C. F-test D. Chi-square test
93. In test of significance for difference of proportions of two populations, for large samples, one can apply
   A. t-test B. F-test C. Chi square test D. Normal test
94. In test of significance for single mean, for large samples, one can apply
   A. Chi Square test B. T test C. F –test D. Normal test
95. In test of significance for the difference of standard deviations of two populations, for large samples, one can use
   A. t –test B. X^2-test C. Normal test D. F-test
96. The chi-square and the t-distribution are both
   A. always symmetrical distributions
   B. used for hypothesis testing
   C. dependent on the number of degrees of freedom
   D. (B) and (C) but not (A)
97. Every hypothesis testing problem, which type of error is always present
   A. Type I error only B. Type II error only
   C. Both Type I and Type II errors D. Neither Type I nor Type II error
98. Fundamental assumption of' F ' test are not
   A. That the population is abnormal
   B. That observations are independent and the samples drawn are random samples
   C. That there is no measurement error
   D. That known as variance ratio test
99. Which of the following is not probability sampling?
   A. Stratified Sampling B. Systematic Sampling
   C. Cluster Sampling D. Quota Sampling

100. Identify the demerit of simple random sampling
   A. There is less chance for personal bias
   B. More representative
   C. Sampling error can be measured
   D. This requires a complete list of the population
101. Identify the demerit of cluster sampling
   A. It introduces flexibility in the sampling method
   B. It is helpful in large-scale survey where the preparation of list is difficult
   C. It is less accurate than other methods
   D. It is valuable in countries where no accurate framework is available
102. Identify the non - essentials of sampling
   A. It must be representative
   B. Sample size should be infinite
   C. Selected samples from the universe should have similar nature
   D. A good number of items are to be included in the sample
103. Which of the following is not non-probability sampling?
   A. Purposive Sampling B. Quota Sampling
   C. Convenience Sampling D. Stratified Sampling
104. Which of the following is not an advantage of RBD?
   A. Accuracy
   B. Flexibility
   C. Ease of analysis
   D. Not suitable for large number of treatments
105. The inherent variability in the experimental material to which treatments are applied is called as
   A. Experimental error B. Design error
   C. Standard error D. Treatment error
106. Which principle is used to average out the influence of the chance factors on different experimental units?
   A. Replication B. Randomization
   C. Local control D. (B) and (C) but not (A)
107. Analysis of variance of two-way classified data is analogous to analysis of variance of which design?
   A. CRD B. RBD C. LSD D. Lattice design
108. Which principle is used in the allocation of treatments to experimental units?
   A. Replication B. Randomization
   C. Local control D. (A) and (C) but not (B)

109. The empirical relationship between the plot size and the plot variance is known as

A. Fisher's variance law
B. Yate's variance law
C. Das and Giri variance law
D. Fairfield Smith's variance law

110. The least significant difference at the given level of significance is known as

A. Critical Difference
B. Mean Difference
C. Standard Difference
D. Absolute Difference

111. Let TSS = Total Sum of Squares

SST = Sum of Squares due to Treatment

SSV = Sum of Squares due to Variety

SSE = Sum of Squares due to Error

In analysis of variance for CRD, identify the correct equation

A. TSS = SSE
B. TSS = SSE + SST
C. TSS = SSE + SST + SSV
D. TSS = SST + SSV

112. Let TSS =Total Sum of Squares

SST = Sum of Squares due to Treatment

SSB = Sum of Squares due to Blocks

SSE = Sum of Squares due to Error

In RBD, which of the following equation is correct?

A. TSS = SSE + SST
B. TSS = SSE + SST + SSB
C. TSS = SSE + SSB
D. TSS = SST + SSB

113. Analysis of variance is based upon a comparison of two estimates of the

A. Variance of the overall population that contains all sample
B. Mean of the overall population that contains all samples
C. Mean deviation of the overall population that contains all sample
D. Range of the overall population that contains all samples

114. Sampling error decreases with

A. Lesser sample
B. Increase in sample size
C. Increase of locality
D. Lesser questions

115. $\Sigma(X_i - \bar{X})$ is always equal to

A. –1
B. 0
C. 1
D. 2

116. There will be only one regression line in the case of two variables if the correlation coefficient is

A. 0
B. +1
C. –1
D. ±1

117. If the sample size n>30 the t distribution tends to

A. Normal distribution
B. F distribution
C. Cauchy Distribution
D. Chi-square distribution

118. Analysis of time series determines
    A. Cyclical, seasonal variations and random fluctuations
    B. Cyclical, seasonal variations and trend
    C. Cyclical, random fluctuations and trend
    D. seasonal variations, random fluctuations and trend
119. The middle value of an ordered series is called
    A. 2nd quartile B. 5th decile
    C. 50th percentile D. All the above
120. With the help of histogram we can draw
    A. Frequency polygon B. frequency curve
    C. Frequency distribution D. All the above
121. Mean and variance of a standard normal variate
    A. μ, $\sigma 2$ B. 0, $\sigma 2$ C. μ, 1 D. 0, 1
122. In Normal Distribution P{$\mu - \sigma < X < \mu + \sigma$} covers----% of area
    A. 68 B. 75 C. 98 D. 99
123. The point of inflection of N(μ, $\sigma 2$) curve are
    A. $\mu \pm \sigma$ B. $\mu \pm 2\sigma$ C. $\mu \pm 3\sigma$ D. $\mu \pm 2/3\sigma$
124. In a normal curve 95% of the observations are included in the range
    A. $\bar{x} \pm 3\sigma$ B. $\bar{x} \pm 1.96\sigma$ C. $\bar{x} \pm 2\sigma$ D. $\bar{x} \pm 0.67\sigma$
125. The Normal Distribution is a limiting case of Poisson Distribution when the parameter
    A. λ tends to 0 B. λ tends to - □ C. λ tends to □ D. λ tends to 1
126. For a set of values the model value can be
    A. Unimodel B. Bimodal C. Trimodel D. All of these
127. The regression coefficients are bi and b2. Then the correlation coefficient r is
    A. b1/b2 B. b2/b1 C. b1b2 D. $\pm\sqrt{b1b2}$
128. For the validity of $\chi 2$ test of goodness of fit, the following constraint on the cell frequencies should be satisfied
    A. $\Sigma Oi > \Sigma Ei$ B. $\Sigma Oi < \Sigma Ei$ C. $\Sigma Oi = \Sigma Ei$ D. $\Sigma Oi \neq \Sigma Ei$
129. For the validity of $\chi 2$ - test of goodness of fit, the minimum theoretical cell frequency should be
    A. 5 B. 10 C. 30 D. 50
130. If the sample size n is small, to test the significance of an observed multiple correlation coefficient, which test is used
    A. $\chi 2$ –test B. F-test C. normal test D. Z test
131. The mode is defined as
    A. Highest frequency
    B. Largest observation

C. The observation with largest frequency

D. None of these

132. Which of the following is an objective of classification?

A. To condense and simplify the data

B. To bring similarities and dissimilarities

C. To help comparison

D. All the above

133. The height of 25 students of class five are taken. If the height of a 6 feet teacher is also included then which measure of central tendency is most affected

A. A.M. B. G.M. C. H.M. D. Median

134. Which of the graphical method is used to find mode?

A. Pie diagram B. Frequency polygon

C. Ogive curve D. All the above

135. From a less than Ogive curve which of the following can be determined

A. First Quartile B. Second Quartle C. Third Quartile D. All the above

136. When the observations are equal the relation between AM, GM and HM is?

A. A.M = GM = GM B. AM < GM < HM

C. AM > GM > HM D. AM < HM < GM

137. The Standard deviation is minimum if deviations are taken from

A. Zero B. Mean C. Median D. Mode

138. Which of the following measures of dispersion depends on all the observations?

A. Standard deviation B. Mean deviation

C. Coefficient of variation D. All the above

139. For any discrete distribution the standard deviation is

A. Not less than mean square deviation from mean

B. Less than mean deviation from median

C. Equal to mean square deviation from mean

D. Not less than mean square deviation from zero

140. Which measure of location will be suitable to compare sale of shirts with different collar size?

A. Mean B. Mode C. Median D. None of the above

141. If two independent series of observations are to be compared, which is most suitable measure from the following

A. Standard deviation B. Mean deviation

C. Range D. Coefficient of variation

142. The coefficient of variation has

A. No unit

B. Same as unit of mean of observations

C. Same as unit of observations

D. Same as unit of variance of observation

143. The width of class interval depends on

A. The range of data

B. The number of class interval to be made

C. The type of data

D. All the above

144. While classifying the data into different class intervals we should take care of

A. Number of class intervals should neither be large nor small

B. No open ended interval as far as possible

C. Characteristic of the variable

D. All of the above

145. To make continuous class interval of a discrete data

A. We substract ½ from upper limit

B. We substract ½ from lower limit

C. We substract ½ from upper limit and add ½ to lower limit

D. We substract ½ from lower limit and add ½ to upper limit

146. Whether classification is done first or tabulation

A. Classification follows tabulation B. Tabulation follows classification

C. Both are done simultaneously D. None of the above

147. Which of the following is not based on all observations?

A. Mean deviation B. Standard deviation

C. Range D. All of the above

148. Which of the following is not a two dimensional diagram?

A. Square diagram B. Rectangular diagram

C. Pie diagram D. Bar diagram

149. For testing independence of m × n contingency table the degree of freedom of a chi-square will be

A. m × n B. (m – 1) × n

C. (m – 1) × (n – 1) D. M X(n – 1)

150. Diagram obtained by plotting (xi, yi): i = 1,2,3....n is called

A. Bar diagram B. Scatter diagram C. Pie diagram D. Histogram

151. Which is the positional average?

A. Mean B. Median C. Mode D. Harmonic mean

152. If you are told that a certain group of observations has mean 5 and variance zero, then which of the following is correct

A. There is mistake in calculation

B. Each observation equals to 5

C. There is only one observation in the group

D. There is no observation in the group

153. Which of the following is not a stable measure?

A. Standard deviation B. Coefficient of variation

C. Mean deviation D. Range

154. Suppose 100 students write a test and a student got 63 (out of 100) with 73rd percentile this means

A. At least 73 students got 63 or better

B. At least 27 students got 73 or better

C. At least 27 students got 63 or better

D. At least 73% student got 63 or better

155. Second decile divides the data in the ratio

A. 1 : 1 B. 2 :1 C. 1 :4 D. 1 : 3

156. Suppose every man marries a woman exactly 0.5 younger than him, which of the following is correct

A. The correlation coefficient between their ages is –0.5

B. The correlation coefficient between their ages is +0.5

C. The correlation coefficient between their ages is +1

D. The correlation coefficient between their ages is –1

157. The 50th percentile of a data is same as its

A. Mean B. Median C. Mode D. None of the above

158. If A and B are independent events then

A. P(AB) = 0 B. P(AUB) = P(A) + P(B)

C. P(AUB) = P(A) + P(B) – P(A) P(B) D. P(AB) = P(A).P(B)

159. If A and B are independent events then P(A/B)

A. 0 B. P(A) C. P(B) D. P(A) P(B)

160. The complete enumeration of all the units of population is called

A. Sample survey B. Census

C. Vital statistics D. None of the above

161. The correlation of X and Y is

A. The degree of linear relationship between them

B. The extent of functional relationship between them

C. Both 1 and 2

D. Neither 1 or 2

162. The correlation coefficient of X and Y in terms of regression coefficients Byx and Bxy is

A. Average of Byx and Bxy B. Geometric mean of Byx and Bxy

C. Harmonic mean of Byx and Bxy D. None of the above

163. The unit of correlation between X and Y

A. Is same as unit of X B. Is same as unit of Y

C. Is same as unit of X, Y D. There is no unit

164. If the correlation coefficient of X and Y is □, then the correlation coefficient between X-a and Y-b

A. > □ B. < □ C. – □ D. None of the above

165. If the correlation coefficient of X and Y is positive, then the correlation coefficient between -X and –Y

A. Positive B. Negative C. Zero D. Both 1 and 2

166. If X and Y have same values for all the observations, then the correlation coefficient will be

A. □ > 0 B. □ < 0

C. □ = 0 D. Cannot be determined

167. The correlation between the circumference and diameter of a circle is

A. + 1 B. –1

C. Zero D. Non sense correlation

168. If regression coefficient Byx is positive, then Bxy will be

A. Positive B. Negative

C. Zero D. Nothing can be said

169. If regression coefficient Byx >1, then Bxy will be

A. Bxy > 1 B. Bxy < 1

C. Bxy = 1 D. Nothing can be said

170. If Byx is regression coefficient of Y on X then regression coefficient of Y-a over X (a>0) is

A. No change B. Byx + a C. Byx-a D. Byx/a

171. Byx is defined as

A. Change in Y per unit change in X

B. Change in X per unit change in Y

C. Change in Y by a unit resulting change in X

D. None of the above

172. In regression of Y on X, the X is known as

A. Independent variable B. Regressor

C. Explanatory variable D. All of the above

173. If correlation coefficient=0 then the two regression line will be

A. Coincident B. Parallel

C. Perpendicular D. None of the above

174. To get idea of fertility conditions of a field, the method used is

A. Local control B. Replication C. Randomization D. Uniformity trials

175. Local control is used to
    A. Increase the number of plots
    B. To reduce the number of replications
    C. To reduce the degree of freedom of total sum of square
    D. To reduce error variance
176. In RBD homogeneity should be there in
    A. Between blocks B. Between treatments
    C. Within block D. None of the above
177. Probability of rejecting H0 when it is true is
    A. Alpha B. Beta C. 1-alpha D. 1-beta
178. To test the hypothesis about equality of two variances of two normal populations. Which test is used
    A. Normal test B. Chi-square test C. T test D. F test
179. To test the hypothesis about means of four normal populations. Which test is used
    A. Normal test B. Chi-square test C. T test D. Anova test
180. To test the hypothesis about the variances of normal population is equal to some assumed value. Which test is used?
    A. Normal test B. Chi-square test C. T test D. F test
181. To test the hypothesis that the mean of two normal populations (with common unknown variances) based on small sample. Which test is used
    A. Normal test B. Chi-square test C. T test D. F test
182. Geometric mean of a given series is always less than its
    A. Harmonic mean B. Weighted mean
    C. Arithmetic mean D. Weighted harmonic mean
183. If any items of a series is zero, which mean of central tendency become zero
    A. Harmonic mean B. Geometric mean
    C. Arithmetic mean D. Weighted Harmonic mean
184. Mean deviation is the least when calculated about (minimal property)
    A. Arithmetic mean B. Median
    C. Mode D. Geometric mean
185. Unit less measure of dispersion is
    A. Standard deviation B. Mean deviation
    C. Range D. CV
186. Arithmetic mean and variance are always equal in
    A. Normal distribution B. Poisson distribution
    C. Binomial D. All of these

187. The probability of committing type I error is known as
A. Test of significance B. Standard error
C. Level of significance D. Sampling

188. For paired observation, which test is used for testing the significance of a mean difference
A. Paired t test B. T test C. F test D. Chi square test

189. The variance of Chi-square ranges from.... and is always positive
A. –1 to +1 B. 0 to 2 C. 0 to infinity D. None of these

190. For calculation of chi-square, N should be at least
A. 20 B. 30 C. 40 D. 50

191. The validity of correlation coefficient is tested by
A. Z test B. T test C. Chi-square test D. Both 1 and 2

192. The range of multiple correlation coefficient lies between
A. 0 to 1 B. –1 to 1
C. 0 to infinity D. –infinity to + infinity

193. Coefficient of determination is a square of
A. Correlation coefficient B. Regression coefficient
C. Both 1 & 2 D. None of these

194. When the fertility gradient of the field is in one and two directions simultaneously, which design is most appropriate?
A. Augmented design & CRD simultaneously
B. CRD & RBD simultaneously
C. RBD and LSD simultaneously
D. SPD & RBD simultaneously

195. In RBD, how many treatments can be adopted without any loss of efficiency
A. 10 B. 15 C. 18 D. 20

196. When several factors with different levels to be experimented simultaneously with same precision, Which design is most appropriate?
A. CRD B. LSD C. RBD D. Factorial scheme

197. When experimental material is very less and number of genotypes are very large, which design is most suited?
A. Split plot B. LSD
C. RBD D. Augmented design

198. The minimum error degree of freedom should be at least
A. 6 B. 12 C. 18 D. 20

199. Which transformation is appropriate for percentages?
A. Square root B. Arc sine C. Logarithmic D. All of these

200. Which design has no of rows=no. of colums=no. of treatments

A. SPD B. Strip plot design

C. LSD D. RBD

ANSWERS

1. D	2. B	3. C	4. C	5. B	6. C	7. B	8. C
9. C	10. A	11. B	12. C	13. B	14. B	15. A	16. B
17. B	18. C	19. C	20. C	21. A	22. B	23. D	24. C
25. A	26. C	27. D	28. C	29. A	30. D	31. D	32. B
33. C	34. B	35. C	36. D	37. D	38. C	39. B	40. C
41. C	42. B	43. C	44. A	45. D	46. D	47. B	48. A
49. A	50. C	51. C	52. C	53. B	54. A	55. D	56. C
57. A	58. B	59. C	60. B	61. C	62. A	63. A	64. A
65. C	66. A	67. A	68. B	69. A	70. D	71. D	72. B
73. C	74. A	75. B	76. A	77. B	78. D	79. D	80. C
81. C	82. *	83. D	84. B	85. B	86. C	87. B	88. A
89. B	90. B	91. D	92. A	93. D	94. D	95. C	96. D
97. C	98. B	99. D	100. D	101. C	102. B	103. D	104. D
105. A	106. A	107. B	108. B	109. D	110. A	111. B	112. B
113. A	114. B	115. B	116. D	117. A	118. B	119. D	120. D
121. D	122. A	123. A	124. C	125. C	126. D	127. D	128. B
129. A	130. B	131. C	132. D	133. A	134. B	135. D	136. A
137. B	138. D	139. A	140. B	141. D	142. A	143. D	144. D
145. B	146. A	147. C	148. D	149. C	150. B	151. B	152. B
153. D	154. C	155. C	156. C	157. B	158. D	159. B	160. B
161. A	162. B	163. D	164. D	165. B	166. D	167. A	168. A
169. B	170. A	171. A	172. D	173. C	174. D	175. D	176. C
177. A	178. D	179. D	180. B	181. C	182. C	183. B	184. B
185. D	186. B	187. C	188. A	189. C	190. D	191. B	192. A
193. A	194. C	195. D	196. D	197. D	198. B	199. B	200. C

Chapter

10

AGRICULTURAL ECONOMICS

- Economics is popularly known as the **"Queen of Social Sciences".**
- The **terms micro and macro economics** was given by **Ragnar Frisch** in 1933.
- Micro-economics and macro economics term is derived from the **Greek word** micro and macro, respectively which means small or a millionth part and large. Micro deals with **"price theory"** and macro economics deals with the **income theory.**

S. No	Economists	Definitio	Remark
1	Adam Smith	Wealth	Father of Economics
2	Alfred Marshall	Welfare	
3	Robbins	Scarcity	
4	J M Keynes	Growth	Father of Modern Economics

- **Agricultural economics** deals with the principles of choice are applied in the use of scarce resources such as land, labor, capital and management in agricultural operations.
- **Goods:** It is defined as anything that satisfies human wants or needs.
- **Utility:** Means the power to satisfy a human want. Any commodity or service which can satisfy a human want is said to have utility.

  Types of utility

  1. **Form Utility**: The change in the form. For example: Processing of paddy into rice.
  2. **Place Utility**: The utility obtained by movement of the goods from one place to another is termed as place utility. *e.g.,* UP potatoes are transported to all parts of the country, thereby increasing the utility of potatoes.
  3. **Time utility**: storage
  4. **Possession Utility**: possession or transfer of ownership of the commodity *e.g.* Agriculture land sold to real estate for plots would increase the utility for the same piece of land.

- Value expressed in monetary terms is **Price**

Wealth- It consists of all potentially exchangeable means of satisfying human wants

- It should possess utility 2. It must be scarce 3. It must be transferable 4. It must be external to person.
- **Law of diminishing marginal utility (LDMU)** – was given by **H.H Gossen Marshall**
- **Marginal Utility**: The addition to the total utility by the consumption of the last unit considered just worthwhile.
- MUa = TUa – TU(a-1)
- The law of demand, the law of equi-mariginal utility and the concept of consumers' surplus are based on LDMU.
- When the TU is max, MU will be Zero
- **Law of equi marginal utility**- A consumer is in equilibrium when he distributes his given money income among various goods.
- Consumer's surplus- introduced by **Marshal** in 1895 in his publication **"Principles of Economics"**.
- Consumer's surplus is the difference between what the consumer is willing to pay and what he actually pays.
- Demand curve has a negative slope
- Price effect = Income effect + Substitution effect
- Substitution effect is always negative.
- For normal goods income effect is positive.
- For giffen goods price and income effect is negative.
- Law of diminishing marginal utility– It is the basic cause of the law of demand
- **Giffen goods**: These are those inferior goods Demand varies inversely with price with ceteris paribus.
- Whose quantity demanded decreases with decrease in price of the good.

Elasticity of Demand

- Elasticity of demand is the degree of responsiveness of quantity demanded of a good to a change in its price.

  Elasticity = % change in quantity demanded / % change in price

  $= \Delta q / q \dagger \Delta P / P$
- The concept of cross elasticity of demand is used for measuring the responsiveness of quantity demanded of a good to changes in the price of related goods.

  Exy = Percentage change quantity demanded of good X
- Percentage change in price of good Y

- The economists grouped various degrees of elasticity of demand into five categories.

  Infinitely elasti = infinite

  Perfectly inelastic = zero

  Unit elasticity = 1, rectangular hyperbola

  Relatively elastic = >1

  Relatively inelastic demand = <1
- **Supply is directly proportional to price**
- At market equilibrium Qs= Qd

  Market types and its characteristics

Market competition	Number of sellers	Entry barrier	Pricing power	Product
Monopolistic	Many	low		Heterogeneous
Monopoly	One	Very high	Price maker	
Oligopoly	Few (3-10)	high		
Duopoly	Two	high		
Perfect market	Many	Very low	Price taker	Homogenous

- **Monopsony** only one buyer
- **Bilateral monopoly** is a market structure in which a single seller faces a single buyers
- **Perfect market :** A market is said to be perfect when all the sellers and buyers are aware of the prices at which transaction take place and all the offers made by other sellers, and buyers, and when any buyer can purchase from any seller and conversely.
- **Imperfect market**: A market is said to be imperfect when different prices prevail for the same commodity at the same time.
- **National Income** is the value of all goods and services in a year in a nation.
- **Gross Domestic Product** (GDP): Gross Domestic Product (GDP) is the total market value of all final goods and services currently produced **within the domestic territory** of a country in a year.
- **Gross National Product** is the total market value of all final goods and services produced in a year.
- Net National Product (NNP) is the market value of all final goods and services after providing for depreciation
- NNP at Factor Cost = NNP at Market Price – Indirect Taxes + Subsidies.
- **Personal Income** = National Income – Social Security contributions – Corporate income taxes – Undistributed corporate profits + Transfer payments.

- **Disposable Income** = Personal income – Personal taxes.
- National income can be measured by three methods:
  1. Output or Production method/value-added method.
  2. Income method and
  3. Expenditure method.

Tax

- It is a compulsory contribution from a person to the state to defray the expenses incurred in the common interest of all, without reference to special benefit conferred.
- **Proportional & Progressive Tax** - A proportional tax is one in which, whatever the size of income, same rate or percentage of tax is charged.
- **Direct tax** - income tax, wealth tax, property tax *etc.*
- **Indirect tax** - sales tax, central excise duty, custom duty, recreational tax, *etc.*

Inflation

It indicates the rise in price of a basket of commodities on a point-to-point basis. It basically suggests an increase in the cost of living over a period of time. Lacks of supply, high demand, and more money supply are main reason of inflation.

- It is an index of the prices of all final goods and services constituted the GDP or GNP.

Measures to control Inflation

(a) Credit Control: reduce the supply of money.
(b) Demonetization of Currency
(c) Issue of New Currency
(d) Reduction in Unnecessary Expenditure
(e) Increase Taxes
(f) Increase in Savings

Economics of farm management and agricultural production

- Agriculture and allied sectors accounted **17.4 % of the GDP** in the year **2015-16.**
- **Law of equi-marginal returns**- A limited input should be allocated among alternative uses in such a way that the marginal value products of the last unit are equal in all its uses.
- **Opportunity cost-** Opportunity cost is defined as the returns that are sacrificed from the next best alternative. Opportunity cost is also known as real cost or alternate cost.

- **Principle of product substitution-** we should go on increasing the output of a product so long as decrease in the returns from the product being replaced is less than the increase in the returns from the product being added. MRS < PR
- The procedure for determining the future value of present sum is called compounding. **FV = P (1 + i)n**

  Where, FV = Future value; P = the present sum, i = the interest rate, n = years.
- The procedure to find the present value of future sum is called **discounting**.
- PV = $P/(1+i)^n$ (P where, PV = Present value P = Future sum i = rate of interest n = years.

Types of farming

- **Specialized farming**: contributes more than 50% of the total farm income.
- **Diversified farming**: no single enterprise contributes 50% of the total farms income.
- **Mixed farming**: It is the type of farming under which crop production is combined with livestock raising.
- **Ranching**: The practice of grazing animals on public lands is called ranching.
- **Dry farming**: Cultivation of crops in regions with annual rainfall of less than 750 mm.
- **Dry land farming**: Cultivation of crops in regions with annual rainfall of more than 750mm.
- **Rainfed farming**: Cultivation of crops in regions with an annual rain fall of more than 1150 mm.

Systems of farming

1. **Capitalistic farming-** Land is held in large areas by private capitalists, corporations or syndicates.
2. **State farming-** State farming as the name indicates is managed by the government. Here land is owned by the state. The operation and management is done by government officials.
3. **Collective farming-** Collective management of land where in large number of families or villagers residing in the same village pool their resources.
4. **Peasant farming-** This system of farming refers to the type of organization in which an individual cultivator is the owner, manager and organizer of the farm.
5. **Co-operative farming-** Co-operative farming is a voluntary organization in which small farmers and landless laborers increase their income by pooling land resources.

Farm planning and budgeting

- **Farm Planning**- future course of action with respect to the farming.
- **Farm budgeting** is a method of estimating expected income, expenses and profit for a farm business.
- **Risk** is a situation where all possible outcomes are known for a given management decision and the probability associated with each possible outcome is also known.
- **Uncertainty** refers to future events where the parameters of probability distribution cannot be determined empirically. Uncertainty is not insurable.
- **Operational holding**: It refers to the total land area held under single management for the purpose of cultivation. It excludes any land leased to another person.
- The resources whose use remains the same regardless of the level of production are called fixed resources *e.g.* Lands, machinery, farm buildings, equipment *etc.*,
- The resources whose use varies with the level of production are known as variable resources. Seeds, Fertilizers, Plant protection chemicals, FYM, feeds, medicines *etc.*, are examples of variable resources.
- Transformation period (or) production period: The time required for a resource to be completely transformed into a product is referred to as transformation period.
- **Cost of cultivation**: It refers to the cost of various inputs and input services used for raising a particular crop. It includes all the operations from land preparation to threshing, cleaning and taking the product from the field to home. Cost of cultivation always refers to unit area (acre or hectare).
- **Cost of production**: It refers to the cost of various inputs and input services used to produce a unit quantity of output of a commodity.
- **Short Run Production Function** (SRPF): Production Function in which some inputs or resources are fixed. Y= f (X1 / X2, X2,… Xn) *e.g.* Law of Diminishing Returns or Law of Variable Proportions
- **Long Run Production Function** (LRPF): Production function which permits variation in all factors of production. Y = f (X1, X2, X3 …Xn) Eg: Returns to scale.

Cost

- Cost is the function of output. *i.e.* C= f(Y)
- The shape of the fixed cost (also called as overhead cost/sunk cost) is horizontal to x axis, Variable cost is saucer shape, AC and AVC is u shape and AFC is rectangular hyperbola.
- The Marginal cost MC will cut the AC and AVC below the line.
- Breakeven point is called where TC=TR

Factor-Product Relationship - 'how much to produce?'

- There are three stages in this production function. In region I or stages I the TP is increasing, MP is more than AP and elasticity is more than one. It is called as irrational region.
- In region II the TP is increasing at decreasing rate, MP is less then AP and elasticity is zero to one. It is called as rational region.
- In region III the TP is decreasing, MP is negative and AP is being positive but decreasing.
- When total product remains constant, the marginal product is zero.
- When total product decreases, marginal product is negative. As long as marginal product increases, the total product increases at increases at increasing rate.
- When marginal product is more than average product, average product increases. When marginal product is equal with the average product, average product is maximum. When marginal product is less than average product, average product decreases.

Elasticity of Production

- Ep = Percentage change in output/Percentage change in input. = MPP/APP
- An isoquant represents all possible combinations of two resources (X1 and X2) physically capable of producing the same quantity of output.
- **Characteristics of Isoquants** are 1. Slope downwards from left to right or negatively sloped. 2. Convex to the origin. 3. Nonintersecting 4. Isoquants lying above and to the right of another represents higher level of output. 5. The slope of isoquant denotes the marginal rate of technical substitution (MRTS).
- Marginal Rate of Technical Substitution (MRTS) refers to the amount by which one resource is reduced as another resource is increased by one unit.
- Slope of price line shows the price ratio.
- MRTS= Number of units of replaced resource/Number of units of added resource
- The isoquants for perfect substitutes is negatively sloped straight lines. The MRTS is constant *e.g.* family labour and hired labour, farm produced and purchased seed *etc,*
- **Perfect Complements**: The isoquant in this case is of a right angle. *e.g.* Tractor and driver, Pair of bullocks and laborer.
- **Isocost Line** (price line, budget line) all possible combinations of two resources (X1 and X2) which can be purchased with a given outlay of funds.
- **Iso-cline** - A line or curve connecting the least cost combination of inputs for all output levels is called isocline.

- **Pansion Path**: many isoclines, the isocline which is considered to be the most appropriate over a production period is known as expansion path or scale line. At any particular time, only one expansion path is possible.
- Ridge lines represent the economics limit of the production
- Production Possibility Curve represents all possible combinations of two products that could be produced with given amounts of inputs.

Types of Product-Product Relationships or Enterprise Relationship-

- **Joint Products**: These are produced through single production process. *e.g.* wheat and straw, paddy and straw, groundnut and hulms , cotton seed and lint, cattle and manure, butter and buttermilk *etc.*
- **Complementary enterprises**: The two enterprises do not compete for resources but contribute to the mutual production by providing an element of production required by each other. The marginal rate of product substitution is positive (> 0) *e.g.* Cereals and pulses, crops and livestock enterprises.
- **Supplementary enterprises**: when increase or decrease in the output of one product does not affect the production level of the other product. The marginal rate of product substitution is zero *e.g.* small poultry or dairy or piggery enterprise
- **Competitive enterprises**- when increase in output of one product, with resources held constant, results in the decrease of output of other product. The marginal rate of product substitution is negative (<0)
- Marginal rate of the product substitution refers to the absolute change in one product associated with a change of one unit in competing product.
  MRPS = Number of units of replaced product / Number of units of added product
- **Iso Revenue Line** - It represents all possible combination of two products which would yield an equal (same) revenue or income.
- **Returns To Scale** - In returns to scale, all the necessary factors of production are increased or decreased to the same extent so that whatever the scale of production, the proportion among the inputs remain the same.

AGRICULTURAL FINANCE AND COOPERATION

- **Short–term loans**: repaid within a period of 6 to 18 months *e.g.* All crop
- **Medium – term loans**: repayment period varies from 18 months to 5 years. purchasing implements, electric motors, milch cattle, sheep and goat, *etc.*
- **Long – term loans**: repayment from 5 years to more than 20 years or even more. Construction of farm buildings, purchase of tractors, raising of orchards, *etc.*

Secured loans:

- **Personal security** - borrower himself stands as the guarantor. Loan is advanced on the farmer's promissory note.
- **Collateral Security** - The movable properties of the individuals like LIC bonds, fixed deposit bonds, warehouse receipts, machinery, livestock *etc*.,, are offered as security.
- **Chattel loans** - credit is obtained from pawn-brokers by pledging movable properties such as jewellery, utensils made of various metals, *etc*.
- **Mortgage** - immovable properties are presented for security purpose *e.g.* land, farm buildings, *etc*.

  3 Rs of credit, which are as follows 1. Returns 2. Repayment capacity 3. Risk-bearing ability
- 5 Cs of credit viz., character, capacity, capital, condition and commonsense.
- **7 Ps of farm credit** - 1. Principle of productive purpose. 2. Principle of personality. 3. Principle of productivity. 4. Principle of phased disbursement. 5. Principle of proper utilization. 6. Principle of payment and 7. Principle of protection
- 19th July 1969 nationalization of 14 commercial banks has taken place. Six more banks were nationalized on 15th April 1980.
- 1969, **Sri. F.K.F Nariman** committee **recommended the formulation of "Lead Bank Scheme".**
- On the recommendation of Narashimham Regional Rural Bank has been established in 1975. Regional Rural Bank has the sharing of 50:15:35 center, state and sponsoring bank respectively.
- **Kisan Credit Card scheme** by banks during **1998 -99**. The scheme was designed by NABARD. The KCC should normally valid up to 3 years and subject to annual review.
- **Differential Rate of Interest Scheme (DIR)** on the recommendations of RBI committee under the chairmanship of **Dr. B. K. Hazare** for the weaker sections at 4 % loan facility.
- **The International Bank for Reconstruction and Development (IBRD)** also called as World Bank was established in the year 1945
- IBRD, IDA, IFC and MIGA are collectively called as World Bank Group.
- **International Monetary Fund (IMF)**: The International Monetary Fund (IMF) is an international organization. Its headquarters is located at Washington, DC., USA.
- **The motto of co-operation is "Each for all and all for each."**

Co-operative Credit Organization	
ST & MT Loans **(Three tier system)**	LT loans (Two tier system)
State Cooperative Bank (SCB)	Central Land Development Banks (At state level)
District Cooperative Central Bank (DCCB)	Primary Land Development (At district level) (PLDBs) (Erstwhile taluka level)
Primary Agricultural Cooperative Credit Societies (PACS) (At village level)	

AGRICULTURE MARKETING

- **Spot or Cash Markets** - A market in which goods are exchanged for money immediately after the sale is called the spot or cash market.
- **Forward Markets** - A market in which the purchase and sale of a commodity takes place at time "t" but the exchange of the commodity takes place on some specified date in future *i.e.* t+1.
- **Grading** means the sorting of the unlike lots of the produce into different lots according to the quality specifications laid down.

  Types of Grading

  1. **Fixed Grading / Mandatory Grading**- goods sorting according to the size, quality and other characteristics which are of fixed standards.
  2. **Permissive/Variable Grading** - goods are graded according to standards, which vary over time.
  3. **Centralized/Decentralized Grading** - Based on the degree of supervision exercised by the government agencies on grading of various farm products.
- **Market Intelligence** - This includes information relating to such facts as the prices that prevailed in the past and market arrivals over time.
- **Commission Agents or Arhaṭias** - A person operating in the wholesale market who acts as the representative of either a seller or a buyer.
- **Brokers** - Brokers render personal services to their clients.
- Speculative middlemen take the title to the product.
- **Food Corporation India (FCI)** was born on **January 1,1965-**
- **AGMARK** is the quality certification marks under the Central Agricultural Produce **(Grading and Marketing) Act, 1937**. This label **indicates that the purity and the quality** of the product on the basis of the standards.
- The marketable surplus is that quantity of the produce which can be made available to the non-farm population of the country. **MS = P - C**
- Marketed surplus is that quantity of the produce which the producer farmer actually sells in the market, irrespective of the requirements for family consumption, farm needs and other payments.

- "**Market integration** as process which refers to the expansion of firms by consolidating additional marketing functions and activities under a single management"
- **Price Spread** - The difference between the price paid by the consumer and price received by the farmer.
- **Commission for Agricultural Costs and Prices** (**CACP**) Agricultural Price Commission, APC was established in **1965** on the **recommendations of Food grains Policy committee under the chairmanship of L.K. Jha.**

  Minimum Support Price (MSP): Role in the agricultural commodity

  1. **Price fixe** by the government to protect farmers against excessive fall in prices.
  2. **Procurement Price -** Refers to the price at which government procures from producers to maintain buffer stocks and feed Public Distribution System.
  3. **Issue Price** - Price at which the commodity is made available to consumers at fair price shops. It is always higher than procurement price.
- **Maximum Ceiling Prices** - fixed by govt. to protect the consumer from unwarranted price.
- **Speculation** - Purchase or sale of a commodity at the present price with the object of sale or purchase at some future date at a favorable price.
- **Hedging** - It is a trading technique of transferring the price risk.
- **Contract Farming/Marketing** - Contract farming is evolving an institutional arrangement of alternative marketing in India. One of the brilliant ideas that are being thrown around is that of contract farming.
- **Free trade** is no restrictions on trade.
- WTO was established on 1st January, 1995 with head quarters at **Geneva**.
- WTO functions as Dispute settlement system is faster and more automatic aims at solving trade problems.

Some formulas

- **Linear production function** $y = a + bx$ where y is dependent variable (output), a is constant, b is coefficient, x is independent variable (input)
- **Cobb-Douglas (non linear production function)** $y = ax^b$ where y = dependent variable, a constant, b coefficient, x independent variable
- **Quadratic function** $y = a + bx - cx^2$ where y = output or yield dependent variable, a constant, c & b coefficient, x input (independent variable)
- **MPP** = Delta Y/ Delta X

Some curves of demand and supply

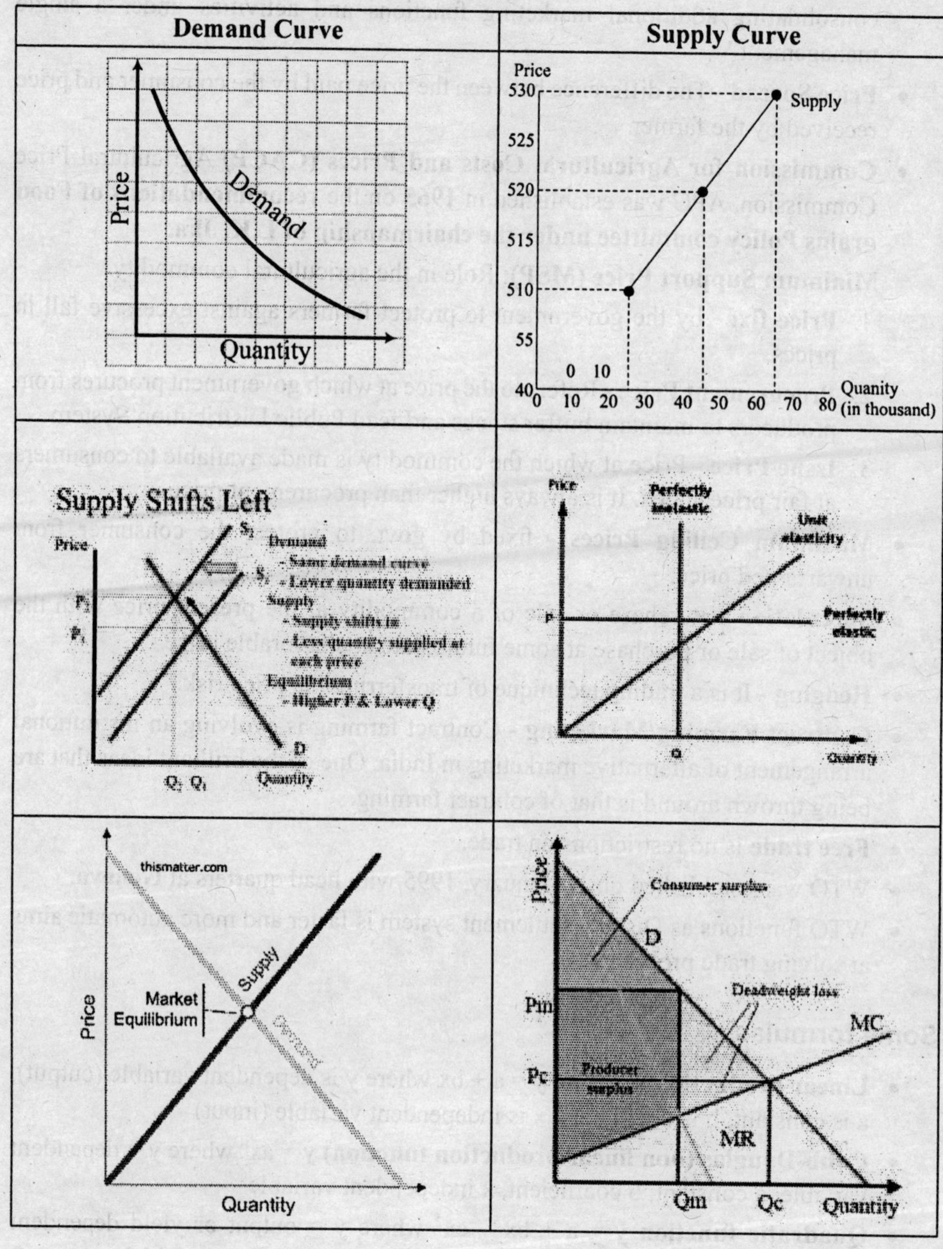

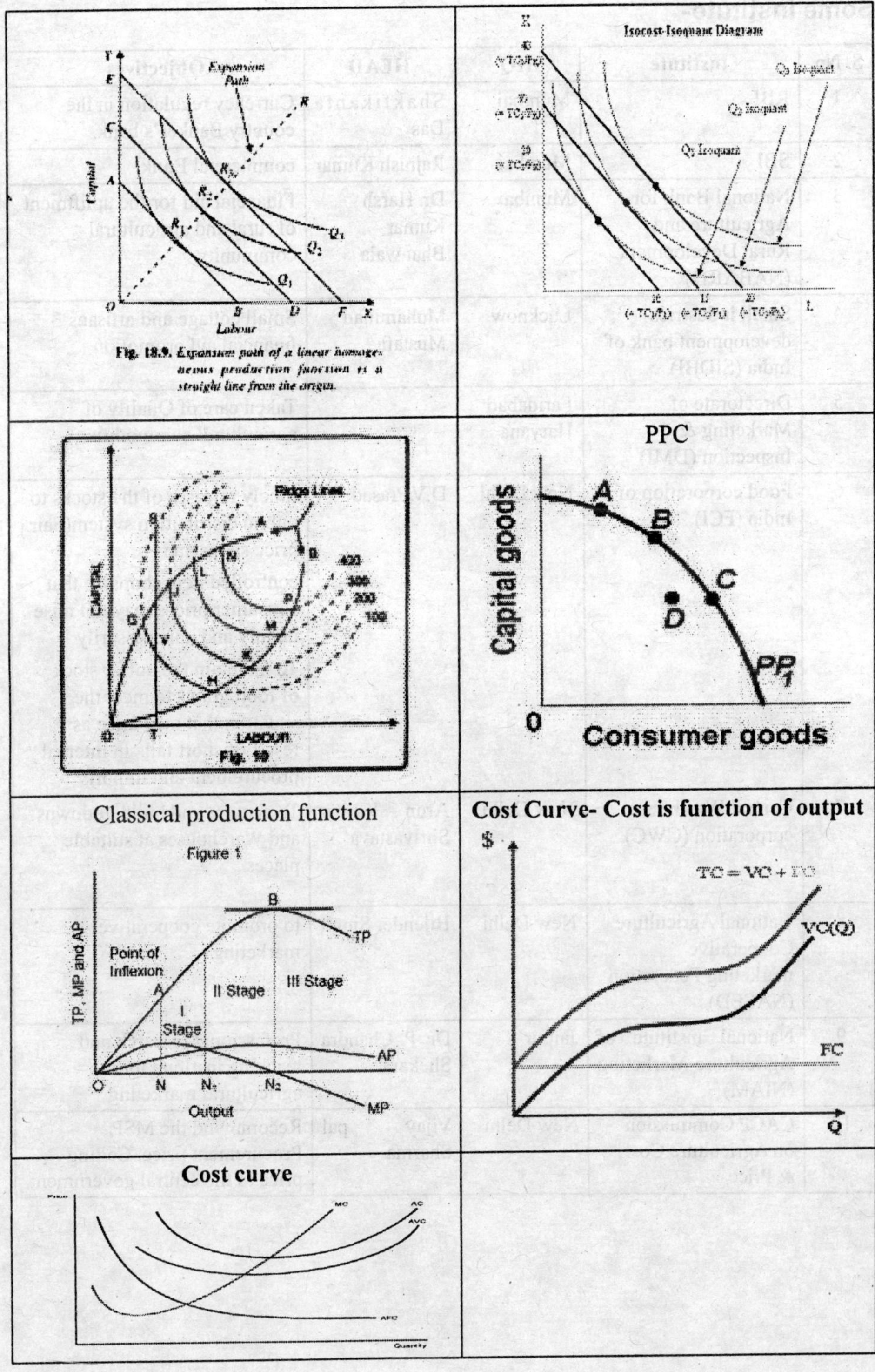

Fig. 18.9. Expansion path of a linear homogeneous production function is a straight line from the origin.

Fig. 10

Some Institute-

S. No.	Institute	HQ	HEAD	Objectives
1	RBI	Mumbai	Shaktikanta Das	Currency regulation in the country Banker's bank.
2	SBI	Mumbai	Rajnish Kumar	commercial Bank
3	National Bank for Agriculture and Rural Development (NABARD)	Mumbai	Dr. Harsh Kumar Bhanwala	Financial aid for the upliftment of rural and agricultural community
4	Small industries development bank of India (SIDBI)	Lucknow	Mohammad Mustafa	Small cottage and artisans financial aid promotion
5	Directorate of Marketing & Inspection (DMI)	Faridabad Haryana	-	Taken care of Quality of agricultural commodities
6	Food corporation of India (FCI)	New Delhi	D.V.Prasad	timely releases of the stocks to public distribution system(Fair price shops and controlled item shops)so that consumer prices may not raise unduly and unnecessarily To maintain the buffer stock of food grains to meet the situation that may arise as result of short falls in internal procurement and imports
7	Central Warehouse corporation (CWC)	New Delhi	Arun Kumar Shrivastava	To acquire and build godowns and Warehouses at suitable places
8	National Agriculture Cooperative marketing Federation (NAFED)	New Delhi	Bijender Singh	to promote cooperative marketing
9	National institute of Agriculture Marketing (NIAM)	Jaipur	Dr. P. Chandra Shekara	Professional research and teaching institute for the agricultural marketing
10	CACP Commission on Agriculture Cost & Price	New Delhi	Vijay pal Sharma	Recommend the MSP, Procurement price, Ceiling price to the central government

Important Objective Questions (MCQ)

1. An exception of law of demand
   A. Free goods B. Economic goods
   C. Giffen goods D. Mono period goods
2. The book "principle of economics" was written by
   A. Adam smith B. Alfred Marshall C. J.M. Keynes D. Aristotle
3. The price below which the seller refuses to sell any of the commodities is
   A. Normal price B. Maximum price C. Reserve price D. Demand price
4. Price taking is the policy in
   A. Perfect competition B. Monopoly
   C. Monopolistic D. Oligopoly
5. Kinked demand curve is a feature of
   A. Monopoly B. Oligopoly
   C. Monopolistic D. Perfect competition
6. Which of the following is not included in national income
   A. Indirect taxes B. Subsidies
   C. Net export D. Personal consumption expenditure
7. Cost of production plays a dominant role in
   A. Normal price B. Short run price C. Market price D. Secular price
8. When more quantity of a good is supplied at the same price it is called
   A. extension of supply B. Contraction of supply
   C. Increase in supply D. Decrease in supply
9. The most immobile factor of production is
   A. Land B. Labour C. Capital D. Organization
10. Remuneration paid to liquid capital is
    A. Rent B. Wage C. Profit D. Interest
11. When commodity changes hands during transaction process it will have?
    A. Time utility B. Possession utility
    C. Form utility D. Place utility
12. Zamindari system was introduces in India by
    A. WilliumWadderburn B. Benting
    C. Lord Cornwallis D. None of the above
13. The price of diamond is higher when compared to water as
    A. Total utility of water is higher
    B. Marginal utility of water is higher
    C. Marginal utility of diamond is higher
    D. Total utility of demand is higher

14. In case of perishable commodities the stock is
   A. Equal to demand B. More than supply
   C. Equal to supply D. Less than supply
15. Partial repayment plan is also called as
   A. Straight end repayment plan B. Balloon repayment
   C. Amortized even repayment plan D. Future repayment plan
16. Cooperative credit structure in India is based on the lines of
   A. Nicholson model B. Raiffeisen model
   C. William Webber model D. Dupernax model
17. The major consideration while fixing the scale of finance for the term loans
   A. Total cost of investment B. Incremental income from investment
   C. Gross income from investment D. Net income from investment
18. To overcome the distress sales by the farmers, farmers are provided with
   A. Crop loans B. Chattel loans C. Market loans D. Term loans
19. Three maxims of cooperation were given by
   A. Nicholson B. Horrace Plunkett
   C. Calvert D. Robert Owen
20. Two products are said to be supplementary when MRPS is
   A. > Zero B. < Zero C. Equal to zero D. None
21. Cost minimization is the goal of
   A. Product – Product relationship B. Factor – Factor relationship
   C. Factor -Product relationship D. Input- Output relationship
22. If MRS< Price ratio, cost can be reduced by using more of
   A. Added resource B. Added product
   C. Replaced resource D. Replaced product
23. The cost of self-owned and self-employed resources are known as
   A. Cost of production B. Implicit cost
   C. explicit cost D. Operating cost
24. The system of farming under which agricultural operations are carried on jointly by the farmers by retaining the right of ownership of land is
   A. Cooperative farming B. Peasant farming
   C. Capitalistic farming D. State farming
25. The following ratio measures the immediate solvency position of a business
   A. Gross ratio B. Current ratio C. Net capital ratio D. Equity ratio
26. a holding which allows a farmer a chance of producing sufficient to support himself and his family in reasonable comfort after paying his necessary expenses
   A. Family holding B. Economic holding
   C. Optimum holding D. None

27. All farm supplies are valued at
    A. Cost or market price B. Net selling price
    C. Income capitalization D. Cost – depreciation
28. The partial budget is used to analyses the profitability of
    A. Input substitution B. Product substitution
    C. Scale of operation D. All
29. In the short run , farmer has to stop the production when the expected selling price is
    A. More than ATC B. Less than ATC C. Less than AVC D. Less than MC
30. Within the ridge lines the products are
    A. Antagonistic B. Complimentary C. Competitive D. Supplementary
31. In the case of law of decreasing returns, the shape of the total product curve is
    A. Vertical B. Horizontal C. Convex. D. Concave
32. Law of diminishing returns implies
    A. Decreasing APP B. Decreasing MPP
    C. Decreasing TPP D. Increasing TPP
33. The goal of product - product relationship is
    A. Cost minimization B. Resource use efficiency
    C. Minimization of profits D. Minimization of products
34. The type of operatorship is individual and that of ownership is collective in the case of cooperative
    A. Tenant farming B. Joint farming
    C. Collective farming D. Better farming
35. Production possibility curve PPC is otherwise known as
    A. Transformation curve B. Isoresource curve
    C. Opportunity curve D. All
36. Total physical product is maximum when elasticity of production is
    A. More than unity B. Less than unity
    C. Less than zero D. Zero
37. In third zone of production
    A. The technical efficiency of variable resource declines
    B. The technical efficiency of fixed resource declines
    C. Marginal product become negative
    D. All
38. Retailers fall under
    A. Agent middle men B. Speculative middle men
    C. Facilitative middle men D. Merchant middle men

39. The measures which have a minimum impact on trade and tariff is
    A. Blue box B. Green box C. Amber box D. S & D box
40. One of them is not a characteristic feature of agriculture commodities
    A. Seasonality B. Perishability C. Bulkiness D. Regular supply
41. Advantage of contract farming
    A. Reduces risk B. Assured price
    C. Quality consciousness D. All the above
42. The financial statement which shows the resources and obligations of a business along with owners equity is
    A. Balance sheet B. Income statement
    C. Profit and loss account D. All the above
43. Patents copyrights and trademarks of business firm are its
    A. Liabilities B. Tangible assets
    C. Intangible assets D. Current assets
44. Competition oriented pricing is popularly known as
    A. Markup pricing B. Target return pricing
    C. Going rate pricing D. Perceived value pricing
45. The sum of the net benefits during the life period of the project is
    A. BCR B. IRR C. NPV D. N/K
46. The Warehousing scheme was launched in public sector based on the recommendation of
    A. AIRCSC B. AIRCRC C. RBI D. RCA
47. MVP=MIC in
    A. Stage I of the production function
    B. Stage II of the production function
    C. Stage III of the production function
    D. All the stages of production function
48. The slope of its revenue line indicates the ratio of
    A. Factor prices B. Product prices C. MPP to APP D. Input prices
49. Cost of production is indicated by
    A. AVC B. AFC C. ATC D. MC
50. Average physical product indicates
    A. Technical efficiency B. Allocative efficiency
    C. Economic efficiency D. Price efficiency
51. Creation of utility
    A. Consumption B. exchange C. Distribution D. Production
52. The terms micro and macroeconomics were first coined and used by
    A. Ragnar Frisch B. Marshall C. Keynes D. D. Smith

53. Power to satisfy a human want

A. Goods B. Utility C. Welfare D. Value

54. Storage aids in creation of

A. Form B. Time C. Place D. Possession

55. The concept of consumers surplus was introduced

A. Keynes B. Marshall C. Smith D. None

56. Tabular representation of the quantity demanded of a commodity at various prices.

A. Demand Schedule B. Demand Curve

C. Individual Demand D. Demand law

57. These are those inferior goods whose quantity demanded decreases with decrease in price of the good

A. Giffen goods B. Normal C. Comfort D. None

58. The goods are graded under this method according to standards, which vary over time

A. Fixed B. Mandatory C. Permissive D. Centralized

59. large number of buyers and sellers are taking place for homogenous product-

A. Perfect B. Imperfect C. Regulated D. Monopolistic

60. Maintaining satisfactory level of operational and buffer stocks of food grains to ensure National Food Security is work of

A. FCI B. CWC C. NAFED D. RAJFED

61. Information relating to such facts as the prices that prevailed in the past and market arrivals over time.

A. Market information B. Market intelligence

C. Market news D. Market function

62. Surplus is that quantity of the produce which can be made available to the non-farm population of the country.

A. Marketed surplus B. Marketable surplus

C. Market deficit D. Surplus drawback

63. Those middlemen who take title to the product with a view to making a profit on it are called

A. Broker B. Commission agent

C. Speculative middleman D. Facilitative middleman

64. Any single activity performed in carrying a product from the point of its production to the ultimate consumer may be termed as

A. Marketing integration B. Standardization

C. Structure a marketing function D. None

65. When the quantity demanded of a good changes by actly the same percentage as price, the demand is said to be?
A. Less Elastic B. More C. Unitary D. Perfect elastic

66. Difference between price paid by consumer and price received by farmer is called as
A. Marketing margin B. Price spread
C. Marketing channel D. Price drawback

67. Commission for Agriculture Cost and Price (CACP) is recommended by which committee
A. L K Jha Committee B. Shashtri Committee
C. Ram Committee D. Shivraman Committee

68. e NAM stands for
A. Electric National Agriculture Market
B. Electronic National Agriculture Market
C. Electric National Agriculture Marketing
D. Electronic National Agriculture Marketing

69. IMF and World Bank came into existence
A. Bretton wood conference B. Geneva conference
C. Doha round D. Rio de jenerio Conference

70. India's rank in Human Development Index in year 2019 is
A. 129 B. 130 C. 138 D. 182

71. Which one is first R among the 3 Rs?
A. Risk bearing ability B. Return from investment
C. Repayment Capacity D. Revenue generation

72. Cash in hand is
A. Current assets B. Current liability
C. Intermediate liability D. Intermediate assets

73. How to produce
A. Factor-Factor relationship B. Factor product relationship
C. Product-Product relationship D. Input Output relationship

74. When Marginal Product is equal with the Average Product, Average Product is?
A. Maximum B. Minimum
C. Zero D. More than 1 elasticity

75. ....budget is best adopted to analyzing relatively small change in the whole farm plan.
A. Partial budget B. Whole budget
C. Enterprise budget D. Complete budget

76. The profit maximizing output level is at the point where
A. MC = MR B. MC = AR C. MIC = MPP D. APP = MPP
77. Social Banking came into existence
A. 1959 B. 1965 C. 1969 D. 1975
78. The additional income from selling additional unit of output.
A. Marginal Cost B. Average Income
C. Marginal Revenue D. Marginal Physical Product
79. Elasticity of demand for food grains is
A. Inelastic B. Perfectly elastic
C. Perfectly inelastic D. Elastic
80. Markets are of permanent nature
A. Retail B. Long term C. Secular D. Seaboard
81. Partial repayment plan is also called as
A. Straight end repayment plan B. Balloon payment plan
C. Amortized even repayment plan D. future repayment plan
82. The slogan Final Raiffeisen is associated with
A. Frederick Nicholson B. B P Patel
C. Narasimham D. Maclegan
83. The second cooperative societies act was passed in the year of
A. 1884 B. 1902 C. 1912 D. 1904
84. An example of collateral security
A. LIC Bonds B. Fixed deposits C. Jewellary D. All of the above
85. To overcome the distress sales by the farmer, farmers are provided with
A. Crop loans B. Chattel loans C. Market loans D. Term loans
86. The committee that advised the setting up of NABARD
A. AIRCSC B. AIRCRC C. CRAFICARD D. Law committee
87. When Total Product decreases, Marginal Product is?
A. More than one B. Less than zero C. One D. Zero
88. Match the following:

Methods	Particulars
1. Under Cover of a Cloth (Hatha System)-	A. By sample of the produce
2. Private Negotiations	B. By pressing/twisting the fingers of each other under cover of a piece of cloth.
3. Quotations on Samples taken by Commission Agent	C. Different lots is mixed and then sold as one lot
4. Dara Sale Method	D. A verbal understanding between buyers and sellers without any pre-settlement of price
5. Moghum Sale Method	E. Prices are fixed by mutual agreement
6. Open Auction Method	F. By offer bids loudly.

A. BEACDF B. BCAFDE C. EDAFCB D. EABCDF

89. GST came in existence
A. 2012 B. 2016 C. 2017 D. 2018
90. Are used to guide one's thinking process during the planning or decision-making Stage
A. Policies B. Strategy C. Programme D. Goal
91. A statement of financial position of a firm at a particular point of time.
A. Balance sheet B. Cash statement
C. Income statement D. Profit loss statement
92. Costs at current market prices are
A. Real cost B. Nominal costs C. Historic cost D. Intrinsic cost
93. Next best alternative is refers to
A. Equilibrium cost B. Nominal cost
C. Opportunity cost D. Real cost
94. Canon of taxation was given by
A. Alfred marshal B. Adam smith C. J M Keynes D. Robbins
95. Considered as the BRAIN in the body of management skills.
A. Coordination B. Organization C. Planning D. Staffing
96. Torque of wheel in management is
A. Planning B. Motivation C. Organization D. Communication
97. Compulsory contribution from a person to the state to defray the expenses incurred in the common interest of all, without reference to special benefit conferred.
A. Fee B. Charges C. Tax D. Fine
98. Measurement of National Income
A. Output or Production method B. Income method
C. Expenditure method D. All the above
99. Total market value of all final goods and services produced in a year
A. GDP B. GNP
C. PI D. Disposable income
100. Kinked demand curve occur in
A. Monopoly B. Oligopoly C. Monopolistic D. Duopoly
101. A market structure in which there are only few buyers.
A. Oligopsony B. Monopsony C. Oligopoly D. Oligopsony
102. Shape of unitary elastic is`
A. Parabola B. Hyperbola
C. Rectangular hyperbola D. Saucer shape

103. The percentage change in the demand of one good as a result of the percentage change in the price of another good.
A. Price elasticity B. Income elasticity
C. Cross elasticity D. Unitary elasticity

104. If the price of gasoline goes up the quantity demanded of automobiles will go down
A. Substitute B. Complements C. Supplements D. All

105. pollution problems comes under
A. Real cost B. Nominal cost C. Social cost D. Hedonic cost

106. With the fall in price of a commodity, the purchasing power of consumer increases. Thus, he can buy same quantity of commodity with less money or he can purchase greater quantities of same commodity with same money.
A. Substitution effect B. Income effect
C. Price effect D. Demand effect

107. The demand for goods which are needed for further production
A. Derived demand B. Autonomous demand
C. Demand effect D. Cost derived demand

108. The famous diamond –water paradox of Smith is related to
A. LEMU B. LDMU
C. Demand Law D. Indifference curve

109. A new product progresses through a sequence of stages from introduction to growth, maturity, and decline is known as
A. PLC B. PPC C. LCC D. MC

110. The discount rate that makes the NPW of the incremental net benefit-stream or incremental cash flow equal to zero.
A. NPW B. Payback period
C. IRR D. Sensitivity analysis

111. Risk comes under which principle
A. Principle of payment B. Principle of protection
C. Principle of productivity D. Principle of phased disbursement

112. Encouraged by the success of first spell of nationalization of banks, six more banks in the private sector were nationalized on15th April 1980. The criteria was
A. 100 crore deposit B. 150 crore deposit
C. 200 crore deposit D. 300 crore deposit

113. Lead Bank Scheme was given by which committee
A. Nariman Committee B. Narshimum committee
C. Shivraman committee D. Ram committee

114. RRBs came into existence by which committee

A. Nariman Committee B. Narshimum committee
C. Shivraman committee D. Ram committee

115. An indicative cost taken as base cost depending on which the amount to be financed to a farmer is fixed

A. Cost of cultivation B. Cost of production
C. Scale of finance D. Cost of secondary goods

116. When KCC was launched in the year of?

A. 1995-96 B. 1996-97 C. 1997-98 D. 1998-99

117. Who gave Differential Rate of Interest Scheme (DIR) scheme?

A. B K Hazare B. Narshimum committee
C. Shivraman committee D. Ram committee

118. the Comprehensive Crop Insurance Scheme (CCIS) was introduced by GIC in all the states

A. 1973 B. 1983 C. 1985 D. 1989

119. RBI nationalized in the year

A. 1935 B. 1949 C. 1955 D. 1969

120. International Development Association (IDA) head quarters

A. Washington D. C. B. New York
C. Mumbai D. Delhi

121. The headquarter of Asian Development Bank at

A. Washington D. C. B. New York
C. Mumbai D. Manilla

122. A phase was considered as "Golden Era" for the cooperative movement in India

A. Initiation phase (1904-1911) B. Modification phase (1912-1918)
C. Expansion phase (1919-1929) D. Restructuring phase (1930-1946)

123. Chairman of All India Rural Credit Survey Committee (AIRCSC)

A. Vaikataipaiyaa B. Horrece Plunkett
C. Gorwala D. Nariman

124. National Bank for Rural Development (NABARD) was established in which plan

A. Fourth Five Year Plan B. Fifth Five Year Plan
C. Sixth Five Year Plan D. Seventh Five Year Plan

125. NCDC stands for

A. National Cooperative Development Corporation
B. National Credit Development Corporation
C. National Corporative Development Corporation
D. National corporative Development Cooperation

126. The apex credit organizations existing at the state level

A. SCB B. DCCB C. PACs D. RCB

127. A link between state cooperative banks and primary agriculturalcooperative credit societies.

A. SCB B. DCCB C. PACs D. RCB

128. Farmers Service Societies came into existence on

A. 1970 B. 1971 C. 1978-79 D. 1980

129. Large-Sized Adivasi Multipurpose Cooperative Societies (LAMPS) was recommended by which committee

A. Bawa Committee B. Kanda Committee

C. Nariman Committee D. Shivraman committee

130. Single window system cam in existence

A. 1985 B. 1981 C. 1978-79 D. 1987

131. The procedure for determining the future value of present sum

A. Compounding B. Discounting C. IRR D. Payback

132. When a farm is organized for the production of a single commodity and this commodity is the only source of income, the farm is said to be?

A. Specialized farming B. Diversification

C. Ranching D. Dry

133. A farming is a voluntary organization in which small farmers and landless laborers increase their income by pooling land resources

A. Peasant farming B. State farming

C. Cooperative farming D. Collective farming

134. Linear programming was developed by

A. Dantzing B. Jethro Tull C. Mishellies D. Domin

135. Where all possible outcomes are known for a given management decision and the probability associated with each possible outcome is also known

A. Uncertainty B. Risk C. Probability D. Ambiguity

136. The total land area held under single management for the purpose of cultivation

A. Agricultural Holding B. Operational Holding

C. Total Holding D. Farm Holding

137. A period of time during which one or more of the production inputs is fixed in amount and cannot be changed

A. Short Run B. Long Run

C. Economic Run D. Agriculture Run

138. This is obtained for those inputs which can be split up in to smaller units

A. Continuous Production Function

B. Discontinuous or discrete Production Function

C. Short Run Production Function

D. Long Run Production Function

139. Each successive unit of variable input when applied to the fixed factor adds more and more to the total product than the previous unit

A. Law of increasing returns B. Law of constant returns

C. Law of decreasing returns D. Fixed return

140. A ratio of total product to the quantity of input used in producing that quantity of product.

A. TP B. MP C. AP D. MR

141. When Marginal Product is less than Average Product, Average Product?

A. Decreases B. Increases C. Fixed D. Less than zero

142. All possible combinations of two resources(X1 and X2) physically capable of producing the same quantity of output

A. Isorevenue B. Isoquant C. Isocost D. PPC

143. It refers to the amount by which one resource is reduced as another resource is increased by one unit

A. MRPS B. MRTS C. PPC D. Iso-cost

144. A line or curve connecting the least cost combination of inputs for all output levels is

A. Isocline B. Isocost C. Isoqusnt D. Ridge lines

145. All possible combinations of two products that could be produced with given amounts of inputs

A. Isorevenue B. Isoquant C. Isocost D. PPC

146. Wheat and Straw, paddy and straw, groundnut and hulms are example of

A. Joint products B. Complementary

C. Supplementary D. Competitive

147. It represents all possible combination of two products which would yield an equal(same) revenue or income.

A. Isorevenue B. Isoquant C. Isocost D. PPC

148. The behavior of production when all factors (inputs) are increased or decreased simultaneously in the same proportion.

A. Law of variable proportion B. Return to scale

C. Substitution D. Short run

149. $Y=aX^b$ represents

A. Cobb Douglas PF B. Quadratic PF

C. Linear PF D. Short PF

150. A family labor income is calculated by

A. Gross income – Cost A1 / A2

B. Gross income – Cost B

C. (Gross income – Cost C) + (Cost B – Cost A)

D. Cost A1 + Rent on leased in land

ANSWERS

1. C	2. B	3. C	4. A	5. B	6. C	7. C	8. C
9. A	10. D	11. B	12. C	13. C	14. C	15. B	16. B
17. A	18. C	19. B	20. C	21. B	22. C	23. B	24. A
25. B	26. B	27. A	28. D	29. C	30. C	31. C	32. B
33. B	34. A	35. D	36. D	37. D	38. D	39. B	40. D
41. D	42. A	43. C	44. A	45. C	46. A	47. B	48. B
49. C	50. A	51. D	52. A	53. B	54. B	55. B	56. A
57. A	58. C	59. A	60. A	61. B	62. B	63. C	64. D
65. C	66. B	67. A	68. B	69. A	70. A	71. B	72. A
73. A	74. A	75. A	76. A	77. C	78. C	79. A	80. C
81. B	82. A	83. C	84. D	85. A	86. C	87. B	88. A
89. C	90. A	91. A	92. B	93. C	94. B	95. A	96. B
97. C	98. D	99. B	100. B	101. A	102. C	103. C	104. B
105. C	106. B	107. A	108. B	109. A	110. C	111. B	112. C
113. A	114. B	115. C	116. D	117. A	118. C	119. B	120. A
121. D	122. C	123. C	124. C	125. B	126. A	127. B	128. B
129. A	130. D	131. A	132. A	133. C	134. A	135. B	136. B
137. A	138. A	139. A	140. C	141. A	142. B	143. B	144. A
145. C	146. A	147. D	148. C	149. A	150. B		

Chapter

11

AGRICULTURAL EXTENSION EDUCATION

EXTENSION EDUCATION

- Extension is that type of education which is str*etc.,*hed out to people in rural areas far and near, beyond the limits of the educational institutions to which the formal type of education is usually confined. The extension is derived from **Latin** word
  - **Ex** means **Out**
  - **Tension** means **Stretching**
- The French word for extension is **Vulgarisation.**
- The Dutch word for extension **'Voorliching'** means **lighting the path**
- Education is an integral part of extension.
- **Education** is production of desirable changes in human behavior (change in knowledge, skill and attitude).
- Three types of education are informal, formal and non formal.
  1. **Informal Education:** is the lifelong process by which every person acquires and accumulate knowledge, skills, attitudes and insights from daily experiences and exposure to the environment at home, at work, at play *etc.,*
  2. **Formal Education:** are the highly institutionalized, chronologically graded and hierarchically structured education system, spanning lower primary school and the upper reaches of the university.
  3. **Non-Formal Education**: is an organized, systematic, educational activity carried on outside the framework of the formal system to provide selected types of learning to particular sub-group in the population, adults as well as children according to their needs.
- **Extension education** is a Non formal education
- Non formal education is a **learner centered, less bureaucratic control and more decentralized**
- **Emancipatory extension** means when aim at empowering or uplifting the poor.
- **Formative extension** means emphasis on human resource development.

- **Persuasive extension** means extension when used as a policy instrument for achieving societal objective.
- **Informative extension** means extension as an instrument for helping people to make well-considered choice
- Term extension was originated in England (1866)
- Term extension education was first time used in 1873 by Cambridge University
- **Extension Education-** as an applied sciences consisting of content derived from research, accumulated field experiences and relevant principles drawn from the behavioural sciences synthesized with useful technology into body of philosophy, principles, content and method focused on the problems of out of school education for adults and youth. (Leagans,1961)
- The Father of University Extension is **James Stewatt.**
- Father of Extension Education is **J P Leagans.**
- Father of extension education in India is **K N Singh.**
- In USA, the first agricultural society established in **Philadelphia (1875).**
- To keep rural people in rural areas **"out of school education"** should be given to them, for this committee was sign known as **Philadelphia Committee**.
- Land Grant Colleges in USA came into existence as part of **Morill Act (1862).**
- Experimental stations came into existence as part of **Hatch Act (1887).**
- Cooperative extension service is a consequence of **Smith Liver Act (1914).**
- The first use of the word extension was done by **Vorhees.**
- Indian Society of Extension Education was established dated on **22 June 1964.**
- Directorate of extension education was set up in the year of **1971.**
- **Philosophy of extension** is the pursuit of wisdom, a body of general principles or laws of a field of knowledge.
- According to **Kelsey and Hearne** (1967) is the basic philosophy of extension education is to **teach people how to think, not what to think.**
- The concept of Extension Educational Process was developed by **Leagans** (1967).
- 5 steps involve in extension educational process are **SOTER** (Situation, Objectives, Teaching, Evaluation, Reconsideration)
- **Teaching** is the process of arranging situations in which the important things to be learned are called to the attention of the learners, their interest developed, desire aroused and action promoted.
- **Learning** is the process by which an individual, through one's own efforts and abilities changes the behavior.
- **Elements of Learning Situation** are **5 (LTSTP)** as Learner, Teacher, Subject Matter, Teaching Materials and Physical Facilities).

- Dale's Cone of Experience was given by **Edger Dale**.
- Cone of experience as modified by **Sheal (1989).**
- We tend to remember as
  - Reading - 10%
  - Hearing - 20%
  - Seeing - 30%
  - Hearing & seeing - 50%
  - Saying - 70%
  - Saying & doing - 90%
- Steps in extension teaching is presented by **Wilson and Gallup (1955)**
- Steps in extension teaching are **AIDCAS** (Attention, Interest, Desire, Conviction, Action, and Satisfaction).
- **Objectives of Extension:** Objectives are expression of the ends towards which our efforts are directed. The fundamental objective of extension is to develop the rural people economically, socially and culturally by means of education. The general objectives of extension are:
  1. To assist people to discover and analyze their problems and identify the field needs.
  2. To develop leadership among people and help them in organizing groups to solve their problems.
  3. To disseminate research information of economic and practical importance in a way people would be able to understand and use.
  4. To assist people in mobilizing and utilizing the resources which they have and which they need from outside.
  5. To collect and transmit feedback information for solving management problems.

Principles of Agricultural Extension

- **Principle of cultural difference** - Differences which exist between groups of farmers as well as between extension agents and farmers regarding their habits customs values *etc.,*
- **Principle of Grass roots** - Extension programmes should start with local groups, local situations and local problems involves *e.g.* establishment of **Panchayati Raj bodies** at various level fulfills
- **Principle of Leadership** - "Never do anything that you get someone to do for you"
- **Principle of indigenous knowledge** - Instead of ignoring the indigenous knowledge system as outdated, the extension agent try to understand them involves.

- **Principle of learning by doing** - People should learn what to do, why to do, how to do and with what result:
- **Principle of participation** - People should willingly cooperate and participate in identifying the problems, solving it then implementing the projects in getting the desired results.
- **Principle of adaptability** - Extension work and extension teaching methods must be flexible and adapted to suit the local conditions.
- **Principle of satisfaction** - The end product of extension work should produce satisfying results for the people.

Extension Approaches

- Book "guide on Alternative Extension Approaches" written by **G. H. Axinn**.
- Number of extension approaches according to Axinn are 8
- **Project approach** - The extension approach wherein short change is the measure of success. *e.g.* **IVLP, NAIP etc.**
- **Cost Sharing Approach** - The extension approach wherein success is measured in terms of farm people willingness and ability to share some cost.
- **Educational Institution Approach** - The extension approach wherein measure of success is the farming people's attendance at and participation in the school's agricultural extension activities.
- **Commodity Specialized Approach** - Extension is fairly centralized and is oriented towards one commodity or crop and the agent may have many functions. *e.g.* **Coffee board, Silk Board** *etc.*,
- **Training & Visit Approach** - The extension approaches wherein the success is measured in terms of production increases of the particular crops covered by the programme.
- **Farming System Development Approaches** - The extension approaches wherein success is measured by the extent to which farming people adopt the technologies developed by the programme and continues using them overtime.
- **General Agricultural Extension Approach** - The extension approach wherein success is measured in terms of the rate of take-up of the recommendations, and increases in national production.

RURAL SOCIOLOGY AND PSYCHOLOGY

- **Sociology** is the science of human relationships.
- **Father of sociology** is **August Comte.**
- **Rural sociology** is involves the study of human relationships in rural situations.
- **Society** is a group of people who have lived together long enough, sharing common values and general interests, to be considered as a social unit. Rural

societies found rural in the rural areas like villages and urban societies in the urban areas like towns, cities, industrial and commercial centers.

- **Socialization** is the process by which an individual is conducted into one's social and cultural world.

Family

- **Family** is a group defined by a sex relationship sufficiently precise and enduring to provide for the procreation and upbringing of children.
- **Family** is the **basic unit of society** and **extension work.**

Different types of families on the basis of several criteria:

A. According to blood relations living together:

1. **Nuclear family (primary)** - consists of husband, wife and their unmarried children.
2. **Joint family** - is an aggregate of more than one primary family, on the basis of close blood relationships and common residence.

According to numerousness of husbands and wives:

1. **Polygynous family** - is the family where a man marries more than one wife.
2. **Polyandrous family** - is one in which a women is married to more than one husband.

According to rule of lineage:

1. **Patri-lineal family** - is one where determinations of descent and property inheritance take place along the male line.
2. **Matri-lineal family** - is one in which determination of descent and property inheritance take place along the female line.

According to rule of residence:

1. **Patri-local family** - is one where the married couple and their offspring put up with the husband's family or in a new household set up by the husband.
2. **Matri-local family** - is one where the husband goes to live with the family of his wife and the family line goes with the mother.

Group

- **Group** is a unit of two or more people in reciprocal communication and interaction with each other. There are several classifications of social group as given below-
- **Primary group** is characterized by intimate face to face interaction, informal personal relationships and a define 'we feeling' on the part of the members. *e.g.* Family, neighbor.
- The term primary group was given by **Cooley.**

- **Secondary groups** are characterized by formal, contractual, utilitarian and goal-oriented relationships. *e.g.* Cooperative society, political party, club *etc.*,
- **Formal groups** are organized, have a membership roll, definite roles, rules or procedures of operation and rigidly enforced behaviour of its members. *e.g.* Panchayati, registered societies *etc.*,
- **Informal groups** are not formally organized and rules and procedures are not rigidly enforced. *e.g.* friendship groups, play groups *etc.*,
- **Gemeinschaft** - A society in which most of relationships are both traditional or personal *e.g.* Zamindari system or Talukdari system
- **Gesellschaft** - A society in which there is neither personal attachment nor the importance of traditional rights, obligation and duties *e.g* Business organization or association of wholesalers.
- **Reference group** is a group with which the individual feels identified, the norms and objectives of which he accepts.
- **Clique** is tightly knit group based on friendship or common interest and having very strong primary group feelings.
- **Social group** a collection of two or more individuals in which there are psychological interaction and reciprocal roles based upon durable contacts, shared norms and interest
- **Community** is refers to group of mutually dependent people, living in a more or less compact continuous geographical area, having a sense of belonging and sharing common values, norms and some interest and acting collectively in an organized manner to satisfy their needs through a common set of organization and institutions.
- **Association** is a group organized for the pursuit of an interest or group of interest in common.
- **Neighbourhoods** are limited geographic areas in which the individual and families are known to each and carry on intimate associations together.
- Tendency of a group to be unity while working towards a goal or to satisfy emotional need of its member is **group Cohesion**
- The **concept of group development** in **5 stage** model developed by **Bruce Tuckman**
- Five stage model of group development are:
  - Forming - Uncertainity
  - Storming - Intergroup conflict
  - Norming - Close cohesiveness
  - Performing - Winding up
  - Adjoining - Enhance activities

Social Control

- **Social Control** is Pattern of influencing the society individuals and groups to maintain order and establish rules in the society.
- **Folkway is** Informal rules of behavior based on habit and tradition.
- How to eat, what to wear, shaking or folding hands, table manner are included in **Folkways**
- **Mores** is socially accepted way of behavior that involve morals are rearing a particular animal, saluting flag, standing national anthem monogamy, policy of women and children
- **More** -Things that ought to be done.
- **Taboo** - Things that one not ought to be done.
- **Norm** is established behavior pattern for the members of a social society *e.g.* acceptable and unacceptable form of behavior.
- Norm – More = Folkways
- Codified system of mores is a **Laws.**
- **Anomie** a French word means **'lack of rules'** or **normlessness.**
- **Social control** is the ability to control the action of others or the degree to which an individual can influence or control the actions of others.
- Forms of social power are **(EOUL)** Elite, Organizational power, Unorganized mass power and Law.

Social stratification

- **Social stratification** is the arrangement of individuals or groups of people into hierarchically arranged strata in a community.
- The gap between the extremes of social pyramid is highest in **Urban society**
- Basic form of social relationship is **Class and caste**
- **Caste** - is the form of social relationship which is ascribed or from birth and can't be change.
- **Class** - is the form of social relationship which is decided on the basis of economic condition.
- Elements of social stratification are **(WAFRB)** Wealth, Ancestry, Functional utility, Religion, and Biological characteristics
- **Social mobility** – is the movement of individual or group from one stratum of society to another.
- **Sanskritization-** is the process by which caste placed lower in the caste hierarchy seek upward mobility by emulating the rituals and practices of the upper or dominant caste
- Sanskritization termed was given by **M. N. Srinivas** (1950).

Social interaction

- **Social interaction** - is the dynamic interplay of forces in which contact between persons and group result in modification of attitude and behavior of the participants.
- Cooperation, accommodation and assimilation are **positive social interaction.**
- Competition and conflict are **negative social interaction.**
- **Cooperation** is a working together toward common objectives or goal.
- **Accommodation** is the process of getting along despite differences a sort of compromise, tolerance, arbitration, truce *etc.,*
- The form of accommodation which serves as "safety valves for release of hostility" is: **institutionalized.**
- **Assimilation** is fusing or blending whereby a cultural difference tends to disappear.
- **Competition** is the process in which person and group compete with each other for resources or to reach a goal.
- **Conflict** is the struggle in the competing parties, attempting to reach a goal, try to eliminate, abolish, wipe out or make the other party ineffectual.
- **Acculturation** is a process of burrowing cultural items from another culture.

Social Institution

- **Social Institution** - The **'Crystallized mechanism'** created by society involving systematic pattern of behavior so organized and directed as to meet the basic social needs.
- Types of function of institution are **manifest and latent.**
- **Manifest Function** - Those function which are deliberately intended to be function of institution in accordance with the purpose which it is intended to serve. *e.g.* the provision of basic education imparting KAS to children so as to influence desirable change in behavior.
- **Latent Function** - Function of institution which is not intended. *e.g.* In school development in children of loyalties to certain ideologies.
- Five basic social institution are **(FRGEE)** Family, Religion, Government, Economy and Education
- **Family** is the basic social institution providing for procreation, upbringing and socialization.
- **Religion** is a Sets of believe regarding the ultimate power in universe, ceremonial ways of expressing the beliefs.
- **Educational institution** is a socializing person into basic values and practices of society is characteristics function.

- **Economy Institution** is an institution provides basic physical subsistence for society and meets basic needs for food, shelter, clothing and other necessities.

Culture

- **Culture** is the complex whole which includes knowledge, belief, art, moral law, custom and other capabilities and habits acquired by people as a member of society.
- **Cultural trait** is the smallest unit of culture.
- Types of cultural traits are **material and non-material traits**
- Material traits of culture *e.g.* **dhoti, sari, bamboo pen etc.**
- Non-material traits culture *e.g.* **pressing the palm together, bowing slightly in respecting elders, etc.**
- **Culture Complex** is a group or cluster of related culture traits. *e.g.* the thread ceremony in a Hindu Brahmin home.
- **Culture pattern** is a group of culture complexes.
- **Cultural lag** is the time difference between some technological change and resulting change in culture.
- **Cultural Inertia** is the resistance to change by people in a culture.
- **Counter culture** is the sub-culture that adheres to a set of norms and values that sharply contradict the dominant value and norms of the society.
- **Ethnocentrism** is a tendency of people to consider their own culture of high value and superior to all others.
- **Xenocentrism** is an opposite of ethnocentrism what is foreign is best, one's own life style products and ideas are inferior to those of others watch, cars *etc.,*
- **Cultural relativism** is the belief that cultures must be judged in their own terms rather than by the standards of another culture.
- Culture change in behavior of people occurs through d**iscovery and invention** within the society and **diffusion and burrowing** outside the society
- **Social role** is an each person in a society is expected to behave in accordance with socially prescribed patterns of behavior.
- Types of Social role are **ascribed and achieved role.**
- **Ascribed role -** Roles are assigned by culture and society to an individual. *e.g.* to become the head of family
- **Achieved role -** Roles are earned or architect by the people through individual effort and competition. *e.g.* elected leaders.

Social change

- It involves alteration in the structure and function of the society.

Process of social change

1. **Invention**: process of development of new ideas or existing ideas in new shape or form.
2. **Diffusion**: process by which new ideas are communicated in the social system over a period of time.
3. **Consequences**: change occurring within the system as a result of adoption of rejection.

Leadership

- It is an activity in which efforts is made to influence people to cooperate in achieving a goal viewed by the group as desirable.

Classification of leadership

1. **Autocratic leader** - This type of leader cannot trust people. He assumes that his workers are idle, not doing what they should do. He thinks that an employee must work since he is paid for it. *e.g.* **Hitler and Sadam Hussain.**
2. **Democratic leader** - This type of leader share with the group members in the decision making process and the planning of activities. The participation of all is encouraged. He works to develop a feeling of responsibility on the part of every members of the group.
3. **Laissez-faire leaders** - This type of leaders believes that if you leave workers alone, the work will be completed in a better way. He seems to have no confidence in himself and withdraws from his responsibility of work group.
4. **Professional leaders** - The professional leader is one who has received specialized training in the job by which he earns his livelihood. *e.g.* village level worker, subject matter specialist.
5. **Lay leader** - These leaders may or may not have any specific training. They are paid for work and generally work only part time with local group organization. *e.g.* Youth Club President, Gram Sahayak or Gram Saathi. Lay leader are also called volunteer leader or local leaders.
6. **Functional leader** - They are recognized as leaders because of their specialized knowledge and function in the society. *e.g.* Headmaster of school, doctor, scientist, agriculture innovator *etc.*,
7. **Political leader** - They arise out of the political system. *e.g.* M.L.A., M.P., Chief Minister, Prime Minister *etc.*,

Psychology

- Science of human behavior is psychology.
- **Personality** is the unique, integrated and organized system of all behavior of a person.
- **Attitude** is a mental state of readiness, organized through experience, exerting a directive and dynamic influence upon the individual's response to all objects and situations with which it is related.
- Components of attitude are **Cognitive, Affective and Conative (CAC)**
- **Cognitive** *e.g.* thoughts, belief and ideas about something.
- **Affective -** Feeling or emotion that something evokes *e.g.* fear, sympathy and hates.
- **Conative** – a tendency or disposition to act in certain ways towards something.
- **Emotion -** A state of being moved, stirred up or aroused and involve impulses, feelings and physical and psychological reactions
- **Emotion** is derived from Latin word **Emovere** means **to push or excite**
- Prejudice means **Prejudgement**
- **Prejudice** - Judgment before due examination and consideration of facts and based on certain assumption.
- **Stereotype** - Fixed images formed in one's mind about people, practices or various other social phenomenons' on the basis of experience and attitude.
- The action which leads to a desirable outcome are likely to be repeated in similar circumstances is **the Law of Effect.**
- **Frustration -** A conditioning in which an individual perceives the wish goal blocked or unattainable.
- **Rationalization -** When a person unconsciously explain the situation to oneself by reasoning that, after all the individual never did really wish to achieve the goal.
- **Sour grapes rationalization** - The form of rationalization, when one fails to achieve his goal he denigrates the goal and reasons he did not want it anyway since it was not worthwhile.
- **Sweet lemon rationalization** - The form of rationalization, when individual reasons that what he has got is really what he wanted after all, even though it is not.
- Frustration of dominant motive is **Aggression**
- **Projection** is transferring one's emotion and ascribing the source of emotion to other object.
- The individual enters an imaginary world in which the person's all wish goals are realized are **Fantasy or Day Dreaming**

- **Substitution -** When a goal is substituted for a goal which is blocked.
- **Sublimation -** When substitution involves moral consideration.
- **Regression -** People adjust to the present by recalling and reliving the 'good old days'.
- **Repression -** The mechanism by which wishes are not allowed to come out of the unconscious or thrown down into the unconscious.
- **Classical Experiment on Conditioning** was done by **Pavlov.**

COMMUNICATION

- Communication derived from Latin word **"Communis"** meaning **common.**
- **Communication** is the process by which two or more people exchange ideas, facts, feeling or impressions in ways that each gains a common understanding of the meaning, intent and use of message (Leagans, 1961).
- Basic functions of communication are **(ICII)** Information function, Command or instructive function, Influence or persuasive function and Integrative function.
- Purpose of communication are **(IPE) Information** (appeal to mind**), Persuasive** (appeal to soul and emotion**) and Entertainment**
- Most basic function of communication is to **influence**
- The art of communication is **listening**
- Level of communication are **(IIII) Interpersonnel, Intrapersonnel, Interorganisation and Intraorganization**

Models of communication

- **Aristotle Model - (SSA)** Speaker -Speech -Audience
- **Shanon-Weaver Model** (1949) - **(STSRD)** Source – Transmitter – Signal – Receiver – Destination.
- **Berlo Model** (1960) - **(SEMCDR)** Source- Encoder-Message-Channel-Decoder- Receiver.
- **Schramm Model** (1961) **(SESDD)** Source- Encoder- Signal-Decoder-Destination.
- **Leagans Model** (1963) - **(CMCTAR)** Communicator-Message-Channel-Treatment- Audience-Response.
- **Rogers and Shoemaker model** (1971) - **(SMCRE)** Source –Message- Channel-Receiver- Effect.
- Basic elements of communication are **(SMCR)** - Source, Message, Channel, Receiver.
- Attitude of source should be self**, audience** and **content.**
- **Credibility** means trustworthiness and competence.

- **Fidelity** means faithful performance.
- Person in any social system, sought out for information and advice on general or specific topics is **Key Communicator or Opinion leadership.**
- Another name of key communicator is Spark plugs, style-setters, taste-makers, influencers, information leaders and fashion leaders.
- **Sociometric method** - The method in which members of social system are asked to name the person to whom they go for advice and information on particular topic.
- **Sociogram** - Diagrammatic representation of internal structure of choices of the members of group.
- Most popular and best method is **Sociometric method**
- **Information rating** - The method in which some members of a social system are randomly selected and asked to designate person giving advice and information on a general or particular topic.
- **Self-designating method** - The method in which each selected person is asked a series of questions designed to determine the degree to which he perceives himself to be a key communicator.
- Dimension of Message was given by **Berlo**.
- Dimension of Message are **Code, content and treatment.**
- Basic characteristic of message are **Timeliness, Specificity & C edibility.**
- Message Distortion was explained by **Kirk**.
- **Systematic distortion** means if no information is lost it is recorded in systematic way.
- **Mirage** means the distortion type wherein the loss of message is maximum.
- **Fog distortion** - The distortion type wherein unwanted information are supplied along with message.
- **Channel** is a Physical barrier in communication process.
- **Personal localite** - Local leaders and local people who belong to the receivers' own social system.
- **Personal cosmopolite** - Channels of communication from outside the social system of the receiver *i.e.* Extension agent.
- **Impersonal cosmopolite** - Channels of communication are from outside the social system of the receiver and at the same time no personal face-to-face contact is involved *i.e.* Mass media.
- Ultimate objective of any communication function is a **audience response**
- **Feedback** means to carry some significant response back to the communicator.
- Concept of Feedback was given by **D Berlo**.
- Communication methods/ extension teaching methods were classified based on use and form by **Wilson & Gallup**.

Extension methods

Extension methods based on use

Individual Method	Group Method	Mass Method
Farm and home visit	Result demonstration	Circular letter
Telephone calls	Method demonstration	Farm publication (leaflet, folder *etc.*,)
Personal letter	Lecture	Exhibition
Official letter	Meeting	Campaign
Farm clinic	Field day	Newspaper
Minikit trial	Study tour	Radio
Flag method	Peripatetic team meeting	Television

Extension method based on form

Written	Spoken	Visual	**Spoken & visual**
Personal letter	Meetings	Poster	Result demonstration
Circular letter	Office calls	Chart	Method demonstration
Leaflets	Radio	Exhibition	Television
Bulletins	Farm &home visit	Flash card	Movie
Farm journals	Flag method	Bulletin board	puppets show
	Farm clinic		Campaigns

- **Audio-visual aids -** are instructional devices which are used to communicate message more effectively through sound and visuals.
- The instructional devices through which the message can only be heard are known as audio aids.
- The instructional devices which help to visualize the message are known as visual aids.

Classification of audio- visual aid

Audio aids	Visual aids	Audio- visual aids
1. Tape recorder 2. Telephone 3. Public address system	**Non- projected**	**Non- projected**
	1. Chalk board 2. Bulletin board 3. Poster 4. Flannel graph, Flash card, Flipchart 5. . Picture and Photograph 6. Diagram, Map, Chart and Graph 7. Specimen, Model, Diorama 8. Translide	1. Puppet show, Drama, Talking doll
	Projected	**Projected**
	1. Slides 2. Filmstrip 3. Overhead projection 4. Opaque projection	1. Video 2. Motion picture (cinema)

Some concepts relating to communication

- **Perception** - as the process by which an individual maintains contact with the environment. Perception of the same situation may differ from individual to individual due to differences and ways of looking into it.
- **Communication fidelity** - is the faithful performance of communication process by all its elements: communicator, message, channel and receiver. Noise and fidelity are two sides of the same coin. Eliminating noise increases fidelity, the production of noise reduces fidelity.
- **Communication gap** - refers to the differences between what was communicated by the extension worker and what has actually been received by the audience.
- **Empathy** - is the ability on the part of one person to understand the other person's internal frame of mind and reference and accept the same.
- **Homophily** - the degree to which pair of individuals who communicate are similar in certain attributes, such as beliefs, education, and social status.
- **Heterophily**- is opposite of homophily and is the degree to which pairs of individuals who interact are different in certain attributes.
- **Result demonstration** - It is a method of motivating the people for adoption of a new practice by showing its distinctly superior result. The demonstrations are conducted in the farm or home of selected individuals and are utilized to educate

and motivate groups of people in their neighborhood. This is a very effective method for the transfer of technology in a community.

- **Method demonstration** - It is relatively short-time demonstrations given before a group of people to show how to carry out entirely new practice or an old practice in a better way. It is essentially skill training, where the emphasis is on effectively carrying out a job, which shall improve upon the result.

DIFFUSION AND ADOPTION

- **Diffusion** - is the process by which an innovation is communicated through certain channels over time among the members of a social system.
- Diffusion include four elements are **innovation, channel, time, social system**
- **Innovation** - is an idea, practice or object that is perceived as new by an individual or other unit of adoption.
- **Relative Advantage** - The degree to which an innovation is perceived as being better than idea it supersedes. *e.g.* providing subsidy to the farmer for introduction of farm machinery.
- **Compatibility:** The degree to which an innovation is perceived as consistent with existing values, past experience and needs of potential adopters. *e.g.* when a new breed of livestock is in agreement with farmer's beliefs and values and new crop variety advocated to farmers suit the agro-climatic condition then it indicates the is in cultural compatibility and situational compatibility respectively.
- **Complexity** - The degree to which an innovation is perceived as relatively difficult to understand and use. *e.g.* adoption of high yielding technologies requires adoption of balanced nutrition practices, appropriate protection technology and better management methods to get best result.
- **Trialability** - The degree to which an innovation can be experimented with on a limited basis. *e.g.* providing free seeds and fertilizers to farmer.
- **Observability** - The degree to which the result of an innovation is visible to other. *e.g.* effect of nitrogenous fertilizer is more prominent in plants as compare to effect of phosphate and potassium fertilizer
- **Predictability** - The degree of expecting certain benefits from adoption of innovation
- **Preventive Innovation** - An idea an individual adopt at one time in order to lower the probability that some future unwanted event will occurred. *e.g.* contraceptives, smoking, buying insurance *etc.,*
- **Innovation Decision Process** - The process through which an individual passes from first knowledge of an innovation, to forming an attitude towards the innovation, to a decision to adopt or reject, to implement and use of new ideas and to confirmation of this decision

- Innovation decision process can shortly called as **KPDIC** (Knowledge, Persuasion, Decision, Implementation, Confirmation)
- **Adoption** is a decision to make full use of an innovation as the best course of action available.
- The adoption process is includes: **5 stages (AIETA)** Awareness, Interest, Evolution, Trial, and Adoption.

Adopter categories:

- **Normal & bell shaped curve -** frequency of adopters when plotted over time follow.
- **S shaped curve** - cumulative frequency of adopters over time follows.

  5 categories of adopter (IEELL)

  1. Innovators - 2.5 %, Venturesome
  2. Early adopter - 13.5 %, Respectable, "check with", Legitimizer
  3. Early majority - 34%, Deliberate,
  4. Late majority - 34%, Skeptical and cautious
  5. Laggards - 16%, Traditional, point of reference is past
- **Rate of Adoption** is the relative speed with which an innovation is adopted by members of a social system.
- **Overadoption** is a sometimes it may happen that people continue to adopt an innovation, rather vigorously, when experts feel that it should not be so done. *e.g.* Indiscriminate sinking of shallow tube wells and excessive use of pesticides

EXTENSION SCHEMES

Pre- independence extension programmes in India:

- 1920: **Gurgaon project** by F. L.Brayne
- 1921: Sriniketan project by Ravindra Nath Tagore
- 1921: Marthandam project by Dr. Spencer Hatch
- 1923: **Sewagram project** by Mahatama Gandhi
- 1945: Indian Village Service by T. Mosher & B.N. Gupta
- 1946: Sarvodaya by VinabaBhave
- 1946: Firka Development Scheme by T Prakashan
- 1947: **Grow more food** Campaign by Thakurdass committee

Post- independence

- 1948: **Etawa Pilot Project** by Albert Mayor
- 1948: Nilokheri Experiment by S.K. Dey

Developmental Programmes and Institutes

- 1952: Community Development (CD)
- 1953: National Extension Service (NES)
- 1959: **Panchayat Raj Institute**
- 1960: Intensive Agriculture districtprogramme (IADP)
- 1964: Intensive Agriculture Area programme (IAAP)
- 1966: High Yield Variety Programme (HYVP)
- 1974: Training & Visit System (T&V)
- 1978: Integrated Rural Development Programme (IRDP)
- 1989: Jawahar Rozgar Yojana (JRY)
- 1995: Intensive Village Linkage Programme (IVLP)
- 1999: Swarnajayanti Gram SwarozgarYojana (SGSY)
- 1999: National Agriculture Technology Project (NATP)
- 1998: Agriculture Technology Management Agency (ATMA)
- 2000: PradhanMantri Gram SadakYojana (PMGSY)
- 2006. National Agriculture Innovation Project (NAIP)
- 2009:Mahatama Gandhi National Rural Employment Guarantee Act (**MNREGA**)
- **Community Development Programme:** based on feedback of the **GMF Enquiry committee** report as well as the experiences gained through the **Etawah pilot project** experiment, it was planned to start 15 pilot projects with the financial assistance from the Ford Foundation. Further, the Indo-US technical cooperation agreement was signed in January, 1952, assisted in initiating 55 Community development projects for a period of 3 years. The programme was initiated on 2 October birthday of Mahatma Gandhi. The programme aimed at **socio-economic transformation of rural people** and as such the administrative arrangements were designed.
- **Intensive Agricultural District Programme (IADP)**

  With a focusing on increasing productivity and production IADP was launched in July 1960 in 7 selected districts of various states in first phase and later extended to another 9 districts more in second phase during 1963-64. This programme was aimed at **integrated and intensive approach** to solve the problems of agricultural production **through adoption of package of improved practices.** The achievements accomplished through this programme were tremendous and it really made a dent into the rural poverty for its eradication through agriculture. This programme was popularly known as **Package Programme.**
- **Intensive Agricultural Area Programe (IAAP)**

  To meet the demand of food for the vast population of the country, it was decided that at least 20 to 25 percent of cultivated area of the country should be earmarked and selected for intensive agricultural development. Accordingly in 1964, IAAP came into operation in 114 districts of the country. The IAAP's main objective was to bring about the **progressive increase in production of main crops** in selected areas by intensive and coordinated use of various aids to production.

- **High Yielding Varieties Programme (HYVP)**

  The IADP and IAAP were concerned with the package approach and intensive agriculture and these programmes increase the food production to some extent. But in the later period, when yield was stabilized, it could not meet the demand of food production and this experience directed the need of HYVP. The HYVP was launched in Kharif 1966-67 with a sole objective to increase the total food production by utilizing high yielding seeds of selected crops. The selected crops were **paddy, wheat, bajra, jowar and maize**. The salient features of HYVP were: (1) Supply of inputs like seeds, fertilizer and plant protection chemicals, (2) Supply of credit, (3) Cooperative marketing, (4) National Demonstration and (5) Training.

- **Training and Visit (T & V) System of Agricultural Extension**

  The Training and Visit (T&V) system, the brain child of **Dr. Daniel Benor** (World Bank Consultant) was introduced in India in 1974 for all round development of agricultural extension system in the country to introduce observation, training and technology transfer to the farmers and extension workers so as to enable them achieving greater productivity and production in the agricultural sector.

- **Training of Rural Youth for Self-Employment (TRYSEM)**

  The Centrally sponsored scheme TRYSEM was launched by Government of India under Department of Rural Development on August 15, 1979. TRYSEM is a facilitating component of the IRDP. The major and significant objective of TRYSEM was to **provide technical skills to the rural youth** from the families below poverty line to enable them to **take up self-employment** in the broad fields of agriculture and allied activities, industries, services and business enterprises.

- **JawaharRozgarYojana**

  Jawahar RozgarYojana to provide employment to the rural poor was initiated in 1989. National Rural Employment Programme (NREP) and Rural Landless Employment Guarantee Programme (RLEGP) were merged into this programme. The expenditure under the programme was shared by the Centre and the States on 80:20 bases.

- **Krishi Vigyan Kendra (KVK)**

  KVK is designed to impart need based and skill oriented vocational training to the practicing farmers, in-service field level extension workers and to those who wish to go in self employment.

Mandates of kvk:

1. On farm testing: to assess the location specificity of agricultural technologies under various farming system.
2. Front line demonstration: to established productivity potential of technologies on the farmer's fields.
3. Capacity development of farmers and extension personal to update their knowledge and skills on modern agriculture technologies.
4. To work as knowledge and resource centre of agricultural technologies for supporting initiatives of public,private and voluntary sector in improving the agricultural economy of the district.

- The first KVK was established in India : 1974 at Pandicherry (TN)
- The first KVK was established in Rajasthan: 1976 at Fatehpur (Sikar)
- Total number of KVK in India :716
- Total number of KVK in Rajasthan: 44
- Total number of ATARI :11
- ATARI (Agricultural Technology Application Research Institute)

Recent Programmes in Agricultural Extension

Agri-Clinics and Agri-Business Centres (ACABC) SCHEME -2010

- To supplement efforts of public extension by necessarily providing extension and other services to the farmers on payment basis or free of cost as per business model of agri-preneur, local needs and affordability of target group of farmers;
- To support agricultural development and to create gainful self-employment opportunities to unemployed agricultural graduates, agricultural diploma holders, intermediate in agriculture and biological science graduates with PG in agri-related courses.
- National Institute of Agricultural Extension Management (MANAGE) will be responsible for providing training to eligible candidates, through Nodal Training Institutes (NTIs) and motivating them for setting up of Agri-Clinics and Agri-Business Centres.

Agricultural Technology and Management Agency (ATMA)

Extension Reforms in India were pilot tested in **28 Districts of 7 States** under Innovations in Technology Dissemination **(ITD)** component of World Bank funded National Agricultural Technology Project **(NATP)** during the period from November, **1998 to April, 2005**. This successful experiment served as a basis to launch the Scheme **"Support to State Extension Programmes for Extension Reforms"** in its first phase since **2005-06.**

The institutional mechanism for extension under the Scheme has hitherto been as follows:

1. **SAMETI** is the State level institution catering to the training and HRD needs of extension functionaries.
2. Agriculture Technology Management Agency **(ATMA)** is responsible for coordination and management of agricultural extension related work in the District.
3. Block Technology Team **(BTT)** consisting of line department representatives in the Block.
4. Farmers Advisory Committees at Block level comprising a group of farmers to advise and provide inputs to the BTT.
5. Innovative support through a **'Farmer Friend'** at Village Level @ 1 Farmer Friend per two villages;

The revised composition, functions and mandate of ATMA

1. **State Level:** Inter- Departmental Working Group (IDWG)
   SAMETI Executive Committee
   State Farmers Advisory Committee (SFAC)
2. **District Level:** ATMA Governing Board
   ATMA Management Committee
   District Farmers Advisory Committee (DFAC)
3. **Block Level:** Block Technology Team (BTT)
   Block Farmers Advisory Committee (BFAC)

National Mission on Agricultural Extension and Technology (NMAET)

The Modified Extension Reforms Scheme was introduced in 2010 with the objective of strengthening the extension machinery and utilizing it for synergizing the interventions under these schemes under the umbrella of Agriculture Technology Management Agency (ATMA). NMAET has been envisaged as the next step towards this objective through the amalgamation of these schemes. National Mission on Agricultural Extension and Technology (NMAET) consists of 4 Sub Missions:

(i) Sub Mission on Agricultural Extension (SMAE)
(ii) Sub-Mission on Seed and Planting Material (SMSP)
(iii) Sub Mission on Agricultural Mechanization (SMAM)
(iv) Sub Mission on Plant Protection and Plant Quarantine (SMPP)

(i) SMAE: Sub-Mission on Agricultural Extension will focus on awareness creation and enhanced use of appropriate technologies in agriculture & allied sectors.

(ii) SMSP: Adoption of quality seeds is the most cost effective means for increasing agricultural production and productivity

(iii) **SMAM:** There is a strong co-relation between farm power availability and agricultural productivity. Therefore, Sub Mission on Agricultural Mechanization will focus on farm mechanization

(iv) **SMPP:** Sub-Mission on Plant Protection included in NMAET envisages increase in agricultural production by keeping the crop disease free using scientific and environment friendly techniques through promotion of Integrated Pest Management.

National Mission for Sustainable Agriculture (NMSA)

NMSA derives its mandate from Sustainable Agriculture Mission which is one of the eight Missions outlined under National Action Plan on Climate Change (NAPCC). The strategies and programme of actions (POA) outlined in the Mission Document, that was accorded 'in principle' approval by Prime Minister's Council on Climate Change (PMCCC) on **23.09.2010**, aim at promoting sustainable agriculture through a series of adaptation measures focusing on ten key dimensions encompassing Indian agriculture namely; 'Improved crop seeds, livestock and fish cultures', 'Water Use Efficiency', 'Pest Management', 'Improved Farm Practices', 'Nutrient Management', 'Agricultural insurance', 'Credit support', 'Markets', 'Access to Information' and 'Livelihood diversification'. During XII Five Year Plan, these measures are being embedded and mainstreamed onto ongoing/proposed Missions/ Progammes/ Schemes of **Dept. of Agriculture & Cooperation (DAC)** through a process of restructuring and convergence.

NMSA will cater to key dimensions of **'Water use efficiency**, **'Nutrient Management'** and **'Livelihood diversification** through adoption of sustainable development pathway by progressively shifting to environmental friendly technologies, adoption of energy efficient equipment, conservation of natural resources, integrated farming, *etc.,* To optimize utilization of water resources through efficient water management to expand coverage for achieving **'more crop per drop'**

NMSA has four (4) components:

1. **Rainfed Area Development (RAD):** RAD will adopt an area based approach for development and conservation of natural resources along with farming systems.
2. **On Farm Water Management (OFWM):** OFWM will focus primarily on enhancing water use efficiency by promoting efficient on-farm water management technologies and equipment.
3. **Soil Health Management (SHM):** SHM will aim at promoting location as well as crop specific sustainable soil health management including residue management, organic farming practices by way of creating and linking soil fertility maps with macro-micro nutrient management, appropriate land use based on land capability, judicious application of fertilizers and minimizing the soil erosion /degradation.

4. **Climate Change and Sustainable Agriculture: Monitoring, Modeling and Networking (CCSAMMN):** CCSAMMN will provide creation and bidirectional (land/farmers to research/scientific establishments and vice versa) dissemination of climate change related information and knowledge by way of piloting climate change adaptation/mitigation research/model projects in the 5 domain of climate smart sustainable management practices and integrated farming system suitable to local agro-climatic condition.

Attracting and Retaining of Youth in Agriculture (ARYA)

- Under the recommendation of ICAR, in the XII Five-Year Plan, it has been proposed to initiate the programme
- The initiative aims at **checking the rate of migration of youth from rural areas** and **building confidenc** among rural youth that they can lead a decent life through farming rather than chasing jobs in cities

'Farmer FIRST'

An ICAR initiative to move beyond the production and productivity and to privilege the complex, diverse & risk prone realities of majority of the farmers through **enhancing farmers-scientists** contact with multi stake holders-participation

Student READY

A one-year composite programme, the Rural Entrepreneurship and Awareness Development Yojana (READY) is proposed with the objective to **develop professional skills for entrepreneurship and knowledge** through meaningful hands-on experience in project mode; confidence through end to end approach in product development; and enterprise management capabilities including skills for project development and execution, accountancy and national/ international marketing.

Government schemes and incentives for promotion of entrepreneurship:

- **Prime Minister's Employment Generation Programme** (PMEGP) - It is a credit-linked subsidy programme introduced by the government of India in 2008. PMEGP is a merger of two schemes, namely, Prime Minister's RojgarYojna and Rural Employment Generation Programme. This program focuses on generating self-employment opportunities through micro-enterprise establishments in the non-farm sector by helping unemployed youth and traditional artisans.
- **Rajiv Gandhi Udyami Mitra Yojana:** RGUMY was launched in 2008, is aimed at providing financial assistance to the selected lead agencies *i.e.* Udyami Mitras for rendering assistance and handholding support to the potential first generation entrepreneurs through '**Udyami** Helpline' (a Toll free Call Centre for MSMEs on 1800-180-6763) support, guidance and assistance

to first generation entrepreneurs as well as other existing entrepreneurs is also provided to guide them regarding various promotional schemes of the Government, procedural formalities required for setting up and running of the enterprise and help them in accessing Bank credit *etc.,*

- Schemes for women entrepreneurs:
- **Mahila Udyami Yojana:** IDBI Bank has set up special fund under this scheme with corpus fund of Rs. 5 crore to provide seed capital assistance to the women entrepreneurs.
- **Priya Darshni Yojana:** it is implemented by Bank of India.
- **Stree Sakthi Package**: is a unique scheme run by the SBI, aimed at supporting entrepreneurship among women by providing certain concessions. An enterprise should have more than 50% of its share capital owned by women to qualify for the scheme.

PROGRAMME PLANNING

- **Extension Programme** - is a statement of situation, objectives, problems and solutions.
- **Programme Planning** - is a decision making process involving critical analysis of the existing situation and the problems, evaluation of the various alternatives to solve these problems and the selection of the relevant ones, giving necessary priorities based upon local needs and resources by the cooperative efforts of the people both official and non- official with a view to facilitates the individual and community growth and development.
- **Situation** - is a statement of affairs that includes the cultural, social, economic and physical conditions in which a particular group of people find themselves at a given period of time.
- **Aim** - is generalized and broad statement of directions with respect to given activities.
- **Goal** - is the distance in any given direction one expects to go during a given period of time.
- **Need** - there is gap between what is-actual situation and what ought to be- the desirable situation.
- **A Plan of work** - is an outline of activities so arranged as to enable efficient execution of the programme. Plan of work indicates 5W & 1H (What, where, who, whom, why & how)
- **A Calendar of work** - is a plan of activities to be undertaken in a particular time sequence.
- **Steps in Extension Programme Planning:** 8 steps

  1. Collection of facts

2. Analysis of situation
3. Identification of problems
4. Determination of objectives and goals
5. Developing plan of work and calendar of operations
6. Follow through plan of work and calendar of operations
7. Evaluation of progress
8. Reconsideration and revision of the programme

IMPORTANT OBJECTIVE QUESTIONS (MCQ)

1. Which of the following is not a feature of formal education?
   A. Sequential B. Hierarchically oriented
   C. Non- authoritarian D. Subject oriented
2. To correct sequence of steps in extension teaching is
   A. Interest, desire, attention, conviction, action, satisfaction
   B. Attention, desire, conviction, interest, satisfaction
   C. Attention, interest, desire, conviction, action, satisfaction
   D. Attention, conviction, interest, action, satisfaction
3. Extension education basically produces
   A. Better understanding of farm technology
   B. Greater knowledge of science in agriculture
   C. Change in skills for handling improved technology
   D. Change in human behaviour
4. It is also known as "single window delivery system"
   A. ATMA B. ATIC C. NAIP D.KVK
5. The relatively non-mobile transfer of technology (TOT) programme of Indian Council of Agriculture Research
   A. KVK B. LLP C.ORP D.AICPND
6. The concept of "Village Guide" was given by
   A. S K Dey B. Albert Mayer C. Spencer Hatch D. F L Brayne
7. The concept of " BratiBalika" was given by
   A. Mahatma Gandhi B. VinobaBhave
   C. Ravindranath Tagore D. Albert Mayer
8. These are elements of diffusion, expect.......?
   A. Innovation B. Adoption C. Social System D. Time
9. Which one is the most important focal point of teaching and learning process?
   A. Learner B. Physical Facilities
   C. Teacher D. Subject Matter

10. Which of following are informal rules of behaviour?

A. Mores B. Laws C. Folkways D. Norms

11. Which of the stages of adoption process is the " mental trial"

A. Awareness B. Evolution C. Trial D. Interest

12. "Technology assessment and refinement" (TAR) took place though

A. T&V B. IVLP C. PMRY D. ATMA

13. "Entrepreneur" is loanword from

A. French B. Amharic C. Arabic D. Latin

14. How many farmers in one group were to be selected "contact famers" in case of T& V system?

A. 5 B. 10 C. 15 D. 20

15. Adoption is essentially a:

A. Physical process B. Social process
C. Mental process D. Psychological process

16. The process of using a new ideas on continuous basis is

A. Diffusion B. Innovation C. Adoption D. Utility

17. The process of providing situation in which learning take place is

A. Learning B. Learning experience
C. Teaching D. Physical facilities

18. The first KVK was set up

A. Hyderabad B. Pondicheri C. Fatehpur D. Nilokheri

19. In India ATMA has been initially introduced in

A. 28 Districts of 6 State B. 22 Districts of 6 State
C. 24 Districts of 8 State D. 24 Districts of 6 State

20. Panchayat Raj first introduced in which the following states?

A. Punjab B. Rajasthan C. UP D. Orissa

21. Which of following is not an element of communication process?

A. Fidelity B. Response C. Channel D. Message

22. T& V system of extension was the brainchild of

A. Neils Rolling B. Robert Chamber
C. Daniel Benor D. Swanson

23. Method demonstration works on principles of

A. Learning by Believing B. Leaning by Practicing
C. Learning by Teaching D. Learning by Doing

24. Who is the father of "modern theory of management"?

A. Terry B. Henri Fayol C. Franklin D. None of these

25. The National Agricultural Higher Education Project (NAHEP) was launched by ICAR, New Delhi through the support of-
    A. World Bank   B. Govt. of India
    C. World Bank and Govt. of India   D. Govt. of India and State Govt.
26. In which year, programme on training of rural youth (BPL farmers) for self employment aimed at providing technical and entrepreneurial skill has been commenced?
    A. 1979   B. 1990   C. 2013   D. 2017
27. A continuous and cooperative activity , involving the local farmers and extension staff in which objective of the work are set forth and action planned is known as
    A. Programming   B. Programme planning
    C. Planning   D. Programme planning process
28. Visual aid for transfer of technology include
    A. Audio teleconference   B. E- mail
    C. Cultural programme   D. All of these
29. Extension education is related where
    A. Audience is more or less homogenous
    B. Teaching is mainly vertical and curriculum centered
    C. Teaching is mainly horizontal and need based
    D. Pre-planned and pre-decided
30. The three characteristics of extension are
    A. Teaching-Informal-Flexible
    B. Teaching –Formal-Voluntary
    C. Teaching-Curriculum based- Compulsory
    D. Teaching-Cooperation-Formal
31. Which of these cannot be classified in "Mass contact method" of extension teaching?
    A. Leaflets   B. Bulletin board
    C. Front line demonstration   D. Film shows
32. Which Agriculture University got section of first Krishi Vigyan Kendra?
    A. TNAU, Coimbatore   B. KAU, Trishur
    C. APAU, Hyderabad   D. UAS, Bangalore
33. The expenditure under JRY is shared the between central and state govt. under in ratio of :
    A. 50 : 50   B. 60 : 40   C. 70 : 30   D. 80 : 20
34. In the Programme planning process, second steps is
    A. Analysis of situation   B. Determination of objectives
    C. Collection of facts   D. Identification of problems

35. Directorate of Adult education of National Literacy Mission –India is located at

    A. Mumbai B. Chennai C. Kolkata D. New Delhi

36. Which one of following is a training institute

    A. KVK B.DRDA C.ATIC D.ATMA

37. The full form of IVLP

    A. Institution Village Livestock Project

    B. Institution Village Linkage Program

    C. International Value Linked Price

    D. Institutional Village Livelihood Development Plan

38. ....is an example of farm publication

    A. Television B. Motion pictures C. Folder D. Radio

39. Method demonstration does not include-

    A. Seed treatment with rhizobium culture

    B. Use of improved variety seed

    C. Taking soil samples for its testing

    D. Budding in fruits plants

40. Which is following sentence is not true about KVK?

    A. Kvk organize long term vocational training for rural youth for generating self employment

    B. Kvk organize trainings for subject matter specialists of agricultural universities

    C. Kvk organize trainings for extension functionaries and farmers

    D. Kvk organize front line demonstrations to generate production data and feedback

41. Agriculture clinic is helpful to farmer for -

    A. Testing soil sample

    B. Testing irrigation water quality

    C. Testing seed viability

    D. Providing specialized guidance in writing

42. How many components were there in the national agricultural innovation project?

    A. Three B. four C. Five D. Nine

43. Which of the following is not communication effectiveness approach?

    A. Attention - perception -comprehension – acceptance

    B. Comprehension - validity- utility

    C. Commonality of comprehension between sender and receiver

    D. Communicator- message- channel- audience

44. Goal of extension education is?
    A. To promote income of farmers
    B. To promote production of crops
    C. To promote new crops
    D. To promote scientific outlook
45. Which one is a non- projected aid?
    A. Cinema
    B. Overhead projector
    C. Slides
    D. Pictures
46. People should realize that the task of extension education is their own work. The above sentence depicts which principle of education?
    A. Principle of need and interest
    B. Principle of cooperation and participation
    C. Principle of cultural difference
    D. Principle of democratic approach
47. Programme Planning starts with
    A. Identifying problems along with people
    B. Transmission of information and ideas
    C. Selecting suitable strategy or action
    D. Studying the resources
48. Under RastriyaKrishiVikasYojana , Government of India provide financial assistance to the states, From 2015-16 Govt. of India reduce their contribution from
    A. 100 per cent to 80 per cent
    B. 100 per cent to 60 per cent
    C. 80 per cent to 60 per cent
    D. 100 per cent to 50 per cent
49. From which district of Rajasthan, the National Rural Livelihood Mission was launched by Mr. Sonia Gandhi on June 3, 2011?
    A. Udaipur
    B. Dungarpur
    C. Bhilwara
    D. Banswara
50. KisanMandal System under T& V system was initiated in which year?
    A. 1992
    B. 1993
    C. 1994
    D. 1996
51. According to Kelsey and Hearne, the basic philosophy of extension education is
    A. To eradicate poverty of people
    B. To provide technical help of people
    C. To teach people, how to think and not what to think
    D. Helping people to helps themselves
52. Which of the following project is implemented in consortium mode, with competitive funding in India?
    A. NAIP
    B. NATP
    C.IVLP
    D. NARP
53. Which of the following pairs not correctly matched?
    A. Individual contact- Personal letter
    B. Group contact- Campaign
    C. Mass contact- Farmers fair
    D. Group contact-Field trip

54. First 55 community development projects were organized in India for a period of
    A. 3 years B. 5 years C. 10 years D. 15 years
55. Which of the following not a style of leadership according to classification given by "white"?
    A. Autocratic B. Professional C. Democratic D. Laissez- faire
56. Which of the following pairs not correctly matched?
    A. Leagans model of communication- 5 elements
    B. Berlo model of communication – 6 elements
    C. Aristotle model of communication – 3 elements
    D. Rogers and shoemaker model of communication- 5 elements
57. Development of Plan of work is part of
    A. Programme Planning B. Programme Implementation
    C. Programme Monitoring D. Programme Evaluation
58. Pradhan Mantri Fasal Bima Yojana (PMFBY) was launched in
    A. 2015 B. 2016 C. 2017 D. 2018
59. Pramparagat Krishi Vikas Yojana (PKVY) was started in India in the year
    A. 2013 B. 2015 C. 2017 D. 2018
60. Who have greater access to land, capital, credit-input, experts, information network
    A. Laggards B. Early adopters C. Majority D. Innovators
61. The most important group teaching that give quick adoption
    A. Group discussion B. Method demonstration
    C. Results demonstration D. Informal discussion
62. Pradhan Mantri Krishi Sinchai Yojana (PMKSY) was launched in
    A. 2013 B. 2014 C. 2016 D. 2017
63. Who is chairmen of ATMA at district level
    A. Chief Executive Officer- Zilaparishad
    B. District Collector
    C. Deputy director Agriculture
    D. Deputy Director ex-officio project director
64. To motivate farmers importance should be given to
    A. Need based approach B. Training
    C. Participation D. All of these
65. Which of the following teaching aids is used for mass communication
    A. Flash card B. Radio C. Chart D. Flip card
66. From which level was should start our planning
    A. Village level B. Panchayat level
    C. Zilaparishad level D. State level

67. At village level to whom one will contact first for transfer of technology
    A. Vikashadhikari B. Gram sarpanch
    C. Chairmen cooperative society D. Agriculture supervisor
68. Result demonstration works on principles of
    A. Seeing is believing B. Learning by doing
    C. Learning by Teaching D. Learning by practicing
69. The expenditure under SGSY is shared the between central and state govt. under in ratio of
    A. 50:50 B. 60:40 C. 75:25 D. 80:20
70. National Food Security Mission was launched in
    A. 2004 B. 2005 C. 2008 D. 2007
71. Father of modern psychology
    A. Meston B. August Comte C. Sumner D. Horton
72. Anomie means
    A. Lack of rules B. Anomaly C. Anonymous D. None
73. Communication competency concept given by
    A. Wiemann and Backlund B. Hymes
    C. Shannon and Weaver D. Thayer
74. The most basic function of communication
    A. Information B. Instruction C. Integration D. Influence
75. The basic logic applicable in extension education
    A. Deductive logic B. Inductive logic
    C. Horizontal logic D. Lateral logic
76. Aggregate of fully established and accepted rules, procedure and norms
    A. Values B. Institution C. Folkways D. Community
77. The term self -actualization coined by
    A. Maslow B. Kurt Goldstein C. McGregor D. Herzberg
78. Empathy is greatly needed in
    A. Brainstorming B. Group Discussion
    C. Roleplay D. Sensitivity Training
79. Science which deals with adult learning is
    A. Androgogy B. Pedagogy C. Pedology D. Edaphology
80. Social stratification is a consequence of
    A. Differentiation and evaluation B. Differentiation and classification
    C. Classification and satisfaction D. Classification and tabulation
81. How is the T&V system of Extension characterized as
    A. Management system B. Supervisory system
    C. Integrated system D. Broad based system

82. Media forum originated from
A. Canada B. Russia C. USA D. India
83. Hierarchy of need theory was proposed by
A. Maslow B. Freud C. Heizberg D. Pipperman
84. How does human eye moves while reading any visual?
A. Y like pattern B. Z like pattern C. W like pattern D. C like pattern
85. Indian type of family characterized by
A. Polyandrous B. Matrilineal C. Conjugal D. Consanguinous
86. Linear model of communication was given by
A. Berlo B. Leagans C. Schramm D. Weaver
87. Files to field programme has been introduced in
A. Ahmednagar B. Sirmour C. Ahmadabad D. Sambalpur
88. Language in which the word family have its origin
A. Roman B. Greek C. Latin D. British
89. The gangplank idea is found in the works of
A. Fayol B. Follen C. Barnard D. Simon
90. At which level, communication is fact & problem oriented?
A. Explorative level B. Participative level
C. Conventional level D. None
91. Extension approach in India is mainly
A. Prescriptive B. Perspective C. Evaluative D. None
92. An agricultural innovator is generally viewed as a
A. Opinion leader B. Social deviant C. Skeptical D. Traditional
93. Objective of education decided by
A. Biological science B. Normative science
C. Physical science D. Social science
94. Society is considered as
A. Abstract B. Concrete C. Both 1 & 2 D. None
95. Execution of plan should be carried out
A. Without change B. With change
C. In presence of leader D. In absence of leader
96. In Iowa, seed corn train were organized in 1904-05 by
A. Dr. Seaman A Knapp B. Prof. P G Smith
C. Prof. Holden D. Prof. AC True
97. In which equipment, direct principle of projection works?
A. Overhead projector B. Slide projector
C. Opaque projector D. All

98. In which, social calendar is a tool most commonly used?
    A. PRA B. PLA C. RRA D. FSR
99. Agri-Clinics and Agri-Business Centres (ACABC) Scheme started in
    A. 2010 B. 2004 C. 2000 D. 2008
100. INTERPAKS model given by
    A. Horton B. Swanson C. Stewart D. Thompson
101. Farmer-First-Farmer-Last (FFFL) Model given by
    A. Chambers and Ghildyal B. Scones and Thompson
    C. Rhoades and Booth D. None
102. Beyond Farmer First (BFF) Model given by
    A. Chambers and Ghildyal B. Scones and Thompson
    C. Rhoades and Booth D. None
103. Farmer-Back to-Farmer (FBT) Model given by
    A. Chambers and Ghildyal B. Scones and Thompson
    C. Rhoades and Booth D. None
104. ....involves a top –down approach that delivers specific recommendations to farmers about the practices they should adopt *e.g.* Training and Visit system
    A. Technology Transfer B. Advisory Work
    C. Human Resource Development D. Facilitation For Empowerment
105. Farmer Field School (FFS) or participatory technology development (PTD) comes under which paradigm of extension
    A. Technology Transfer B. Advisory Work
    C. Human Resource Development D. Facilitation For Empowerment
106. Government organizations or private consulting companies respond to farmers enquiries with technical prescriptions is
    A. Technology Transfer B. Advisory Work
    C. Human Resource Development D. Facilitation For Empowerment
107. Concept of Agricultural knowledge system first given by
    A. Uwe Nagel B. Swanson
    C. Chambers and Ghildyal D. Thompson
108. A system of beliefs, cognitions, models, theories, concepts and other products of mind in which the experience of a person or group with respect to agricultural production is accumulated is
    A. Agricultural knowledge System B. Agricultural Information system
    C. INTERPAKS model D. Extension Science
109. The body of knowledge which accumulates experience and research findings with respect to extension and burrows insights from other disciplines and fields of endeavour which seem pertinent to extension
    A. Agricultural knowledge System B. Agricultural Information system
    C. INTERPAKS model D. Extension Science

110. The process by which an innovation is communicated through certain channels over time among the members of a social system is
A. Diffusion B. Innovation
C. Social System D. Rate of adoption

111. The ability on the part of one person to understand the other person's internal frame of mind and reference and accept the same
A. Empathy B. Sympathy C. Both D. None

112. The degree to which an innovation is perceived as relatively difficult to understand and use
A. Compatibility B. Complexity C. Predictability D. Trialability

113. The Dutch word for extension 'Voorliching' means
A. Lighting the path B. Advisory work
C. To promote D. Str*etc.,*hing out

114. Extension when aim at empowering or uplifting the poor
A. Emancipatory extension B. Formative extension
C. Persuasive extension D. Informative extension

115. Extension that emphasis on human resource development
A. Emancipatory extension B. Formative extension
C. Persuasive extension D. Informative extension

116. Extension when used as a policy instrument for achieving societal objective
A. Emancipatory extension B. Formative extension
C. Persuasive extension D. Informative extension

117. Extension as a instrument for helping people to make well-considered choice
A. Emancipatory extension B. Formative extension
C. Persuasive extension D. Informative extension

118. Extension that moves beyond the traditional crop production technologies to include land based
A. Emancipatory extension B. Formative extension
C. Persuasive extension D. Broad based extension

119. Book" guide on Alternative Extension Approaches" written by
A. Axinn B. Berlo C. Leagans D. Chamber

120. Expression of ends towards which our efforts are directed
A. Objective B. Aim C. Goal D. None

121. The process by which an individual, through one's own efforts and abilities change the behavior
A. Teaching B. Learning C. Process D. Behavior

122. The process of arranging situations in which the important things to be learned are called to the attention of the learners, their interest developed, desire aroused and action promoted
A. Teaching B. Learning C. Process D. Behavior

123. The relatively enduring changes in overt and covert responses as a result of perceived stimulus
    A. Teaching B. Learning C. Process D. Behavior
124. Teaching-Learning Process explain by
    A. Axinn B. Berlo C. Leagans D. Chamber
125. The learning curve in teaching follows a
    A. S shaped B. L shaped C. Z shaped D. Reverse S shape
126. Steps in extension teaching presented by
    A. Wilson & Gallup B. Berlo
    C. Leagans D. Chamber
127. Difference between what is (actual situation) and what ought to be (desirable situation)
    A. Need B. Objective C. Training D. Problem
128. 'Extension Science' book written by
    A. Rolling B. Berlo C. Leagans D. Chamber
129. 'Challenging the Professionals' book written by
    A. Rolling B. Berlo C. Leagans D. Chamber
130. Jounalism should be
    A. Attractive, brief and clear B. Accuracy, Bravity and Clarity
    C. Both D. None
131. Decision to reject an idea in order to adopt a better idea that supersedes it is
    A. Replacement Discontinuance B. Disenchantment discontinuance
    C. Discontinuance D. None
132. A collectivity acting with some continuity to promote a change or resist a change in the society or group of which it is a part is
    A. Social Movement B. Propaganda
    C. Publicopinion D. None
133. The 'Crystallized mechanism' created by society involving systematic pattern of behavior so organized and directed as to meet the basic social needs is
    A. Social Institution B. Social organization
    C. Committee D. Latent Function
134. Socializing persons into basic values and practices of society is characteristics function of
    A. Educational institution B. Family
    C. Economy Institution D. Religion
135. Such institution provide basic physical subsistence for society and meet basic needs for food, shelter, clothing and other necessities
    A. Educational institution B. Family
    C. Economy Institution D. Religion

136. A systematic display of models , specimens, charts, photographs, pictures, posters information *etc.,* in a sequence around a theme to create awareness and interest in the community is

A. Mela B. Campaign C. Exhibition D. Mediaforum

137. A temporary collection of people reacting to stimuli, whatever may be the stimuli is

A. Crowd B. Audience
C. Public D. Social movements

138. The collective opinion of the public and constitutes the composite point of view arrived at the public concerning an issue is

A. Public opinion B. Propaganda
C. Social Movement D. Public

139. Working together toward common objectives or goal is

A. Competition B. Conflict C. Cooperation D. Compromise

140. When contending parties do not settle differences among themselves, then third party is require as a mediator who is capable of studying the issue objectively is

A. Assimilation B. Arbitration
C. Accommodation D. Assimilation

141. The process of burrowing cultural items from another culture is:

A. Acculturation B. Accommodation
C. Assimilation D. Competition

142. The complex whole which include knowledge, belief, art, moral law , custom and other capabilities and habits acquired by people as a member of society is

A. Religion B. Perception C. Thinking D. Culture

143. The resistance to change by people in a culture is called

A. Cultural Inertia B. Cultural lag
C. Counter culture D. Culture complex

144. Core of norms and values that characterize a sub-culture and give it a "feeling" or "tone" is term as

A. Belief B. Ethos C. Thinking D. Value

145. The PRA technique that represents the degree of importance of individuals and institution in decision

A. Venn Diagram B. Cause effect diagram
C. Preference Ranking D. Trend Analysis

146. Curriculum in training programme is

A. Heart B. Arteries and Vein
C. Both D. none

147. Training method in training programme is

A. Heart B. Arteries and Vein
C. Both D. none

148. Systematic, regular collection and occasional analysis of information to identify and possibly measure changes over a period of time
A. Monitoring B. Evaluation C. Training D. None

149. Measuring performance against pre-determined goal is
A. Monitoring B. Evaluation C. Training D. None

150. A package of exercise organized at the beginning of any programme to psychologically stimulate the trainees is
A. Microlab B. Rapport building
C. Training D. Curriculum

ANSWERS

1. C	2. C	3. D	4. B	5. A	6. D	7. C	8. B
9. A	10. C	11. C	12. B	13. A	14. B	15. C	16. C
17. C	18. B	19. D	20. B	21. A	22. C	23. D	24. B
25. C	26. A	27. D	28. B	29. C	30. A	31. C	32. A
33. D	34. A	35. D	36. A	37. B	38. C	39. B	40. B
41. D	42. B	43. B	44. D	45. D	46. B	47. A	48. B
49. D	50. B	51. C	52. A	53. B	54. A	55. B	56. A
57. A	58. B	59. B	60. D	61. C	62. C	63. B	64. D
65. B	66. A	67. D	68. A	69. C	70. D	71. B	72. A
73. B	74. D	75. B	76. A	77. A	78. C	79. B	80. A
81. A	82. A	83. A	84. B	85. D	86. B	87. C	88. A
89. A	90. A	91. A	92. A	93. B	94. A	95. A	96. C
97. B	98. A	99. A	100. B	101. A	102. B	103. C	104. A
105. D	106. B	107. A	108. A	109. D	110. A	111. A	112. B
113. A	114. A	115. B	116. C	117. D	118. D	119. A	120. A
121. B	122. A	123. B	124. C	125. A	126. A	127. A	128. A
129. D	130. B	131. A	132. A	133. A	134. A	135. C	136. C
137. A	138. A	139. C	140. B	141. A	142. D	143. A	144. B
145. A	146. A	147. B	148. A	149. B	150. A